PENOBSCOT NARROWS
PUBLISHING

THE ADVENTURE, JOY, AND VALUE OF SCIENCE:

Lessons from a Field Geologist in Antarctica, Mount St. Helens, and the American West

Peter D. Rowley

PENOBSCOT NARROWS PUBLISHING

PENOBSCOT NARROWS PUBLISHING

7 Third Street
Bucksport, ME 04416

This book is an autobiography by Peter D. Rowley.
Views mentioned in this book represent only those of Peter D. Rowley, and they may not be the views of the publisher or people mentioned within.
For some of the author's research listings, see Notes.

COVER AND BOOK DESIGN BY
SCOTT ROWLEY, SCOTT DESIGN

Set in Times New Roman and
Margin MVB

ISBN: 979-8-9960467-0-6 (hardback)
ISBN: 979-8-9960467-1-3 (paperback)
ISBN: 979-8-9960467-2-0 (epub)

Library of Congress Control Number: 2026915781

DEDICATION

This book is dedicated to all persons who are mentioned here.
This book is their story too!

CONTENTS

PREFACE

Creators of movies and fiction commonly depict scientists as nerdy, shy, skinny, and rather unattractive, glasses-wearing persons. Or as wearing long white lab coats and sitting indoors fiddling with numbers or computers. And if that is not bad enough of a portrayal of individuals who make up most of my friends, the public vision of what they do—namely science—is even worse: something that may be occasionally interesting but is predominantly boring and largely irrelevant to what an average person does.

Not so! For a scientist—and for me as a geologist—almost every day is exciting and fun, with the goal of our daily work being publication or data releases of the results, thus creating new knowledge! Most conclusions, when published or otherwise released, have practical value to the citizen taxpayer, although some value may not become fully evident until years later. For us scientists, an "aha" moment can occur on an ordinary day in the office or lab during the task of compiling data. Or it can occur out in the "field," as it often has done for me. Geologist Chuck Thorman, a friend, describes geology as a great detective story, as one assembles facts on route to reaching conclusions. Or, as to quote the lyrics to one of my favorite songs, "Life's Been Good" by Joe Walsh of the Eagles, "they say I'm crazy," but "I'm just looking for clues at the scene of the crime." Collaboration with others is always a major part of the job, for doing science requires that we necessarily, happily, and unselfishly share data and ask for and value input from others, whether critical ("Are you nuts?") or supportive. I have found that most of the highly productive (in this book I use "productive" for the ability to produce lots of publications or other data releases to the public) scientists have a great sense of humor, which, as you'll see, shows up as practical jokes on others, laughter over beers together, and an almost universal practice of kidding and "putting-down" our closest colleagues—and they of us.

This memoir records a career that is typical of most in science, a story intended for curious folks of all ages but particularly youth to give them an idea of what to expect if they choose science, and to assure them that such work is worth doing. My career, however, is atypical in that I have been fortunate to experience more adventure than most, in fascinating areas of great beauty, a happy result of luck during a simple attempt to benefit society while pursuing my curiosity. Fortuitously, my story has some important lessons to the non-scientist reader on how science operates, in an era of massive societal transformations (and ignorance) as climate change reaches a critical level and denial abounds. Therefore, this memoir is not only about my scientific career and personal things, but the times in which I lived and their impact on science. In some chapters I make observations on world events and how they influenced others and me. For example, environmental movements surged in the 1970s, then waned, and although now the topic of climate change has become the world's most pressing problem, world leaders are doing little about it and most citizens are too busy to care. In other chapters I write about the Vietnam War, changes in perceptions about science, and of course American politics. Regarding politics, the past eight years (as I wrote this book) have witnessed the greatest collective stupidity in the history of the USA: first it was the election and presidency of Donald Trump and, during and since then, the infantile reactions by citizens to masks and vaccines for COVID-19, to the "Big Lie" that Trump did not lose the election, to the January 6 insurrection and its denial, and to truth in general. During this time, the Republican party has shown a gross contempt for democracy, for facts, and for science.

There is a saying that it is not the things you do but the things you do not do that you later regret. I have been fortunate enough to have acted on my dreams and ambitions, and this book will tell you of most of them, and how they turned out. Action requires making decisions. I made most of these decisions easily. Perhaps at the time my choices seemed natural because of the serendipity of having a choice in the first place. Furthermore, most positions in which I found myself seemed to be the result of strange coincidences. I chose the fork in the road that seemed the most interesting or exciting, or that would teach me something.

Although most decisions in my life were easy, writing this book manuscript was *not*. There always seemed to be something more pressing to be done. What finally got me moving was a June 2018 Nevada Public Radio interview with a Hells Angels biker who had been recently released from twenty years in prison for killing a person or two, whereupon he had written a memoir. I figured that if he could justify a memoir, so could

I. I started typing the next day!

I've mentioned loving science. What made my career more enjoyable was the soundtrack that's accompanied my life, especially that of rock and roll. Many persons throughout time have found experiencing music to be a deep need within them, and they have had their own musical preferences to meet it. My favorite bands are Credence Clearwater Revival/John Fogerty, The Rolling Stones, The Band, and Bob Dylan. Yet other types of great popular music also characterized this period of the 1960s to 1990s. Early country and western music, with Willy Nelson, Waylon Jennings, and Emmylou Harris, was also terrific and crossed into rock-music formats. And I like Bluegrass music, such as that of Bill Monroe, and folk songs by Gordon Lightfoot, Kingston Trio, Nitty Gritty Dirt Band, and Arlo Guthrie, and the gospel/country/rock music of Aretha Franklin, Ray Charles, Mahalia Jackson, and Dolly Parton. The music and lyrics of these and other artists have inspired me, have been an escape from boredom or stress, have provided memories, and have been part of life's fun. Therefore, you'll find woven into most chapters a few words from some of my favorite songs that bear on that chapter. Some songs (e.g., "Wooly Bully" and "The Weight") have mostly nonsensical lyrics but are happy. My life, like that of most scientists, has been satisfying and amusing, and music has been and remains an integral part of it.

The most important aspects of this memoir are truth and accuracy. Most events concerning my personal life are based on memories, yet I found that these are quite accurate, even in remembering names of persons whom I knew for only a month or so. Beyond memories, I rely on letters from people with whom I have stayed in contact, and on fifty years of my Christmas newsletters, some of which I saved. Events in my professional life I recorded and published or wrote down in many dozens of field notebooks (averaging two per year), that include descriptions and conclusions about rock outcrops I visited each field season, plus some personal information. Of my five field seasons in Antarctica, part of a long career with the U.S. Geological Survey (USGS), I also kept journals and wrote lots of letters to my wife and kids, Mom and Dad, and friends, and many letters were saved and returned to me. I also retain the originals of the field notes of all Antarctic participants and copies of all draft geologic maps we compiled (in 2017, I sent copies of all these field notes and maps to the Ohio State University Polar Library and to the British Antarctic Survey). The field notes, journals, and letters contain our geologic observations, with dates and locations. Similarly, events during three years of working at Mount St. Helens were recorded in detailed field notebooks, and the events of the first hundred days of the eruptions are in *USGS Professional Paper 1249*, cited below in the Notes. If my recollections are inaccurate despite all these records, I take full responsibility and apologize for that.

This memoir is written as my way of saying thanks, of repaying a debt, by crediting in detail the many remarkable persons I have met who have not only inspired and mentored me, but whose friendships have kept me going. Some have been my family. Some have been my teachers. Most others have been collaborators on scientific problems. Still others have contributed significantly in various ways to the enjoyment of what I did each day. All have had a major role in instructing me in science and in life. I have stolen many of their stories and adventures, and these purloined tales are also given below. Their stories provide additional examples of how science operates. My references to persons in the text indicate my deep gratitude to each, so those persons are not specifically credited in the acknowledgments. Nonetheless, I have kept in touch with most of them over the years in letters and in annual Christmas cards and newsletters. And those persons who still live when I was writing this memoir knew about it and agreed to contribute to it in this way.

INTRODUCTION

Louis Armstrong's song, "What a Wonderful World", is to me the best and most joyful popular song ever written and sung, and an optimistic way to start the story of a lucky life. That life has been founded on curiosity. Curiosity forms the foundation of science, yet discovery forms the excitement of science. Of the two, the former presumably derives early on from parental influences and, if it survives the hormones of junior high school, it is the main reason a person enters professions in science. Productivity (writing) in science, which commonly follows discovery, similarly requires favorable genetics and early skills, but it is more likely the result of mentoring, training, and good luck. Most personal scientific discoveries are mundane (many scientists are merely figuring out what some in previous generations knew already), but the mental excitement can be profound. Most discovery comes from the always necessary and sometimes enjoyable evaluation of reams of data collected through laborious work conducted in the field, lab, and literature review, and it occasionally results in "aha" moments. A few discoveries, however, are unique and unexpected, a "eureka" moment, although these rarely happen unless considerable background work has already been done.

Science consists of branches of study that apply objective scientific methods to understanding the physical universe. The key to science is how it is done, namely the scientific method, which is systematic accumulation of data by observation, measurement, and experiment. This results in an hypothesis that is first proposed, then tested, then modified as necessary. Commonly multiple working hypotheses are developed to explain a phenomenon, then whittled down impartially to one or more. Scientists cannot discard data simply because they are troublesome or outliers, so any hypothesis must explain all data. Rising above the level of an hypothesis is a theory, which to a scientist is something that is accepted even though it cannot be proved correct (because science is discovery of data, and new data keep being added, so conclusions are never final, thus proved). And, once an hypothesis or theory is proposed, as Albert Einstein once stated, "no amount of experimentation can ever prove me right; [yet] a simple experiment can prove me wrong." Of course, because no hypothesis can ever be proved correct, scientists can sound as wishy-washy as a group of lawyers, even for concepts as foundational as gravity or evolution, for which nothing has been found in hundreds of years to prove them wrong. Objectivity is critical: first, the objectivity of the scientist herself, then the objectivity of peer reviewers and editors who consider for publication the papers describing the scientist's findings.

In contrast to such objective science is something I call biblical "science," in which the hypotheses are proposed first, perhaps by the Bible, and only then does the "scientist" look at data, specifically trying to prove that each favored hypothesis is true. They do that by accumulating evidence that supports the hypothesis and by rejecting evidence that does not support the hypothesis. In other words, there is no objectivity. Citizen "scientists," some highly educated (though rarely in the topic on which they are speculating), and even trained scientists need considerable self-discipline not to lose their objectivity when proposing an hypothesis. Of course, biblical science is not science; it is a far cry from how Einstein operated! For conclusions of this type coming from the literal interpretation of the Bible, for example, the "evidence" may be passages from the Bible itself because the "scientists" consider those passages to be the word of God and therefore true. Yet we see many more examples of biblical science that are not related to fundamentalist religion. This kind of "science" includes reports of aliens/UFOs/alien abductions, the supposed location of Noah's Ark, crystals, Atlantis, astrology, soothsayers/psychics, a hollow earth, Immanuel Velikovsky's many nitwit ideas, reasons against evolution, opposition to vaccination, homosexuality is a choice, and so forth. Particularly appalling examples include "scientific studies" that purport to show that people of color are mentally inferior to whites, ignoring the fact that all of us are *Homo sapiens* who have cross-bred to varying extents over many tens of thousands of years. Conspiracy theories, so beloved by Republicans, result from this same non-objective thinking. Most ads on TV are at best non-objective and at worst intentionally deceptive, including those that claim that "clinical trials" or "scientific studies" support their product; such studies are in fact typically done by the very company that made the product, thus hardly passing a test of objectivity. The key to avoiding loss of objectivity is to seek out rigorous peer reviews by experts in the field about which one is theorizing.

I attempt in this memoir to put faces (my own and many others') of scientists doing their job in one branch of science—geology—that I witnessed. I want to share with you the joy of doing science, a profession that requires one to think creatively. I want to show you that a career in science is exciting, not only mentally, but also physically through the field work we scientists do. In addition to objective accumulation of data and their interpretation, science requires writing up of results for peer review and publication or for oral presentation, and this step is fun too. Beyond that, no scientist works alone, and collaboration with others adds a much higher level of enjoyment.

Finally, science benefits society. That benefit may not be as immediate as the benefits from professions such as nursing or policing. But even if long in coming, science must result in the accumulation of knowledge that can be turned into a societal benefit at some future time. Moreover, most benefits from scientific professions become evident both practically and, in many cases, relatively quickly, at least in the time it takes for scientific conclusions to be presented or published. For example, scientists and analysts working on the COVID-19 pandemic of 2020–2023 daily provided valuable internet information and press briefings. Scientists such as doctors and nurses (treating patients), statisticians (analyzing tests and tracing cases), and lab technicians (doing testing and exploration) gathered, analyzed, and parsed information to try to find a vaccine or other form of relief; results of those studies were made public daily, changing behaviors and doubtless saving the lives of those who took the time to listen.

Science is one of the few professions in which the object is to think thoughts and have ideas that nobody has ever thought or had before. In whatever discipline or field that a scientist is in, each project is one of exploration. Such exploration does not have to be with a dog sled into unvisited lands. It can be exploration of the gene, where one of the main tools is a microscope. Nor does such exploration necessarily entail something visible, for it can be in the mathematics of a feature, as for example that of an orbit of a planet or of an atom. Or it can entail interpretation of a geophysical pattern regarding what the interior of the crust, mantle, or core of the earth looks like, with the knowledge that you will never see anything about which you are speculating. Regardless of the scientific discipline, a scientist uses whatever has already been discovered as a stepping stone to discovering something new. Indeed, scientists advance by climbing on the shoulders of predecessors, always acknowledging such previous work. Whatever the discipline, whatever the subject, or whatever the method, in all cases, practicing science is an adventure!

This memoir focuses on my own scientific specialty—that of field geology, which concerns the study of geology in the out-of-doors, where all the rocks and surficial sediments are. Excelling in field geology requires a person who does her best thinking in the field. That person is more often a generalist than a specialist (a generalist knows a little about almost everything, and a specialist knows a lot about almost nothing). Rocks have different lithologies (that is, descriptions, including mineral types, color, degrees of consolidation, etc.), which in college Introductory Geology (Geology 101) are commonly subdivided into sedimentary, igneous, and metamorphic rocks. Unconsolidated young sediments at the surface (surficial deposits) also have different lithologies and origins, such as from recent landslides, stream deposits, and sand dunes. Furthermore, the rocks and young sediments have different ages, depending upon when they were formed or deposited, ranging from yesterday when a flash flood spread new stream gravel, to a lithified sandy beach formed more than a billion years ago (for with time, pore spaces between grains of clay, sand, and pebbles are filled with minerals deposited by groundwater, like the white calcium crusts that accumulate on a faucet or shower head). The relative age of sedimentary rocks, which were originally deposited in horizontal layers, can be determined by stratigraphic position, that is, older rocks are below and relatively and successively younger rocks are progressively in higher up layers. Fossils in these sedimentary rocks provide the start of determining absolute age (in years), for paleontological study of similar or identical fossils allows correlation with sedimentary rocks in other parts of the world that have been dated by isotopic methods. Isotopic dating provides numbers, generally in millions (Ma) or billions (Ga) of years before present, that can be assigned to some rocks by measuring relative amounts of isotopes (an element with the same number of protons in its nucleus but differing from another in its number of neutrons) in minerals in those rocks.

The main thing that a field geologist produces is a geologic map, which shows how various rock types are distributed across a particular area, generally shown by different colors, and the relationship of those rock types to each other and to towns and other cultural or geographic features in the area. Making a map requires two actions—observation and interpretation—and a map contains both. A geologic map shows contacts (boundary lines) between different rock types and sediments. Rocks and sediments may have been deformed by structures such as folds and faults, and the locations of these are also plotted. Putting a map together is known as geologic mapping. Making a map is not a routine or entirely descriptive exercise (although some persons have disparaged it in these ways) but requires interpretation, for contacts and structures may be partly covered by soil and vegetation. Foot traverses of the area make up most of the effort, although one first examines flight lines of vertical aerial photos of a similar scale to the map. Library and lab work follow the field work. Lastly, in addition to the map, a field geologist must include or produce later (generally both) a report that discusses her observations and interpretations.

Maps are made in quadrangles, block by block. A quadrangle is a rectangular map bounded by lines of latitude (parallel east-west lines) and longitude (north-south lines that meet at the poles). These are published in a map series of a prescribed scale. The map scale that one uses depends on the objective of the mapping and determines the amount of effort and time spent. The USGS publishes paper copies belonging to three main series of topographic maps (quadrangles), in which topography is shown by contour lines. The most famous (used by hikers and engineers) and the most detailed scale of published topographic quadrangles by the USGS is at 1:24,000 (1 inch on the map equals 24,000 inches on the ground; or, put another way, 1 inch on the map is almost 0.4 miles on the ground), and the paper map (quadrangle) stands about a foot and a half wide and nearly two feet vertically. The map represents 7.5 minutes of latitude and 7.5 minutes of longitude. It represents an area (in southern Utah, for example) that is about 7 miles east-west and 9.5 miles north-south, about 67 square miles, but because latitude lines converge at the poles, a quadrangle is smaller toward the poles. The Topographic Division of the USGS published 55,000 of these 7.5-minute quadrangles between 1947 and 1992 in the lower 48 states. You may also buy an intermediate-scale topographic quadrangle, at 1:100,000 scale (1 inch on the map is about 1.6 miles on the ground), which covers 30 minutes of latitude and 1 degree (60 minutes) of longitude, and such a quadrangle stands almost 3 feet wide and 2 feet high. This 1-degree quadrangle (or 1-degree sheet) represents an area that is about 55 miles east-west and 35 miles north-south in southern Utah, or about 1900 miles^2. The third map sold, considered to be reconnaissance scale, is a 1:250,000-scale quadrangle (1 inch on the map is a little more than 4 miles on the ground), which covers 1 degree of latitude and 2 degrees of longitude, and it stands about 2.3 feet wide and 1.4 feet high. This 2-degree map represents an area that is about 108 miles east-west and 68 miles north-south in southern Utah, or about 7300 square miles.

Geologic mapping on the three types of quadrangles requires quite different amounts of detail/accuracy such that each type can be published after a reasonable amount of field work (two or three summer field seasons). James Gilluly (1896–1980), an eminent USGS geologist (a member of the National Academy of Sciences, employed by the USGS from 1922 to 1966) that I used to see regularly in the USGS library in Denver after he retired, had a rule of thumb for fieldwork, regardless of scale: he suggested accomplishing about a square inch of geologic mapping per day. Therefore, it was most fun to map a 7.5-minute quadrangle, for it gave you the time to visit most outcrops. And it was least fun to map a 2-degree sheet, where a square inch covered a great deal of land and you did much of the work preparatory to mapping by working on aerial photos in the office before you went to the field, then field checking your office work by driving all the dirt and paved roads in your field vehicle but not being able to spend much time on outcrops except those near the roads; each 7.5-minute quadrangle (of which there are 128) within the 2-degree sheet needed to be done at a rate of almost one per day! That is why it is known as reconnaissance geology. This was our mapping scale in Antarctica, about which more later.

Undertaking field work is of course much more than whacking on rocks. The person who does field geology, being a general geologist, is curious about things besides the rocks and sediments in her field of study. Moreover, knowledge about soil and vegetation commonly allow one to make better interpretations of underlying rocks, and one must also apply knowledge about geophysics and chemistry. And the experience of field work is made better by the beauty and quiet of the hills and mountains traversed, the smell of sagebrush and pinyon, waiting out a cool downpour under a tree or cliff, the wild animals one will see or hear, and many other experiences. Even the screams of peregrine falcons on thermals above, and the howl of coyotes make an all-day traverse entirely worthwhile. For a field geologist, the rocks are the most interesting of these experiences because each outcrop visited provides a new discovery and a piece in a jigsaw puzzle that allows us geologists to project rock distributions and interpret what happened to the rocks. The next outcrop to which one hikes becomes the next piece in the puzzle, and so on. Geology is the perfect profession for the curious!

A person is not doing science unless she releases the data to the public, generally by publishing her findings and interpretations. Publication must reveal everything (tables of supporting data, followed by interpretations) she has discovered, with no withholding of secrets, no company-confidential data, and no leaking of information to a company or person who might profit from a discovery. For example, USGS geologists may discover rich mineral deposits during their mapping and related studies, but they are happy merely to write about them, with no eye to profiting from the discovery. When they publish the discovery, people who read the report first are the ones who stake the claims and who develop the deposit by mining it. Publication of results allows scientists who come later to replicate or advance what the earlier geologist has done. If a geologist merely records stuff or works out problems on their own, with no written record that went through peer review or release in an objective, established journal, we would say that that geologist has a hobby or is a journalist, but she is not doing science.

Although I have published more than two hundred reports, chapters, and small books, almost all my writing has been for peers in my field, using a lingo that my peers will understand but the average citizen likely will not. This memoir, however, has a far more difficult writing task: to communicate my love of science to citizens.

Most scientists are employed by city, county, state, and federal governments. Whether working at a low level or a high level, they gather data, interpret it, and pass on to businesses, peers, and the public any advice that comes out of their findings. Scientists include many teachers in public schools, private colleges, and state universities. In other words, these scientists who assemble and interpret facts are mostly working for the taxpayer; they are public servants. However, significant scientific research is paid for by the private sector, such as pharmaceutical companies that develop drugs, oil companies that find oil and gas, mining companies that exploit mineral resources, etc. None of these are jobs that will allow a person to get rich, but then money rarely motivates scientists: the discovery and the information do. Many members of the taxpaying public—perhaps you too—are unaware that they pay for most science, not only that of public servants but also by the prices they pay for drugs, metallic objects, fuel for their car, etc. Most technologies we now enjoy, such as computers, software, fiber optics, satellite technology, and the internet, were developed during science projects in research and development (R&D) laboratories run by the federal government (see the book *Concrete Economics* by Stephen S. Cohen and J. Bradford DeLong, 2016). So too were vaccines (such as for COVID-19, influenza, and malaria). Such government-paid scientists are almost overwhelmingly competent, dedicated, and hard-working; they are proud of what they do, and they have every reason for this pride. Society could not function without them. Yet many persons today think that government workers are primarily bureaucrats, engaged merely in pushing papers or fomenting regulations, and that therefore they are not important and this type of work force should be shrunk to save money. Republicans have long peddled this tune, and it grew worse under Trump.

In late 2018 and early 2019, Trump ordered a 35-day shutdown of the federal government. Only so-called "essential" federal workers were allowed the privilege of working, but they were not paid until later. The whole fiasco—a political blackmail attempt by Trump against the Democrats—was a huge hardship on millions of people inside and outside the government, all of whom were innocent targets of Trump's wrath! A beautiful article ("Portrait of an inessential government worker") by Michael Lewis in the October 21, 2019 issue of *Bloomberg Opinion To-*

day (also summarized in Lewis' 2019 edition of his book *The Fifth Risk*) describes one such "nonessential" scientist, an oceanographer named Art Allen, who worked in the Search and Rescue Division (the lone scientist there) of the U.S. Coast Guard, reaching a grade of GS-14 before retiring recently. This person, unassuming and unheralded (except by many in the Coast Guard), whose only career research topic was analyzing winds and currents and their complicated effects on the routes of people and overturned boats in the water in order to rescue them, used his science to save the lives of many persons lost at sea, and he ended up spearheading the development of SAROPS (Search and Rescue Optimal Planning System), a software that analyzes many types of data to find and rescue people in the water. All government employees, including scientists, can and should identify with Art Allen, for he is the model of a public servant but hardly atypical from many I have known, some of whom I describe in what follows.

Ours is a time in which many people disparage science—and indeed truth—perhaps in part because we scientists did not try hard enough to reach members of the public, who after all pay our salaries. Scientists do more than describe things and accumulate facts: they are also responsible for interpreting why and what those facts mean (the fun part of science), based on their experience. Those interpretations may not always be correct (think of how astronomers like Nicolaus Copernicus, Galileo Galilei, and then Stephen Hawking increased understanding of the solar system, but only incrementally, over time, with more discoveries undoubtedly to come in the years ahead), but at least scientists make those interpretations honestly and impartially to explain things using the best evidence available to them. This evidence includes that of previous workers as well as what each scientist discovers on her own, and the conclusions are published to allow the next generation of scientists to come along to accept, modify, or render outdated what each scientist has done. To some non-scientists any criticism may seem to be an onerous price to pay for a scientific career. But for scientists it is not, even if one's conclusions turn out to be dead wrong. Some conclusions/interpretations are weak and may be scientifically controversial, but peer review and new data will eventually sort out the correct interpretation, if not in this generation, then in future ones. The strength of each scientific interpretation is directly proportional to the strength of evidence behind it.

Some important interpretations, including at times those that have become politically controversial, are based on evidence whose strength is huge. These include evolution, human-caused climate change, bad health consequences from smoking, acid rain, the ozone hole, and many others. The "controversy" of these, even though the evidence for them is overwhelming, arises because corrupt—typically meaning profit-driven—companies or politicians argue against them and commonly will pay equally corrupt scientists to side with them. To the non-scientist public, such a "controversy" thus becomes a contest between arguing scientists, even though the corrupt scientists are very much in the minority.

Another aspect of science upon which I will expand below is the practical nature of science. Most non-scientists know in their hearts that scientific discoveries have led to inventions, new drugs, behavior changes, etc., and that these have resulted in benefits to many lives. Geology, the most practical of the sciences, discovers potable and farmland groundwater, oil and other energy resources, metallic and non-metallic minerals, and information with the help of which we can counter natural hazards such as earthquakes, flooding, and landslides. Yet especially since President Reagan, some politicians with axes to grind or profits to make have told their public that all scientists are just a bunch of nerds with thick glasses living on the public's dime and pursuing curiosity-based research that has no value to anybody, and worse, that many of their data and conclusions are wrong or falsified. These are lies that foster skepticism of science and truth—exactly as such politicians intended.

My story is one of a shy, introverted, yet curious kid who moved through life, seemingly effortlessly in retrospect, by way of lucky adventures that came from doing science out of doors. My talents have been field geology and a love of the outdoors, plus the enjoyment from writing reports and stories about them. Along the way, I have met and made many friends of extraordinary talents and accomplishments, and I tell many of their stories too. The one thing that we all have in common is a love of doing our craft. Certainly, my family had to endure my absence beyond the typical forty-hour work week while I enjoyed doing geology. My passion took place against the backdrop of the times in which we lived and worked, not all of them pleasant. These times included the rise and fall of the USGS, a formerly important federal agency that lost much of its usefulness largely due to its own actions and inactions, including its lack of courage in the face of our national ultra-partisan politics. We also witnessed a rise and fall of science in general. Partisan politics by the Republican political party has worsened, not least through its encouragement of formerly nascent right-wing religions, but also through its condoning if not glorifying the ignorance generally associated with fundamentalist members of these religions. Such ignorance is overwhelming in those who buy into the literal translation of the Bible, Quran, etc., because that buy-in also assumes rejection of data and objective science. Of course, Democrats have been inept in countering the onslaught, perhaps for the reason given by the poet Robert Frost, namely that "a liberal is a man too broadminded to take his own side in a quarrel."

The Trump administration disparaged science, objectiveness, and the truth in the US, and it hired sycophants who went along with the Trump party line and refused to refute his lies. The people who elected this emperor, Donald Trump, were blinded to the reality that he wore no clothes, to use the well-known parable. Thus, until the election of Joe Biden as President in November 2020 and the subsequent attempt to return to normalcy, the US slowly slid toward a corrupt autocracy that resembled classic banana republics. Before and since Trump, we also witnessed events of international significance, including increasing political turbulence, migration, diseases, and genocide in response not only to leaps in world population, but also to corrupt countries that survive by exploiting and dividing people, and to climate change. In the absence of global calamities, it might have been possible for an incompetent politician to muddle through a four-year presidency. But the accelerating effects of climate change have highlighted the hazards of electing a politician such as Trump, who preferred to turn back the scientific clock and to reject eleventh-hour realities. And as the year 2020 dawned, we also had COVID-19, an exceptionally contagious virus for which no cure existed (see chapter below).

A lyric in Bob Dylan's song "Subterranean Homesick Blues" reminds us that "you don't need a weatherman to know which way the wind blows." The greatest change in the world is that of climate, and it does not take a scientist to detect climate change. Anyone who goes out of doors and can remember things just a decade ago can see differences in temperatures, storm frequency and strength, and other extremes of weather and the concomitant rise and fall of flora and fauna that, taken together, clearly show that something horrific in the climate, and hence life on our planet, is taking place and accelerating in strength. Because all these things are so interrelated, this book is therefore not only about geology, but about objective observations in climate, life, and the politics of science as well. But it's also about how geology (and science more broadly) became part of my life, and to that I now turn, beginning with my beginnings—an abbreviated family tree. In the words of Louis Armstrong, let's see how the red rose bloomed for you and me!

CHAPTER 1

EARLY AWAKENINGS

Peter DeWitt Rowley. The name Rowley came from my father's grandpa, William A. (Armstrong) Rowley (and of course from his forebears), born in Belfast, Northern Ireland. He was a man who "wore orange"—that is, he expressed loyalty to the British crown—and he was an educated son of a well-off family that made textiles, apparently linen. Presumably, the prospect of spending his life producing fabrics did not appeal to the young man, and he ran off to sea, eventually becoming a captain of a sailing ship. After a career on the high seas, at the turn of the twentieth century he settled in northeastern Massachusetts and married Sarah Hooper. They had six children, four of whom lived beyond infancy and one of whom was James H. (Henry) Rowley, my father's dad.

Now my father's grandpa William was a feisty guy who enjoyed the bars and got into fights with Irish Catholics who had settled in abundance in Massachusetts. My dad barely knew his grandpa William, for he died at 56, leaving Sarah not only penniless but having to return to work in one of the many textile mills powered by rivers in Lawrence. She married a second William—William Stansfield, a Welchman and a wonderful person ("a sainted man" according to my dad); they had one child. In the absence of his own grandfather, his step-grandfather William S. became a major influence on my dad's upbringing. Jim, my grandfather and Dad's dad, married Marian Carter, who had descended from the Hemenways, a family that reached America in the 1600s. Jim worked his way up from a teller in a bank in Lawrence to becoming a bank manager for twenty years, but I never knew him because he also died early, coincidentally also at age fifty-six. Jim and Marian had two kids, my dad, Arthur C. (Carroll) Rowley (born February 23, 1917), and sister Ruth, five years his junior. They lived in nearby Methuen, and our family often drove north to visit Grandma Rowley and Aunt Ruth. Though the family grew up as Quakers, later my dad observed that organized religion was not his life's priority, although he was always proud of one of the main credos that characterizes Quakers, that of honesty. His main reason for leaving the Quakers (now known as the Friends) took place when World War II broke out; Dad had no use for religious objectors to military service.

The name DeWitt, my middle name and that of my brother Clark, was my mother Barbara's maiden name (Barbara Jean DeWitt, born September 25, 1918). She grew up with her younger brother, Richard C. DeWitt (1922–2006) in Utica, New York, where my grandfather, Guy C. (Clark) DeWitt, was a civil engineer for the City of Utica. Guy married Angie Mae Humstone in 1916; both were born in 1893. We called them Gamie and Gampie, for we drove to visit them as often as we went to Methuen. Gampie, a quiet but loving, thin six-footer, majored in engineering at Cornell University, and early in their marriage they lived in New York City, where he worked on the tunnels beneath the city (Mom was born in nearby Yonkers). After a brief stint in a Buick plant in Flint, Michigan, Guy accepted the aforementioned engineering job in Utica, not far from where the two had grown up. They built a home in a suburb at the end of a trolley line from downtown, at 11 Bonnie Brae. Mom spent happy summers with her grandparents Humstone in the small town of Vernon, west of Utica, helping them with chores on their dairy farm. Her Grandpa DeWitt was a veterinarian in Oak Hill, a small community southwest of Albany, where he and his wife lived in a pre-Revolutionary War farmhouse and raised sheep and chickens.

Long, long before that, brothers Johan and Cornelis DeWitt were statesmen who dominated Dutch politics during the Dutch Golden Age, when the Netherlands was the main European sea power. But politics changed and both were lynched in 1672 by a mob loyal to the House of Orange, which had just gained power. I have a scrapbook put together by one of the DeWitts that traces the family name back into the early 1600s to the two brothers and others. Apparently, being on the wrong side of politics drove offshoots of this family to the Plymouth colonies in the New World after the voyage of the Mayflower. My mother and father lived in Rhode Island and then Connecticut, the states where Clark and I grew up. Clark was named after William Clark, of Lewis and Clark fame, a family friend of previous generations of DeWitts. The useless fact regarding the Mayflower was recognized early in Mom and Dad's marriage by busybodies at the Daughters of the American Revolution (DAR), who invited Mom and Dad to join their exclusive "club" made up of people who thought that those with ancestors who arrived earlier in America were progressively better than those whose ancestors arrived later. Mom ignored all such racist invitations (they came from time to time) and had nothing but contempt for people who thought this way. Because of similar examples of arrogance, commonly manifested by a lack of acceptance of newcomers to their town or state, my parents tended to paint most New Englanders with at least mild disdain, and they were happy to leave when they eventually retired to Tucson, Arizona, where they found people to be friendlier.

Art (Dad) got a bachelor's and master's degree in chemical engineering from the Massachusetts Institute of Technology (MIT), finishing the latter degree in 1939 (see Notes) as war clouds loomed across the world. Mom graduated from Laselle Junior College in Auburndale, Massachusetts, not far away. Both were members of a glee club in their colleges, and they met at some joint concerts by these schools. They got married in August 16, 1941. Mom was shy and introverted, but beautiful and mentally strong. Dad was not particularly good looking but more extroverted. And he was brilliant. As World War II approached, Dad specialized his studies in artificial-rubber technology, for the Japanese Empire was soon to conquer and annex the rubber plantations in Indonesia, where most commercial rubber trees were grown. Upon graduation, the U.S. Rubber Company in Providence, R.I. hired him. He was too valuable to serve as a soldier in the war, he noted with significant regret. He was in Research and Development there, helping invent artificial rubber. In short, he was a scientist, too.

We lived first in an apartment in Seekonk, a suburb of Providence. I was born there on December 6, 1942, just ahead of the "baby boomer" generation. I spent my first days ("little bitty baby") being rocked by mom in an old family cradle, as the old song "Cotton Fields", one of my favorites, brings home. One of my earliest memories is of Dad returning to MIT with me on weekends to borrow a two-man scull (rowing shell) to row so Dad could get some exercise; he plunked me down in the second seat while he rowed up and down the Charles River. He had rowed as a varsity collegiate sport while at MIT.

The postwar years brought prosperity to all. By about 1943, the three of us moved to the suburb of Barrington, seven miles southeast of Providence and on the eastern side of Narragansett Bay. Here Mom and Dad refinished a small farmhouse on a couple of acres, with a big garden out back. Clark was born there on August 3, 1944, and the first time he slept through the night coincided with the hurricane of 1944, which caused significant damage in other parts of Rhode Island. Our first new (!) car was a 1950 light-blue and white Chevrolet.

A couple of miles away from our new home, Mom found in Molly Nye Tobey a mentor and close friend. Mrs. Tobey was a thin, high-intensity and high-intelligence, no-nonsense grandmother and also a well-known artist with a studio in her house. (Some of her art is in the Metropolitan Museum of Art and the Smithsonian American Art Museum.) She did all types of paintings, but her specialty was another medium altogether: rug hooking. Her set of fifty "Statehood" rugs done between 1946 and 1961 to commemorate the states became famous and are now displayed in the Shelburne Museum in Vermont. One of her sons was the CEO of Amica Insurance Company, another the CEO of Chase Bank. Mom saved her letters from a correspondence that lasted decades until Mrs. Tobey died in 1984, and Molly gave Mom several hooked rugs, some woodcut prints, and a half-dozen watercolor paintings that depict floral designs and U.S. Navy sailing ships, plus a mobile of geese in flight, all of which decorate a guest bedroom in our present home.

We kids, however, found a less inspirational friend, Dutch Vetterlein, on the next block from us, where he and his wealthy parents rented a big Victorian house with a wraparound deck, adjacent to which were big shrubs. Somewhat of a spoiled hell-raiser, he inspired us to climb up on the deck railing and jump into the bushes, something we would not be allowed to do at home. Back then, we had a comfortable, middle-class life, although Mom was strict. Clark was better looking, a thin six-footer like Gampie, not so bookish (did not need glasses) or introverted, and with a better personality, so he got away with more trouble than I did. Newport, where really rich people had their estates, was another twenty-five miles to the south, as were many long stretches of empty beaches on Narragansett Bay. Our family spent many memorable hours walking those beaches, and if a storm came up on a weekend, we rushed down to a beach in the early morning to collect quahogs (round hard-shelled clams used for clam chowder) and oysters that washed up, and soft-shelled clams that we dug up at their "blow holes." We put them in gunny sacks, and Dad shucked oysters when we got home, to be eaten raw. The soft-shelled clams were steamed and served in butter, with a side dish of corn on the cob. I have memories of neighborhood clambakes on the beach that consisted of the same ingredients, some of the best food ever prepared!

While Clark and I were jumping into neighbors' bushes, Dad spent eleven years in research with U.S. Rubber, then in a similar capacity for three years with General Electric. He made inventions and patented them, and he wrote publications about some of his discoveries. But he left me no records of these, and I have been unable to turn up anything. I often wonder why I did not press him more about his career. He modestly noted once that he "made some contributions" to rubber technology, so I assume that his contributions were minor compared with those of many others. Nonetheless, artificial rubber was available for the war effort and perfected in subsequent years. He spent the remainder of his career employed by R.T. Vanderbilt, a large chemical supply company based in Norwalk, Connecticut, where he was a traveling consultant in wire and cable technology until retirement. Although Dad loved and was good at inventing things and doing research, he did not have a Ph.D., and there was a growing number of young people who got that degree after the War. So that was how he came to Vanderbilt, eventually to be put in jobs that involved solving problems (consulting for his employer) and even in sales. He was hardly thrilled to sell chemicals or chemical technology, but he was accommodating; Mom recognized the frustration. This later was an important lesson for me: if you want a career in research, make sure you get the damned Ph.D.!

Mom and Dad put their primary life emphasis on Clark and me. But it was subtle, and we were not doted on or pampered, so it took me at least fifty years to realize their motivations. They were always supportive and they trusted us in our decisions, urged us to figure things out for ourselves. I do not remember them ever doing something without us, other than occasionally visiting neighbors and leaving us with a babysitter. They both—but particularly Dad—encouraged us to be curious, and they made sure we had ready access to interesting people, libraries, and ideas. Dinner conversation seemed to be centered on concepts rather than people, politics, or finances; and we knew to eat everything on our plates (a stay-at-home parent, Mom was a good cook)! We were never protected or urged not to do things that might be too dangerous. Exploring on our bicycles or on foot, to the library and occasionally going miles away from home, was a common pursuit of ours. Weekends and vacations consisted of visiting state or out-of-state parks or museums, camping with our German Shepherd dog Vera, and things that were educational or interesting. Studying and chores had to be done before fun, and fun included reading lots of books. We lived in various suburbs in Connecticut, including Stratford, Bridgeport, Woodmont, and New Haven, and later (after I left for college) Newtown. We stayed the longest in New Haven, not far from the Yale Bowl, where we attended private day schools in junior high and high school. Clark is smarter than I am, so he did not have to work as hard at his lessons as I did. At Newtown, my parents bought a small one-story house on two treed acres, which was the guest house that had been broken up from an estate owned by the actor Paul Muni; it also had a guest house (bedroom and bathroom), where Gampie spent his last years, cared for by Mom and Dad.

During several Christmas breaks, the family drove all the way from Connecticut to a beach at Clearwater, Florida, to escape the cold weather. My most enduring memory of these vacations was not the swimming but a disturbing incident in a small town in Georgia, where we had stopped for gas on a trip south. I recall walking down a narrow sidewalk adjacent to shops on the main street. An older black man was walking toward me and, as we got to within twenty feet of each other I was about to move off the sidewalk to let him pass. Instead, this adult got off the sidewalk to let me, a roughly nine-year-old white kid, go by. I was disturbed, and my consternation and questions precipitated a long conversation as we continued driving, a discussion about the Jim Crow South and the abuses and insults (and lynchings) that blacks still suffered then (and now, even at the hands of the police).

Both Clark and I had curious, adventurous spirits. During the summer between his junior and senior year of high school, Clark hitchhiked across Canada to Vancouver (I drove him up to Quebec, leaving him on a highway outside Montreal), then south through California and to Yosemite (he even climbed Mt. Lyle) and Sequoia national parks and other places, then east to Las Vegas, then southeast to Houston to visit Uncle Dick and Joyce, then New Orleans, and back to Connecticut. He said he lived on Oreo cookies and quarts of milk. An interesting lesson he took away from this, among many others, was that few friends or schoolmates wanted to quiz him during his senior year about the huge adventure, presumably because they could not relate to his experiences. Typically, the only persons who will participate in conversations about unique experiences are others who have had similar encounters. Military veterans will tell you the same. This was the case with my Antarctic adventures too. Many of my best friends are persons who have had similar experiences.

What sparked such early adventures, or the later Antarctic ones? Perhaps we were driven to it, for despite Clark's and my objections, for years our parents refused to own a TV. Later, while we were in high school, Mom and Dad relented and bought one, but they severely restricted our viewing, generally to Sunday nights and then only if our homework were done. One of our favorite shows was *Maverick*, with the actor James Garner. But that was the exception. Usually it was homework, chores, and outdoor play that occupied us.

Then, after a conversation Dad had with a colleague, we started a weekend hobby: collecting minerals. Connecticut and nearby states had pegmatite and other types of quarries that had been mined for abrasives and metals, and what we considered to be the beautiful minerals were now in the discard piles ("dumps'). Getting to the quarries or other spots required hikes, but all of us needed such exercise. This hobby was so much fun that the entire family commonly spent summer vacations visiting areas to look for minerals or fossils, plus we also visited nearby national parks and tourist spots in such places as Ontario, Michigan, Wyoming, Colorado, Utah, and South Dakota. By then we had a gray and white 1957 Chevrolet station wagon, and its tail end was dragging from boxes of rocks by the time we got home.

On one such weekend day while digging in a dump at a quarry, an adult fellow rock hunter nearby asked me the typical adult question of what I wanted to be when I grew up. I had not given much thought to a career, even though I was well into high school. While I hesitated, Dad volunteered: "Why, he's going to be a geologist!" Even though by then we had been collecting rocks for several years, for some reason it had never crossed my mind that I could make a profession out of this!

Decisions on college finally came around. Mom urged me to get out of evil New England, so that took the Ivy League off my list. Instead, our attention focused on small liberal-arts colleges in the Midwest. I was a dutiful child. On spring break of my junior year in high school, Mom and I drove the '57 station wagon around to visit colleges in Ohio and Minnesota that had a geology department. I remember that Oberlin College was ugly, but Carleton College was beautiful. Besides, its president was a guy named Laurence McKinley Gould, formerly a geologist. Two professors staffed its Geology Department, one a kindly older man named Duncan Stewart, who brought his large Airedale terrier, Bobby Marvel, with him to his office, and the other a fast-walking, dynamic young athlete (a wrestler who just missed making the Olympics once), Eiler L. Henrickson. Carleton was the clear favorite after the trip and, following a stressful application period and a worse wait, I was fortunate enough to be accepted to study there.

CHAPTER 2

Laurence M. Gould
c/o Condo'tel
P.O. Box 738
Pinetop, AZ 85932

4/8/83

Dear Pete,

I was greatly pleased to have your warm-hearted letter and the items with it. I have read all 3 with an increasingly high opinion of a lad named Pete Rowley. I am sorry indeed that I did not know you as an undergrate at Carleton.

As a former member of the National Science Board who was involved in the development of USARP I still get the Antarctic Journal and other Antarctic items. I was immensely pleased to see a reference to one of your early Antarctic papers with reference to the fact that you were a graduate of Carleton College. The name seemed familiar but it was not until I was reading the imposing list of your publications and came across the name of Mackin that I suddenly realized why. Some time after I joined the Dept of Geology at Univ of Arizona I had a delightful letter from Mackin telling me what a fine department of Geology we

... impressive one indeed for one ... young. I learned that your ... record was much more ... than I had realized ... course I am delighted to ... that it will continue. ... interested in your extensive ... Utah. I did the geology of the ... 1921+23 for my doctorate. And ... I had an old ... to see "The ...) and immediately ... Albright, Head of the ... of Utah ... one of the ... and persuaded them ... "The Windows". The ... the creation of ... to become Arches ... me much that ... "the Arches" ... a contribution.

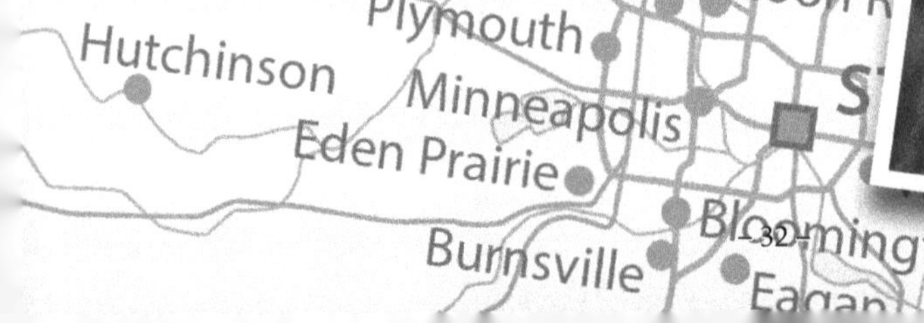

Carleton College, in Northfield, Minnesota, had/has an enrollment of about 1,300 to 1,800 students. Going there was a great decision, for it is an excellent college with a beautiful campus, in a bucolic rural setting; indeed, the sign marking the city limits noted that it was a place of "cows, colleges, and contentment." This farming town straddles the Cannon River and lies forty miles south of Minneapolis/St. Paul. The town had its five minutes of fame in 1876 when the James-Younger outlaw gang attempted to rob the First National Bank. The bank clerk on

LARRY GOULD AND UNDERGRADUATE STUDIES

duty was killed when he refused to open the vault, the gunfire alerting townspeople who then fought back, although one died, as did two outlaws. Two weeks later a posse caught up with the gang, killed another of them, and captured the three Younger brothers. Frank and Jesse James had escaped but the Youngers were convicted of murder and sentenced to life in prison. The bullet holes in the bank were still visible when I enrolled at Carleton. The college is on a hill on the eastern edge of town. Another excellent college, St. Olaf, on a hill about a mile to the west of town, was our major rival in football, basketball, and other sports, but it was a close partner in spirit and in some shared classes.

I was a decent student, although my sophomore year was hardly stellar. My main memory was that I studied day and night to keep up. I was still shy and introverted, so did not date for nearly the entire four years. The constant study and shyness meant that I did not particularly enjoy college, and I was certainly ready to get it over with, to invoke the same spirit of future longing in Emmylou Harris' lyric line "one of these days" in her wonderful song of the same name. My sport was soccer, and working out, making friendships, and the camaraderie of the varsity training table resulted in some added muscle and friends. In the first game of my senior year, however, during a tackle for the ball, I blew out the ligaments and cartilage of my right knee, ending my soccer career. The ligaments came back by the end of the year through rehabilitation and jogging, but the cartilage was never operated on and bothered me from then on. But it became an incentive to walk distances and to exercise every day, so the injury perhaps was a good thing.

As in high school, my worst classes were languages, and the college requirement was to take two semesters of a language (Spanish in my case)—my nemesis, even though I liked and valued professor Antonio Obaid. Cars were banned for students at Carleton, so we were all forced to get along with each other and to study. All of us lived in one of a half dozen dormitories. Applicants to Carleton were chosen largely on their academic achievements, so the women and men (about 50 percent each) were smart, hard-working, and dedicated. At Carleton, I had some good friends and we all did the standard college things, including practical jokes, debates with classmates about the turbulent times, and complaining about the meals (actually, they were quite good). My best friend was my roommate for four years, Lowell A. Quenemoen, a handsome biology major who eventually became a highly successful surgeon in Montana. His supportive and intelligent parents were Minnesota farmers who stressed academics, and Lowell used that as his stepping stone out of that world.

As study-obsessed college students, we were largely insulated from the outside world, but as Bob Dylan sang, "it was a-changin." The civil-rights movement slammed the South on the side of its head, countering the Jim Crow South years that white Southerners adored. Martin Luther King, Jr., during his peaceful march on Washington in August 1963, gave his "I Have a Dream" speech from the steps of the Lincoln Memorial to a crowd of 250,000. King received the Nobel Peace Prize in 1964 for his nonviolent resistance to oppression of blacks and other minorities. I remember with great shock and sadness the assassination of President John F. Kennedy in November 1963, while the civil rights movement was going on. I recall being with other geology majors in our small "office" (two rows of drafting tables) when the shots rang out, and we huddled around a radio to get the details. America changed forever and for the worse with that one event. He was one "elder" person, a politician no less, whom we young people could believe in, but now he was gone because of hate. Five years later, more assassinations destroyed our hope and idealism for the future even more, as I will relate.

I said that I was with other geology majors when JFK was assassinated, but in truth there were not many in each graduating class back then. In fact, only two other geology majors were in the class of '64, Alain Kahil and Terry E. Tullis, and the three of us became good friends and colleagues as we struggled to keep up with the work load. Both did well after Carleton. Alain, born in Alexandria, Egypt, (once he quipped with irony "you love your president; I love my dictator" [Nasser]) left his country when he was accepted at Carleton and ended up living and working in Canada; he got a Ph.D. in groundwater hydrology and worked in it and field geology in business (coal-bed methane) and as a consultant. Terry dated a geology major in the class of '65, then married her (she became Jan A. Tullis) when she graduated from Carleton; they both got their Ph.D.s in "Structural Geology" at UCLA, then both became professors in the Dept. of Earth Sciences at Brown University, and unfortunately later divorced. (Structural geology is a branch of geology that deals with the structure [e.g., folds and faults] and their analysis and cause.) The '63 class had only two majors, one of whom was Wendell A. Duffield, who later had a stellar career with the USGS as a field geologist and geochemist in volcanology, notably at the Hawaii Volcano Observatory, and after retirement became a novelist. During my return from my first trip to Antarctica, Duff showed me around an active pit crater and lava flow along the East Rift of Kilauea volcano, to see a favorite rock type (basalt) in action! The pit crater was about a hundred feet in diameter with vertical walls, and about forty feet below us was red-hot magma swirling around; the nearby flow, about fifteen feet across, was moving at about five miles per hour, with solidifying black chunks breaking off the leading edge, over which the red flow moved. We continued to stay in touch. In the '62 class, there were about five majors, including Mel A. Kuntz and Walter Alvarez. Mel, after earning a Ph.D. from Stanford University, first taught at Amherst College. Then our branch of the USGS, Central Regional Geology, hired him while I was the Assistant Branch Chief under Ed Ruppel, and Mel and I ended up as close friends. He is a great field geologist and volcanologist, and I brought him into my project mapping 1980 pyroclastic flows at Mount St. Helens, and later with the Nevada Test Site (more about these exciting projects below). In 1981, the Carleton geology department invited Mel and me to come to Carleton and give a slide talk about Mount St. Helens, which we did to the Carleton students and faculty in the big chapel, where Larry Gould and other luminaries had held forth hundreds of times. Who would have anticipated such an honor!

> ***Walt Alvarez*** *became a Professor at the University of California, Berkeley, where he and his father put forward his famous theory about the extinction of the dinosaurs by meteorite impact 66 million years ago. With Cretaceous dinosaurs and many other plants and animals extinct in a flash, mammals and other species proliferated and this new age was termed the Tertiary. For this, Walt received the Penrose Medal from the Geological Society of America, the most prestigious prize in the geologic profession. As years went by, I concluded that the Carleton Geology Department was the best among small liberal-arts colleges, if not most or all major universities. That conclusion was based on how most graduates of the department turned out, for many geology departments had or have a Carleton graduate on their faculty, and the USGS also had many of our grads.*

After my freshman year (1960–61), I had a summer job in Gouverneur in upstate New York, doing labor for a talc mill there. (Talc is a clay mineral, the softest mineral, which is used as a thickener or lubricant in paint, ceramics, and roofing materials; ours was not used in talcum powder, a use for purer varieties than ours.) I became friends with a couple of welders, and they occasionally invited me to their home for Sunday dinner with their families. One of the welders had a small farm, and I volunteered on weekends to help the family bale hay for their dairy cows. That was probably the toughest job I ever had, and it was great to drill that into the mind of a youngster, in case that person ever had doubts about getting a college education! The tractor that my welder friend was driving towed a baling machine, behind which was a flat-bed wagon, where I stood. As the bale popped out of the baler, I had to pick it up and stack it on the wagon. As soon as I had stacked it, I turned to find another bale waiting for me. They just kept coming!

My parents bought me a car before I left for my sophomore year. It was a second-hand, about ten-year-old, slightly beat up Peugeot 403 sedan that had been wrecked, although they did not recognize that (I found the repaired places later). This was the payoff for my parents bribing me to not smoke by the time I started college. The car, made in France, had several good points, the main one being that the back of the front seat would go completely down to a horizontal level against the back seat, meaning I could sleep in the car! I just had to pull off by the side of the road each night and throw my sleeping bag on top of the lowered seats. The other good thing was that the car was mechanically simple, so even I could fix it when it broke down, and new parts did not necessarily have to be made by Peugeot. Cars for students, however, were illegal at Carleton. But I drove it out there anyway (it being too far to walk from Connecticut) and found an old garage that I secretly rented from a farmer a couple miles from Carleton. I, of course, used it to drive back and forth from Connecticut, but I also occasionally used it (against Carleton's rules!) to go to the Twin Cities (Minneapolis/St. Paul) or elsewhere. By the time I was in my early twenties, I had visited all states except Hawaii and Washington, much of it with this car.

During my sophomore year, this wanderlust took me to Alaska. In the spring, I applied for numerous jobs up there, but the few that responded suggested I apply in person. With my friend, Dave Field, we decided to split gas costs and go together. So, when school was out, I found a mechanic to put foam pads on the bottom of the gas tank to protect it from rocks, and we headed off on a true adventure, up the ALCAN highway to Anchorage. The ALCAN officially starts (Milepost 1) at Dawson Creek, British Columbia, which is about 400 miles northwest of Edmonton, Alberta. The first hundred miles of the ALCAN were paved. After that, we had 1,100 miles of gravel road (50 mph recommended driving speed) through sparsely populated country before we passed from Yukon

Territory into Alaska. Along the way, just past Whitehorse, the capital of Yukon Territory, we took a gravel road south 150 miles to Haines, in the panhandle of Alaska. In Alaska, the ALCAN and other main roads were paved, and the ALCAN itself officially went to Fairbanks. So, we followed it northwest only a hundred miles past the border, to Tok, from where we took a left onto the Glenn Highway for another three hundred miles to Anchorage, the big city of about 100,000 people (this information comes from a 150-page travel guide that I still have). We slept in the car most nights. The country was beautiful and we soon came to love Alaska and Canada. There were virtually no towns except Whitehorse (population 5,000), but gas stations spaced at regular intervals consisted of clusters of log-cabin buildings that included a café and lodge, and areas for free camping, where we stayed; most of these establishments made their money by guiding people on fishing and hunting trips. It was fun to talk to the owners and workers at these places, for they truly lived isolated lives. They all had exciting stories, especially about encounters with bears. The story I remembered the most (I heard it more than once) was the one about a person who went missing, until they found his boots six months later, with his feet still in them.

After I dropped off Dave at the Anchorage airport to fly home, I began a job search. Almost immediately, I got a labor job, as deck hand and "instrument helper," on M.V. (Motor Vessel) *Grebe*, a narrow, hundred-foot long, wooden-hulled, former submarine chaser built in the Great Lakes and used during World War II. The depth-charge gun had been replaced by another "gun," a thousand-pound submersible seismic rig, which mixed propane and oxygen that were sparked and sent a shock wave down through the sea bottom. We were to look for oil structures (anticlines and such) in Cook Inlet, whose northern end was the port of Anchorage. This was almost a geology job! The tides in Cook Inlet are among the highest in the world, about forty feet between low tide and high tide. We slept on the boat (the *Grebe* was not big enough to call it a ship) when in port. Once on the job, we lowered the gun into the water and did lines of seismic profiles, recording the results on machines (early computers) in a rear hold. This was some of the first exploration for oil in Cook Inlet, which then and later became the sites of many offshore oil rigs.

Harry Furford, the captain of the *Grebe*, had a wealth of harrowing tales from his career as skipper of commercial fishing boats in Alaska. One of his more chilling warnings (pun intended) was that if we ever fell overboard, we would freeze to death from hypothermia before he could turn the boat around—even if he saw us go in the water and even if we had a life vest on! It may have been only a "boat," but it had the turning radius of the *Queen Mary*! As a precaution, the boat therefore had one strand of quarter-inch steel cable strung through a bunch of spindly, sagging three-foot-high metal posts spaced around the edge of the deck as a "railing." Large propane tanks, containing fuel for the gun, were strapped to the outside of the cabin on both sides. Most crew members smoked cigarettes, and one thought picture I have of the *Grebe* was dozens of cigarette butts scattered along the deck, including adjacent to the propane tanks. This obviously was a job where we had to watch our asses and never screw up! But what an adventure! From these salty characters I learned the meanings of such words as port, starboard, marlinspike, belaying pin, and lots of others, some of them clean. Not everything Harry did was above board either. Once, when all the tugs in the harbor were busy, Harry and the *Grebe* pushed a large ship into its berth. Harry billed the ship, not expecting to be paid. When he got a check a month later, Harry took the entire *Grebe* crew out to dinner at a nicer restaurant in Anchorage ("fancy" would not be an adjective I could use for restaurants there at that time).

When our traverses in Cook Inlet were deemed to be done, I had a week's vacation while the crew took the boat west into Bristol Bay in the Bering Sea to shoot more seismic lines. I used the time to explore, taking the Peugeot north about four hundred miles to visit Fairbanks. On the way, I toured Mt. McKinley (later renamed Denali, the Native American name for the mountain) National Park (it rained so I did not see much) and went north of Fairbanks to the end of the road, at Circle, not far south of the Arctic Circle, where huge land dredges were processing river and glacial gravels for gold. Fairbanks, although the capital of Alaska, resembled a frontier town; I remember a drunk cowboy on a horse trying to get himself and his horse through the swinging doors of a bar on main street. Then I returned to Anchorage and we flew to Naknek, at the northeastern head of Bristol Bay. This was our only port, a tiny Indian community of salmon canneries. I got drunk in one of the bars there, and was told that members of the crew had to borrow a wheelbarrow to bring me back to the boat that night. One day, late in the summer, a typhoon (the Pacific word for hurricane) came up suddenly out of the west. We stupidly tried to finish the seismic line we were doing, but within a half hour the waves were so high that we could not winch the gun out of the water. With such a sea anchor preventing us from heading the boat into the wind, we had to cut—by blow torch—the cables holding it and therefore we lost the gun. As it fell into the sea, it pulled out about a square foot of the farthest rear hold (fortunately above the water line), where it had also been attached. One of my jobs in the storm was to drop down into that hold and man the emergency bilge pumps, holding on to the flimsy "railing" to get there! Once the gun was cut, we turned the boat westward, into the wind and waves, and we rode out the storm. Those two days and nights seemed much longer than that. At times the engine flooded, and the engineer had to rush down into the hold and get it started again. The terror of the minute or two of silence each time is etched into my memory! The waves were about thirty feet high, which is near the maximum for large storms. Gordon Lightfoot's popular song, "The Wreck of the Edmund Fitzgerald", recalled that experience to my mind a dozen years later, especially a line about the "tattletale" sound of the wind in the rigging of the ship that foundered.

When the storm was over, we limped back into Naknek. Of course, we could do no more seismic lines, so most of the staff flew home. But Harry kept me on as part of the crew. The trip back southeast through the Bering Sea, then a "pass" between two of the Aleutian Islands, was made the more eventful because one of the volcanoes erupted and covered the deck in ash, and because I had to learn how to pilot the boat, once avoiding hitting a whale by twirling the wheel through the many revolutions it takes to move a boat's direction.

With the job over, my little car and I motored back down the ALCAN, directly to school.

The next winter, while still at Carleton, the great Alaska earthquake, the third largest ever recorded at the time, with a magnitude of 9.2, had its epicenter in Cook Inlet, resulting in much of Anchorage and its western suburbs adjacent to the Inlet being devastated. With much sadness, as I flipped through the pictures in *Life* magazine of collapsed Main Street, I remembered some of the bars we had frequented. One of my best friends later in life, Wallace R. Hansen of the USGS, did the definitive geologic report on the earthquake long before I joined his Branch years later. But back to Carleton.

> *For me, much of Carleton's story concerns* ***Laurence M. Gould****, a hero in many categories, including as an Antarctic explorer. He was born in 1896 on a farm in southwestern Michigan. When he graduated high school in 1914, he had no money to go to college so he took a job as a teacher of all eight grades in a one-room schoolhouse in Boca Raton, Florida, thereby launching his career as an educator. After two years he was able to enroll at the University of Michigan, but when President Woodrow Wilson declared war with Germany in April 2017, Larry enlisted. He was assigned to an Army ambulance unit in Italy for World War I, mustered out, and returned to the University of Michigan for the fall of 1919. That year he took an introductory geology class taught by Professor William H. Hobbs, an eminent geologist and an excellent and witty teacher, and Larry became hooked on geology, with Hobbs his mentor. Larry also impressed other professors, and after receiving a B.S. in 1921, he was appointed as an instructor there. In one of the introductory classes that he taught in 1922, he met freshman Margaret (Peg) Rice. He earned his M.A. in 1923, realized that college teaching was his second love (after Peg), and embarked on his Ph.D. For his dissertation study he mapped the LaSal Mountains in Utah, and his analysis of the laccoliths (plutons that rose*

from an underlying batholith, then intruded between horizontal sedimentary strata so they have a flat floor) that formed the range. He received the Ph.D. in 1925, but Michigan would not let him go and hired him as an Assistant Professor for the 1926–27 academic year.

Before the academic year ended, Hobbs was planning a June to September expedition to western Greenland to study weather patterns and glaciology, and he invited Larry to be his second-in-charge. They reached Greenland by way of a schooner, with both a motor and sail, and endured swarms of black flies and mosquitos to do their science, camped on muskeg (peat bogs formed because of permafrost) in fjords near communities of Eskimos who came to be friends and benefactors. Although logistics prevented them from doing little more than venturing onto the edge of the glacier ice, the lure of polar exploration had grabbed Larry. In the summer of 1927, he accepted the job as assistant director and geographer for another summer expedition, this time to western Baffin Island. The American Geographical Society sponsored this one, and Larry used the same motor schooner. The work was simple exploration to chart coastlines, and though it included several inland treks on ice, for the most part it consisted of small-boat work amongst icebergs and grounded ice to survey bedrock. That expedition began only one year after thirty-eight-year-old Navy Lieutenant Commander Richard E. Byrd had completed a flight across the North Pole, for which he was fêted with a New York ticker-tape parade. Thirty-year-old Larry's expedition started only a few weeks after Charles A. Lindbergh had made the first solo flight across the Atlantic Ocean. Byrd had planned to make the first solo transatlantic flight, but he crashed during a test flight and broke an arm. He ended up making the third transatlantic solo flight, for which he received a second ticker-tape parade, while Larry was exploring Baffin Island.

Now Byrd announced plans for a trip to fly to the South Pole of Antarctica, and influential friends and backers of both Larry and Byrd made sure they met each other, with the objective being for Larry to join Byrd's expedition. They met, got along well, and Byrd made Larry his geologist, effectively the chief scientist. The Byrd Antarctic Expedition was privately funded, and Byrd took on the main job of fundraiser. By the summer of 1928, Larry moved to New York to work on final preparations, amidst a huge publicity campaign and daily newspaper headlines. Along with four airplanes and five hundred tons of supplies and materiel, the expedition was to consist of eighty-two men. Among them was nineteen-year-old Eagle Scout Paul A. Siple, who later joined four other Antarctic expeditions, got a Ph.D., and had a long career with the Army Scientific Office in cold-weather research, specifically determining what was the most effective clothing in such conditions. He also invented the name and concept of wind chill. Paul died in 1968. With funds raised and supplies and men gathered, and once under way, Byrd announced that Larry was to be his second in command. In late December they arrived at the Bay of Whales, an indentation in the eastern Ross Ice Shelf that was recognized as the closest "port" to the South Pole, and they established their base, called Little America. The Bay of Whales was where Norwegian explorer Roald Amundsen had established his base, Framheim, from which he launched the successful first attempt (by dog sled) to reach the South Pole in 1911 (discussed further below). Flights started soon after, and led to the discovery a hundred miles to the east of previously unknown mountains, which they named the Rockefeller Mountains after a benefactor. The entire coastline of the continent to the east of the ice shelf was named Marie Byrd Land, after Byrd's wife.

Preparations were made for forty-two of the men (including Paul Siple) to winter over; the disappointed remainder shipped home in late February. However, on March 7, right before the winter set in, Larry made a rookie mistake (it was too late in the season for this!): he persuaded Byrd to let him take one of the planes into the Rockefeller Mountains to do some geology. He, the pilot, and the engineer moved into a tent near the plane, but the next morning a storm formed that intensified over the next days to wind gusts of more than 100 mph according to the plane's airspeed indicator, but likely significantly higher than that. The wind resulted in the plane going airborne for at least a half mile, to its destruction. With the radio gone, the news flashed across the US of three missing men, and it was not until March 22 that Byrd piloted a rescue plane and flew them back to Little America. Now the expedition was down to two aircraft for a dangerous polar flight from Little America to the Pole of about eight hundred miles each way. The expedition no longer had sufficient planes to support geologic work or to lay caches of emergency food, fuel, and oil along the route to the South Pole.

Predictably, Larry had a bright idea: he proposed a sledge trip that would precede the polar flight and go to the base of the Queen Maud Mountains. This range was part of a vast string of mountains that spanned Antarctica and that Warren Hamilton of the USGS later called the Transantarctic Mountains. The plan was for Larry and his men to leave a southern supply depot at the edge of the Queen Maud Mountains, provide weather reports to guide the flight, and of course do geology. The route would be essentially the same as the one that Amundsen took in 1911, except that it would stop at the base of the Axel Heiberg Glacier, rather than climbing up it through the range to the Polar Plateau.

The planning and preparations took place over the long winter. Then, as the spring approached, advance sledge parties went out the first two hundred miles, laying depots every fifty miles for Larry. Black Tuesday, marking the stock market crash that prompted the Great Depression, took place on October 29. On November 4, 1929, Larry and five others headed south the 450 miles to the mountains, with forty-six dogs distributed in five teams of two sledges each. On November 28, when they were in sight of the mountains, they reported favorable weather to Little America. The larger of the two planes, piloted by Bernt Balchen and with Byrd the copilot, left at 3:30 pm and returned on the morning of the 29th, a successful flight that was immediately broadcast to the world. With their primary job of air support over, Larry and his colleagues made it past some large fields of crevasses to the Queen Maud Mountains a few days later and started doing geology. They climbed into the mountains to map, measure, and sample the rocks, then continued east along the edge of the mountain range, skirting more crevasse fields as they continued with the geology along the mountain front. This advance into unknown land continued until December 20, when the conditions and their supplies required that they begin to trace their steps back. They returned to Little America on January 19, having covered about 1500 miles. With that last event on the exploration schedule completed, everybody packed up and returned to New York by ship, arriving in mid-June.

Larry and Peg got married in August 1930. For a two-year leave of absence, they lived in Manhattan, writing book manuscripts and publishing the trip's scientific data, primarily their new geographic discoveries and the geology. To recoup past and present expenses, Byrd and Larry were on a lecture circuit, with Larry giving about sixty-two hour-long lectures around the country. The one at Carleton College so impressed the president and others at the college that they began recruiting Larry to establish a geology department there. In the summer of 1932, as the Dow Jones Industrial Average bottomed out, Larry was hired to teach geology, having accepted a full professorship and the highest paid salary of any Carleton prof! He was then a mere thirty-seven years old. A wonderful teacher with a great sense of humor, his classes were well attended and his students loved him, then and throughout his career. Peg also was a well-loved part of the Carleton team. Larry hired Duncan Stewart in 1933 (more about Dunc later). In 1947, Larry hired Eiler Henrickson (a Carleton geology major, class of 1943) as an instructor, following duty with the U.S. Marines, USGS, and U.S. Army; this was, of course, the World War II era. From there, Eiler worked his way up to his Ph.D., and later full professorship, and retired in 1987; in "retirement," Eiler became chairman of the Geology Department at Colorado College and built up that department.

In May 1945, a few days after Germany surrendered, Larry became Carleton's fourth president. In this role, he made Carleton one of the top liberal-arts colleges in the country. Besides his attributes of hard work, courage, common sense, and brains, his personality was ideally suited for his role as college president: he was a gifted speaker, he was charismatic and empathetic, he attracted donors and their funding, and his great sense of humor, including always wearing a red tie, made the college a happy place. Larry's philosophy was strong on academic freedom, the importance of a liberal education, the necessity of science, and the value of small liberal-arts colleges. He considered his most important task to pick the strongest faculty possible. He could be quite outspoken on politics, education, and local events, including his abhorrence of Nazi Germany. Later, although an Eisenhower Republican, he expressed admiration for President Kennedy and disgust with Presidents Nixon and Reagan. Through his many public speeches (he received twenty-five honorary degrees) and his leadership role on many national committees, he was an important voice in academia and was well known to the press.

Larry clearly loved and made students a major part of his day, unlike some college presidents who were hired to spend most of their time doing administration and fund-raising work. One morning, when I was walking to classes across an open area (known on campus as the "Bald Spot," part of it flooded deliberately annually to become the hockey rink) in the center of the campus, yet only one month (!) into my freshman year, I passed Larry going the other way. He grinned and said, "Good morning, Pete." That he knew my name knocked my socks off, and I immediately knew I was in the right school! He was the primary inspiration to all 1,300 students, and I never met any of them who did not love him. At his request in his "Welcome Freshman" speech, every one of them called him "Larry" from day one. As told (p. 530) in Larry's biography by Eric Hillemann, Larry was asked in 1984 what it takes to be successful in life. His reply: "Discipline—it's indispensable to any successful thing, isn't it? The other essential is integrity. And you can't be lazy. If you have those, and you have any gifts at all, well there you go!"

It was a great sadness to all Carleton students when Larry retired in 1962 at age sixty-five. But he became a boon to the University of Arizona, for in February 1963 the Geology Department there hired him. More important, he was key to upgrading the department and bringing in money. Partly under his leadership, the university hired new faculty of unusual competence. He and the people he hired turned the Arizona Geology Department into one of the best large state-supported departments in the country. One of these was the hiring of James H. Zumberge in 1968, who as department chairman in turn hired George H. Davis. Zum-

berge had led several geological and glaciological expeditions to Antarctica, and later became a college president, including at the University of Southern California. As told below, I was the first to step foot on one of the features in Antarctica named after him, Cape Zumberge. And George was one whom I would meet my first day in graduate school and remain friends with since then (more on George later). As at Carleton, Larry was an inspiration to Arizona students and faculty. That and his ability to solicit private contributions led to a new geology building, the Gould-Simpson building, in 1986. He taught his last course (Glacial and Quaternary Geology) in 1979. Peg died in 1988, whereas Larry lived to 1995, aged ninety-eight years. At Larry's death, Tony Obaid, my Spanish professor, quoted a student at the University of Arizona saying that "[i]t just isn't fair for any department to have one of its members be so popular and admired." Antarctica has three places named for Larry: Mt. Gould, Gould Bay, and the Gould Coast. Was he the inspiration for my later career in Antarctic geology? Certainly not exclusively, but he was the first of a series of people and coincidences that led to it, and the next one happened during the spring of my junior year.

I have a three-page handwritten letter that Larry wrote to me on April 8, 1983, in response to my letter to him accompanied by the obligatory resumé and several reprints of reports I had written about Antarctica. It seems that he had maintained his familiarity with the literature on Antarctic geology and had seen my name before but had not remembered our Carleton connection. He wrote,

"[Your] name seemed familiar but it was not until I was reading the imposing list of your publications and came across the name of Mackin that I suddenly realized why. Sometime after I joined the Dept. of Geology at Univ. of Arizona, I had a delightful letter from Mackin [the story of Hoover Mackin appears in the next chapter] telling me what a fine department of Geology we had at Carleton and how well our students compared with others in graduate school. Naturally he mentioned you. I have always been sorry that I didn't know this remarkable man better."

Larry mentioned our mutual work in Utah (where he also did his dissertation study), including the news that when he was doing this Ph.D. work, he "had an old cowboy take me on horseback to see 'The Window,'" an arch that inspired him "immediately [to] get in touch" with several senators in Washington, leading to the creation of Arches National Monument. He also commented on our mutual work at Mount St. Helens, which he climbed in 1941.

The best part of Carleton was the professors because, as is typical of a small liberal arts college, they prided themselves on being teachers. It was clear that they valued this role beyond all else. Although all of them did research and published books and reports, they differed from some more egocentric professors at research universities by being student oriented. Those in the geology department were in this mold: good field geologists but, more importantly, supportive, dedicated, likeable, and demanding teachers. Duncan Stewart (1905–1969) and Bobby Marvel returned to the office most evenings, for Dunc was the foremost petrographer (specializing in the microscopic study of thin sections [rock slices cut thin so that a light bulb below them will allow light to pass through them and minerals in the rock to be identified]) of Antarctic rocks, on which he published extensively. That he was also a geologic consultant, doing field work during the summers, seemingly meant that he was available night and day to all students, to whom he was devoted. Dunc had a dry sense of humor, as reflected in the classroom and in all personal contacts. No wonder that in a framed picture of him that I've had on my desk for much of my life he is wearing a sombrero and an impish half smile, suggesting that he has discovered a secret that life is good! I remember him in my first introductory geology course (Physical Geology, Geology 101) drawing a *Brontosaurus* on the blackboard, presumably so that we could distinguish it from *Tyrannosaurus*, and without a word or fanfare finished the picture off by putting a smile on its face; few students noticed but I, sitting in the back row high up in a big lecture hall, had to stifle my laughter. Though Dunc had done his Ph.D. dissertation (University of Michigan, when Larry was there) on the petrographic description and geologic interpretation of the rocks that Larry had brought back from Antarctica from his sledge journey, he himself had never been to Antarctica. It so happened that Larry, who had maintained a close association with (the leader at) the United States Antarctic Research Program (USARP), had arranged a VIP tour for Dunc in the austral summer of my freshman year. Thus, Larry taught about half the Geology 101 course that I took, when Dunc was not there to put smiles on the faces of dinosaurs! It was a special honor and joy to see Larry in action.

Eiler Henrickson was not technically a good or well-organized teacher, but he made up for it with his incredible enthusiasm for geology and geology students, and he piled on the work until, but not beyond, our breaking point. One of his adventures was a summer job (1958) looking for mineral deposits on the shores of Lituya Bay, a fjord at the opening to the Gulf of Alaska, in the panhandle of Alaska. One Friday night, after being lifted off by helicopter when the day's traverse had ended and after being taken to a nearby town for the weekend, an 8.4-magnitude earthquake shook the bay like a bathtub, and a giant wave (1,900 feet high, the highest ever recorded) from the bay topped the fiord and produced a path of destruction (including through their thankfully deserted camp), fortunately resulting in no loss of human life.

Eiler was as much a lifetime field geologist as Dunc. I came to admire both men, even love them. We all went by first names, or at least we used theirs after we passed our comprehensive exams late in our senior year. Their emphasis was on field geology and specifically getting students to think in the field, a process that requires understanding the implications of what they observe and what and where to go next to work out remaining questions of age and origin of exposed rocks. Minnesota is not blessed with exposed rock, for Pleistocene (a geological Epoch of time characterized by ice ages and interglacials, from about 2.58 million years ago to 11,700 years ago, the end of the last ice age; as opposed to the Epoch we now live in, the Holocene, from 11,700 years ago until today, and brought the domestication of plants and animals, thus farming, towns, and civilization) continental glaciers had deposited, when they melted, their unconsolidated debris and covered up all real rocks. So virtually every holiday of three days or more, we students and Eiler visited and camped out wherever he could find bedrock in adjacent states, Ontario, the Upper Peninsula of Michigan, the Black Hills, and other places. That perfectly suited Eiler's and my energy levels. Thus, I got D's in Spanish but A's in all my geology courses. I became a field geologist, or at least aspired to be one. There was never any doubt after my sophomore year that I would be moving on after Carleton to grad school and a Ph.D.

Another Carleton graduate (1952, history) was ***John J. Anderson*** *(1930–2017). Graduation was followed by a Fulbright Scholarship to the University of Copenhagen, Denmark, in 1952–1953. Born in Port Arthur, Texas, he and his older brother Poul (later to become a prolific science-fiction novelist) moved to Northfield as young children with their mother Astrid, after their dad/ husband died in an automobile accident. I had met Astrid before I knew John, for she had become a reference librarian at Carleton's library (later named for Larry Gould). An emigrée from Denmark before her marriage in Port Arthur, she was a thin, intense, attractive, highly-intelligent person with a delightful personality. John was headed toward a career in political science but in 1954 he was drafted into the U.S. Army, with his duty station (one year) as a Personnel Specialist in Iceland. Happily, perhaps due to the active volcanoes in Iceland, John discovered geology. When his Army tour was over in 1956, he enrolled in the Geology Department at the University of Minnesota to pursue a Master's degree. His thesis advisor was J. Campbell (Cam) Craddock, the pre-eminent US specialist in the geology of Antarctica. Later, during my Antarctic career, I came to know Cam well, and considered him one of the finest and kindest geologists I ever met—as did John.*

John's thesis was to be on the geologic mapping and analysis of part of the Ellsworth Mountains, the highest range (more than 16,000 feet in elevation) in Antarctica, and previously unvisited. But the timing of the trip was delayed a year by the funding agency, the National Science Foundation (NSF), so John used his his-

tory background to do a library thesis on the "Bedrock Geology of Antarctica," which in 1965 was published as a long and important summary article. Nonetheless, the trip of a lifetime to Antarctica was scheduled from October 1961 to February 1962, the plan being for Cam to lead a joint program of students from the Universities of Minnesota and Wisconsin. Unfortunately, Cam became sick at the last minute and John—as Cam's Research Fellow—was appointed to lead the field party instead, even though he had never previously been to Antarctica. The team lived in two-man Scott tents and traveled by motor toboggans (the predecessor of snowmobiles). In later years, John often marveled that it was impossible to replicate such a professional and personal high as had been his work in this beautiful, huge range. Anderson Massif, a twelve-mile-long mountain mass in the Ellsworths, was named after him in 1968. He described, with Cam and the others, the geology of the area in an article in the prestigious journal Science, *including a dramatic photo on the cover of this issue of one of their tent camps at the base of a mountain swathed in clouds. An enlarged framed version of the photo thereafter hung in his office or home. Cam led other trips to the Ellsworths and, with John as co-author, went on to publish other articles on the mountain range, including the discovery of the fossil plant* Glossopteris *of Permian age, also released in* Science, *that was used to demonstrate the then hotly debated concept of "continental drift" because the other southern continents also contained this fossil and therefore had to have been joined together then, as the former supercontinent Gondwanaland.*

From there, in 1962, John entered another grad school, this time the Department of Geology at the University of Texas, where he selected J. Hoover Mackin as his advisor. For his first year or two, John supported himself with teaching, including Historical Geology (Geology 102), an introductory course for undergrads. (Interestingly, that course in the spring of 1963 was the second that student Richard B. Waitt, Jr. ever took—and even back then John was an excellent teacher, according to Richard—on his way to a bachelor's degree in geology at Texas. Coincidentally, Richard became one of Hoover's last grad students and later, having joined the USGS, accompanied me on one of my Antarctic trips, and worked on the 1980 eruptions of Mount St. Helens, among other projects.) Like Cam Craddock, Hoover was the best in his field, which was geomorphology and field geology, and he was a member of the National Academy of Sciences. The year before, Hoover had been lured away from the University of Washington to occupy the first endowed professor chair in the geology department of the University of Texas, for a heretofore unheard-of professorial salary.

While at Washington, Hoover and his students did geologic map-

ping in southwestern Utah, in thesis areas that moved progressively farther out from the Iron Springs mining district, which Hoover had mapped during World War II (details given below). So, Hoover and John visited Utah for several days and toured a large area, ending their hunt at a previously unmapped area in the northern Markagunt Plateau, on the Colorado Plateau and the southern flank of the great Oligocene- and Miocene-aged Marysvale volcanic field. Hoover also suggested that John extend this area to the next range to the east, the Sevier Plateau, but John was able to talk him out of that. As John told it, Hoover's parting line after the grand tour was "There it is, John. Do what you want with it!" What an invitation to become a professional! John settled on an area of about 225 square miles—and determined that he needed a field assistant. That spring of 1963, while on a visit home to see Astrid, he stopped at his alma mater *and talked a junior in the Geology Department into joining him for most of the upcoming summer. That junior was me! John was to become one of my most important mentors.*

With field support from one of Hoover's National Science Foundation (NSF) and National Aeronautics and Space Administration (NASA) grants, John then bought a large, old panel truck that he gutted and fitted with beds and a Coleman stove. He named it Detgaav'nok, Danish for "the beast will probably run." Thanks to the beast, John and I were able to map all day without interruption, then pull off the road at the end of each day (most of the area is in the National Forest), throw a couple cans of chili, spam, or "Chicken in Golden Gravy" into boiling water for dinner, and call it good. One day, during a sudden flash flood while we were mapping terrain in a low-lying area, John maneuvered the brute off the road and on to the top of a low hill, where we were soon surrounded by water for a half mile in most directions, and we learned first-hand how pediments (gently sloping landforms akin to alluvial fans) formed.

Early in that season, I recall that John needed a bar of soap. He stumbled into Bulloch's Drug Store in Cedar City, just west of his thesis area, where he met Linda Jones, a beautiful co-ed at the University of Utah, who had a summer job at the drug store. Of course, by mid-summer, John had become totally smitten with Linda, and my departure was hastened! I was happy for them and anyway had my senior thesis—mapping Precambrian rocks in the Upper Peninsula of Michigan—to research and write, gratefully supported by one of Carleton's NSF grants, and that thesis led to my first published report. As for Linda and John, they married in 1964, John graduated from Texas with a Ph.D. in June 1965, and he and Linda headed off to Kent State University in Kent, Ohio, where he had been offered a job as Assistant Professor in the Department of Geology. John's story will continue through several more chapters below.

CHAPTER 3

This is to Certify that

Peter D. Rowley

UPON COMPLETION OF SUFFICIENT FLIGHT INSTRUCTION FROM

RAGSDALE AVIATION

AUSTIN, TEXAS

DEPARTED AND RETURNED TO AUSTIN MUNICIPAL AIRP

AS PILOT IN COMMAND OF A 150-C ON 10/24/

THEREBY COMPLETING His FIRST SOLO FLIGHT.

Robert Ragsdale

HOOVER MACKIN AND GRADUATE STUDIES

I've already told you about a host of bright and engaging geologists. But in my life's many wanderings, I've run into only one genius: J. Hoover Mackin. Mackin was the advisor of my Ph.D. dissertation while he was the Stamps-Farish Professor in the Geology Department at the University of Texas (Austin), and he served on the National Academy of Sciences, which required trips to and from Washington, D.C.

How does one recognize a genius? In many ways, one just knows it. If you take the time to listen to Mozart, you will recognize his genius; it seems to me that all Mozart's works (more than six hundred before he died at the age of thirty-five) that he composed as an adult (he also composed many as a child) come as close to perfection in music as anything manmade has ever been. Hoover Mackin likewise produced a lifetime of excellent symphonies, concertos, and chamber pieces—in his case published geologic reports and newly minted geologists—which had a great impact on the geologic activities of geomorphology (the specialty of geology dealing with the description and origin of landforms) and geologic mapping. One of these reports, certainly of symphony caliber, was on "the graded river," which explained—as only a genius could—how rivers and streams move their sediment under the constraint of a fixed volume of water. This topic, more complicated than you may think but certainly not sexy, when successfully understood had tremendous influence on the engineering of waterways, water diversions, dams, penstocks, flood-control structures, etc. Because of these practical implications, Hoover built a robust consulting career alongside his research and teaching career. Hoover's genius in the classroom included being able to define perfectly a scientific mystery and then point to what was needed to solve it. His insight was therefore in zeroing in on the crux of a problem through the maze of ephemeral theories and of rough data that can be interpreted several ways. As he once said in class, "define the problem and you're halfway home." And his defining of a problem made it almost self-evident what the answer was. As Paul L. Williams, another student of his, noted, only idiots refused to come around to Hoover's interpretations, so his papers generated little controversy.

Although known as a geomorphologist, ***Hoover*** *in fact was a field geologist and a geologic mapper, one of the best. He was born in north-central New York in 1905. He had polio and his father died early, so Hoover had to work his way through New York University, where his first course in geology established the trajectory of his career. In 1930 he both received his B.S. in geology and married Esther Fisk, a strong, intelligent, and beautiful businesswoman from Brooklyn who ran a collection agency, therefore managing to put Hoover through grad school during the Great Depression. He received his master's degree (1932) and Ph.D. (1937) from Columbia University, where his mentors were famous geomorphologists Douglas Johnson and William Morris Davis. Hoover's dissertation study was the erosional history of the Bighorn Basin, Wyoming. Esther and he spent their first summer in the field together, mapping by day and sleeping in their Model-T Ford by night. Esther packed a pistol on her hip and killed twenty-eight rattlesnakes that season. In 1934, Hoover accepted a job as instructor with the Geology Department at the University of Washington. Esther liked to recall that, as they were driving down from Snoqualmie Pass into Seattle, Hoover turned to her and said words to the effect that "you are not going to see much of me for a while, for I have a career to make." Esther always concluded the anecdote with the words "and I didn't." But make a career he did! And of course, they did find the time to have two children, Barbara and Robert, on whom they doted. Barbara, a brilliant and beautiful woman, in 1964 married young University of Texas geology professor Daniel S. Barker, had daughters Molly and Amy ("considered their finest achievements," Dan later said) before the marriage failed in 1984. Barbara went on to considerable fame as Barbara Barker Warner, a professor of dance at the University of Minnesota, leading it to national prominence that culminated in 1999 with the dedication of the $4.5-million University of Minnesota Barbara Barker Center for the Dance. She died far too early three years later, from cancer. At the Universities of Washington and Texas, Hoover was a classic absent-minded professor; before he went off to classes each morning, Esther made sure his shoes matched, that he had the right lecture notes, and so on.*

Besides his publication on the graded river, Hoover's work in the Bighorn Basin resulted in several other published conclusions. Then he moved on to studies of areas closer to Seattle, including its glacial history and engineering geology. He became a prolific scientist and teacher and, thanks to his humorous and outgoing, larger-than-life personality, passed on his love for his field to every student who took his two classes, Map and Photo Interpretation and Geomorphology. Most students in his classes, I among them, found him to be the best teacher they ever had. To the students closest to him, meaning his graduate students who pursued master's and Ph.D. degrees, he was extremely supportive and loyal, yet he encouraged them to finish their degrees as soon as possible, to get on with life before they became burned out, as professional students are prone to be. He chose his grad students carefully, then devoted most of his time to them, despite his frenetic schedule.

The door to Hoover's office was always open and his desk faced that door; if he were on the phone, he would wave you in when you appeared at the door, then abbreviate the phone call. But his devotion entailed much more than that, for he had a singular warm and caring personality for colleagues, even those on different sides of a scientific argument, and especially for students.

He was a person of great principle. John Anderson told a story of the time he revealed to Hoover that he was having difficulty writing up the conclusions section in his dissertation. Hoover gently stated "You know, John, it's very difficult to state something that is scientifically wrong." This was Hoover's beautiful way of telling a budding geologist that more than likely his interpretations sucked! John went back to his office and rethought his science, and then the paragraphs flowed more easily.

Hoover was an excellent writer, and he passed that skill on to others, along with his life philosophy. Though Hoover's scientific writings could be considered beautiful prose, in the sense of a great writer, like a Hemmingway, that prose did not come easy, and he could spend an entire day finishing a paragraph, not merely to write beautifully, but to communicate ideas accurately and in a way that did honor to his field and its forbears. He once stated that "I never write anything without the feeling that Gilluly is watching over my shoulder" (James Gilluly, who died in 1980, was another great USGS geologist, and was also on the National Academy of Sciences).

Some of Hoover's best writings were for students and concerned both how to interpret geologic maps and on the logic of multiple working hypotheses in geology. That concern for students extended not only to how he wrote for them, but to shaping them as writers. To that end, he dedicated enormous amounts of time to carefully and conscientiously editing multiple drafts of theses that these students gave to him to review. Through such careful editing, he not only covered these theses with probing red-ink comments that, with luck, decreased with each new draft, he also taught his grad students how to write well and clearly—in addition to teaching them how to do geology.

No wonder Hoover was much loved and respected by peers and students! His wife Esther likewise was loved for her counsel and empathy. Truly, they were a two-person team that turned students into scientists. They built a beautiful home, complete with swimming pool and two German Shepherd dogs, on a ridge with a beautiful view of Austin, and I remember spending delightful weekends there doing yardwork for them.

As part of the war effort during World War II, Hoover joined the

Strategic Mineral Program of the USGS. Though this Program included broadly providing more field data to mercury deposits in Washington and iron deposits in Idaho, Hoover's main effort was mapping ore bodies in the Iron Springs iron mining district of southwestern Utah. Working in conjunction with geophysicists using surveys by magnetometer (instruments that measure the magnetic field, commonly by towing the device behind aircraft) and geologists of U.S. Steel Corporation and Utah Construction Company, he mapped and demonstrated that huge hematite ore bodies surrounded the edges of plutons (23 to 21 Ma) that he recognized as laccoliths (most others vehemently disputed this, but he was proved right). This study had huge practical consequences for the war effort, as these deposits were subsequently mined and the ore transported by train to blast furnaces that were set up in Provo, Utah, and Pueblo, Colorado. The iron became steel for ships, tanks, trucks, armaments, etc. So vast were the deposits there that this district soon became the most important iron-producing district in the West. After the War and well into the early 1980s, the iron mines continued to be the primary employer in Cedar City. Once the war was won, publication of the geologic maps and reports about the district began.

The curiosity-driven academic consequences of Hoover's mapping were also great, but their study had to wait until the 1950s. For example, thanks to his mapping skills and general genius, Hoover came up with an innovative hypothesis for the origin of the ore deposits in which the iron came from the breakdown (called deuteric alteration) of dark iron-rich minerals in the plutons, after which iron-rich solutions moved outward to pluton margins along early-formed joints (fractures) in the consolidating magma. Hal James, his friend and an equally renowned geologist (and a member of the National Academy of Sciences and a Chief Geologist of the USGS), stated in his memorial (eulogy) to Hoover that this hypothesis was "one of the finest contributions to the science of ore deposits of the past three decades." It was not until 1995 that Dan Barker was able to articulate the geochemistry that made Hoover's hypothesis a likely explanation. Dan took such a long time because he knew he would have trouble being objective, for he not only loved Hoover as we did but of course Dan had married Hoover's daughter Barbara!

Next, Hoover recognized that thick stacks (stratigraphic sections) of formerly horizontal Miocene (an Epoch of 23.03 to 5.33 million years old) and Oligocene (an Epoch of 33.9 to 23.03 million years old) volcanic rocks adjacent to the laccoliths and previously considered to be mostly lava flows were, in fact, ash-flow tuffs. Ash-flow tuffs are resistant beds deposited as volcanic ash that is generally more than a thousand degrees Fahrenheit, with subordinate mineral crystals (the first minerals to settle out of its underground magma chamber, before eruption) that form broad rock sheets, each from several tens of feet to more than a thousand feet thick, that spread out as much as a hundred miles in all directions from their vents. The sudden explosive eruption of huge volumes of ash from the vent area evacuated the underground magma chamber, so that the surface above subsided to form a large circular depression called a caldera, which could be tens of miles in diameter. Yellowstone National Park, for example, is largely underlain by several huge calderas that formed following eruptions that deposited ash beds as far east as Nebraska. In contrast to ash-flow tuffs, lava flows result from quiet eruptions of molten lava and are mostly confined to less than several tens of miles from their vents (volcanoes).

The study of the ash-flow tuffs and their use in interpreting the geology also began after the War. In 1954, Shell Oil Company discovered the first large oil field in Nevada, at Railroad Valley, about 150 miles northwest of Cedar City. The oil was in Eocene (an Epoch of 56.0 to 33.9 million years old) lake beds and Oligocene volcanic rocks, and other oil companies soon rushed in to exploit the natural resources. John Osmond, District Geologist for Gulf Oil Company and a former Mackin student, contacted Hoover for help in understanding the stratigraphy of the volcanic rocks so that additional oil fields could be found. (Few were.) With Gulf paying his salary and expenses, beginning in 1955 this was the excuse that Hoover, along with his students Earl F. Cook and Paul L. Williams, needed to range far and wide across eastern Nevada and western Utah to look at the volcanic rocks. Earl, then a geology professor at the University of Idaho, had been a former field assistant of Hoover's during the Iron Springs mapping and later (1954) finished his Ph.D. under Hoover, mapping the Pine Valley Mountains (just west of New Harmony, where we now live) south of Iron Springs. Paul became Earl's field assistant. In the summer of 1955, Hoover, Earl, and Paul traced a series of ash-flow tuffs westward from one range to the next for at least a hundred miles west of Iron Springs into Nevada. They found that individual tuffs in the stacks were distinctive, especially in their amounts and proportions of mineral crystals. Paul and Earl's microscope examination of thin sections thus allowed tracing each tuff from range to range. An additional conclusion from this study was that tuffs were the main Tertiary rock type in the entire Great Basin. By such broad mapping, Hoover, his students, and others started to understand the geology of the Great Basin. And for one of the students, Paul Williams, the support from Gulf Oil to identify ash-flow tuffs over a broad area funded his Ph.D. dissertation, begun in 1964 and completed in 1967, which I consider one of the best I ever read. His field work also reported on thickness variations in the tuffs, leading to conclusions about their caldera sources. (I tell the story of Paul later.)

Because there were no topographic maps of the Iron Springs area during the war, Hoover had to "survey in" the topography and geology (that is, make a topographic base map first, then the geologic map) using a plane table and alidade (a horizontal stand on a tripod and a sighting device, commonly a monocular, that were used for surveying and, even now, can be seen in use by road crews), on the latter of which he sighted to a measuring rod held by an assistant. One area that needed mapping was a ranch owned by the Isom family. A particularly curious horse there spent part of its day looking over Hoover's shoulder and scratching itself against the plane table. Hoover formally named a prominent ash-flow tuff sequence on that ranch the Isom Formation, after first (as required by the USGS and Board on Geographic Names, geologic names must come from nearby named geographic features) naming a nearby topographic feature Isom Creek.

Hoover believed, as I now do too, that geologic mapping is how you do geology, so all grad students mapped a large area. When Hoover resumed his academic career at the University of Washington, he put his graduate students on mapping projects adjacent to, although progressively moving farther away from, Iron Springs. Because each ash-flow tuff was distinctive, represented only an instant of geologic time, and was spread over a broad area, these tuffs helped Hoover and his students decipher the geologic history of this part of southwestern Utah. In the Great Basin, the mountain ranges are oriented north-south and they alternate with basins (valleys) that also are north-south. The basins and ranges were formed by north-south normal faults, which are faults that have mostly vertical offset such that basins moved down along the faults and ranges moved up. The main U.S. expert on ash-flow tuffs, who did a classic study of a caldera and its ash-flow products in New Mexico, was Robert Smith of the USGS. Hoover invited Smith to join him on a field trip to the tuffs of the Iron Springs area. When Smith saw the rock units in the Isom Formation, he declared them to be lava flows. Yet later mapping and microscope examination of the rocks demonstrated that they were indeed ash-flow tuffs, as Hoover had suggested. Although informally and good-humoredly, Hoover commonly called those tuffs the "Bob Smith ignimbrite." Such an incident is in fact typical of how scientists treat other scientists whom they value and respect, for laughter and practical jokes are endemic. When one scientist catches another good scientist making a minor mistake, the needling and kidding can go on for months or even years. All of us have given, and all of us have taken such good-natured ribbing!

When Hoover accepted an endowed chair at the University of Texas in 1962, he also received a new batch of graduate students to work with him in southwestern Utah, the first being John Anderson. From the mid-1960s, however, he suffered from heart problems that curtailed strenuous field work. Instead, he concentrated on teaching, writing, giving invited speeches (he was an excellent speaker), and participating in national committee meetings. His best graduate course was Map and Photo Interpretation, in which each week he would give us a topographic map or geologic map and tell us to define our own problem, then solve it and write it up. The classes that followed the completion of each weekly assignment consisted of vivisecting (or complimenting) those write-ups picked for having the most teachable moments. Hoover was less interested in a "correct" answer than an analysis of the situation, even if the conclusions were incorrect. His criticisms were made with such humor that few resented the method. What he was trying to counter was bad science and bad thinking. This exercise also taught us to think in four dimensions, for we needed to visualize what a landscape or fault looked like one, five, or twenty million years ago. At the end of such a class, each of us would walk out mumbling "Why didn't I think of that?" or, better, "Why didn't I think like that?" This was a class in thinking and in discovering that getting something right was fun! As Hal James said in his memoir to Hoover, "as a teacher of earth science, Mackin was almost without peer during his lifetime." Hoover, of course, used the same approach in thesis work by his grad students, and he once stated, "if there's one thing I can't stand, it's a graduate student who has to be led by the hand." So, as with John Anderson's thesis study, we defined the problem, then solved it. He always had confidence that we would.

Unfortunately, Hoover's publications are by and large no longer familiar to geologists, let alone to the public. Most scientists today are familiar only with the last two decades or so of the literature in their field. Progress in science is rapid, although generally in fits and starts, as new workers on a problem cite pieces of ideas from previous credited publications, then continue on either established or on new paths of the research, resulting in better and more refined hypotheses. Some of these newly published ideas have barely any evident connection to their ancestor ideas, but nonetheless all of us have at one time stood on the shoulders of earlier scientists to march our science onward. So even though few of his ideas are cited nowadays, Hoover's career does have a legacy of ideas—and of course a legacy of the students he produced (master's and Ph.D.), who in turn spawned their own students and hypotheses.

What was my connection with Hoover? During my summer as field assistant to John Anderson, John recognized that I had some budding but raw geologic talents. Earlier, Hoover had wanted John to map the Sevier Plateau, in addition to part of the Markagunt Plateau. So John suggested the Sevier Plateau to me as a topic for a possible master's thesis. Viewed east from the Markagunt Plateau, the east-tilted Sevier Plateau had an imposing (5000-foot) west-facing fault scarp, and climbing it sounded like fun to me. That fall, John asked Hoover about the idea, and he responded positively. During Christmas break, I drove to Austin and talked with Hoover, who agreed to take me on as an advisee and even to support me with grant money. The Chairman of the Geology Department, Ronald K. DeFord, approved the arrangement and, because I was already familiar with the stratigraphy (rocks) from John's area, they agreed that I could start next summer rather than taking a summer camp in field geology, as required for most grad schools.

So it was that upon graduation from Carleton in June 1964, I spent the summer tackling the Sevier Plateau. I found a place to rent, an old mobile home out behind the red-brick house of the Utah Highway Patrolman for this area. Dean J. Pierson and his wife Sheren lived in the small town of Junction, along north-south U.S. Highway 89 in the Sevier Valley west of the plateau. Junction, the county seat of Piute County, now has a population of about two hundred—as it did way back then. Not much has changed. Then as now, tourism and raising hay and cattle were the chief occupations along this high (valleys have elevations above 6,000 feet), beautiful, secluded transportation corridor. I thus became part of the Sheren and Dean Pierson family, with pre-teen daughters Tad and Annette, junior high-school son Rick, and high-school son Lynn. The Piersons were jack Mormons, the best kind in my opinion and common in rural Utah. This designation means that they did not pay their 10-percent tithings and that they did drink alcoholic beverages; Dean's favorite was Coors Light. Dean, born in 1925, was a handsome, dashing, extremely competent patrolman, but he was also a former smoker who had contracted throat cancer, which had required a tracheotomy. Therefore, he had to talk by gulping in air, a method by which he was perfectly understandable but that was probably disconcerting to the motorists to whom Dean dispensed speeding tickets! Sheren owned and was chief cook in a café in Circleville (population now about five hundred), the town about five miles to the south, and she became my surrogate mom. Dean and the boys did a lot of fishing in the nearby reservoirs, and they showed me their favorite spot at the Piute Reservoir a few miles to the east, where thereafter I could always catch my dinner of a half-pound Rainbow trout. Ricky was a troublemaker, and we occasionally went out together to shoot jackrabbits on the dirt airport runway near Piute High School. What else was there to do, after all? The closest thing to a town that contained amenities such as a food store, bar, multiple cafes, and a mechanic was Panguitch (population now about 1,600), the county seat of Garfield County and about twenty-five miles south of Circleville, also in Sevier Valley. Panguitch is now a popular launching point for visitors to Bryce Canyon National Park, less than twenty miles to the southeast of it, and to other parks, so it swarms with tourists each summer.

The Sevier Plateau, whose southern end is just north of Bryce Canyon National Park, was a beautiful place to work, with sagebrush and pinyon and juniper trees in the valleys, and ponderosa and aspen trees and elk herds at the eleven-thousand-foot top. For the first time I experienced the solitude and beauty of day-long traverses up and down hills. I'd locate myself on vertical aerial photos, map the geology along the way, and in the evening plotted the relationships (contacts) of rock units and structures on a base map.

When alone in the field—as I have been almost my entire career—I had to watch out, for there were dangers. For example, that first season I killed about a half dozen rattlers, which I did by holding the head of the snake down with a stick, then cutting it off with my Swiss army knife. I did that for years, although now I feel badly about having too little respect for the life of another creature. During that first field season, however, there were times when I mixed up the different rock units, and I wondered if I would ever get things right. After a while, of course, the procedure became easier and my love of doing field work increased. When time came for me to map the Sevier Plateau fault scarp, with the Pierson's eldest son Lynn, I'd drive my car west up the dirt road along the eastern tilt slope to the top of the big west-facing fault scarp. He would drive my car back home while I hiked west off the scarp to the bottom, then west five to eight miles across the valley flats to north-south U.S. Highway 89, while mapping the surficial sediments of the Sevier Valley. There I would hitchhike back north to Junction. I did about eight or ten traverses off the scarp over the three summers of field work, and now I wonder why I was not killed doing it. For at times, as I worked my way down the gulch in the sheer rocks, I was "cliffed," meaning I could not get down a cliff below me. With no way down, I'd go back up, then out of the canyon, and around the canyon rim. But that route was not always a safe option, so there were times when I had no choice but to slide down the cliff into the gulch on my butt and hope that there were not a worse cliff below me that I could not yet see. If that happened, I would not have been able to get back up.

When September rolled around each of those three years, I'd pack up and head back to Austin. I was now in the second largest state in the US, where the state ego was just as big, as in the 1978 song Texas ("When I Die") by Tanya Tucker. The Department of Geology was in fact the largest geology department in the country, for most undergraduates and graduate students found employment in the nearby oil industry in Dallas, Houston, and Midland. The Department oriented the new grad students, perhaps twenty or so that fall of 1964, and we picked and enrolled in classes. I found that I had achieved a certain status in already having an advisor and in having completed my first field season on my thesis topic before I even enrolled. As with most incoming graduate students, I was offered a Teaching Assistantship (TA), for which I taught laboratory classes to undergraduates who had wisely decided to take introductory geology for their science elective. The TA, which included tuition costs, had a salary that was fully satisfactory for paying for a rental apartment (if shared with roommates), for food and clothes, and of course for beer. I found a roommate, Russ Burmester, and we soon were under way! All grad students were given offices in the attic of the old Geology Building and we quickly settled into our routines of taking classes, teaching labs, and starting master's theses or Ph.D. dissertations. Nearly all grad students were smart, fun, and compatible, and we had many parties. Most members of the faculty also entertained students and other staff. I overlapped for one year with John and Linda Anderson. John received an award that year as the best grad student, and by then he was supported by fellowships. There were many organized and informal field trips to see the geology in parts of Texas and Mexico, and at times we accompanied a professor who was visiting a field area in which one of his students was working. Field studies were emphasized throughout the program.

All professors were of high quality, and as the hiring of Hoover showed, the department was actively upgrading itself, under the department chairman, Ronald K. DeFord. For DeFord knew a secret: when early Texas governors and legislatures established the state universities, the agricultural school of Texas A&M received prized agricultural land in East Texas, whereas the University of Texas received land in West Texas that at the time was deemed somewhat worthless. However, the West Texas land contained oil, and therefore the University of Texas was suddenly well endowed. (Things are different now, with all state universities sharing the largess, but with Republican governors each year decreasing allotments for college and all other education.) In addition, because so many Texas grads got rich in the oil industry, annual alumni funds were substantial. DeFord used these funds to attract world-class geological educators. While I was there, a large grant from a rich alumnus in the oil business enabled a new geology building to be constructed. And in 2001, long after I'd left, the grandiose Jackson School of Geosciences was established.

These funds allowed other kinds of expansion, too. Hoover's first office had been a tiny ground-floor office. His desk had soon become piled high with teetering stacks of reports and papers that he was reading or editing, yet despite the apparent chaos, when he needed one, remarkably he knew where to find it. Being absentminded, he commonly locked himself out of this office, but fortunately he never latched his window, so he could simply open the window and climb back in the next morning! The new endowments meant that suddenly he had a huge office, in the center of which was a large boardroom-sized conference table. However, within a few months that table also became piled high with teetering stacks of reports and papers! Other stacks included aerial photographs of the Moon and Mars, for the NASA grants he received were for his help as a geomorphologist and volcanologist to identify volcanic versus meteor-impact landforms. Whenever students visited him, he would regale them with these photos. His enthusiasm, curiosity, and interests had no bounds: at about this time he had heart surgery to repair a bad heart valve, and he would press willing students' heads to his chest so that they could hear the valve click!

Such intimacy and enthusiasm were common, and friendships made between students and faculty were many and typically lasted the rest of our lives. As at Carleton, I was consumed by my studies and didn't have much of a social life, so my grades were virtually all A's. Within a year or two, however I finally did have some dates and some fun, not all of it on campus. One spring break I reconnected in Houston with Uncle Dick DeWitt, Mom's black-sheep brother, and his wife Joyce, and we drove to New Orleans for Mardi Gras, hitting Cajun restaurants and bars across the Mississippi delta. Dick was a six-foot-tall, movie-star-handsome, former oil geophysicist (University of New Mexico graduate, in archeology I think) who had managed field geophysical surveys in the Lake Maracaibo oil fields of Venezuela, and had lived in Caracas with Joyce. The big oil company for which he worked after a few years asked him to pay off political figures under the table; Dick refused and was fired. Joyce and Dick were not permitted to take their savings out of Venezuela, so upon his termination they toured northern South America. That included driving south (some rivers had to be crossed by ferrying the car on small boats) in their 1956 Chevy on a true adventure south through the Andes all the way to Lake Titicaca and La Paz, Bolivia. Dick had the engaging personality and wit sufficient to keep you laughing most of the time. In fact, it seemed impossible for him to be serious for even a moment. Life was a joke to him, and he made friends easily and drank a lot. He owned Picaro Microfilm for thirty years, selling microfilm in Houston. At the time of our trip to New Orleans, he was a host of a local morning TV interview/talk show in Houston, apparently like "Good Morning America."

When I was not acing tests, Hoover introduced me to some of his former grad students, for almost all of them had done their theses or dissertations akin to mine. Many of them now worked for the USGS, and many of these were mapping in Antarctica. The most influential person to my career was Paul L. (Lincoln) Williams, then with the USGS in Denver, whom I first met as my second Sevier Plateau field season began. He was heading to Utah, where he was to finish up field work on his dissertation, then to continue mapping on a project of 1:250,000-scale (2-degree sheets) maps of the Colorado Plateau and its uranium deposits; his Moab and Salina quadrangles were two of the best reconnaissance geologic maps ever done. He knew I was heading west too, and he suggested we meet at the bar (the only one, on Main Street, now a parking lot for a supermarket) in Cedar City on such and such a date. I was his field assistant for over a week and in the process learned about the major Utah tuffs on which he and Hoover had been working. Some of those tuffs likely had flowed into my own thesis area. Paul also regaled me with Antarctic stories—stories that would stay with me and that rekindled my interest in that area. As at Carleton, I was anxious to get grad school behind me, especially if there were employers like the USGS and jobs out there that involved mapping in Antarctica! I'll take up Paul's story again later.

Student years rolled by. I made lots of friends, and we socialized at Scholtz' Beer Garden or each other's apartments or, quite commonly, at homes of the faculty. Most of those fellow grad students had exceptional careers.

*One, **Roberta (Robbie) Rice Gries**, who got her master's degree while I was there, had already become "first woman" in many things, such as her B.S. in geology at Colorado State University, and soon she began accumulating seemingly dozens more, including jobs in the oil business and memberships and positions of leadership in many geological organizations. Wikipedia online tells us that she was the daughter of a shrimper on the Texas Gulf coast, where she was fascinated by marine fauna she saw on her dad's boat and from there to rocks! Bright, strong, and attractive, she had a good sense of humor and made friends easily. She married John Gries, another classmate who earned a doctorate at Texas before they went off to his teaching position at Wichita State University. Before she divorced John, they had a daughter named Lynn, who is now a medical doctor in Tucson. Lynn and my daughter Jill used to ski together when we later all lived in the Denver area. After her divorce, Robbie became an oil geologist, publishing some innovative and well received reports on the structural geology of the Rocky Mountains. She soon became a consulting geologist in oil and gas, and founded a small oil company, Priority Oil and Gas LLC. She is best known as a past President of the American Association of Professional Geologists (AAPG), the main research society for oil geologists. Thanks to earnings from her oil and gas company, she had the time to research and write a book on women stars in oil geology, of which there are many. She parlayed her talent for organization and working with others by being picked (at the time of the start of this writing) to be President of the Geological Society of America (GSA), the largest and best geological society in the world. She is a leader and a role model for women scientists, one of the most important of our generation. She and her husband, David Bailey, lived in Lakewood, Colorado, and in Tucson until, sadly, David died of a virulent form of leukemia in February 2021.*

*Another fellow student, **George H. Davis**, entered the University of Texas the same semester I did, with his beautiful wife Merrily. Like me, they had just earned B.A.s from an excellent small college, theirs being Wooster College. That B.A., plus later life*

> *experiences, including a close friendship with Larry Gould, made George an outspoken advocate of a liberal arts education. He got a master's at Texas, then went on to a Ph.D. at the University of Michigan. George was an early but significant product of Larry Gould's upgrade of the department, then he continued with his hiring of exceptional teachers and researchers. George has had a long career at the University of Arizona, where he became an exceptional teacher as he climbed the ladder to Professor of Geology, Chairman of the Geology Department, and subsequently Provost and Vice President there, interrupted by a couple years as President of the University of Vermont, where he ran into old-time New England politics and resigned (my interpretation). At Arizona, he and his graduate students did fundamental research on how some large mountain ranges formed during extreme extension (pulling apart), resulting in huge low-angle ("detachment") faults, and he found and studied a bunch of those ranges in Arizona, starting with the Catalina and Rincon Mountains on the north and east edges of Tucson. He wrote the best geology textbook I ever saw,* Structural Geology of Rocks and Regions, *great in part because it contained humor, which I have not seen before or since in a serious textbook! Like Larry, George has received many honorary doctorates, including one from Carleton. George preceded Robbie in being President of GSA and, while there, established a GSA award for geologic mapping, with my help. He and Merrily remain in Tucson, where they are philanthropists. Only in 2022 did he stop teaching a field-based geology course as Emeritus professor. In 2023, he senior-authored a magnificent report by GSA on the Catalinas and Rincons; he, colleagues, and students had been working on this range for more than fifty years, a time period that is commonly required in science before major problems can be addressed with some degree of accuracy; of course, no matter how "magnificent" a report is, future geologists are likely to add to or modify its conclusions.*

I put in my second field season in Utah, again staying with the Piersons. When Hoover visited me to see what I was doing, he realized that the Sevier Plateau was a bigger job (eventually I mapped six hundred square miles at 1:62,500 scale) even than John's area, and he proposed that we turn the work on it into a dissertation. And, even better, that we forgo the master's degree and skip straight to the Ph.D. I was delighted. After that, I had a third field season with the Piersons. My final two years were supported by a University Fellowship. Besides Hoover, members of my dissertation committee were Robert E. Boyer, a young professor of structural geology, and Daniel S. Barker; Paul Williams was my outside (USGS) member. From Bob Boyer, I took an excellent course in tectonics, which is the study of continent-scale structural features. Those were days in which, after years of being laughed at, the concept of continental drift was finally receiving the data support to make it not only viable but increasingly likely. It soon came to be called plate tectonics as researchers better defined which parts of the crust were moving around. Most of these new data came from geophysics and the geology of ocean basins, a new field. (The story of plate tectonics is expanded on later in my 1986–1987 season in Antarctica in Chapter 6). I recall debating the pro- and con-evidence in the classroom. It made for an exciting course.

> *I became even closer to* ***Dan Barker****. Dan was born in 1934 and raised in rural Maine. He graduated from Yale University (magna cum laude) in 1956 and received a Ph.D. from Princeton University in 1961. He was hired by the University of Texas in 1963 and "stuck," as he once said. For thirty-six years he did research in igneous rocks in the US and around the world, with his many grad students. Perhaps more importantly, he was one of the best teachers in the Geology Department, earning teaching awards in multiple years from both the department and the university, as well as awards for research. Some of his undergrad and graduate students referred to him as Obi-Wan from the Star Wars movies because of his wisdom and mentorship. Like Hoover, Paul Williams, and Bob Boyer, Dan Barker was extraordinarily supportive of students. Dan, about ten years older than I was, seemed at times more like one of the grad students, so he was inspirational to more people than me. Dan taught Thermodynamics, the toughest course in the department, and one that intimidated almost all grad students because most of us were not that good in chemistry. I had only a little fear, because I knew from his Igneous Petrology course that he was a good teacher and could somehow drill the subject into my sorry brain! Somehow, I got an A+ in either the final exam or the course (I cannot remember which). Dan also once gave me the second-best advice I ever had (after that of my brother Clark, as reported in the next chapter): "Just get the damned Ph.D. ASAP and move on to something more important." I stayed in touch with Dan over the years, including collaborating with him on Iron Springs research. He published an igneous-geology textbook, guidebooks, encyclopedia articles, and more than fifty reports. He retired in 1999 but continued to do research until his death in 2015. As he noted in an obituary for himself that he wrote before the actual event, "he enjoyed puns, even good ones, and tried to hide Down-East humor that 'was so dry it made your nose bleed.'"*

> ***William R. Muehlberger*** *became Chairman and continued upgrading the department where Ronald DeFord had left off. Bill was a big guy with a square jaw and a balding head; his booming laugh could be heard two hundred yards away. No wonder he was a favorite with students. Because of that and because he was a structural geologist, I took all his courses and participated in every field trip he ran. Later he became very involved in training astronauts, including the famous ones, in preparation for their*

trips to the Moon. For NASA wanted all astronauts to make intelligent geologic observations and collect samples. Hoover worked with Bill in interpreting (using vertical aerial photos) what lunar landforms the astronauts were likely to see and visit. Bill retired in 1992 and died in 2011.

Another exceptional professor was ***Robert L. Folk****, who later received the Penrose Medal of GSA. After Hoover, Folk was the best teacher I ever had, and a favorite among students for his unorthodox teaching methods, including plenty of humor, in his classes in sedimentary petrology (i.e., limestones and sandstones and what they looked like under the microscope). Because of his teaching ability, I took all his classes, even though I was not heading toward those specialties.*

In my spare time, in the fall of 1966, I began taking flying lessons through the Longhorn Flying Club, affiliated with the university and thus cheap for a student. I earned my pilot's license in December. The trainer we used was a Cessna 150, a high-wing two-seater. The aircraft was small and light, and I remember one particularly hairy solo training flight. I was nearing the approach to land to refuel in the small town of Junction, Texas, during a significant wind, when I discovered that I was competing for airspace with large numbers of buzzards circling in the thermals and the breeze and therefore swishing right past the cockpit (and within flapping distance of the propeller, more to the point!). The buzzards appeared to be as big as the plane!

I loved the freedom of flying and even considered giving up my studies and becoming a professional pilot instead. In early 1967, I checked out how to fly a Cessna 172, a four-seater. Hoover, who had a trip to West Texas planned in June with a couple students who were starting theses out there, heard about this and suggested I fly them there. One of these students was Charles ("Chip") Groat, one of Hoover's last grad students (along with Richard Waitt), who later became the Director of the USGS, perhaps the second-worst Director in its history. Taking off from Austin with four on board was fine, but as we approached our destination in the thin air of the relatively high (4,700 feet), small town of Marfa and with a load with which I was unfamiliar, I experienced one of my most embarrassing moments in life. As I touched down on the runway, the aircraft bounced, and kept on bouncing as I approached the end of the runway, which was surrounded by a high chain-link fence that I still remember. Mortified, I applied power, managed to clear the fence, and went around again. Like a rubber ball, the plane did the same thing but this time I just applied the brakes until the bouncing stopped, then motored up to the one-room "terminal" with a face that must have been bright red! Hoover never said a word.

In the early fall of 1967, after a field season in which I had not practiced flying at all, I took a 150 out from Austin to brush up on what to do if the engine stalls, in this case high above the tiny farm town of Pflugerville. In a stall, with the throttle pulled back (power killed), the nose does not necessarily drop straight down but commonly tilts off to one side or the other. That is easily corrected (straightened out) by applying opposite rudder (the foot pedal opposite to the turn), but for some reason I had forgotten this and applied same-side rudder, which accentuated the turn. So I was suddenly in a deep turning dive. I realized what I had done, applied power and opposite rudder, and pulled back hard on the wheel. A plowed farm field was below me as I fumbled, losing at least a thousand feet of altitude, and I watched the clumps of overturned soil getting clearer and closer, until the plane, verrrry, verrrrrry slowly, came out of its dive, this a scant two-hundred feet from the ground according to the altimeter (I memorized the elevation, then compared it to that when I landed)! I climbed back up to altitude and did a couple more practice stalls, correctly this time, before heading back to the Austin airport. Flying was not quite as much fun after that, and geology once again looked better as a career.

My last year was spent entirely on the dissertation, doing lab work, writing, and assembling the geologic map on the second year of a full fellowship. At some time before this, probably in the spring of 1967, I had to pass my written and oral comprehensive exams for the Ph.D., with questions not only from my committee but also from a few other, presumably more objective, professors. The written questions went fine and the orals were going well for a while—until one of the external members of my committee asked me an oddball question on a fringe topic of geology. I probably said "I don't know" but I do not remember my exact transgression or how I responded, except that I became flustered after that until one on my permanent committee pitched me a soft ball to get me back on track. Some of the outside members wanted to reject me for the Ph.D., I think. But Hoover pointed out that he and Esther had a hobby, thanks to which they knew quite a bit about eighteenth-century furniture and he threatened to ask those same committee members any number of basic questions on that topic, knowing they wouldn't have a clue. I passed with "conditions," which meant I had to read a book on the topic of the oddball question. It was a good lesson for me on not getting too high an opinion of myself, for there are lots of people out there who may think otherwise.

I sent the first draft of the dissertation to Hoover in late 1967; he covered it with red ink. The second draft had fewer red marks. The third draft went also to my dissertation committee, Dan Barker, Bob Boyer, and Paul Williams. So, because of these three people, by the time I repaired my final draft, I had learned the fundamentals of scientific writing! In June 1968 I graduated, having gotten a Ph.D. in the shortest amount of time (four years) of anybody in department history and, with Susan Longacre who also graduated in June 1968, we were the first in the department to go straight for the Ph.D. without a master's degree in between. Soon after graduation, I packed my car and headed north, never to return to Texas. I am not sure why I never came back, for liberal and educated Austin was like a diamond in the middle of a sand box that was Texas, but perhaps my impression of the sand box out-weighed that of the diamond.

John offered me a temporary job at Kent State University, where we collaborated and started the publication process on our joint research. He had received a research grant from the National Science Foundation (NSF) for the summers of 1968 and 1969. To facilitate field work, I returned to Connecticut to borrow my parents' green 1963 CJ Jeep, which they used to plow their winter driveway in rural Newtown. John extended his mapping northward from the Markagunt Plateau into the southern Tushar Mountains. I started mapping in the Black Mountains west of the Markagunt Plateau, then continued south into the Iron Springs mining district. We were attempting to tie together an area of several thousand square miles. Hoover flew into Cedar City in late July 1968 to visit us for several days to see what we had found, about two weeks before he had to return to the heart specialists at Houston Hospital to repair his leaking heart valve. These specialists included Dr. Debakey, an egomaniac heart surgeon who had been on the cover of *Time* magazine, and because Hoover was going to be in such good hands, we all thought little of this upcoming surgery. After the field visit, the three of us had a steak dinner and much laughter at the best restaurant in Cedar City, Milt's Stage Stop Steak House. The next morning, we put Hoover on a plane for Austin and returned to mapping, with John returning earlier to Kent than me. About two weeks later, I received a letter from Bill Muehlberger, which I opened with significant apprehension. As gently as he could, Bill told me that Hoover had died on the operating table in Houston. I wept. When I wrote to Esther the next day, I told her that Hoover Mackin was my standard of excellence, both as a scientist and as a human being. That is still true. He is always with me, and I am deeply grateful for his example and friendship. Esther maintained their beautiful home, with its view of Austin, and at age sixty-one went back to work fulltime. She continued until she fell and broke her hip, at the age of eighty-four. We corresponded until she died in 1998.

Lao Cai
Red R.
Thac Ba
VIETNAM
MOON LANDING USA - ONE SMALL STEP FOR MAN-ONE GIANT LEAP FOR MANKIND
COMMEMORATING
FIRST MEN
ON THE MOON
ASTRONAUTS
ARMSTRONG
COLLINS
ALDRIN
APOLLO 11
JULY 20, 1969
FIRST DAY OF ISSUE
12 АПРЕЛЯ-ДЕНЬ КОСМОНАВТИКИ
ПЕРВЫЕ СПУТНИКИ
ПОЧТА СССР 4к
33 USA
"I Have a Dream"

CHAPTER 4

A single life is strongly influenced by the times in which it occurs. This chapter attempts to discuss where my little life fit in to these times—between about 1960 and now—the best and the worst of times. Three long-term events began in my teen years and continued to play out for the rest of my life and those of my contemporaries: (1) both the golden age and the decline of science, (2) the Vietnam and other wars, and (3) climate change. I also will try to put my life in the context of the various presidential administrations of this era, for these have had a significant

TIMES ARE CHANGING

influence on the attitudes, knowledge, and economies during the three long-term events. Covered first will be the rise of science, which in the modern era began during and after World War II but accelerated during the space race. The space race elicited a huge interest in science and prompted a blossoming of science careers, with positive effects for the country and the world. Concurrent with the space race were the far more negative effects of various wars involving the US, starting with the Vietnam War. I address in Chapter 12 the environmental movement and its focus on climate change, which began with Rachel Carson's magnificent science and her courage to present it. Unfortunately, it is also a story of neglect and denial of science, not only by all of us inasmuch as every run-of-the-mill household pollutes, but by corporate polluters and politicians paid off by those polluters to suppress true science and reality. Climate change has now reached a decisive and critical phase, for its human-caused effects have now become obvious to all but the corrupt, stupid, ignorant, or uncaring among us.

The rise in emphasis and funding for science is in large part what prompted my career. Investments and inventions in scientific technology, including nuclear weapons, radar, and others, can claim that they won World War II, but it was the huge production of weapons, supplies, and means of transport by the US, the third most populous country in the world, that must get much of the credit. For during the war, virtually every American played a part in the war effort, and fortunately America had political leadership and an innovative democracy that encouraged inventions. The postwar years expanded on the victory by still greater development of science and technology, noteworthy among which was free education (the GI bill) for returning veterans, many of whom chose science for their careers. The US, credited by many with victory in World War II, became the leader of the free world. Its prosperity increased rapidly, and its political leaders had the wisdom to share their wealth by rebuilding the devastated infrastructure and economies of its former enemies in Europe and Japan, as well as by helping and building up its allies. The Cold War followed soon after the post-World War II good times, of course, but the US had enough wealth to combat that without hurting its economy. The Soviet Union could not keep up, although that was not apparent until the Ronald Reagan presidency.

On October 4, 1957, amidst the post-war good times and to the surprise of the US, the USSR launched Sputnik 1, which stayed in orbit for three weeks. This was the start of the space age, and suddenly the complacency that derived from our post-war wealth and status turned into a mild panic. President John F. Kennedy provided the leadership necessary for easing the anxiety when in May 1961 he challenged the public and the US Congress "before this decade is out, [to land] a man on the Moon and return […] him safely to earth." His leadership was particularly impressive because the main reason he gave for doing it was simply the challenge. Then he got Congress to fund it. Investing in science and technology was deemed the answer, and a golden age began for all types of science in the US. Of course, this golden age came not from any vast recognition by the public or its politicians about the merits of science, but only because now we had to catch up with a communist adversary. Politicians did it by funding everything that had the name "science" attached to it. National Science Foundation grants were relatively easy to win, and scholarships and teaching assistantships were available to nearly anybody who wanted graduate degrees in science. Such funding allowed me, as well as most of my contemporary science students, to be self-sufficient once we started graduate work. The golden age lasted only several decades, but that was enough to get many careers in science going strong and to produce a phenomenal number of new discoveries.

The culmination of Kennedy's challenge was, of course, Apollo 11, now over fifty years in the past. Using the Saturn V rocket, Apollo 11 was launched on July 16, 1969, and it splashed down in the Pacific Ocean on July 24. In between these dates, of course, Neil Armstrong and Buzz Aldren landed the lunar module, which separated from the command module that Michael Collins piloted in orbit. The landing took place at about 4 pm of July 20, 1969, eastern daylight time with commander Armstrong stating "Houston, Tranquility Base here. The Eagle has landed." Those words still thrill me. Armstrong and Aldren stayed in the lunar module for six and a half hours, then stepped onto the lunar surface with Armstrong's statement: "That's one small step for a man, one giant leap for mankind." Aldren soon followed, and they spent about two hours on the surface, bouncing around in the low gravity, leaving tributes and plaques to Soviet and American astronauts, and planting an American flag. Then they loaded up almost fifty pounds of rocks and sand, locked the hatch, pressurized the cabin, and slept for seven hours. Their sleep gave people on earth time to worry whether the main engine would fire up and lift them off the Moon, for there was no backup to that, and speeches had been written ahead of time just in case the worst happened. But after two hours of engine preparations and chatter with Houston, Armstrong uttered "Roger, I understand we're number one on the runway," fired up the engine with no problem, lifted off to rendezvous with the command module, and returned to earth.

I returned early that day from mapping in the Black Mountains of Utah, and my new family and I watched the landing on a TV in the house we were renting in Beaver. As usual, I was exhausted from a hot day and I doubt that I waited for the space walk. Chet Wallace and his then-wife Britann watched the landing at a friends' house in Santa Barbara on a TV they had just given their friends, in which all images were in shades of green. Chet and Britann were packed up and ready to leave for Americus, Georgia, where they had just accepted a teaching position at Georgia Southwestern State University after completing the Ph.D. Before this auto trip, they had been approved to report for work with the USGS in Menlo Park in a month, until President Nixon declared a government-wide hiring freeze. It took a couple years for them to get to the USGS, and then it was with Central Regional Geology in Denver, my branch!

Art Ford noted that, on the day of the Apollo 11 landing on the moon, he was on a field trip to South Africa to discuss the similarity of the huge Dufek intrusion in the Pensacola Range of Antarctica, which he had worked on in 1965, to the huge, economically important Bushveld igneous complex in South Africa. He recalls "that July 21, 1969, evening is indelible. Following another steamy day searching the veldt for platinum, cool pints were raised in the Salisbury … pub to the work of Walter Cronkite, and as Apollo 11 was landing, [a] Rhodesian farmer … stood, raised his pint, and told us 'you bloody Yanks can do anything.'" Certainly, every American felt pride and incredible optimism for the future. In fact, as the years went by, the lunar landing grew in importance, such that most of us looked back on it as one of the most important events of our lives. What a contrast to the fiftieth anniversary of this monumental event, with President Trump telling non-white American citizens that they should "go home" to their country of origin, even if they and their relatives have been here for generations!

The aborted flight of Apollo 13 in April 1970, when an oxygen tank exploded on the shuttle's way to the moon, demonstrated the courage and American knowhow of James Lovell and his crew, and the public's excitement about space exploration. But the public and the US Congress then grew bored, and the last flight to the moon took place in December 1972 with Apollo 17.

The presidency of Richard M. Nixon, which oversaw the beginning to the end of moon visits, can be credited with establishing the Environmental Protection Agency (EPA) and the National Oceanic and Atmospheric Administration (NOAA), as well as the technical "end" (by way of the Case-Church Amendment, which terminated direct involvement) of the Vietnam War in August 1973. (The fall of Saigon in April 1975 marked the actual end of the war.) But Nixon and his administration were corrupt. The most visible demonstration of the corruption was the break-in to the headquarters of the Democratic National Committee at the Watergate office complex in Washington, D.C., during the start of the campaign for Nixon's second term as President. In August 1974, about eighteen months into his second term, Nixon resigned in disgrace to avoid impeachment.

Coinciding with the good days for science, sadly, was the Vietnam War, which lasted nineteen years. It started in 1955 but escalated in 1960 under President John F. Kennedy. With the bogus Gulf of Tonkin Resolution in 1964 under President Lyndon B. Johnson, it moved to a whole new level of combat and enlistments. Television news brought the public images of the war that were both graphic and horrifying. The war's main death and destruction began as I entered the University of Texas, as did the protests. Most of us opposed the protests, for at that time we did not know the full scope of the lies that the government was telling us about the progress of the War. Credence Clearwater Revival/John Fogerty penned and sang the protest song "Fortunate Son", which noted the hypocrisy and racism of those who did the fighting ("I ain't no Senator's son, son"), and their protest song, "Déjà Vu", although written for a later war, describes the horror many felt ("another Mamma's crying").

How did the war affect me? More than me directly, it affected my brother, Clark. Clark had graduated in 1966 from Tulane University with a degree in journalism. At Tulane, he had met his soon-to-be wife, Karen West, a gorgeous fellow student who was born and raised in New Orleans. Anticipating that he would be drafted, Clark had decided to go as an officer instead and enlisted in military training (Reserve Officers' Training Corps, or ROTC) his last year or two in college. As expected, upon his graduation the US Army came calling. By luck, he was sent to Korea to be in the Demilitarized Zone (DMZ) between North and South Korea, rather than being sent to Vietnam. Still, things were hairy there, for both sides sent patrols into the DMZ and fired at each other. (After military service, Karen and Clark settled in the historic district of St.

Louis, Missouri, where they bought and restored a large house. Karen was a manager and Clark became a jack of all trades, including restoring historic buildings. They retired to South Carolina, but sadly Karen passed away on April 14, 2024. Their son Christopher W. [West] Rowley earned a Ph.D. in biochemistry from Washington University in St. Louis, and married a beautiful woman, Yelena, who was born in Uzbekistan. He runs a lab in St. Louis that does genetics, and she translates Russian into English.)

Meanwhile, back to Texas, where we graduate students, in contrast to Clark, had student deferments (exemptions) and so had only minimal fears of being drafted. In my third year, as the grim toll of the war hit the US public harder and harder, I had three roommates, Robert S. Kier, David W. Sipperly, and Doy L. Zachry, with whom I often spent evenings debating the war and whether we as individuals should have a role in it, specifically by enlisting. We had much guilt and idealism. And in fact, Sip did join the Marines just after he got his master's degree in 1967. I never heard from him again, although a recent Google search showed that he made it back and had a career as a geologist with Atlantic Richfield Company. Much later, after I had joined the USGS in Denver, I became a close friend and colleague of Lawrence W. Snee, who had also suffered from guilt and idealism at about this time; he quit college and became a Marine officer, and engaged in combat in Vietnam. Still another USGS geologist colleague, David W. Moore, was a combat helicopter pilot in Vietnam. Both lost their idealism in a hurry! Although neither was injured physically, both suffered from PTSD and other emotional problems thereafter.

At the time, of course, we in Austin had no friends like Larry and Dave to guide us through our reflections and decisions. However, I happened to mention our debates in a letter to Clark while he was in Korea. A few months later, while on leave back to the States and with little announcement, he made a visit to see me. There, after a few pleasantries, he had to sit me down to tell me, so to speak, how the cow ate the cabbage. Most of it was in the form of profanities (learned in Korea, and mostly continued from then on, although Karen must have been a little taken aback—as I was—when we first encountered them)! Clearly, Clark had lost any idealism he might have had about the conflict! The theme of his talk with me was that the Army was not likely to make proper use of any of my talents (he suggested that they would turn me into a cook or clerk) and because of my compromised eye sight might not even allow me into combat. He told me that if I signed up despite this good advice, that would simply convince him that I was an idiot! His argument was so forceful and clear, so lacking in equivocation, that I never again considered signing up with the military. Clearly this was the best advice I ever received, and I realized it at the time! Later, when I got to know Larry

Snee and Dave Moore (more on them later), I realized that their idealism, like my brother's, had also been dashed quickly, but not quickly enough to prevent lasting damage to them.

During the time that I fortunately remained in my academic cocoon, the country convulsed in war, corruption, and turmoil. In February 1968, Walter Cronkite, the most prominent TV newscaster in the US, took an extended tour of Vietnam, during which he declared that a US victory was not possible. Richard M. Nixon declared his intention to run as a Republican for the Presidency soon after. His candidacy was given a substantial boost when President Lyndon Johnson announced in late March that he would not seek another term.

What else was happening in this era? Martin Luther King, Jr. was assassinated in April 1968 in Memphis, as racial hatred in America continued. Cities and states, starting in 1971, began to create Martin Luther King, Jr. days as holidays, but most Republican states, including Utah, refused. Finally, such a national holiday was instituted by federal legislation in 1986, which President Ronald Reagan signed. In June 1968, Robert F. (Bobby) Kennedy, brother and chief advisor to President John Kennedy, was assassinated while celebrating his Democrat primary victory in California for the Presidency, a job he probably would have landed. Something was terribly wrong with our country. Racism was a large part of it. Small minds who could not face the possibility of change was another. The protest song "Abraham, Martin, and John" by Dion expressed the grief that most people, especially the young, felt.

The Democratic National Convention in late August 1968 showed the generational divide that the Vietnam war had partly created. Inasmuch as the war was now taking place during the presidency of Lyndon B. Johnson, he got the blame. The Convention was in Chicago, where the secure political machine of Democratic Mayor Richard Daly and his enthusiastic policemen felt that attacking the young war protestors with excessive force was what his older generation should do, and riots raged for weeks between protestors and cops. Hubert Humphrey (Johnson's Vice President), who represented this older generation, was nominated for President at the Convention. (The fifty-year anniversary of the riots and the angry ballot process occurred as I wrote this manuscript, and it reminds us that not much has changed since then.) Richard Nixon, even more representative of the older generation than Humphrey, ran a campaign of law and order, one pervaded by racism. The main difference between that era and the Trump presidency was that Trump's racism and corruption were out in the open. And more distressing, now about 45 percent of voting Americans like it that way. Back in November 1968, Richard Nixon was elected President. I am ashamed to say that I voted for him, just as my first semester of teaching at Kent State University was beginning.

Under Nixon, the Vietnam War continued with no respite. Although his presidency had some civilian successes (e.g., the origin of the Environmental Protection Agency), all of these were overshadowed by the agonizing daily news reports, pictures, and videos of the war. During this time, I taught at Carleton College, then joined the USGS in August 1970. Late in Nixon's first term, as he prepared for another election, his character deficiencies (paranoia, narcissism, and a belief that he was entitled to use dirty tricks) came to the surface as he orchestrated the Watergate break-in (June 1972). Now the American public had something else that was gruesome to distract them, besides the war, and it soon got worse. Legal wrangling and statements by both parties continued through the late summer and fall, with most Republican elected officials and voters siding with Nixon' excuses and lies. By the time the next presidential election took place, I had figured out that Nixon was guilty of the break-in. By then, I was starting on my second trip to Antarctica. Although I disagreed with most of the policy positions of Democrat George McGovern, I somewhat reluctantly voted for him by absentee ballot in McMurdo Station. Other voters were not paying enough attention, and unfortunately Nixon won by a landslide in November 1972.

The year 1973 was one of the most distressing in the nation's history. Legal wrangling, charge versus counter-charge by the political parties, and official committees concerning Watergate continued nonstop, with astounding news headlines almost daily. In July 1973 it was revealed that Nixon had a secret tape-recording system that taped his conversations in the White House; Nixon's pleas of executive privilege traded the spotlight with partisan calls to make public the tapes. That August, Nixon ended the Vietnam War by decree, terminating all direct involvement by US troops. To top it off, unrelated to his Watergate dealings, Nixon's Vice President Spiro Agnew was found to be guilty of corruption for tax evasion and money laundering while Agnew was Governor of Maryland. This resulted in Agnew's resignation in October 1973, to be replaced by Gerald Ford. Archibald Cox was appointed as the first Watergate Special Counsel, but when he subpoenaed the tapes in October 1973, Nixon fired him in what became known as the "Saturday Night Massacre." Thankfully, after that it was all downhill for Nixon. Nixon thought that the next Special Council, Leon Jaworski, would be his stooge, but he was not. The legal battle for the tapes continued through early 1974. By July 1974, the Supreme Court ruled that all tapes had to be released. By August 1974, Nixon resigned, and it seemed as if the public nightmare had ended. Unfortunately, Vice President Gerald Ford pardoned Nixon in September 1974.

CHAPTER 5

THE START OF REAL LIFE

The city of Kent, in northern Ohio, was dominated in that era by the large but unattractive campus of Kent State University. More than a year later (May 4, 1970), the campus would become infamous for the Kent State shootings, in which police and security guards fatally shot four protesting students and shot and injured eight others during demonstrations over the Vietnam War. But when I drove into town in September 1968, all was quiet. Linda and John had by then settled in at Kent and had two daughters, Janet and Kathy. John's mom, Astrid, had bought a home nearby, and John had become Associate Professor in 1968. At the Department of Geology, I was assigned to teach Introductory (Physical) Geology (Geology 101), Geomorphology, and a seminar in Volcanology. I had been hired as a full-time Temporary Instructor. In our spare time, John and I put together our geologic maps and did lab work and wrote manuscripts for publication on the geology of the High Plateaus of Utah. John contacted people with whom we would like to collaborate, including Bob Fleck of Ohio State University, who would undertake K-Ar ages of the volcanic rocks, and Alan Nairn of Case Western Reserve University, who specialized in paleomagnetism.

It probably was a blind date concocted by a friend, perhaps John, but I started dating a petite young divorcée named Karen Ward, who taught elementary school in Kent. She had a six-year-old son, named Scott (born March 27, 1963). Karen was pretty and giggly. I fell in love. It seemed both shocking and incredible for a bookworm like me to reach this stage, for I previously thought that love was only for the lucky, not me, as in the later (1980) beautiful song "The Rose" by Bette Midler. But was I mature enough to start married life with a fully formed family? We did all the things we needed to do to practice being a family, such as go on picnics, visit the local zoo, make and eat meals and watch movies together, and so on. By late in the spring, we were ready to try marriage. After a few bumps in the road, Scott and I became very close and we maintained that relationship from then on, well beyond that with Karen, unfortunately. We got married in June 1969. Part of the new family was Barney, a black standard poodle who had a gift: he could speak a short sentence, when prompted generally near his dinner time, "I'm hongry." At the time of our marriage, I had a sports car (a 1959 Jaguar XK-150) and I traded it in for a used white Volvo station wagon.

Paul Williams had asked me if I would join the USGS and go south to Antarctica with him, but government bureaucracy and a political hiring freeze delayed the appointment, as is common in Federal government. In the meantime, Eiler Henrickson was going on sabbatical leave to do research in Greece for the next academic year, so I accepted a temporary job with Carleton College as Assistant Professor to replace him. John and I still had money for summer field work from John's NSF grant. So, after our wedding, I adopted Scott, who became Scott E. (Edward) Rowley. Then Karen, Scott, and I headed to Utah, again borrowing my parents' green Jeep. We rented a little bungalow in the small, pretty, but isolated rural town of Beaver, at the western edge of the high Tushar Mountains, which is north of the Markagunt Plateau. I continued most of my mapping in the low, hot, east-trending Black Mountains southwest of Beaver. But it was still mapping, which I loved, and small pleasures like listening to the shriek of peregrine falcons on thermals above me kept me enchanted. When I returned home at the end of each day, Karen and Scott had been rotting in the little town and wanted some action. That, unfortunately, was pretty much limited to playing basketball with Scott, using a hoop attached above the garage door. When I could get away, we saw some of the parks and other beautiful attractions of Utah.

In September 1969 we moved into the large Henrickson home in Northfield, Minnesota, adjacent to campus. The Geology Department had added a third professor in 1966, that of Caryl (Ed) Buchwald, an unusually competent and dedicated geologist, young and very good with students. His wife Cinny was part of a Buchwald team that made the department a great place to get a major. After I left, the department hired Shelby Boardman, the best friend of George Davis when and after they got their doctorates at Michigan, and this put the department on a trajectory of even greater quality. Largely because of Ed (and Shelby after I left), there were far more majors in the department than when I was there. Even better, female majors were approaching about 50 percent of the total. When I arrived, it was good to see Dunc and his wife Graziella. I taught Igneous and Metamorphic Petrology, Environmental Geology, and Structural Geology. Ed and I ran nonstop field trips.

Late in the academic year, the shootings at Kent State University took place, so students from all over, including from Carleton, marched in solidarity and protest, in addition to the general protests over the Vietnam War. In early November, amid the year's chaos of student protests, Dunc died tragically and suddenly, at the age of sixty-four of an apparent heart attack. After the shock of the loss, then the funeral, Ed and I took over Dunc's classes and other duties for the rest of the academic year. Mineralogy was the main additional course I inherited. Karen and Scott did not see much of me that year (nor did Cinny and their kids see much of Ed), as I struggled to keep up with each new class. It was tough to stay on top of lectures and, as at Kent State, I commonly prepared lectures the weekend or the night before the classes in which I gave them.

In the summer of 1970, an Alfred Sloan Foundation grant to Carleton again allowed me to return to Utah. We rented a small house in Cedar City and I extended my mapping from the southern Black Mountains southward to the Iron Springs district. Another basketball hoop above the garage door enabled Scott and me to play a little basketball most days when I came home. Finally, during the field season, Paul Williams told me that the USGS appointment for mapping in Antarctica had cleared the bureaucracy, and the USGS officially offered me a permanent position. In August, the three of us, plus Barney the dog, piled into the Volvo and took the I-70 route back across the Colorado Plateau to Denver, a trip that became commonplace the rest of my career. Then we moved our belongings out of Kent and on to Denver.

Here let us take a break from my personal narrative and discuss how scientific collaborations, interests, and careers are built, topics of interest in our case because it explains how six Antarctic field parties into unexplored lands were put together and completed as a seamless and successful project.

> *The story starts with **Paul Williams**, who was born in 1932 in New Jersey to Paul L. Williams and Marjorie L. Dean. His early main influence was his mother, college educated and an elementary school teacher who taught in New Jersey beginning in 1920. She traced her ancestry to a Walter Deane, who emigrated from England to Massachusetts in 1636 on the Mayflower (as did an ancestor of Art Ford's). Marjorie introduced Paul to books and the natural world of animals, volcanoes, and dinosaurs. His dad, an office worker in New York City, moved the family to Maryland in 1942, when he was employed by the US-Great Britain Lend Lease program out of Washington, D.C. Sadly, Marjorie died of cancer that same year. Paul's dad remarried and the family moved to Seattle when the war ended. A good student who received some excellent career advice from a cousin who was an oil geologist, Paul enrolled in 1950 at the University of Washington as a geology major. And here the name Hoover Mackin again enters the picture, for in his sophomore year Paul took Geomorphology from Hoover Mackin, and in his junior year he took Hoover's Map Interpretation course. Those exciting courses certainly eliminated any doubt about Paul's career choice! However, Paul recalls a comment made by Hoover in humor about one of Paul's homework reports in the Map-Interpretation course, "Williams, if you truly believe*

> *what you wrote here, you should consider getting out of geology!" Fortunately, there were not many of those incidents. Paul received a B.S. degree in geology in the summer of 1954 and attained an M.S. degree in 1957, with a thesis supervised by Hoover Mackin on the Pleistocene glacial deposits in Stanley Basin in central Idaho. His Ph.D. came a decade later from the University of Washington, while he was a USGS geologist.*

At this time the combination of the University of Washington and J. Hoover Mackin led to a surprising number of USGS geologists, many of whom went to Antarctica. The first of note is Arthur B. (Barnes) Ford, like Paul born in 1932 and entering the University at the same time as Paul. They took most of the same courses in undergrad and grad school and became best friends. Although Hoover was on Art's Ph.D. committee, his Ph.D. was supervised by Professor Peter Misch, the second-best teacher and personality in the department and similarly strong on geologic mapping (more on Art below). Another student there then, also benefitting from Hoover's courses, was Charles H. Thorman, who got his master's and Ph.D. there (Art was the lab instructor for one of Chuck's classes), also under Peter Misch. Chuck, the son of a Baptist minister in southern California, finished up his dissertation in 1962 by mapping the Pequop Mountains of Nevada. Following a career with Humble Oil Company, teaching stints at Olympic Community College and the University of Oregon, he eventually became Chief of Central Mineral Resources with the USGS in Denver while Paul Williams was Chief of Central Environmental Geology, and they became best friends. Another was Dwight L. Schmidt, who did a Ph.D. under Hoover to study radioactive minerals in Idaho placer deposits, supported by a USGS grant to Hoover during the uranium boom. Dwight later joined the USGS after Hoover wrote a letter of recommendation that, among other comments, stated that Dwight "was the best geologist since Joseph Barrell and the nicest guy since Jesus Christ"! That work for Dwight partly supported Paul's master's degree. Another contemporary of Paul and Art was H. Richard Blank, Jr., who finished his Ph.D. in 1959, doing his dissertation under Hoover, involving mapping in the Bull Valley Mountains of Utah just west of the Pine Valley Mountains. And Willis H. Nelson, after a bachelor's degree at Montana State University, completed a M.S. thesis in 1950 on the volcanic rocks of the Iron Springs district, supervised by Hoover. All these friends became great field geologists and ended up together with the USGS, all in Denver except for Art and Willy in Menlo Park. Paul, Dwight, Art, Willy, and Dick mapped in Antarctica. Later, all became friends of mine.

Critical to all this was something called the International Geophysical Year (actually, eighteen months) of 1957–1958. A group of the world's top scientists proposed the IGY to the International Council of Scientific

Unions to gather global data in such geophysical topics in the earth sciences as cosmic rays, geomagnetism, gravity, meteorology, oceanography, seismicity, and solar activity, the year being picked to coincide with a surge in sun-spot activity. The program was endorsed by sixty-seven countries, the main exception being China, then continuing a spat over Taiwan that even today might turn into war. Much of the effort was to be in the Antarctic and Arctic, for the inspiration for the proposal was previous international polar years. The 1953 death of Joseph Stalin marked a softening of the Cold War, and since then Russian scientists have been major contributors to the study of Antarctica. Larry Gould, though president of Carleton College, was asked by the president of the National Academy of Sciences to coordinate the Antarctic effort as Chairman of the IGY Antarctic Committee. He chose Paul Siple, Bernt Balchen (the pilot of the flight with Byrd over the South Pole), and Richard B. Black (for which is named the Black Coast, which I was among the first to visit) to that committee. The IGY work in Antarctica became by far the most visible effort to the public, for many expeditions to study geology, glaciology, and other topics were launched. Commercial American television, then taking hold in the 1950s and early 1960s, showed indelible images of field work, storms, Snow Cats falling into crevasses, and other photogenic images. Walt Disney, through the Mickey Mouse Show, provided weekly newsreel updates of these adventures. Most of the bases in Antarctica were established during the IGY and began scientific work by various nations, including the Amundsen-Scott South Pole Station by the US. Shortly after, in 1959, these nations signed the Antarctic Treaty, declaring the continent to be only for peaceful purposes and cooperative science. The achievements over the short "year" were spectacular, including defining the mid-ocean submarine ridges (a major piece of evidence in plate tectonics) and discovery of the Van Allen radiation belts. Yet, as the IGY began, in October 1957, the Soviets launched Sputnik I, which of course led to other satellites and scientific breakthroughs, but started the space race, as discussed earlier.

Interest in Antarctica during the IGY came at a time when geologists were debating the theory of continental drift, although convincing evidence of the theory was about a decade away. Although persons had recognized for hundreds of years that many continents had outlines that could be matched up with others across oceans (e.g., the western coast of Africa with the eastern coast of South America), Alfred Wegener in 1912 put the name "continental drift" to the separation of the matching continents and grouped the continents into a single original supercontinent, Pangaea. Paul Williams observed that the name "drift" sounded incredibly anemic for a theory that launched arguably the greatest revolution in geology! Antarctica formed the central part of Pangaea with respect to other continents and subcontinents of the southern hemisphere, and this fact made it a schoolroom for study by those who favored the theory. And in fact, most geologists in the southern hemisphere were more inclined to like the theory. Unfortunately for Wegener, he could not identify any force that drove the continents to their present positions, and many learned people ridiculed the "idea." Besides, the critics argued, his scientific career was in meteorology and he had no Ph.D. Not only did many continents match geographically, geologists soon started finding that the geologic matches were far more compelling. In 1931, Arthur Holmes and Harry H. Hess suggested that mantle convection (heat) provided the power. This was a major step forward, but where and how still had to be answered, so the debate continued (to be covered in subsequent chapters).

The USGS was an agency that was well suited for full-time Antarctic field work during the austral summer (October through February), something that academic geologists could not do because they were obliged to teach classes during that time. Furthermore, the USGS evolved from the expeditions of early exploration of the American West in the late 1800s, when geologists such as Ferdinand Hayden, Clarence Dutton, and John Wesley Powell studied unvisited areas and determined their resources. For Antarctica, geologists from geological surveys of other countries were likewise enlisted. This entire international crew of scientists became colleagues and friends, and few of us can remember anything except support and sharing of data from all of them. Not much later, of course, geology professors themselves, and their graduate students, such as Cam Craddock and John Anderson, became involved. Antarctica not only presented mountain ranges that had never been visited, but the beauty of these ranges and the excitement of being there was hard to believe. Of course, Antarctica was, and still is, a continent for young persons because of the rigors of working outside in the cold. So, most scientists were necessarily young. After all, there is a good reason why most persons who enter middle age prefer Florida!

Dick Blank, not yet with the USGS, was the first Mackin student to take up an Antarctic adventure. He received a Fulbright grant in 1958 to work with New Zealanders mapping in the Transantarctic Mountains. These mountains (the longest in Antarctica, at more than 2,000 miles) separate the more circular area of old (Precambrian) rocks called East Antarctica from the younger (mostly Cenozoic and Mesozoic), more northern rocks of the Antarctic Peninsula, Marie Byrd Land, and other areas, called West Antarctica. Peaks in the Transantarctic Mountains reach almost 15,000 feet. The Transantarctic Mountains are broken into individual mountain groups, such as the Queen Maud Mountains that Roald Amundsen and Larry Gould explored. Art Ford, after a short stint at teaching at San Diego State University, joined the USGS in 1960 to begin mapping in the Thiel Mountains (the eastern part of the Transantarctic Mountains) during the 1960-1961 season, then led the mapping

(Dwight was in the field party) of the Patuxent Range of the Pensacola Mountains (also an eastern part of the Transantarctic Mountains) in 1962–1963. Dwight led a team that included Willis H. Nelson to map the Neptune Range of the Pensacola Mountains in 1963–1964.

Although not a Mackin or University of Washington student, geologist Warren warrants a mention here.

> ***Warren B. Hamilton*** *joined the USGS in 1952 not long after getting his Ph.D. from UCLA, and he led a two-man field party during the IGY (1958–1959 season) to parts of the Transantarctic Mountains. In fact, Warren coined the name Transantarctic Mountains in a 1960 report he wrote. Warren returned to other parts of the Transantarctic Mountains in 1963 and 1964. From there, his remarkable career found a theme. He visited some of the ranges in southern continents to compare their rocks with those he saw in Antarctica. A self-described "closet drifter" before he got to Antarctica, Warren Hamilton's travels in the southern hemisphere convinced him of continental drift, and he suggested that the Transantarctic Mountains had been formerly connected to the Adelaide fold belt of Australia. His ideas preceded the rapidly incoming data from the ocean floor in the late 1960s, after which only the most stubborn geologists did not accept the theory. With more data, it became apparent that it was not only continents that were moving but that the continents rode on other upper rigid layers (plates of lithosphere), namely the upper mantle (together the continental crust and the upper mantle are called the lithosphere). The name of the concept thus changed from continental drift to plate tectonics.*
>
> *The new data showed that oceanic crust was created at spreading centers of upwelling magma, such as the north-trending Mid-Atlantic Ridge, which locally breached the surface to create volcanic islands, such as Iceland in the North Atlantic and Ascension in the South Atlantic. The upwelling magma beneath, for example, the central Atlantic Ocean, drove "conveyer belts" in opposite directions, one going westward carrying North and South America and the other going eastward carrying Europe and Africa. But there was another part to the picture, because the earth's circumference could not increase (because the earth's mass could not increase) as the Atlantic basin became wider. This was the concept of subduction, whereby the westward moving plates containing North and South America over-ran the lithosphere (including oceanic crust) of the Pacific Ocean basin, which descended in east-dipping subduction zones beneath North and South America. Indeed, the "ring of fire" that surrounded the Pacific basin due to subduction zones resulted in volcanic mountains (such as the Cascades) that were eroded to their intrusive sources (batholiths and plutons exposed in the Sierra Nevada and Andes), with associated earthquakes along faults.*

Running with these ideas, Warren became the world's foremost student and prolific publisher of plate tectonics, known as "arm waving" because it dealt with continent-scale concepts. Yet he had been solidly based in field geology and was a member of the same branch I joined (Branch of Central Regional Geology) in 1970, although he later transferred into the Branch of Geophysics. He later received the Penrose Medal of GSA and was a member of the National Academy of Sciences. I got to know Warren, although not well (I considered him humorless, but perhaps he was just shy), but I respected him greatly, liked him, and was delighted when he was awarded the Penrose. He retired from the USGS in 1996, disgusted with the USGS Reduction in Force (covered in Chapter 8) and joined the Department of Geophysics at the nearby Colorado School of Mines. Here he taught only one or two courses a semester that included introductory geology and geology for the non-geologist (an excellent teacher, according to Margaret Wallace, a banking executive who took the course and the wife of Chester Wallace, covered below) while continuing research on topics of even larger scale, including plate tectonics in other parts of the solar system. Warren died in 2018, aged ninety-three. Almost every geologist who had to summarize in their reports the regional aspects of their little area of study had to cite one of Warren's reports. Whenever I had to discuss the plate-tectonic picture of the areas we mapped in Antarctica, I called upon Warren to be one of my technical reviewers. He was an excellent reviewer, so I picked him many times, but each time I knew I would have to face one of the peculiarities of his style of reviewing: among his many good edits would be sprinkled put-downs in the margin, such as "How could you even think that?" "Ridiculous!," "Are you nuts?," and other personal comments that queried my intelligence. But the good of his reviews outweighed the bad, and he prevented me from writing something boneheaded about plate tectonics. Rest in peace, and thanks, my friend!

Now back to the story of ***Paul Williams****. In 1957, Paul joined the USGS full time. The uranium boom had started in the early 1950s, and the USGS was still hiring. Most deposits were on the Colorado Plateau and the work was run out of an office in Grand Junction, Colorado. The job included a new program that involved mapping ten 1:250,000-scale quadrangles on the Plateau. Paul was assigned to work on the Moab two-degree sheet, which included the Uravan uranium belt. Here he met E. Bart Ekren, nine years Paul's senior and a four-year USGS veteran already, then working on individual uranium deposits. Bart taught Paul the mysterious ways of the USGS. Both were to become great geologists and colleagues who worked with each other on and off for the rest of their careers. Bart was also to have a major impact on my career; his story appears below. In 1958, the office closed and all were transferred to the Denver Federal Center, in Lakewood just west* *of Denver. With completion of the Moab sheet (it was published in 1964), Paul was to start on the Salina two-degree sheet. But, just prior to starting, Antarctica intervened: he was asked to join the 1963–1964 field party led by Dwight Schmidt to the Neptune Range of the Pensacola Mountains. This later resulted in the geographic names Williams Hills and Schmidt Hills in the Pensacolas, and they affixed the name Mackin Table to a nearby plateau. Having gotten the Antarctic virus but mapping the Salina sheet, Paul in 1968 applied to NSF to lead his own field party into Coates Land, at the far eastern end of the Transantarctic Mountains. But NSF deemed that to be British Antarctic territory and turned down the application. The British Antarctic Survey (BAS), however, were perhaps more in tune with the international character of research in Antarctica, and they invited Paul to visit their bases as an Exchange Scientist during the 1968–1969 season to talk about the American program and the field methods the USGS used in mapping. This involved a flight to Punta Arenas, Chile on an ancient C-124 Globemaster cargo plane, where he was picked up by the* John Biscoe*, an ice-strengthened, rather small (1,500-ton) BAS cargo ship that was used to supply BAS bases on the Antarctic Peninsula. Paul brought his banjo, for he was an accomplished musician who, for beers and yuks, had been playing banjo regularly during evenings with a traditional Dixieland jazz band at Your Father's Mustache, a nightclub in Larimer Square, near downtown Denver. Paul is over six feet tall, fit from all his field work, and instantly likeable, with a great sense of dry humor and a ready, infectious laugh. Although generally quiet, his dry humor was such that, as I was to learn later, if you were eating lunch or dinner with him, you had to be careful if you had a mouthful of food when he was about to say something. Otherwise, you might laugh so hard that the food would come out your nose! During the trip, Paul gave slide talks about the geology, logistics, sights, types of air and ground transportation, and mission of the US Antarctic program. Of course, his singing and banjo playing with others in the evenings on the ship and at the bases also earned him many friends, and this paid dividends in later years for him, the US program, and even me! And the next time Paul applied to NSF to do mapping in British Antarctic Territory, he was not denied. But he ate in the galley with the other scientists (most were young, generally in their twenties, who were on a two-year tour), field hands, and crew. From Punta Arenas, they went to Port Stanley, capital of the Falkland Islands, a British dependency, to take on supplies and fuel. Then they headed for the Antarctic Peninsula, first visiting Deception Island, an active volcano off the northern tip of the Antarctic Peninsula. Then the ship passed along the western side of the Peninsula to Adelaide Island Base, an area of almost unbelievable beauty of mountains, ice, and Pacific Ocean shorelines. This route would in later years be part of a regular tour by cruise ships, with most of the passengers being* *wealthy retiree tourists who now paid for an adventure that included eating and drinking to excess, watching whales and killer whales amongst icebergs from the ship, and taking inflatable rubber Zodiacs equipped with outboard motors to visit the mostly bare rocks (there are no trees in Antarctica), penguins, and old bases. From Adelaide Base, Paul flew by Twin Otter aircraft to the small BAS station of Fossil Bluff, on Alexander Island. This was the farthest south he got. Art Ford and I were to replicate much of Paul's trip in 1986–1987, as described below. On the return trip northward, the volcano on Deception Island erupted, damaging a British base and destroying a nearby Argentine base. The ship returned to Punta Arenas and Paul flew home.*

In April 1969, Paul married a beautiful widow, Jean M. Glen-Myers. Jean's former husband, a Navy pilot, had died on a training flight while they lived in Corpus Christi. At the time she became a widow, she had three children, aged four years, two years, and four months. One of the musicians in the band at Your Father's Mustache was Jean's cousin, and he introduced Jean to Paul in 1967. That proved to be Paul's lucky day! When they married, Paul had an instant family, and the kids—Suzie (by then eleven years old), Heidi (nine), and Charlie (seven)— were delightful. On New Year's Eve of that year, Marjorie was added (while Paul was away on The Ice!). Jean was a caring, mentally strong, highly intelligent, wise, and accomplished person, and a great mother who kept the kids in line while nurturing all of them into becoming successful and caring adults and parents in their own right. Paul and Jean were to become mentors, great friends, and two of the main reasons for my growing up and achieving something in geology.

By the time they married, Paul learned that NSF had agreed to fund a 1969–1970 field party to Antarctica, for which he had applied and which he was to lead. This was to the southern Lassiter Coast, a mountainous area near the base of the Antarctic Peninsula and clearly within "British Antarctic Territory." It was a four-man party, with Dwight Schmidt as second in charge. The others were Lawrence E. Brown, a young geologist with an MS degree, and Carlos C. Plummer, a young geology professor at California State University in Sacramento. Carlos turned out to be an expert radio operator and a good cook as well. Topographic Division sent four topographic engineers, headed by Eberhard G. (George) Schirmacher, to provide control for base maps for our geologic maps. A base camp consisting of a thirty-foot long, well-padded (insulated) Jamesway Quonset hut was established on the four-mile-wide Wetmore Glacier (about 74° South), where the field party was put in by Navy C-130 Hercules aircraft. The party worked out of the base camp on traverses of as much as several weeks, traveling by motor toboggans—a precursor to snowmobiles—and living in two-man Scott tents. The motor toboggans were geared low so that they could carry fuel barrels, boxes of food and gear, tents, and rocks on the Nansen sledges that they towed.

The Lassiter Coast has three main mountain ranges, separated by two large glaciers that flowed east into the Weddell Sea. The mapping, at a scale of 1:250,000, disclosed that the rocks were indeed like those in the Sierra Nevada. As with all other field work that the USGS did, whether in Antarctica or elsewhere, later stateside work consisted of lab work, preparation of geologic maps, and writing reports for publication. But with the rush into the following field season (which was my first), they made little progress on the lab and office work; instead, Paul and I ended up doing that in 1971. One of the interesting things these four geologists found during the 1969–1970 field season was a porphyry copper deposit not far from the Jamesway, but it needed more work in 1970–1971. The geologists returned to the US in March 1970.

Rio Grande
Isla de
los Estados
Marysvale
volcanic
field
Map Location
MINERAL MOUNTAINS
Drake

CHAPTER 6

South Shetland Islands
Elephant I.
King George I.
Renaud
Lavoisier I.
Matha Strait
Antarctic
Peninsula
mfs
PETE ROWLEY
U S Geological Survey
Dolleman I.
Steele I.
Alexander
Island
George VI Sound
Palmer Lan
Butler I.
1986 – 1987
Black Coast
1984 – 1985
English Coast
1972 – 1973
N. Lassiter Coast
1969– 1971
S. Lassiter Coast
1977 – 1978
Orville Coast

I joined the USGS as a geologist with the Geologic Division in August 1970. Paul Williams hired me specifically to work in Antarctica. Denver was in a housing boom, with an expensive rental market and, for economy's sake Karen, Scott, and I had to rent a summer cabin well to the south, in the beautiful small town of Palmer Lake at the base of the Front Range. The USGS offices are in the Denver Federal Center in Lakewood, a western suburb of Denver. As the fall approached, it became obvious that the summer cabin was poorly insulated. Dwight Schmidt came to the res-

ADVENTURES IN ANTARCTICA AND ELSEWHERE

cue by suggesting that we rent their house. He and his wife, Carol, and their sons, Lyman and Keegan, were headed to Saudi Arabia, where the USGS had a geologic "mission" in Jeddah to map the geology there. We jumped at the chance, for their house is in the suburb of Green Mountain, only a few miles west of the Federal Center. Carol, Dwight, Paul, and the whole Branch seemed to be helping our young family get going ("everyone is helpful, everyone is kind"), as in the later (1973) song "Shambala" by B.W. Stevenson. Our October trip to Antarctica was coming up fast, so Paul and I had a lot to do, including hiring the rest of the field party. I was to be second in charge of the four-man geology party. Another four-man party of USGS topographic engineers from Topographic Division, again led by George Schirmacher, was to join us to survey the ground control for topographic base maps.

This trip to The Ice was my first of five, together making up the most exciting part of a career that was rarely dull and always fun. Those five trips, plus the previous 1969–1970 trip that included Paul and Dwight, were to explore the last large mountainous part of Antarctica—and therefore of the land surface of the earth. The US Antarctic Research Program of NSF supported this work. Each of the six field seasons mapped areas of between 2,500 to 10,000 square miles.

After the 1970–1971 field season, Paul started a new project of mapping in the Snake River Plain of Idaho, and I took over the Antarctica project. The fact that Paul, Dwight, Art, others, and I were able to complete this mapping effort, and therefore put a bow around the entire project area of more than 50,000 square miles (the size of Illinois) is itself remarkable for its continuity.

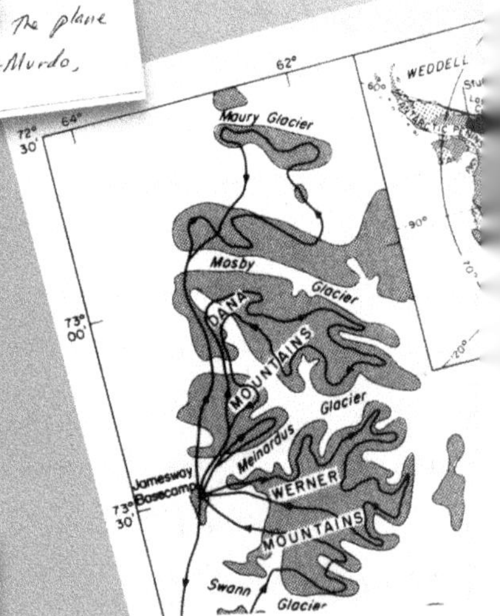

1970 – 1971
S. Lassiter Coast

The 1970–1971 Field Season on The Ice

1

Antarctica is the fifth largest of the seven world continents. Its diameter is that of the east-west distance across the US, from California to Maine. Continents are made up of continental crust, which is old, thick, and relatively silicic rock (therefore lighter, so it floats on less silicic rock that makes up the underlying mantle), as opposed to oceanic crust, which is younger, thinner, and basaltic (the opposite of silicic, thus heavy). Most continents have separated from each other at various ages by movement of the tectonic plate of which they are a part. That movement continues. Yet other debris of continental crust trapped within oceanic crust, debris we know as countries such as Madagascar, New Zealand, and the Falkland Islands, has also moved with respect to the continents, but these blocks are too small to be called continents so they are demoted to "islands."

The place we were going was the central Lassiter Coast, on the eastern side (the Weddell Sea of the Atlantic Ocean) of the southern Antarctic Peninsula. The Antarctic Peninsula is that finger of land that extends north from the main subcircular continent toward the southern tip (also a finger) of South America. The Drake Passage, with the stormiest seas in the World, separates the two fingers. As such, the Antarctic Peninsula is the continuation of the Andes, and both are geologically like the Sierra Nevada and other parts of the west coast of North America.

Few people had been on the ground in the southern Antarctic Peninsula before Paul Williams got there in 1969. A relatively small area called eastern Ellsworth Land (west of the Lassiter Coast) was visited by a group of scientists in large enclosed tracked vehicles (snowcats) of the Antarctic Peninsula Traverse in 1961–1962, led by John C. Behrendt (more about him later). And in 1965, a four-man field party led by Thomas S. Laudon of the University of Wisconsin at Oshkosh, did mapping there after being put in, I believe, by a ski-equipped R4D aircraft (the Navy term for what the Army calls a C-47 and Douglas Aircraft Company—which built it—calls a DC-3). I would later hire Tom to be a part of two of our field parties.

The rest of the southern Antarctic Peninsula was the last large unexplored area in Antarctica because it was the most isolated part of the continent, far from US and UK bases. The nearest US base was Palmer Station, seven hundred miles to the north, whereas the nearest UK bases were Adelaide (later replaced by nearby Rothera) about 250 miles to the north and Fossil Bluff (staffed only in the summer) nearly two hundred miles to the north. But it became logistically possible to explore because of a new type of aircraft, which had begun to be used only a decade earlier. This was the four-engine turboprop C-130 Hercules cargo plane, made by Lockheed Aircraft. By outfitting the landing gear with skis, the US Navy, which supplied the logistics for the US in Antarctica, opened vast areas of the continent for exploration. Production of Hercs began in 1954 and continues to this day. They are the world's best cargo plane, and pilots love flying them because of their powerful engines, sturdy airframe, and dependability. In 1969, they finally reached the southern Antarctic Peninsula, and we were the beneficiaries.

Aircraft had flown over our field area and taken trimet photos (each flight line a strip of overlapping vertical shots, then two strips of side views), which we used to locate ourselves and on which we mapped. We also had 1:500,000-scale "sketch maps" made by the Topographic Division of the USGS from the trimet photos, but a publication-ready topographic base map required the engineers to establish control by land surveying (as it worked out, Topo never delivered on their promise to provide controlled base maps, even though NSF funded them generously). The USGS topographic data, however, were passed on to the British Antarctic Survey, which published the controlled maps.

Maps and photos were easily available, but hiring the correct people to go to Antarctica is a more exacting process. First, the person must be an excellent field geologist, especially in their ability to make geologic maps. Next, the person needs to be in excellent shape, well-coordinated, and in good health; youth helps a lot. They must have a love and respect for working in the remote out-of-doors. It is a benefit to have mountaineering and cold-weather-camping experience, or of course previous trips to Antarctica. A person must be an accomplished skier and have athletic ability; be compatible and be able to get along with others; must be willing to accept a certain amount of danger; and be able to operate under pressure when faced with danger. Danger includes extreme cold and windchill, falling into crevasses, operating vehicles, and sliding down steep ice slopes. Of course, such persons need to have common sense so they do not do something stupid or macho (e.g., be a "hero") that will put them or others in jeopardy. Any person who becomes sick or hurt in the field requires a flight from McMurdo to get out, and if that happened the Navy would simply pull the rest of the party too, ending the field season. Finally, no person we ever hired to do this was luke-warm about this job! All of us were fired up, and were about to get "wooly bully," that is grow a beard—literally or metaphorically—as described in the 1964 song of that name by Sam the Sham and the Pharaohs. (The first woman into the area was Janet Thomson, who joined us on the English Coast; Art Ford hired the first women in Antarctica somewhat earlier to work in the Pensacola Mountains.)

For the 1970–1971 field party, in addition to me, Paul temporarily hired Karl S. Kellogg and Richard L. Reynolds, the former with a B.A. in geology from Berkeley and the latter from Princeton, now going straight for Ph.D.s at the University of Colorado. After they received their Ph.D.s, our branch hired Karl permanently to do mapping in Colorado and the Uranium/Thorium Branch hired Rich to do paleomagnetism (the study of Earth's magnetic fields in the past). Both became unusually prolific authors, with Karl becoming the expert on the geology of Colorado, among other places, including Antarctica. Rich applied expertise in paleomagnetism and mineralogy to studies of many things over a long career, including the environmental effects of atmospheric dust, including microplastics; he retired in 2013 but continues his work free of charge, including supervising graduate students at several universities on these topics. As for Karl, during a trip from his home in Boulder to our office in Denver for an orientation meeting, he wrecked his car on an icy road and broke a leg. So, at the last minute, Paul hired two other USGS permanent employees from Menlo Park, both veterans of Antarctic work. They were Willis H. Nelson and Arthur B. Ford, who had busy schedules but, in this emergency, agreed to split the field season. Willy was to join us first, and Art was to come in on the resupply flight at Christmastime. Willy was then forty-nine years old, by far the oldest member ever in our field parties, but in good shape.

> ***Art Ford**, classmate and close friend of Paul's for more than fifty years, began his USGS and Antarctic career when he was co-leader of an expedition to the Thiel Mountains, an eastward continuation of the Transantarctic Mountains, in 1960–1961, then led the field party there in 1961-62. He then led field parties in 1965–1966, 1973–1974, 1976–1977, and 1978–1979 to the Pensacola Mountains, east of the Thiel Mountains. Among other discoveries in the previously unvisited Pensacolas, they mapped the largest layered mafic intrusion (economically important because these rocks contain platinum) in the world—the Jurassic-age Dufek intrusion in the northern part of the range. By the time Art was unjustly fired by the USGS (in the 1995 Reduction in Force, known as the RIF, described below), he had made twelve trips to The Ice and was the greatest US expert on the geology of the continent. He has the Ford Massif and the Ford Ice Piedmont named after him.*

In addition to the four geologists, our four topographic engineers would be led by Eberhard G. Schirmacher (George), who was at the Lassiter Coast in 1969 to 1970. George had been a young child in Germany just after World War II, before his family emigrated to the US. He remembered post-war hardships, and now he ate all beef steaks well done, which he had been accustomed to as a child to avoid getting sick from bad meat. He was an excellent engineer, leader, and companion. The other engineers were Ronald F. Whiting, Antonio I. Malva-Gomes, and Robert L. Johnson.

As before and in subsequent years, a month or two before we left, the NSF organized a week-long symposium for everyone scheduled to be on The Ice that season. The symposium was held at Skyland Resort, Virginia, a National Park Service complex in Shenandoah National Park that consisted of a lodge, motels, cabins, and lecture rooms. What a beautiful and classy tradition that made all participants feel welcome and part of the team! It was a great opportunity to meet geologists in other projects, biologists, aircraft pilots, logistics managers, and others whom we were likely to see or work with in McMurdo. We met officials from NSF, the Navy, and support groups. Lectures included summaries of all upcoming science activities, safety talks by the Navy and others, and general information and policies for this long and complex adventure.

On October 14, we flew by commercial airlines from Denver to San Francisco, where we connected with Willy, then on to Honolulu. From there we were bused next door to Hickam Air Force Base, with sleeping quarters in officers' military housing (BOQ, Billeted Officers Quarters). The USARP contracted with the US Navy to supply all US logistics, and so from this point on we were in their hands. We were scheduled to take off the next day in a Lockheed C-141 Starlifter, the first military jet-powered transport (production line began in 1965), their workhorse until about 2006 and much bigger than a Herc. But the flight was delayed, so we rented a car and enjoyed un-Antarctic things such as palm trees during two days of touring western Oahu. We took off on the evening of the 16th, loaded like sardines in a small can—at least sixty Navy enlisted men and some scientists, all bound for Antarctica. The aircraft is a noisy boxcar with no windows, webbed seats facing rearward, six seats across in the central aisle and side aisles of four, all our luggage in a net in the rear. In a set-up like that, all you can do is sleep, so everybody did. After a flight of five and a half hours, we refueled at Pago Pago (for some reason, pronounced Pango Pango), American Samoa, in the middle of the night in choking hot humidity, then flew another five and a half hours to Christchurch, New Zealand, losing a day crossing the International Dateline and arriving early on a Sunday morning, after a total of 7,200 miles. As with the Australians, New Zealand (NZARP) has been doing Antarctic science long before any Americans got there. We spent part of the first day at the USARP warehouse, adjacent to the airport, supplying ourselves with cold-weather gear. That included a bright red parka with hood, red plaid woolen shirts, black baggy heavy wind-breaker pants, gloves and mittens, long-john underwear, etc. We got white floppy canvas mukluks, which laced up halfway to our knees, and although those were soft and warm, they provided no foot support suitable for all-day hiking and climbing, so we had previously bought double mountaineering boots (which have a smaller inner boot).

Helping us was New Zealander Margaret Landon. The USARP hired her to run the logistics part of the program, so she was the first person we met in Christchurch. Because of her competence and outgoing and humorous personality, she was always greeted with hugs and happy greetings. She embodied the fact that we were finally close to something fun and memorable. For although geologic research in Antarctica was hardly a large specialty, those involved from many countries became a tight-knit group because of the adventure, and it consequently attracted good people. In fact, the international appeal of all types of Antarctic research was and continues to be remarkable, with close interactions among individuals from several dozen countries during international symposia of many disciplines. Antarctica is indeed a continent of science, probably in part because of the adventure and beauty of the surroundings, but now becoming increasingly important in studies of climate change.

Our clothing issues solved, we checked into Warners Hotel on the downtown square. Christchurch is a beautiful city, ideal for exploring now that it was early spring, with temperatures getting into the 60s during the day. The cathedral for which the town is named is on the square, and a large botanical garden, about a mile away along the Avon River, is a common walking destination. In that era, restaurant food in New Zealand was not very good, and the powerful smell of mutton emanated from all restaurant kitchens. Though it was the most popular local dish on these densely sheep-populated islands, we avoided mutton.

On October 19, we learned that we were scheduled to leave the following day on a Constellation ("Connie"), a four-engine propeller aircraft made by Lockheed. We were apprehensive with that news. For although the Connie had been in production from 1943 to 1958 and was a favorite of commercial aviation and as a military troop and cargo plane, its reign was fortunately nearly over. The military apparently bought up the last ones from the production line, but because of its limited range (see below) they finally determined that this aircraft was not suitable for Antarctica. Nonetheless, until midseason, McMurdo can accommodate wheeled aircraft such as the Constellation by spraying water on the runway, which instantly turns to ice. Later in the summer, the snow and ice surfaces tend to become slushy, and in such conditions only ski-equipped aircraft can be used.

Our destination was to be McMurdo Station, the main US base, at a latitude of 78° South. McMurdo is a port perched on the shore of Ross Island, which is about fifty miles long. The entire island consists of black lava flows, whose source is several jagged (extinct and eroded) volcanic cones, plus more rounded Mt. Erebus, an active volcano 12,500 feet high in the center of Ross Island. Erebus almost continuously erupts puffs of ash and occasionally small stubby lava flows. In October, Ross Island is surrounded by sea ice. On the southern side, this ice is called the Ross Ice Shelf, which extends southward about fifty miles to the mountain peaks of 10,000 to 13,000 feet, which rise virtually from sea level, called the Royal Society Range of the Transantarctic Mountains, beyond which is the Polar Plateau. The Ross Ice Shelf is made up of permanent ice, perhaps two hundred feet thick, floating on sea water but undergoing no changes, at least on the timescale of historic exploration. Seaward, north and west of McMurdo, the permanent ice is fringed by annual ice of the Ross Sea. Annual ice, about six or seven feet thick in October, generally melts during the Antarctic summer. The runway (Williams Field) is on annual ice several miles south of McMurdo; when this ice goes out, planes use the runway known as outer Williams Field—another dozen miles southward—which is on permanent ice. The contact point between the two ice types is at McMurdo, and therefore the summer melting allows ships to dock and offload fuel and large items such as vehicles, lumber, and prefabricated housing.

Despite the plans to leave on October 20, we learned that day that, although several C-130 flights the day before had made it to McMurdo, a trip of 2,400 miles, they had had to turn back because of high winds on the ground. Our flight was scrubbed. We rented bicycles and rode to a nearby town, then returned to Christchurch to shop for Christmas presents, which we sent home. And that day we also got the good news that our flight was now to be by one of these ski-equipped C-130 Hercules, on October 21. One of the advantages of the C-130 is that it has a range that would get us to McMurdo, but if the conditions turn out to be bad on the ground, as they had been for the flights on October 19, the plane had the fuel capacity to return to Christchurch. The value of this became obvious as we circled McMurdo on our approach the following day, for there, just off the skiway was a Constellation—the plane we had first been scheduled to use—its long dimension oriented about 60 degrees to the runway, and one wing and the landing gear broken off. A Constellation has enough range to make it only halfway to McMurdo (and back), so if the crew learns that the weather is bad at the landing site, it can return to Christchurch. The wrecked Constellation had gone more than halfway before the weather turned bad. Without the fuel to return, it had no option but to continue anyway. Fortunately, that stormy landing caused no loss of life, but the use of Constellations in Antarctica soon came to an end, for obvious reasons.

On the day of our trip, October 21, we put on our cold-weather gear before boarding in Christchurch. Three of the four topographers were already in McMurdo, and George was to come on a later flight. In flight, we slept and at times wandered around the cabin, stacked high with food ("freshies" of lettuce and vegetables were especially valued at the McMurdo dining hall), mail (even more highly valued), people, and personal gear. USARP scientists made up most of the people on board. All were clad in the standard heavy woolen shirts and black windproof pants, with their heavy red parkas nearby and ready to don right before landing. From the round windows, we saw open ocean for the first several hours, then icebergs, then ice shelves, then finally the rock of the continent.

After a seven-hour flight, we landed to frigid conditions (14 degrees below zero), and wind. It was early spring, and we were the early birds. The sun was below the horizon only for about an hour (at midnight, resulting in twilight) when we arrived, but we knew that within a week it would be up all the time until February 20. Our first impression was of the beauty of the Royal Society Range of the Transantarctic Mountains to the south. By contrast, McMurdo, to which we were transported by small buses—or, in more adverse conditions, by tracked vehicles such as Weasels—was ugly: a mishmash of thirty-five or forty buildings of different types, colors, and sizes poking out of the black lava flows and separated by snow-packed (and later muddy) roads, with lots of telephone poles clogged with wires. The effect of the wind driving sand and dust from the black lava flows through and onto buildings was to pepper the snow between buildings with black flecks. Near the buildings is Observation Hill, one hundred feet high, which is capped by a large wooden cross erected by the winter-over part of the Robert Falcon Scott expedition in the spring of 1912 after they discovered that Scott and his field party had died. It's at once a sobering reminder of the dangers of life here as well as of the joys from generations of exploration and discovery.

The USARP staff and officials checked us all into McMurdo, then assigned us rooms in a two-story dormitory. Job One after getting situated was finding our field gear, including food, fuel, motor toboggans, tents, and rock boxes located at various warehouses, transporting all of them to central locations, and putting them on aluminum pallets for the Hercules. Downhill skis had to be outfitted with bindings. We selected our motor toboggans at the auto-mechanics' shop and warehouse with care, for they were likely to have been abused or badly worn during the previous season, and winter-over mechanics were unlikely to have serviced or repaired them, meaning they'd require a lot of our attention to fix them. We then serviced and field tested them on the ice shelf and outfitted them with the hardware we needed. The central locations for all the palleted gear were under cover, for the winds associated with any storm that might come in before we were ready to go might bury everything. Gathering up our gear had to be done as soon as possible because the Lassiter Coast was a whopping 1,600 miles off, and leaving for the field area would happen as soon as the weather cleared. It could be tomorrow or two weeks from tomorrow. We decided that Paul was to be our cook,

Willy our mechanic, Richard our radioman, and I our weatherman.

Weather information was hard to come by, typically estimated from satellite views and any reports from bases in the Antarctic Peninsula or from Byrd Station, a refueling stop halfway between McMurdo and our field area. Because we were the farthest field party from McMurdo, thus the most isolated, we were to be one of the first to be put in. Once we were in the field, we were required to send radio transmissions once daily (or much more often if planes were in the air) to let McMurdo know we were ok. The most useful type of message was one with weather data, for those provided information that could be critical for other field parties.

In a letter from Karen that arrived just after we landed at McMurdo, I learned that she was pregnant! Letters that followed told of her morning sickness and the start of visits to the obstetrician. But she, Scott, and Barney were doing fine. Then the Schmidts came through for our family again. Even though Dwight's paperwork had not yet been finalized for their trip to Saudi Arabia, they invited Karen, Scott, and Barney to share their home until their departure date. This, of course, was a godsend to Karen, a stranger in town and popping pills for morning sickness in a cabin with little insulation. And in Keegan Schmidt, Scott had a friend of nearly the same age.

Although all of us were trained in first aid and had prepared ourselves for likely Antarctic conditions, and although others of the field party had been on the continent before or had mountaineering experience, we were required to take a two-day snowcraft school, which taught field and survival techniques such as how to use an ice axe, rig ropes for belays and climbing, wear crampons, etc. An ice axe is an important tool; we carried one most of the time we were on snow and ice as a walking or climbing stick, with the hand grasping the middle of the metal head and, at the other end, a metal spike that is thrust into the snow/ice with each step. The axe head has an adze (a small hoe) on one end that we used to cut steps when climbing. The pick at the other end has been a life saver for most of us at one time or another, for it enables a self-arrest: if you fall and start sliding down a slope, you flip onto your stomach and then, with one hand on the handle and the other on the head of the ice axe, rotate yourself so that your feet are facing downhill, then bring yourself to a stop by slowly digging the pick into the snow/ice.

All of us knew the difference between a square knot and a granny knot, but now we learned how to tie other knots, how to ski or walk while roped to a colleague, and how to extract ourselves from a crevasse or move up or down an ice cliff. We learned how to dig snow trenches and make igloos, and spent a night in one, as this was the first-line defense if we were caught out in a storm with no tent. We also took radio school briefly, learning how to use a short-wave radio, which is powered by a gasoline generator.

Throughout our wait for the right conditions, the camaraderie and spirit between the eight of us and the Navy aircrews and weathermen, and the USARP staff of mechanics, cooks, clerks, and young summer or year-round laborers, was always at a high level, for all of us were in the middle of an adventure that we knew we would remember fondly for the rest of our lives. We understood that our safety depended on one another. There was a lot of joking, laughing, and paying attention to everything the veteran Antarctic folks had to tell us. Truly, we were all having the time of our lives!

Once everything was finally packed, all we could do was wait. We were just as anxious as the Navy to get in the field, so we frequently visited the weather bureau for updates. If the weather were bad over the field area and the wait were likely to be marked in days, we now had time to write letters to our loved ones, shop for Christmas presents in the PX gift shop, write Christmas cards (with a McMurdo post mark!), take afternoon or day trips to explore nearby parts of the island and its wildlife, and visit the New Zealand Scott Base a couple miles away, ice caves along the shore, and the Navy Officers club (that is, the bar, where pilots and aircrew were also waiting out the weather). The smell of diesel exhaust in the cold was always present, and from then on, this smell always brings back Antarctic memories. The wildlife was extra fascinating, and on good days we spent as much time outside as we could to view animals. On clear days, Weddell Seals, as much as a thousand pounds in weight and with a length of eleven feet, sunned themselves on the ice like giant slugs, with newly born pups lying alongside them. Occasionally we would see a penguin, mostly Adelies, but the big rookeries were several hours of travel away, and if we did visit them, it was at the end of the season. Brown, noisy skuas, which are much larger than seagulls and eat penguins, sea food, and pickings from the McMurdo garbage dump, circled overhead. At the end of the season when the annual ice had melted, we occasionally saw an Emperor penguin and the dorsal fins of killer whales who were waiting for tasty penguins just off the edge of the permanent ice.

On October 26 and 27, I volunteered to help our topographic engineers on a short project for some Russian glaciologists in the Dry Valleys—beautiful, north-south valleys in the Transantarctic Mountains that are free of ice and snow. Katabatic winds draining off the Polar Plateau had scoured the valleys clean, unique in Antarctica. The bedrock floors of the valleys were instead covered with ventifacts—boulders and pebbles of rock beveled on several sides by sand carried in the wind. A Huey helicopter took us the sixty miles and dropped us off; and we spent the night in a Jamesway hut that had been erected years before. The Dry Valleys are a popular study area for many scientists, including investigators who looked for bacteria or other forms of life in frigid conditions that might be analogous to those on Mars. A day or two later, the first C-130 flight made it to Pole Station to open it up and bring food, especially freshies and mail. The temperature upon landing was 58 degrees below zero.

The Lassiter Coast is mountainous, with local relief on rock peaks from several hundred to more than three thousand feet. Glaciers three to five miles wide define and separate the mountain ranges. The Lassiter Coast was named after Captain James W. Lassiter, from the US Army Air Force and the chief pilot for the Ronne Antarctic Research Expedition of 1946 to 1948 that wintered over at East Base, a base erected by the U.S. Antarctic Service Expedition of 1939 to 1941 (Richard Byrd's Third Antarctic Expedition) on Stonington Island, about halfway down the western side of the Antarctic Peninsula. Lassiter and Finn Ronne had explored parts of the Antarctic Peninsula well north of us by landing their ski-equipped, single-engine aircraft at various places to get latitude/longitude positions (sun shots with a sextant). The huge Ronne Ice Shelf southeast of the Lassiter Coast was named for Finn, the expedition leader. Most ranges in and near the Lassiter Coast had been named after eighteenth-century geologists, such as Hutton, Playfair, and Werner, whereas the glaciers were typically named after more recent explorers, including pilots, who had visited other parts of Antarctica. One range, just south of our put-in site on the Wetmore Glacier, was the Latady Mountains, named for William Latady, the person in charge of the aerial photography for the Ronne expedition. (As chief of the 1957 winter-over Ellsworth Station, Ronne displayed severe emotional problems, as told in the 1998 book written by my USGS colleague John C. Behrendt.) Paul Williams and his 1969–1970 crew named our dominant rock, mostly a black shale, the Latady Formation after good exposures in the Latady Mountains.

The snow surface on which we now spent most of our time was amazingly beautiful. For example, hoar frost from high-humidity mornings would leave ice crystals on the ground and other surfaces and would gleam the colors of the rainbow when the sun hit them. At times such ice crystals would fall like gentle snow. The squeaking sound that your boots made changed with the temperature. Snow in Antarctica is rarely deep and fluffy, for Antarctica is one of the world's great deserts—here meaning that it gets little precipitation (in the form of snow rather than rain, as in other deserts), except along the coasts; of course, all that falls, stays. The winds tend to blow away any fluffy snow or drive it into low, one- to two-foot-high hard dunes, called sastrugi. Therefore, the snow surfaces over most of the continent are hard packed snow called firn, so one walks on its surface rather than sinking into it. This prevails if the temperature stays below freezing, which it does except on rare occasions during the Antarctic summer.

By contrast, the glacier ice that underlies most of the Antarctic snow surfaces is brittle, yet moving. This movement and changing temperatures result in cracking sounds that at times sound like gunshots. On a larger scale, the movement of the interior of glaciers results in cracks known as crevasses in the brittle upper surface. These extend to depths of roughly two hundred feet, below which ice flows. Falling to such depths in a crevasse, of course, can result in death, so crevasses need to be avoided—hence the safety drills we learn and the equipment we bring. On big glaciers and ice shelves, crevasses can be hundreds of yards across and more, but these are generally visible (meaning open) and easily bypassed. It is the smaller crevasses, typically a few tens of feet across, that are a bigger problem because the cracks at the surface are commonly bridged by wind-blown snow, and thus nearly invisible. When we set out each morning, we'd often remind each other that crevasses are omnivorous, meaning that they eat anything! These small to medium-sized crevasses are especially hard to see, even invisible, on a cloudy day, when shadows are not present on a glacier surface.

Cloudy days result in a whiteout, a phenomenon for which there are two definitions. Most popular is the one to which most people can relate because they have at times been out in it, namely a condition of near-zero visibility due to an intense blizzard. The other definition, used here, has no bearing on visibility but is rarely experienced except by people on wide expanses of snow or in polar regions. The *Merriam-Webster Dictionary* defines it as a surface weather condition in a cloudy snow-covered area in which no object casts a shadow, the horizon cannot be seen (meaning the land surface is indistinguishable from the sky), and only dark objects are discernible. With complete cloud cover, surface definition is absent; the thicker the clouds are, the worse is the whiteout. In such conditions, one's feet appear to be floating in a milky surrounding. Yet visibility itself is not impaired, so bedrock even miles away is visible—but it appears to be floating in a white cloud. It is always difficult to tell distances in Antarctica (there are no trees and few buildings to give perspective), but without a horizon or surface it becomes still more difficult, and something that appears to be miles away may only be ten feet away. I remember one such cloudy day early in the first season, when Willy and I were out on a traverse, and Paul and Rich had gone in another direction. We had stopped in the middle of a glacier, perhaps to take a breather or to change drivers. The wind was blowing, as it normally was. We ate a candy bar as we talked. Then I saw a red thing in the distance. Our motor toboggans were red and I observed "Look, there's Paul and Rich way over there; what are they doing in that direction?" As we stared at the red thing, we suddenly realized that it was a candy wrapper that had been blown about thirty feet away in the wind.

Winds were almost constant but we could do field work in most of them. I exaggerate only slightly when I say that when we could not stand upright in the wind, we knew we would have to take the day off! Storms therefore were our nemesis. And sometimes these storms lasted a week or more, bringing all progress to a standstill. Though we had wind gauges in the kits furnished by the Navy weather bureau to take our daily weather readings, these would break if we tried to take wind-speed readings at those times when we could not stand up, so all we could say with certainty was that the wind was at more than fifty miles per hour. When working outside in such conditions, a short-handled metal shovel would be blown to a horizontal position. Tents would flap furiously, the wind-driven snow would make a whooshing noise moving along the tent's outer surface, and above these sounds would be a high-pitched howl of the storm. Wind speeds of more than two hundred miles per hour have been recorded in Antarctica. At those speeds, the wind blows humans away. The primary type of almost-constant wind is katabatic or gravity-driven wind, in our case coming off the crest (of about 6,500 feet elevation) of the southern Antarctic Peninsula to our west and traveling east down the glaciers. In the US, these are akin to both Santa Ana winds, moving west off the Sierra Nevada of California, and to Chinook winds, east off the Front Range of Colorado. Of course, we also had winds from weather fronts moving through the area, as is common elsewhere.

On those days when such winds or other storm events kept us laid up in our tents waiting for the weather to clear, conversation or reading books was our main source of entertainment. On this expedition, all of us had brought a library of adventure books, which we shared with each other. During one season I digested the Hobbit trilogy, a favorite of the others too. This of course led to us quoting from memory some of the colorful expressions in the books as we experienced our own challenges. Among the books we borrowed and devoured from the extensive library at McMurdo were biographies of Arctic and Antarctic explorers of the Heroic Age (we kiddingly considered ourselves part of that club of OAEs, Old Antarctic Explorers, although we were not in their league). Sometimes we debated the methods, successes, failures, and tragedies of each explorer. Yet, whether they failed or succeeded, we acknowledged the bravery of each one of them.

> *Among those of the Heroic Age that we discussed was British explorer* ***Ernest Shackleton****, who made the first attempt at the South Pole in 1908. Art Ford, in a talk before The Explorers Club in 1998, quoted the following from a recruiting ad said to have been placed by Shackleton in* The Times *of London in early 1907: "Men wanted for hazardous journey. Low wages, bitter cold, long hours of complete darkness. Safe return doubtful, honour and recognition in the event of success." You may know that he and his party had to turn back within ninety miles of the Pole when Shackleton realized that he could not both reach the Pole and survive the return to their base. Norwegian Roald Amundsen, the first to make*

it to the South Pole, on December 14, 1911, along with his five men, was the most resourceful in his understanding and use of the best "technology"—that of dogs—in that as food was consumed and the weight of the loads decreased, the men killed and ate some of the dogs. But he had no interest in (or refused to be distracted by) the science of the continent.

British Captain ***Robert Falcon Scott*** *either did not plan adequately or did not accept available advice. He and four other men followed Shackleton's route, from Ross Island up the Beardmore Glacier to the Pole, arriving on January 17, 1912. His first mistake was that he was averse to sacrificing dogs, so he used horses instead. When the horses died of exposure or fell into crevasses, he and his men resorted to man hauling. Such exertion, and the increased time it took, led to their demise. Scott and his men also lost precious time doing science, primarily geology, and collecting samples in the areas they explored. Therefore, when they did reach the Pole a month after Amundsen, it was late in the tiny window they had for their field season. As winter approached, the entire exhausted, disappointed, and weakened field party died after making it all the way back down the Beardmore Glacier. There, on the Ross Ice Shelf, the first person died on February 18, 1912. The rest struggled on until a storm pinned them down almost within eyesight of Ten Ton Depot, and they ran out of food, heating fuel, and luck; they died on March 17 and March 31. In hindsight, this result would seem to have been predictable.*

Australian ***Douglas Mawson*** *was a special example of the will to survive under tragedy. He and two others were the first to get to the South Magnetic Pole, during a separate trip but as part of the 1908 Shackleton expedition. Having reached this Pole, he then proposed his own expedition, staffed by Australians and New Zealanders, to explore the unexplored Antarctic coast south of Australia (Australian "territory"), which is counter-clockwise around the circular main mass of Antarctica from Ross Island and past the western end of the Transantarctic Mountains. The circular mass of the continent (excluding the Transantarctic Mountains) is an old (Precambrian) "shield," as makes up the core of continents. A geology professor in his home country, Mawson in the 1911–1912 season established his main base at the head of a bay that unluckily happened to be one of the worst funnels for katabatic winds off the polar plateau. Another base was established by ship farther counter-clockwise up the coast. The two bases communicated with each other by radio—until the antennas blew down. From the bases, sledging parties went out in several directions during the 1912–1913 season. Mawson led a three-man team (with two dog-sleds) to the east from the main base. On December 13, about three hundred miles east and inland of the main base, di-*

saster struck when one man and his dog team disappeared into a crevasse—along with the sled that contained most of the dogfood, the tent, spare clothing, and supplies. For several hours, Mawson and his companion called to the lost man but got no reply and could not even see the sledge, so they had to assume he had fallen to his death. The two men fashioned a tent out of the bag that had been used to carry the lost tent, then they and a half dozen dogs headed back in terrible conditions, killing dogs as they traveled to feed themselves and the other dogs. The other man became sick on January 1, and Mawson hauled him on the sledge until he died on January 8, about one hundred miles from the main base. Alone, Mawson struggled on, weak and with frostbite, while hauling the sledge through storms and despite falling into crevasses. But he was up to the challenge. By January 29, with his food all but gone, he stumbled upon a small cache of food and supplies that a party from the base looking for the team had left a few days previously, just in case. This gave him the strength to make it the last thirty miles to the base. He arrived on February 1, only to see the ship on the horizon, heading for Australia. Six men, however, had remained behind to look for Mawson, and they made radio contact with the ship. Unfortunately, weather conditions made it impossible for the ship to return to the base, so the seven wintered over and even made another sledge trip in late November. The entire party was finally rescued when the ship returned on December 13, 1913. Mawson still is considered Australia's greatest hero!

The best of all the Antarctic explorers was undoubtedly ***Ernest Shackleton****, who never lost a man. His explanation to his wife Emily, after he turned back from the South Pole in 1908, was "I thought, dear, that you would rather have a live donkey than a dead lion." Though he didn't achieve his objective of reaching the geographic Pole, his reputation rests on his eight-hundred-mile open-boat journey in the "*James Caird,*" little more than a rowboat with a single sail.*

The adventure started in December 1914 when his ship the Endurance *left Grytviken, a whaling station on the eastern coast of South Georgia Island. The intention of the expedition was to reach the South Pole, then go on to the McMurdo area, from the Weddell Sea (or eastern) side of the continent, south of the Antarctic Peninsula. But by January the Endurance had become locked in pack ice, which froze to annual ice, with no hope of escape. Currents carried the sea ice and beset ship northward along the coast of the Antarctic Peninsula over the course of the winter. Then, as the spring (October 1915) opened leads in the ice, it crushed the ship, which the crew had to abandon before it sank. The entire crew took to several open lifeboats, including the aforementioned* James Caird, *and so they continued (sometimes by man-hauling the boats across pack ice) the rest of the way north along the eastern coast of the peninsula, until in April 1916 they reached Elephant Island, off the northern tip of the peninsula. Not only did the outside world not know they were there, there was little food to support the entire crew, and the Antarctic winter was coming on.*

And with that began, on April 24, 1916, what must be the greatest open-boat voyage ever, following ocean currents and using a small sail to head east along the southern Drake Passage, attempting to hit the relative flyspeck of South Georgia, eight hundred miles away! Through skilled navigation and a great deal of luck, Shackleton and his five companions made it. In high surf they landed on the western coast of this mountainous island on May 10. But the whaling station was on the eastern coast! In more than thirty hours of continuous hiking in terrible conditions, they crossed a pass estimated at 4,500 feet elevation, then descended to the whaling station. The factory manager, who knew Shackleton, asked the three men knocking on his door "Who the hell are you?" The next day, a whaling boat rescued the three men on the western coast. Attempts by Shackleton, employing several ships, to get to the twenty-two remaining men at Elephant Island were rebuffed by the ice conditions and weather, until the steel-hulled Chilean sea-going tug Yelcho *succeeded on August 30. The telling of this, perhaps the world's greatest adventure story, by Alfred Lansing in his 1959 book* Endurance: Shackleton's Incredible Voyage *was my favorite book in high school. The first sentence, which begins the preface of the book, is "The story that follows is true." It is a common saying among people who have been to Antarctica, all of whom know the story, that when you are in great danger, pray for Ernest Shackleton!*

Now, back to our narrative! For us, neither man hauling nor dog sleds were how we traveled in the field. Instead, each two-person team had a FoxTrac motor toboggan, manufactured by the Fox Company in the US. When they were working, we certainly felt fortunate to have them, but many times they were our diabolical nemesis. They were of mid-1960s vintage and powered by a twelve-horse-power Kohler engine. Weighing about 1,100 lbs, each was geared low so that it rarely went more than ten miles per hour and averaged five to eight. This way, each two-person team went for miles and hours on end, pulling more than a ton of supplies, fuel barrels, food, and rocks. These supplies were placed in canvas "tanks" that were lashed to beautiful wooden Nansen sledges; drawstrings closed the tanks, and we tied tents, bamboo trail flags, ice axes, etc. to the top of the load. Named for the nineteenth-century Arctic explorer Fridtjof Nansen, who first used them, these were carefully crafted (by elves, no doubt, and thus expensive) in a northern European country. Each motor toboggan pulled two Nansen sleds when moving camp. On day trips to do the geology, we towed only one lightly loaded Nansen. Behind the rear Nansen sled, we attached a bicycle wheel with an odometer. The motor toboggans, the predecessor to the snow mobile, were our life, for we were in the boonies and would be re-supplied only once, at midseason near Christmas, so each of these vehicles had to be carefully maintained, serviced, and cared for. Unfortunately, many of the motor toboggans in the warehouse at McMurdo had been abused, so at the beginning of each season we carefully inspected and field tested them. The four people in our party were all geologists, but we chose those who had some ability as mechanics. And it was amazing how each one of us became a pretty good mechanic during breakdowns or when machines would not start, when the only other option was to walk!

The USGS was unique among organizations in that it had a method of using ropes for driving motor toboggans and (in later seasons) snow mobiles. Why? To keep us out of crevasses. If a bridge collapsed over an open hole, the motor toboggan might be lost but the human could let go of the rope and survive, at least if he or she were on skis. For a person on skis exerts far fewer pounds per square inch than a person on foot (and a horse far more!). So unless we were on rock or in our tent camps, we stayed on skis. Crevasses are so prevalent that it is best to assume that they underlie even tent camps or other places on snow where we might be tempted to take off our skis. Indeed, occasionally we broke through a bridge while puttering around our tents.

We had perfected our camping and trail methods over many field seasons. Some of our methods can be traced back to Charles W.N. Swithinbank, a British glaciologist on the Norwegian-British-Swedish Expedition of 1949–1952 to Queen Maud Land. Swithinbank passed on those methods to Thomas Taylor, a topographic engineer of the USGS during the Ross Ice Shelf Expedition of 1959–1960. Taylor then taught them to Arthur B. Ford on his trips to the Thiel Mountains. From there, they were relayed to Dwight Schmidt during the 1962–1963 season in the Pensacola Mountains, then to Paul Williams in 1963–1964, and from him to me. Using the USGS method, one person would ski alongside the machine, steering it by ropes hooked to the front skis of the toboggan. To get to this position, a seated driver wearing skis would set the throttle at a constant, appropriate speed, then slide off the seat of the moving machine and stand up, with steering ropes in hand. A rope attached to the rear Nansen would pull along the other team mate, also on skis. The toboggan would trail in the snow a line of quarter-inch parachute cord, attached to the throttle, so that if either person fell or came upon a crevasse, he/she could dive for the parachute cord and pull it, stopping the motor toboggan. As you can imagine, it took some training before each two-person team had ironed out all the bugs to this method, such as inadvertently stepping on the parachute cord, falling in whiteouts, tripping over sastrugi, etc. Early on, we felt like The Three Stooges. But after a while, traveling by this method for hours each day came naturally, and it kept us in great shape—and out of crevasses! Although we published an article on this method, I know of no other organization that used it. Yet each field season, persons would be injured or killed falling into deep crevasses in Antarctica. Clearly crevasses were our greatest hazard, although storms, always fast moving, were a close second.

Most of our "nights" in the field would be spent in two-person, seven by seven-foot, mostly tannish yellow Scott tents, made famous by Robert F. Scott. These are double-walled pyramid tents with a canvas floor and four corner poles that meet at the top, with vents at the peak for smoke and moisture to escape. They are sewn particularly carefully out of high-quality materials and were consequently expensive. The tents are stable in heavy winds, for the floor continued as flaps beyond the tent walls, and stakes and snow blocks anchored these flaps. The tents accommodated a metal-frame canvas camping cot on either side, with a central aisle extending back from the door. On each cot we placed our three-inch-thick foam mattress, then our five-pound goose-down sleeping bag. For heat and cooking, we put a Coleman white-gas stove on a storage box, opposite the door and against the back wall. On one burner, we set a pot for cooking the morning and evening meals, and on the other, a pot for melting snow blocks into water, the blocks gathered by wielding a shovel or snow saw out the front door of the tent, with our caveat being "no yellow snow!"

On November 2, with the weather seemingly clear over the Lassiter Coast, a truck drove the pallets of gear weighing about 19,000 lbs down to the skiway at McMurdo, where others loaded them onto a C-130. Because such a flight was heavily loaded, not only with our gear but also with the aircraft's fuel, most flights departed close to midnight, when the ice runway was at its coldest and hardest (and therefore fastest). On this night, it was 20 below zero, and still light out at 11 pm. The sound and smell of the four turboprops of a C-130 are distinctive, and the scent hangs pungently over the crisp night air as the aircraft positions itself for loading. The four of us did what we could but, to get out of the others' way, we soon boarded and occupied the orange webbed "seats" set up for us. We would need two aircraft for this trip, but the second flight would not take off for hours, depending upon weather reports from the first aircraft and from our short-wave radio and its gas generator that we would set up upon the presumably successful landing of the first aircraft.

The engines then wound up and we were off! The plane shook and rattled as it raced down the rock-hard ice surface, and it took a while to lift off because of the heavy load. The trip was about the distance between Salt Lake City and Washington, D.C. Though we all tried to spend the five- to six-hour flight napping, we instead spent much of the time looking out the circular windows in the cabin at the ice shelves and occasional mountains. Near the end of our flight, we caught a glimpse of the Ellsworth Mountains off to our south, Antarctica's highest range (16,000 feet), that John Anderson had first explored in 1963, then Cam Craddock after that. Twelve-mile-long Anderson Massif, named after John, was down there, although we could not distinguish it. As we approached the Lassiter Coast, the pilots asked Paul to join them in the cockpit to spot the Wetmore Glacier and ranges he knew well from his trip the year before.

As we neared the Lassiter Coast, we found continuous cloud cover, plus lower clouds and ground fog, and we soon had to head back, disappointed. We landed at Byrd Station, eight hundred miles from Lassiter, to refuel. Byrd is in the middle of nowhere at 80°south, with no nearby hills or rocks, on a snowy plateau at five thousand feet of elevation. The base is only for refueling aircraft, and all of it is underground, except for black fuel bladders. Its name suggests that it was formerly near where Admiral Byrd and Larry Gould wintered over, but that was at Little America, far to the north and along the Ross Ice Shelf four hundred miles east of McMurdo. From Byrd we flew back to McMurdo.

The next attempt—this time successful—left at about 2 pm on November 12 (McMurdo time), also with all eight of us. A second flight was scheduled to follow, containing most of the topographers' equipment and most of the food, but it was delayed. We refueled at Byrd Station, a process that takes about fifteen minutes, then continued east, including over a low part of the Ellsworth Mountains, perhaps Anderson Massif. The clouds were again overhead by the time we got to the Lassiter Coast, but pilot Dave Eldridge (with co-pilot Bruce Campbell) found a few holes in the cloud cover and dropped down. They asked Paul and George to come to the cockpit to see if they could spot the old Jamesway, which was to be our base camp/airport terminal for the season, although the odds were that the winds had totally buried it in snow. Fortunately, Paul and George saw the top of the black roof. As they retreated to the cabin, the pilot announced, "Fasten your seatbelts." Within a minute the pilot had put the aircraft down near last season's skiway, heading west, up-glacier and upwind, amidst tremendous shaking and banging as the skis skittered over the sastrugi surface. We hoped the straps would not give way under the swaying, creaking load. We taxied to within about a hundred feet of the Jamesway at an elevation of about 4500 feet. As we scrambled out, Paul asked pilot Dave what the surface definition was like when he landed. He replied, "There wasn't any." The temperature was 5°F and it was 6 am on the 12th, Lassiter time (11 pm on the 12th, McMurdo time; add 17 hours to our time to get McMurdo time, for the International Date Line passes just east of McMurdo). The plane would not leave until we had established radio contact with them and McMurdo, until we had gotten into the Jamesway, and until we had got the Preway oil heater going in it. First, we started the generator for the short-wave radio, then unpacked and tuned the radio, housed in a large metal box, shielding it as best we could from snow driven by nature and propwash. Most of our supplies were offloaded quickly in a process called "freight-training": the rear cargo door was lowered and the rollers beneath the pallets released, then the C-130 wound up its engines again and drove out from under the pallets, one by one. We pried open a plywood box of shovels, tools, and parts and started digging down to the Jamesway door. Though the pilots knew that their remaining fuel was an issue, and though there was only one gas station on the way back to McMurdo, they kept the engines going for nearly two hours until we were set, and then everybody pulled out their cameras to commemorate the occasion. Our work had finally started! Our base camp was about forty feet long, fifteen feet wide, and nine feet high. It had several bedrooms, an office/radio room, and a main room containing a kitchen, a large central table, about a dozen metal folding chairs, and a stove in the middle, which warmed us as we celebrated with a beer. Paul remarked "Home again! It seems like I never left!"

Paul Williams, six feet tall and solid, had a supportive nature and was a natural leader. At radio school in McMurdo, we had learned how to spell out words using the standard short-wave-radio spelling dictionary (e.g., alpha, bravo, charlie for a, b, c). Paul's initials were Papa Lima Whiskey. Whiskey was appropriate, of course, but Papa was more so! I believe it was Richard Reynolds, Paul's tentmate that season, who started to call him Papa. The name stuck, and even years later, we called him Papa almost as often as Paul. Paul had become an instant father of three when he married Jean, and Jean and Paul were probably the best parents I ever ran into, and their kids showed it. Paul had a full beard even without Antarctica, and now we all grew them (the wooly jaw of the song) because of Antarctica.

An Antarctic field season is like your college education: a fixed amount of time for which you paid plenty and out of which you need to get as much as you can. We were familiar with the glaciers and ranges from studying our aerial photos and therefore knew where we would schedule our traverses. But we had much to do and about two months in which to do it. The party chief was under the most pressure, for he knew that we could leave no nunatak (an isolated rock peak surrounded by ice) or mountain range untouched but had to get all or most of them done, at least in some small way. Our job was simple exploration of unknown territory, all while mapping the geology in reconnaissance. Because nobody had been on the ground here before, we did not know whether mineral resources, fossils, or practical or scientific breakthroughs awaited us. We could not tell what obstacles, including storms, we would have to overcome. The Jamesway base camp contained most supplies and its location was locked in the Navy memory (although GPS had not yet been invented at this point) as the resupply and pickup point. We made traverses outward from the Jamesway that generally lasted two or three weeks, until our supplies got low. We generally stayed in each tent camp for three days to a week, unless a storm pinned us down. Because of the pressure of time, we worked every day we could, with no time off. On the trail, we slept in Scott tents and our evening meal was necessarily simple, fast, and nutritious meat-bar stews. Back at the Jamesway, we had cases of lobster tails and tenderloin steaks, frozen corn on the cob and potatoes, and other delicacies that constituted a welcome, wonderful break from the routine stews. Our appetites were huge and we ate large meals with vast amounts of calories from beef, fat, and sweets, all required by our outdoors exertions but frowned upon by any sane stateside cardiologist monitoring average people with an office job. Staying hydrated was an important part of our daily routine, so each day we carried thermos flasks of water and hot tea.

Nutrition experts from the early days of polar exploration had correctly identified high-energy food, notably fats and sugars, as necessary for a person to survive when doing constant physical labor in the very cold outdoors. We ate our meals sitting on the cots, with their sleeping bags, in one of the two tents, as the pots percolated. Drying socks and boots dangled from the peak of the tent. The cook sat next to the food pot. Conversation was animated, with many stories and much laughter. The morning meal was generally oatmeal ("hoosh"), containing raisins or other freeze-dried fruits and almost always bacon, and flavored with jam or honey. At times we had powdered eggs. The beverage was coffee or tea. The conversation centered on where each field party was going—generally in opposite directions as we explored as far from the tents as we could, yet typically within five or ten miles. The evening meal was a stew ("mush") whose basic ingredient was the aforementioned meat bars, which are made of shredded beef and fat compressed into a 3-cubic-inch bar in a foil wrap. To the reader, this staple may not sound appetizing, but we craved it. As a reminder of those good ol' days, I keep one such meat bar in a vegetable compartment of my refrigerator; it is probably still edible. We also added freeze-dried or frozen vegetables and freeze-dried meat to the stew, thickened it with powdered mashed potatoes, and ate it with bread and jam. The cook served up each bowl, and the recipient seasoned it with his favorite condiment from bottles tucked under a cot in the cooking tent. Although we had a system of putting cans of beer in a bag that we tied to the peak of the tent, it was difficult to prevent them from freezing. Most drinks were hot and flavored by tea bags or powdered Jello, coffee, or lemonade, and in the evenings we added hard liquor.

Each evening we compared notes on what we had found, we marked up base maps and aerial photos with station numbers, and drew in the geology we had observed that day. The evening meal was particularly animated, for every party consisted of compatible people, and each day we had significant adventures in addition to recording the previously unexplored geology. I remember the laughter being so loud sometimes that I reflexively checked behind me in case we were bothering anybody sitting nearby. Of course, the only thing behind me was the wall of the tent, and the nearest people were hundreds of miles away, at a British base.

During field work each day, lunches were chocolate bars, primarily Cadbury bars from Great Britain, or meat bars that we munched as we worked. Handfuls of nuts, trail mix, and an early type of inedible fruit bar were also available, but Granola, nut/fruit, and Power Bars had not yet been invented.

During the day, each two-man team took a motor toboggan, to which was roped a nearly empty Nansen sledge. Into that we put emergency supplies, consisting of a Meade tent, food, ropes, sleeping bags, and Coleman stove, as well as boxes for rock samples, jerry cans of gas, and flags on bamboo poles. We added rock hammers, ice axes, crampons, and other geologic supplies. We always wore our backpacks, even during travel with the motor toboggans, for they contained emergency down clothing, meat bars, candy bars, belay ropes, etc. We wore our skis (downhill skis rather than cross-country skis, but the bindings were set so that the heel could rise, to allow us to walk in them, as with cross-country skis). Each person carried a field notebook in a leather case that we attached to our belt, along with our Brunton compass. Each geologist gave a station a unique number, and recorded its rock types and structures (faults or folds) in his or her notebook. When we were ready to leave each morning, we fired up the motor toboggan, put on our skis, and were off. Although it was daylight around the clock, at midnight the sun was near the horizon and temperatures were the lowest, generally dipping to about 10 degrees Fahrenheit below zero; the working part of each day averaged about 10 degrees above zero. When we reached a rock outcrop, we parked the vehicle, removed our skis, and headed up a snow slope to the rock, always aided by our ice axe and wearing our back pack. We collected and loaded all fossils and samples for thin section, chemistry, and isotopic dating into small plywood boxes for transport back to the States. The ice axe that we always carried gave us a sense of confidence when traversing steep snow slopes, when self-arrests might be necessary to prevent ourselves from slipping down them. Occasionally we encountered blue ice (that is, real ice, typically from a past rare above-freezing day), and to be able to walk this required that we attach crampons to the sole of our boots and use an ice axe. We also used our ice axes to probe snow bridges across crevasses, while we stayed on skis.

On November 16, we four geologists headed north on our first traverse from the Jamesway. We crossed the Wetmore Glacier, climbed over a low pass through the six-to-eight-mile-wide RARE (Ronne Antarctic Research Expedition) Range, then traversed the four-mile-wide Irvine Glacier to establish a tent camp two miles southwest of Mount Laudon in the southern Guettard Range. This mountain, one of the highest, sharpest, and prettiest in the Lassiter Coast, was named after Tom Laudon (in Antarctic place names, unlike place names in other parts of the world, the person named for the feature does not have to be dead!). Its elevation is listed on the internet (www.mapcarta.com) at 4,619 feet, but this seems low. During the 1977-1978 and 1984-1985 seasons, I would hire Tom to join our parties!

This trip to Camp 14 lasted the entire day, a total of eighteen miles recorded on the odometer wheel. Willie drove (skied) the whole way while I held on in back, but I fell a lot and I was exhausted for the last few miles. Within a few days I'd found my legs, however, and soon I shared the driving responsibilities. In our tent camps, we had a routine of getting up at about 7 am, being "on the road" by 8 am, then working to about 7 pm, followed by dinner and sleep. No wonder my body was exhausted!

On November 21, two of the topographers visited our camp while we were out doing geology; to our delight, they brought us some mail; the second flight had arrived a couple of days previously and now the topographers were fully outfitted too.

Thanksgiving Day, as recorded in a letter to my son Scott, started as a calm sunny day on which Willy and I were mapping on two different ridges several miles from Camp 15, about ten miles east of Camp 14. The wind picked up and a thick cloud cover dropped the temperature. After an hour or so of mapping, it was apparent that the wind velocity was increasing substantially. Willy fired up the motor toboggan and picked me up halfway along the ridge and we headed back in a full storm and total whiteout. We made it part of the way to camp, aiming the vehicle at a rock landmark near the tents. Then the vehicle caught too much wind-blown snow and stopped running. With our packs, we skied back toward the landmark, locating the tents in a few miles. Inside, we turned on the Coleman stove and prepared hot chocolate. Paul and Rich arrived about an hour later; they too had abandoned their motor toboggan several miles from the tents. Just another day at the office! Dinner was meat-bar stew, but we put a little more poultry seasoning in it to make it taste a little more like turkey, and we toasted Happy Thanksgiving with rum! The next day was calm and clear with a temperature of about 15°F and we worked in our shirtsleeves.

During our time at Camp 16 in the eastern Guettard Range, we experienced alternating good weather (temperatures one day reached 22°F) and windstorms. We worked most of the time. The only vegetation we ever saw on the Lassiter Coast was sparse black lichens that could be seen in joints that trapped the rare meltwater from the few warm days each summer. On November 30, on our way to this camp (Paul and Rich had arrived there a few days earlier), Willy and I were caught in a storm. Erecting the tent took us an hour, for the wind blew down the tent and pulled up the stakes repeatedly. Because we feared that cots, sleeping bags, and the stove inside the sledge cover would blow away if we untied the canvas tank, we spent a cold night in the tent, huddled in our emergency down clothing from our backpacks. The wind had died down by "morning" and we managed to reach Camp 16. A day later, the motor toboggan experienced serious mechanical problems (as opposed to the less serious problem of snow-packed carburetors). A dip stick in the crankcase had rattled apart, with the pieces falling into the engine. Willy, our best mechanic, tried his best to fix it, but we had to abandon it. Fortunately, the topographers had used our one spare motor toboggan to haul a particularly heavy load of equipment into the Guettard Range and had cached the toboggan near Camp 15. We relayed (two trips) the load on December 5 to get all of us and our sledges to the spare. Unfortunately, the spare toboggan had a bad clutch, so Paul, Rich, and I did the fun stuff (the geology) while Willy volunteered for the tough duty (working on the clutch). Willy ended up having a better day than us, for he fixed the clutch, whereas I fell a hundred feet down a snow slope on my ridge and Paul fell thirty feet down a snow slope on his ridge in a severe whiteout. On December 7, following tracks in whiteout, we made it back to Camp 14. The next day, also in a whiteout, we followed our tracks the eighteen miles back to the Jamesway.

Following or re-tracing tracks was a common practice, especially across crevassed glaciers, for we knew that if we had made it over successfully in decent weather, the odds of a safe return, even in bad weather, were better this way. In anticipation of future whiteouts or storms that might obscure our tracks, we flagged our route, at least across glaciers. We tied large orange flags to six-foot bamboo poles and planted them every half mile or so. The Irvine Glacier was severely crevassed, so we stayed on our single flagged route every time we crossed it, knowing from having scouted the terrain that this route was the least crevassed. Yet we knew (from sunny weather only!) that we had to cross or skirt several dozen crevasses. Most seemed to be ten to twenty feet wide and as much as half a mile long. At times the snow bridges sagged under the weight of the motor toboggan and at other times the motor toboggan or Nansen runners, or our skis, broke through snow bridges. Yet when on skis, we felt perfectly safe although we occasionally noted deep holes that opened below the tracks of our skis as we broke snow bridges.

By these methods, we managed to return to the Jamesway on December 8. The four of us were exhausted even though we were now finally in good physical shape. The next few days we recovered by eating huge amounts of steaks, lobster tails, vegetables, and sweets, by sleeping late and goofing off (reading, telling stories, writing letters). We discussed and compiled geologic maps and field notes.

Then on December 12, we left the Jamesway for the RARE Range, the area closer to the Jamesway, because we anticipated the resupply plane coming in before Christmas. This was the best weather of the season, mostly low wind, with temperatures of 15 to 20 degrees. But then we experienced something more sinister: snowfall during low winds, one day about six inches, then four, and so on. Slogging through fluffy snow while mapping on foot was tiring, but traveling by motor toboggan was worse because the machines bogged down, especially when pulling a full load. We had little choice but to break trail with a motor toboggan that carried no load. We finished the RARE Range and returned to the Jamesway on December 23. The topographers soon joined us, the first time we had been back together. With an impending resupply flight, we gave weather reports every three hours, then when the plane was in the air, once an hour. We were waiting for Santa, in the form of a great gray bird, and hoping that the clouds would lift.

On Christmas day, I wrote Scott a letter after I had taken weather measurements that Paul transmitted by short-wave radio: visibility of one hundred yards with light snow falling, 16°, and high clouds (whiteout); the plane, which had been rescheduled for tomorrow, was of course on hold. Next, Paul attempted to get in touch with Palmer Station, the US base seven hundred miles away near the northern part of the Antarctic Peninsula. Palmer would typically relay Christmas messages for remote field parties by going through ham radio operators in the US, but that day we had no luck reaching them. As I told Scott in my letter to him, the previous night we had decorated the Jamesway. First, one of the topographers had borrowed a bed sheet from McMurdo that we put over our dirty wooden dinner table, then he brought out a box that contained a two-foot-high plastic Christmas tree that he put on top of some white paper napkins on which we had drawn Christmas scenes, all placed in the middle of the tablecloth. Next out of the box were some light bulbs for the tree, a straw angel for the top of the tree, small birthday candles attached to the tree with pipe cleaners, and several tiny wooden angels, shepherds, and a crèche for under the tree. From the box we took a package of tinsel and scattered it around the Jamesway, as well as a tinfoil star that we taped to our rusty Coleman stove. Then we got a roll of toilet paper on which we had drawn candy canes, santas, reindeer, and Christmas bells with red magic marker, and we draped it from the rafters. We had no bells, so we dangled from the rafters a pair of crampons, which jingle a bit like bells. We got some reception from our battery-operated radio, including Christmas music from a Spanish station, but the reception was poor and it soon faded away. Then drinks were served! And after a while, we enjoyed a sumptuous feast of steak and lobster tails with a bottle of wine. The next morning, we found that Santa had indeed visited us: each of us had a small present, such as pipe tobacco for some, a candy bar for one, and packages of Life Savers for others!

Just past midnight on December 26, the clouds lifted and we notified McMurdo. The resupply flight arrived at 3 am local time the following day. Out stepped Art Ford, and he brought Christmas presents: another motor toboggan, fuel, more steaks and lobster tails, beer, and wine. The C-130 took away our co-worker Willy Nelson and sixteen boxes of rocks. Willy had been a tent mate who had taught me a ton. But it was a good trade, for Willy was beginning to tell his favorite stories for the third time! Art, my new tent mate, proceeded to teach me still more, and I had not heard his favorite stories before. Art has a booming laugh that you can hear from a hundred yards, even in a ground blizzard. At his home in Menlo Park he drives a Model A that he restored; its vanity license plate is "A Ford": cute! The aircraft also brought letters from home, something of almost unbearable pleasure. We all sat around the table reading letters, and drinking beer for three hours—except topographer Ron Whiting, a Mormon who passed on the beer. I learned that Karen and Scott were faring well and missed me. As any member of the military far from home will tell you, letters from family members always prompt warm thoughts! And, speaking of warmth, that day was the warmest we had seen, at 27°F.

On December 29, we were on the trail, this time into the Hutton Mountains north of the Guettard Range. On New Year's Eve, we led the topographers (they were not as adept at remote skiing as we were, and generally sat on their motor toboggans) across the crevassed Squires Glacier into the small Playfair Mountains, our farthest north. At the end of a hard day, at about 11 p.m. local time, we all had dinner and drinks and celebrated a muted New Years with a few yells of "Happy New Year!" The temperature was zero, although the sky was clear and the sun out. We all thought of our families. New Year's Day was foggy, and Art and I bumped around from nunatak to nunatak, guided by our aerial photos. Most of the day we had about one hundred yards of visibility, with some light snow, but every so often the fog would lift and we would get a compass bearing on the rock we wanted to visit. With a compass bearing, the rear person hanging to the Nansen sled would be far enough from the steel toboggan that he could use his Brunton compass and guide the person steering the motor toboggan with arm signals. At each nunatak, we would get off and study the rock, taking notes, measurements, and samples, then head off into the gloom again to the next outcrop. If we had a long ridge of rock to examine, we would leap-frog our traverse, leaving one of us off at the near end of the ridge, while the other person drove a mile or two down the glacier along the ridge; this person would leave the motor toboggan there, then climb the ridge and continue along it. The first person would reach the ridge adjacent to the toboggan, drop down to it and drive it a mile or two farther before leaving it and climbing up onto the ridge to continue.

On January 3, with a sense of "fall" in the air, we decided to concentrate on the large Hutton Mountains, so we returned southward by our trail across the Squires Glacier, then east across the Johnson Glacier. Fall meant no more "hot" temperatures of 20°F or more, and a return of winds. Fog was common because east of this range was ocean (the Weddell Sea) that opened during the summer, although we never saw it. In contrast, the eastern ranges farther south are bounded by the large Ronne Ice Shelf, and the eastern ranges of the Black Coast to the north are bounded by the Larsen Ice Shelf. On this day, Paul and Rich went ahead eastward to establish Camp 23 near Mt. McElroy, while Art and I stayed back to do the geology of the western part of the range, knowing that we could follow their tracks to the camp. Fog settled in, in addition to high clouds that caused a whiteout. I let Art off on a ridge with a lot of bedrock, as I sat on the motor toboggan and visited bedrock spurs off the ridge. About this I wrote to Karen and Scott: "With a whiteout and about 100 yards visibility, a person is just in a world of white. It is a strange sensation, and a lovely one: all you have is a sputtering motor toboggan and the fog and falling snow. The only sense of direction I was able to retain was that I was hugging a snow slope, and if I stayed on that slope, I knew I would hit bedrock every so often. And sure enough, every once in a while, a dark blob ahead would change into forms of rock as I chugged to within a stone's throw. I would stop, get off the toboggan and make notes and collect samples. Then a half hour later, off I would go into the mist. Even one's sense of time is thrown off, and I would look at my watch constantly and even count numbers out loud to give myself some sense of reality."

At the end of this day of bumbling, we returned to where we began, found Paul and Rich's tracks despite the whiteout, and made it to Camp 23.

A few days later, while working with Rich, the fog came in while we were heading toward a nunatak. Just before it socked us in entirely, we stopped and got a compass bearing on it; it was to the right front of us. Then, the steering ropes got away from me and, by the time I recovered (only a few seconds) we were off the bearing in dense fog. A few minutes later the fog cleared for a moment, and now the same nunatak was to the left of us. We had been so disoriented that we had made a 180° turn. The next day, in good weather, Art and I drove past the site of our confusion, and we laughed at the aimless wandering of our tracks.

On January 9, 10, and 11, we unexpectedly experienced an "Indian Summer," with our highest temperatures of the season—about freezing. Unfortunately, now the snow became wet and sticky and the motor toboggans bogged down, with snow clogging their articulated tracks. We had to stop and clear the snow with shovels and ice axes. We moved into the eastern Hutton Mountains on a compass bearing, hoping to avoid a small crevasse field that was visible on the aerial photos. Unfortunately, once the fog lifted, we found that we had set our tents (Camp 26) right in the crevasse field. With binoculars, we could see open water, with ice bergs, east of here. One day, Paul and Rich saw two Snowy Petrels pass by—the only wildlife (other than the occasional black lichens) we had seen all season! The winds picked up after the warm spell and most days then were cloudy, therefore whiteouts. At Camp 27, at our farthest northeast point in the Hutton Mountains, we broke two of the four engine mounts on one of our motor toboggans. Of course, engine mounts hold the engine to the frame. These were old motor toboggans, and the cold and strain of working every day were taking their toll. We had no spare engine mounts, so we substituted wire and nylon parachute cord and operated the machine with tender care. The increased winds, of course, had scoured the snow surface, so it was harder and had higher sastrugi than we had had at midseason. We moved to Camp 28, back to the west in the Hutton Mountains. We kept in touch with the topographers, who now were in the Ferguson Nunataks, forty miles to the northwest. But since resupply, we had had no word from the outside world; we did not even know how the Super Bowl had turned out! Camp 28 was our last camp, and on January 22 we headed back toward the Jamesway.

It took us two days to return the fifty miles west and south to the Jamesway, as Art and I babied our motor toboggan with the bad engine mounts. But the other motor toboggan was not in much better shape, and we had to stop periodically and adjust its carburetor. We returned by way of our previous routes, for we had cached barrels of gas when we had first gone north. Back at base camp, we slept until noon and gorged ourselves on steaks and lobster tails. Yet we found that we craved vegetables and candy bars even more, a sign that we were both exhausted and weak. For several days, working from the Jamesway, we traveled the seven miles (1.5 hours) to the northwest across the Wetmore Glacier to map in detail and sample the copper deposit found the previous season. We took one motor toboggan, with three persons each hanging onto their own rope attached to the rear Nansen, sharing jokes and stories as we skied alongside each other; we had contests to see who remembered more of the Jabberwocky poem, by Lewis Carroll. After one of these field days, as we pulled in to the Jamesway at 10 pm, the temperature was -8°F and the wind was at ten knots. No wonder we were so chilled!

The copper occurrences just across the Wetmore Glacier were what geologists call a porphyry copper deposit. The rocks were scattered over a half dozen nunataks over two square miles. The outcrops were exceptional, as are most exposures in Antarctica, because they had never been covered by soil or vegetation and therefore are not broken down by chemical weathering. Porphyry deposits are typically of low grade but are now the copper deposit favored for mining around the world because they are of huge volume. When found close to the surface, they are amenable to open-pit mining. Our occurrence might not have the size that would make it commercial in an area other than Antarctica, but it was of scientific interest anyway. In 1975 and 1977 we published articles on it, including a USGS Professional Paper. This was the first mineral deposit discovered in Antarctica that might have been of commercial appeal if it had been found on another continent. This study made us the de-facto experts on metallic mineral resources in Antarctica, a topic of great interest to many countries around the world and a primary reason why they participated in Antarctic research. By the Antarctic Treaty signed by these countries, however, Antarctica was off limits to territorial or mineral claims or to development. Yet many countries informally proposed areas of interest or "territory," such as the Brits, Chileans, and Argentinians in the Antarctic Peninsula, and many others for areas around their bases. The Treaty made Antarctica a continent of research. Yet some persons who had never been to Antarctica persisted in dreaming that great mineral wealth, especially oil and gas, would be found and developed there, as on all other continents. Furthermore, with the price of oil shooting up in the 1970s, countries considered even Antarctica as a potential oil source. Therefore, by the early 1980s, Antarctic treaty countries began renegotiating an Antarctic mineral resources regime, and they needed information. They also needed experts who were aware of the limitations of any development in order to dampen expectations. Inasmuch as many scientists working in Antarctica came from academia and did not have much interest in mineral or energy resources, it was appropriate that the USGS deal with these issues. Therefore, in subsequent years I wrote many summary reports on metallic and nonmetallic deposits of Antarctica, while USGS colleague, geophysicist John C. Behrendt, also of Denver, wrote about oil and gas resources. Neither of us ever reported on anything that could be considered as having remotely commercial possibilities.

McMurdo Station was anxious to get us out, but the topographers were still a hundred miles to the north. A radio blackout, due to upper-atmosphere conditions (sunspots), prevailed for about four days anyway. At 1:30 am on January 31, the topographers pulled in, waking us up. After notifying McMurdo, all enjoying dinner and drinks, then sleep, we spent the rest of the day packing what we needed. On February 1, the takeout flight was on its way toward us, and it finally appeared overhead. But by then a thick cloud cover and ground fog prevented a landing. On February 4, another flight reached our position but could not land either. Finally, on February 5, a C-130 touched down and we loaded what we could, including rocks, tents, Nansens, camping equipment, topo gear, personal effects, and a few stricken motor toboggans. The plane's turboprops never shut down, and in less than a half hour we were ready to go. The crew deemed that JATO (Jet Assisted Take Off) was required, so they attached a bottle (each generates a thousand pounds of thrust for twelve seconds) to each side of the plane, and that is how we got off the ground. The crew also recommended that we all remain in the back of the cabin, due to our strange odor from three months of field work. The Herc was low on fuel by then, but so was Byrd Station, so we landed at Pole Station, about a thousand miles away. The temperature was -25°F, and while the aircraft refueled and took on outgoing mail and some other cargo, we ran around taking pictures of the barbershop pole that marked one position of the South Pole (it moves a few inches a day as the ice moves). Interestingly, the station temperature has never reached above zero, and in winter it is 80° or more below zero.

On our way back, we crossed over the Transantarctic Mountains and the Beardmore Glacier, by which Shackleton, Scott, and their field parties had accessed the Polar Plateau. Even from 30,000 feet up, we could see the open crevasses that marked most of the glacier, and we cringed as we remembered their story. We landed at outer Williams Field, for the annual ice beneath inner Willy Field had already melted out. Though McMurdo was now far uglier than when we had left, the summer melting having exposed more black rock and wind-swept dust, it was a welcome sight to us anyway. Our priority was a hot shower, our first in three months. Then to a chow line that fortunately was not serving meat-bar stew. Letters from home were the next great highlight. And I enjoyed a real treat the following day: after many attempts, a patched ("over," "out") "phone call" via volunteer ham operators in the States to Karen! The joy each of us felt is hard to describe, but I am sure the reader can imagine it. A few beers were needed too, to celebrate.

Over the next few days, we returned whatever we had borrowed from USARP, packed and labeled our wooden rock boxes for transport by ship, and readied ourselves for our trip home. I got a haircut and removed the beard, but could not bear to cut off everything, so I kept the mustache. Just after midnight of February 9, the four of us began the eight-hour flight by C-130 to Christchurch. The sights of green grass, flowers, and women in miniskirts, all in the early fall with a temperature of about 65°F, were truly awe-inspiring! We traded our Antarctic clothes for our civvies with Margaret Landon. On February 12, we had another long C-141 flight to Hickam AFB, with another refueling stop in Pago Pago.

Karen and I had been planning a vacation in Hawaii well before I left for Antarctica. On February 13, she flew in from Denver and I met her at the Honolulu airport. I had rented a car and we toured, slept, and ate our way around Oahu. Then on February 16 we flew to Hilo on the big island of Hawaii and did the same there. On February 22 we flew back to Denver, where Scott and Barney were waiting. Real life began anew!

In summary, during the 1970-1971 field season, we had explored the central Lassiter Coast, an area of about 2,500 square miles, about half the size of Connecticut. The area mapped during the 1969-1970 season, the southern Lassiter Coast, was about three thousand square miles. The geology of both, of course, was much the same, although Cretaceous (a geologic Period, extending from 145 to 66 million years ago and terminating with the meteorite-impact extinction of the dinosaurs) intrusions became much more common as we worked our way north. Most rocks in both areas consisted of black shales and lesser sandstones, the Latady Formation, that had been deposited in shallow to deep sea water and contained marine fossils, mostly oysters and clams, that indicated that they were of Jurassic age (the preceding Period, from 201 to 145 million years ago). The sedimentary rocks were interlayered westward with volcanic rocks, which we called Mount Poster Formation, which formed the former axis of the Antarctic Peninsula; that axis, in fact, was similar in location to the present axis. All sedimentary and volcanic rocks had been thrusted and folded into anticlines and synclines. Then they were intruded by the stocks and batholiths of coarse-grained granites and related rock types that are of Cretaceous age. When these Cretaceous stocks and batholiths were intruded into the two Jurassic rock types, their surface expression in Cretaceous time would have been as volcanoes that erupted lava flows and ash-flow tuffs, but such surface features and rocks have long since been removed through erosion. In fact, in Cenozoic time (66 million years ago to the present time), all Cretaceous, Jurassic, and older rocks had been uplifted and eroded. Uplift and stream erosion (and glacial erosion in late Cenozoic) continued until the present surface came to be, likely cutting several miles into the crust vertically. Jurassic and Cretaceous time was, of course, the maximum extent of the dinosaurs. Early dinosaurs, including *Brontosaurus*, must have been roaming the surface of the ancestral Antarctic Peninsula when the Mount Poster Formation was being deposited on the land surface, while the Latady Formation was being deposited just to the east of the volcanoes in streams and the sea off the eastern coast of this peninsula. Later, during the Cretaceous, different types of later dinosaurs, including *Tyrannosaurus*, must have been roaming among the Cretaceous volcanoes, which were derived from the underlying stocks and batholiths now exposed. And of course, plate tectonics had moved the continents to quite different places than they had been in the Jurassic and Cretaceous.

Granby
Longmont
Boulder
Brighton
Broomfield
DENVER
Central City
Golden
Aurora
Littleton
4400
COLORADO
Gunnison
Blue Mesa Res.
Great Sand Dunes National Park and Preserve
Walsenburg
San Luis
Conejos

2

Between Field Seasons, 1971–11972

Denver is a good place to raise a family. The biggest city in the intermountain west (Rocky Mountains), it lies at the eastern base of the beautiful, massive Front Range, where peaks rise to 14,000 feet. To the east is the Great Plains. Denver is attractive, although sprawling. Ever since Mom, Dad, Clark, and I visited it during our summer vacations, it has been my favorite place in the country. Even its professional football team, the Denver Broncos, was my favorite as a kid. To live there was a dream that had miraculously become reality. Via east-west Interstate 70, Denver's western suburb of Lakewood was the gateway to the mountains, where we spent many weekends picnicking, hiking, skiing, and touring. We rented the Schmidt's house on Green Mountain, which is a 2.5-mile-long, flat-topped hill (more specifically an old dissected alluvial fan) that rises almost seven hundred feet above the "mile-high city" and whose lower eastern flank is suburbanized. By the time Karen and I returned from Hawaii, she and Scott knew Denver better than I did, so I had some catching up to do. Denver also is a friendly city, for most residents were born elsewhere, and we made many friends easily. The enthusiasm of these years seemed to be typified by the band The Rolling Stones, and to take a line from their 1968 song "Jumpin' Jack Flash", our new life was going to be "a gas, gas, gas"!

The Denver office of the Geologic Division is one of three main offices of the USGS (the others are in Reston, Virginia and in Menlo Park, California), although smaller offices are scattered around the country. The Denver office is in the Denver Federal Center, a square mile of mostly single-story, old government buildings in western Lakewood that during World War II made ammunition and explosives, and thus access was restricted to its government employees. Our offices were housed in Building 25, a mostly one-story sprawling building whose corridors zigged and zagged intentionally so that if an explosion occurred, it would not destroy the entire hallway and rooms off it. We all had real offices (no cubicles here!) whose size depended mostly upon the GS (General Schedule, or primary pay schedule of the US Civil Service) grade of the geologist.

I was hired at a grade if GS-12, as is usual for a person with a Ph.D. and two years of job experience beyond the Ph.D. Government grades and promotions to higher grades are determined differently in the USGS than in most other federal government jobs. Most other federal jobs have grades based on the job for which you are hired, which includes what you do, how many people you supervise, etc. To get a job of higher grade, a person must apply for a different job. Back then our grading system, however, was determined by an annual committee of about a half dozen peers who appraised each scientist largely on the quantity and quality of her publications. So, as they say in academia (which also has a peer system to determine Assistant, Associate, or Full Professor) it was "publish or perish" at the USGS. Some grades were determined, of course, by administration positions of supervising others, for which we could apply.

In the Geologic Division, the standard organizational unit then was a Branch, generally determined by geologic specialty and consisting of about twenty-five to seventy-five geologists. The leader of each of these units was a Branch Chief, with a grade of GS-15. The Branch Chief was picked based on her ability as a scientist and her potential as an administrator. Some who occupied the office were lousy at the administration part, but about an equal number were good at it. Their term was four or five years, so the bad ones could not inflict damage for more than that time period. When their term was over, most Branch Chiefs wanted to return to science. The grade of GS-15 could also be achieved by scientific productivity and achievement. A GS-15 (equivalent to a Full Professor or, in the military, to a Colonel or one-star general) was about as high as one could aspire to by doing science in the USGS, although a few "super grades" of GS-16 were awarded to a handful of superstars. Our branch went through various names as it tried to be more responsive to the public or to trends in science, but the name that made the most sense was the one it had when I joined, so I will use only its name, the Branch of Central Regional Geology. All of us were field geologists, in other words those who specialized in geologic mapping, then publishing the results of our research as geologic maps and as reports that interpreted the meaning and implications to the public of those maps. (There was a Branch of Eastern Regional Geology in Reston, and a Branch of Western Regional Geology in Menlo Park.)

Branch Chiefs had their own supervisors of several levels above them, all housed in our main office in Reston, Virginia, and with grades of GS-16 or better. Our Office Chief oversaw the three branches of Regional Geology as well as the branches of Engineering Geology, Special Projects, Paleontology and Stratigraphy, and Astrogeology. There were a half dozen other Office Chiefs who ran branches of other specialties, all headed by the Chief Geologist, who of course oversaw the Geologic Division. Our Office Chiefs and Chief Geologists were generally good, although I rarely had any dealings with them. Two Office Chiefs of Regional Geology were particularly good, as administrators, geologists, and supportive managers, and they came to play a role in my career. One was Douglas M. Morton and the other was John F. Sutter.

> ***Doug*** *was chief from 1980 to 1984, following a one-year stint as Branch Chief of Western Regional Geology. Born in 1935 near Riverside, it was natural that he attend the University of California at Riverside (UCR), which then had an enrollment of only a thousand. He earned his bachelor's degree in geology in 1958 but, perhaps more importantly, he met his future wife Robyn there, resulting in an unusually happy marriage and inseparable partnership that resulted in three children. He did a field and detailed petrologic study of a nearby pluton for his Ph.D. (UCLA, 1962), and from there never left, at least mentally, and became perhaps geology's greatest expert on the geology of southern California, with dozens of geologic maps of all scales and with many reports to his credit, thanks to ongoing collaborations with professors and students in the area. Southern California is an area of unusual complexity, and it contains the San Andreas fault. This, of course, is a seismically active area of high population, and his mapping and related studies were highly practical in understanding the seismic and landslide hazards and many other topics of value to taxpayers. Before joining the USGS in 1972, he had been a geologist with the California Division of Mines and Geology, with a one-year leave to be a professor of geology at Pomona College and Whittier College, which launched his life-long mentorship of students and young professors. Until 1978, in fact, his USGS office was at UCR, but his supportive, caring nature, scientific talents, sense of humor, and ability to work with others called him into his Branch Chief job (Western Regional Geology in Menlo Park), then to Office Chief in Reston. Few office chiefs were more loved or better managers. He built up the mapping program financially and in terms of staff (with more than five hundred geologists and staff in the office) as never before by emphasizing its practical nature; few managers sold programs to outside funding agencies better than he. His tenure could be thought of as the high point in the office, and through his collaboration with Director Dallas Peck (more about him later), perhaps as the high point of the USGS as well. But by 1984 he and Robyn were back in the Riverside area, where he ran several large mapping projects again out of UCR; one of them was with Gary Dixon, whom we will meet later, and they sold mapping to US Air Force bases and the like to fund their work. Doug officially retired in January 2005, but he continued mapping and publishing maps as an Emeritus Geologist with the USGS and teaching as Adjunct Professor at UCR. He died at the age of eighty-five in September 2020 of complications related to diabetes. Even a caregiver during his last illness came to love him and expressed this as follows: "I have only had the privilege of knowing Mr. Morton and Robyn for a short time. I was first taken by the love they have for one another, something rare and beautiful. He and*

> *Robyn are authentic and humble. The salt of the earth. He was my patient for one and a half years but I'm forever enriched."*

> ***John F. Sutter*** *was Office Chief from 1993 to 1997, following a stint as Assistant Office Chief for an Office Chief who was largely incompetent; before that, John was chief of Eastern Regional Geology. John received his Ph.D. from Rice University at about the time I did, where his advisor was B. Clark Burchfiel, a legendary structural geologist and good guy who did great early work mapping in southern Nevada and later joined the faculty of MIT to study the Himalaya Mountains. Before joining the USGS, John took over the K-Ar Lab from Bob Fleck at Ohio State University when Bob joined a similar lab with the USGS in Menlo Park. There John was a professor of geology, and the dissertation advisor to Laurence W. Snee, who will enter our picture later. John joined the USGS K-Ar lab in Reston in 1979, turning it into a* $^{40}Ar/^{39}Ar$ *lab (a modern method, giving more accurate ages); he retired from the USGS in 2005. In between, he played a significant role in securing the National Geologic Mapping Act (more about the Act later) from Congress. To do that, he worked closely with directors of state geological surveys, who were to be major beneficiaries of the funding and therefore were staunch supporters (through their members of Congress) of the Act. This wise move to bring state directors into the study proved to be the clincher in getting the Act passed (as John knew); therefore, the Act continues today. In so doing, John built bridges to state surveys that had been destroyed by his predecessor. During the USGS Reduction in Force (RIF) of 1995 (see Chapter 8), John was one of the lone voices for sanity, although it did little good.*

My heroes aside, let me return to my story.

Doing our field work in Antarctica was only the first step in my job. The second step included working with the samples we collected, many of which were submitted for chemical analyses (that is, determining the weight percent of elements in a rock, a method to describe rocks or correlate them with other rocks) or for thin sections that I was to study and interpret, and some of which were submitted for isotopic dating by a geologist whose specialty was to use amounts and ratios of isotopes to determine the rock's age, in other words the number of millions of years ago when that rock was formed. The third step is writing a report for publication to interpret what one has found in one's research. Writing this report, however, is commonly preceded by preparing an oral talk, plus publishing an abstract (one paragraph in length), on one's preliminary conclusions, and presenting such a talk and abstract at a domestic or foreign scientific meeting. Then, later, the manuscript of the final report goes through peer review by other geologists in the Branch, specialists in other branches, or specialists in domestic or foreign academic institutions or geological surveys. If technical reviewers are fine with most of one's conclusions and if one has adequately addressed their helpful suggestions for revision, an editor of a book or journal will determine whether to allow publication of the manuscript.

As a new hire, I took a week-long course on how to write, taught by professional writers and some of the senior geologists in the USGS. I took another week-long course on how to do technical reviews, for all of us were expected to be called upon to drop everything and do an objective peer review of somebody else's manuscript. The job of a technical reviewer was to be positive and helpful to the author and always to offer ideas of how to improve the clarity and conclusions in the report, unless said report was total garbage. On the rare occasions of the latter case, a manuscript can be rejected by the Branch Chief. The Branch Chief has an Assistant Branch Chief, one of whose jobs is to pass judgment on manuscripts and advise the Chief.

My working days were taken up by compiling the field notes of the 1969–1970 and 1970–1971 field parties, doing the lab work, and preparing reports. I worked with and was supervised by Paul. Although field work is by far more fun, I found great pleasure in doing lab and office work, then writing. Writing requires you to ask yourself some hard questions about your data and preliminary hypotheses, so writing enables you to refine conclusions and is an important part of the discovery process.

The Branch was made up of veterans, all of them ten to forty years older than I was, nearly all highly regarded as scientists and with large bibliographies of published reports. Many were household names in the profession. Most took me under their wing, for I was one of the youngest. The Branch Chief assigned me to work with some of them on their projects, and thus I was mentored by one after another. At least half the Branch had their GS-15 standing, demonstrating the respect that their peers had for them. I think it was the best branch in the USGS, and I was delighted and honored to ask their advice, work with them, or simply pass them in the hall and grin. To get expert advice on almost any geologic topic, all I had to do was walk down the hall and visit one of them, their office doors being always open. The Branch was a place of great camaraderie and much laughter, including practical jokes. Everyone loved their job.

My first Branch Chief was Arthur B. Campbell, a well-respected and well-loved administrator and scientist. But, sadly, within the year Art died of a heart attack. The successor in 1971 was Edward T. Ruppel. Like others in the Branch, he was a well published field geologist. He had been born in Twin Bridges, Montana, and he would retire there. He earned a B.A. at the University of Montana, an M.A. at the University of Wyoming, and a Ph.D. at Yale. He spent his career mapping in southwestern Montana, east-central Idaho, and the Yellowstone Park area. In addition to being an excellent and productive field geologist, he was a superb administrator, liked and well respected, and the Branch flourished under him. He had been a lieutenant in the US Navy during World War II. He and his wife Phyllis were highly moral and rather religious church-goers (Presbyterian, I think), and they raised two daughters and two sons. Both Phyl and Ed were warm, supportive persons who loved the outdoor life. Tall and thin, Ed resembled a Montana cowboy, and he had certainly used horses during some of his field work, though perhaps not granting them the respect they might deserve (horses were "thousand-pound chickens" to him). Phyllis was petite alongside Ed, but no pushover; she once shot a bear that had wandered into their summer field camp. Ed retired from the USGS in 1986 to become State Geologist and the Director of the Montana Bureau of Mines and Geology until 1994. They subsequently returned to Twin Bridges, although he continued with his mapping and publishing on Montana geology. I stayed in touch with them until he died in 2014 at the age of eighty-eight.

In my first year with the USGS, I bounced the idea off Paul that I finish the maps that Hoover Mackin had started at Iron Springs and get them published. There were at least a half dozen 7.5-minute quadrangles in which Hoover had put lines on paper maps but had not taken the steps to get more than parts of one quadrangle published. Paul liked the idea, and Art and Ed were supportive. Besides, it gave me two field seasons (in our summer and in the austral summer) each year, as well as returning me to Utah to finish up a lot of work. This project eventually morphed into being one of my primary projects with the USGS, in the huge Marysvale volcanic field of Utah.

But my first project, the geology of the Iron Springs mining district, encroached upon the tasks (economic geology) of the Branch of Central Mineral Resources, then run by Daniel R. Shawe. So administratively and financially, I was put partly (for three months per year) in Dan's Branch. Dan was another terrific human being and administrator, with a long, productive career of mapping and field geology, mostly in Nevada mining districts. Thus, when the summer of 1971 came up, I was off to Utah to do more mapping. Hoover once kidded me that "if anybody changed a line on any of my maps, I'll break his arm." I ended up changing a lot of his lines, although I am sure he would have approved. For though his mapping was classic, there were some holes in the coverage. To complete the coverage, I ended up remapping or at least walking the contacts and faults that he had put down. Virtually all held up, and I learned a lot following his mapping, but I made a few changes in his geology, as well as adding all surficial sediments ("dirt"), and then completing his maps to the quadrangle boundaries. I identified him as the senior author on most of the maps, and over the next few years they were published.

The year of 1971 had a great personal milestone: the birth of our daughter Jill E. (Elizabeth) Rowley on June 23. We were still in the Schmidt home, for we had a contract of a year. The baby made a big change in all our lives, making us happier and more contented. All, in fact, was blissful! Scott was totally in love with Jill, and they have remained close friends since then. Karen was a stay-at-home mom, devoted to the kids. Jill was healthy and energetic, and it was soon obvious that she was going to be a beautiful child and adult. All of us pampered her, and she was a little on the bratty side. Barney had another family member to tease him. Now that we had two children, the replacement number for Karen and me, I got a vasectomy two months later! The Schmidts returned from Saudi Arabia at the end of 1971, so in December 1971 Karen and I bought our first house about a mile south of them, at 1050 South Foothills Drive, Green Mountain. It had a view east of Denver and the Great Plains. Our neighbors were about our age, except for retired Army Colonel Peter Dragalin and his wife Arlene across the street. We were all compatible and it seemed as if we had dinner or barbeque parties with each other, plus occasional summer block parties, at the drop of a hat. It was the friendliest neighborhood to which I ever belonged, and we all stayed in touch with each other even when we or others left. Karen and I had settled into a typical middle-class life.

Paul and I applied to the Office of Polar Programs of NSF for a third season in the Lassiter Coast, and it was approved. But NSF funding and Antarctic logistics did not necessarily balance, and they postponed the field season for a year. Soon Paul moved on to map the geology of a large geothermal resource south of the Snake River Plain, in Idaho, and I took over the Lassiter project. Paul mapped a series of four fifteen-minute quadrangles at 1:48,000 over a broad area that included the Cassia Mountains.

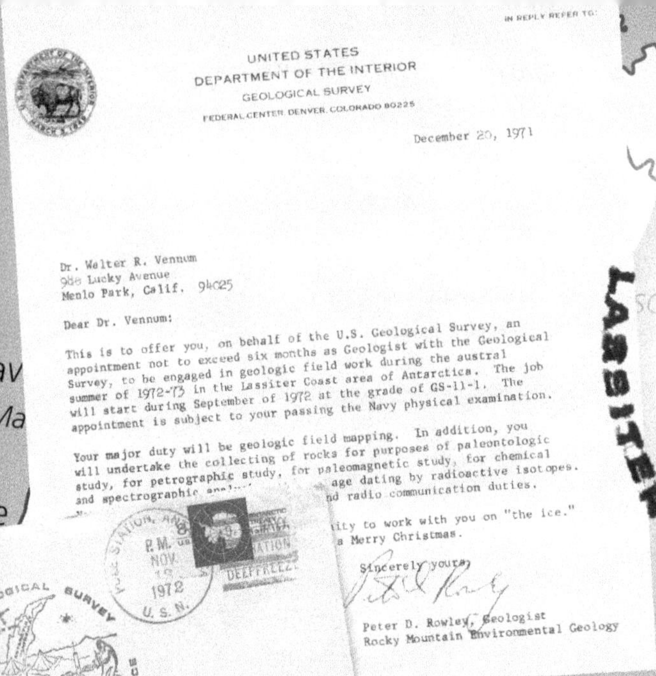

IN REPLY REFER TO:

UNITED STATES
DEPARTMENT OF THE INTERIOR
GEOLOGICAL SURVEY
FEDERAL CENTER, DENVER, COLORADO 80225

December 20, 1971

Dr. Walter R. Vennum
988 Lucky Avenue
Menlo Park, Calif. 94025

Dear Dr. Vennum:

This is to offer you, on behalf of the U.S. Geological Survey, an appointment not to exceed six months as Geologist with the Geological Survey, to be engaged in geologic field work during the austral summer of 1972-73 in the Lassiter Coast area of Antarctica. The job will start during September of 1972 at the grade of GS-11-1. The appointment is subject to your passing the Navy physical examination.

Your major duty will be geologic field mapping. In addition, you will undertake the collecting of rocks for purposes of paleontologic study, for petrographic study, for paleomagnetic study, for chemical and spectrographic anal[illegible] age dating by radioactive isotopes. [illegible]nd radio communication duties.

[illegible]ity to work with you on "the ice." [illegible]s Merry Christmas.

Sincerely yours,

Peter D. Rowley, Geologist
Rocky Mountain Environmental Geology

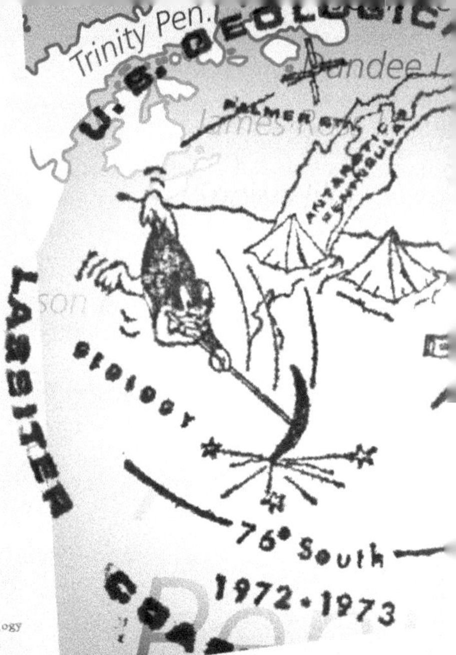

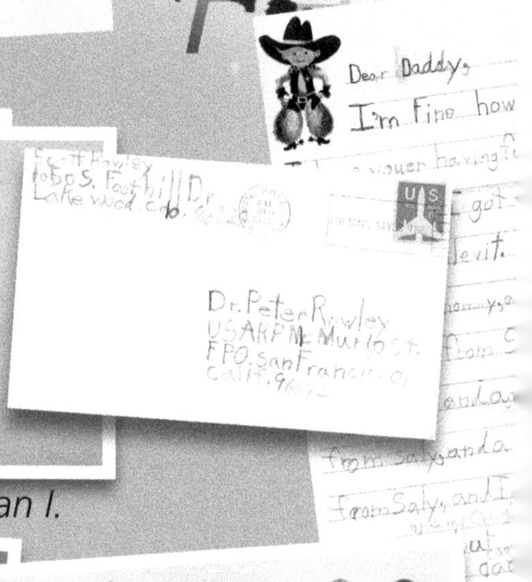

Dolleman I.

1972 – 1973
N. Lassiter Coast

3 The 1972–1973 Field Season on The Ice

As I planned for the upcoming season on the northern Lassiter Coast, I knew that such a leadership role was a tremendous responsibility, but I was excited about the challenge. Years later, one of my favorite songs—"Centerfield" by John Fogerty (1985)—expressed my thoughts at the time with the line "oh, put me in, Coach!" Karl Kellogg was by then a permanent geologist with our branch, with a specialty in structural geology, so I asked him the first. Karlogg, as we called him, had earned a B.A. from the University of California, Berkeley in 1966 and a Ph.D. from the University of Colorado in 1972. His return to get his Ph.D. was guided by his love of Boulder, Colorado, where he, from then on, made his home. For he had grown up there in the place where his dad, William W. Kellogg (1917–2007), was a famous meteorologist/climatologist with the National Center for Atmospheric Research (NCAR) and a pioneer in weather satellites and climate change. Karl's trip to The Ice would be the first of three of my field parties on which he would join me, a source of stability and experience that would serve the project well. In addition, before the third one of those field parties (to the English Coast), Karl participated in a party to the Ohio Range, in the eastern Transantarctic Mountains. Karl's career with our branch was mostly spent doing outstanding and productive mapping in complex terrains in Colorado, where he was equally good at 7.5-minute and 30' x 60' sheets.

It was common for the NSF to support a foreign exchange scientist, and even though the Cold War was still on, such tensions did not much affect the harmonious international cooperation of Antarctic work. And, in fact, there had been a recent thaw or détente in US-Soviet relationships. So I asked Eugene N. Kamenev, a thirty-five-year-old Russian geologist from the Research Institute of the Geology of the Arctic, in Leningrad, to join me; he became my tent mate. As I later learned, Gene was soft spoken, had good mastery of English, owned both an apartment and a country house, but had only a motorcycle for transport, and drank well; we got along great. He was a veteran of five Antarctic field seasons by Russia, plus he "wintered over" at McMurdo before our season started. He had co-authored a 658-page book published in 1972 on Antarctic geology. He gave me a copy, which I retain today, but I have spent little time on it as I cannot read Russian!

Among the other applications I received for the Lassiter position, one by Walter R. Vennum caught my eye. Walt, not two years older than me, was an Assistant Professor in the Geology Department of Sonoma State University, Rohnert Park, California, about fifty miles north of San Francisco. He was a field geologist, specializing primarily in Sierra Nevada igneous rocks, on which he had done his Ph.D. while at Stanford University. He would be at home with Lassiter geology, which was much the same. After talking with him, I hired him sight unseen to a temporary position and to be second in charge of the field party (and after that, for two subsequent field seasons). He had been an athlete (football, track, etc.) in high school in southeastern Washington state, with a poor academic record, but he enrolled in the University of Montana, intending to be a forestry major until he ran out of money and enthusiasm. Dropping out, he drove to Alaska and found a job for six months with an oil exploration company in the Gulf of Alaska. He returned to school and switched to a geology major because, he said, the geologists with whom he had been working "flew around in helicopters a lot [and] drank truly huge amounts of beer." The field work of geology turned him around and he graduated in 1964 with honors. Then he took two years off from school to return to Alaska during summers, this time as a Bureau of Land Management (BLM) smokejumper into remote areas to fight forest fires; for the other months he worked construction and in lumber mills. Then, from 1966 to 1971, he went straight through for his Ph.D. in geology at Stanford.

While at Stanford, he discovered the Stanford Alpine Club, a bunch of "long-haired, bearded, irreverent, dope smoking, dissidents who were among the best climbers in America." This became a life obsession, resulting over time in ascents of rock walls such as El Capitan and Half Dome in Yosemite, a winter ascent of Granite Peak (in the Beartooth Mountains of Montana), Mt. Denali (formerly named Mt. McKinley, at 20,000 ft the highest peak in North America), the south tower of the Golden Gate Bridge (three times), Cheops Pyramid at Cairo, the high point in all fifty states (a joke in that only a few are technical climbs), and most of the highest mountains (and stratovolcanoes) on all continents, including some in the Himalayas. He once described himself as an "adrenaline-seeking outdoor adventure junkie," but on later reflection denied that he climbs for the adrenaline rush (because that would mean he would be scared all the time, hardly pleasant); instead, he admitted to liking the ambiance and the atmosphere of the effort and the places he reaches. However, Walt, showing great sense, never attempted to climb Mount Everest (29,000 feet high, as most readers know), which he called a "cluster fuck." Jon Krakauer's book *Into Thin Air* admirably exposed the sordid details of Everest's modern-day "adventurers," many of them rich but with sparse climbing abilities. The book's conclusions spurred a lot of conversations when we were laid up in our tents during storms.

Sonoma State hired Walt in 1971, and he stayed with them until he retired in 2010. It was his base of operations, which allowed him to do mountain climbing and all the other things, the most enjoyable of which was Antarctic science. His tales of the other faculty in later years made it sound as if most of them were incompetent; his publication record of more than forty published reports, not including abstracts, exceeded all the others in the department combined. Although the students were generally ok, he bemoaned that at times the college seemed to be a combination of "an advanced day-care center [and a] Betty Ford Rehab Facility." He clearly needed real jobs that combined field work and thinking, and was therefore more than amenable to working in Antarctica, resulting eventually in five complete field seasons, as well as four field seasons in southeastern Alaska and one in Saudi Arabia with the USGS, and two cruises on the R/V Glomar Challenger. Of these, the first one to Antarctica with us was his most enjoyable, he told me later, and it made his career (as it did mine). I came to rely on his stability and counsel more than anybody else, during our three field seasons together, none of which were picnics! Next, between 1990 and 2012 he spent eight Christmas/New Year's holidays giving geology lectures on cruise ships visiting Antarctica, including "driving zodiacs [big rubber boats] full of plump elderly people with large disposable incomes around to molest penguins and seals." All these temporary "jobs" (actually, the cruise ships provided only free lodging, food, and drinks) were squeezed in on sabbatical leaves, summers, or by various types of subterfuge. In 2005 he began his second marriage, to Lois Rice, another exceptional athlete and adventurer, and together they "immediately began to orbit the planet like a pair of teddy bears in a clothes dryer," including visiting all seven continents. They continue to climb, kayak, and cave; each of his Christmas newsletters itemizes one scary, stupid, or silly feat after another, all with his remarkable sense of humor that you can imagine from the aforementioned quotes.

I interviewed Stephen J. Boyer, who had finished a master's degree in geology at the University of Colorado. Steve was a marathon runner and iron-man athlete. When he was working on his undergraduate geology degree at Yale, he was friends and ran the Boston Marathon with Frank Shorter, who won that race once, while Steve was not that far behind. After our field season, Steve earned an M.D. degree and spent most of his career in the Emergency Room at the Kaiser Permanente Hospital in Portland, Oregon (his wife Priscilla also is an M.D.). Art Ford later hired Steve for a temporary job of mapping in the Pensacola Mountains, but to get Steve a decent salary Art hired him as a doctor, even though Steve did geology 99.9 percent of the time. Steve once tackled Mount Everest as the M.D. on an expedition. He climbed to the last base camp on the mountain, but gave his oxygen bottle to a colleague and never quite made the summit himself. Steve's iron-man competitions continued throughout his whole career, although in "retirement" he mostly does biking.

Finally, I hired Richard B. Waitt, Jr., who was a master's student of Hoover Mackin. I got to know him during my last few years at the University of Texas. He later earned his Ph.D. and became a permanent geologist with the USGS, serving with distinction in mapping and studying Quaternary geology in the northwestern US. We became reacquainted after the 1980 eruption of Mount St. Helens, where he did studies that I will document below, and after which his office was at the Mount St. Helens Volcano Observatory in Vancouver, Washington.

Three USGS topographic engineers were to join us on the field party. Because George Schirmacher and his colleagues had already got much of the data they needed in the northern Lassiter Coast during the 1970–1971 season, the 1971–1972 group would concentrate on areas south of the Lassiter Coast. They were to be landed in the Sweeney Mountains of the Orville Coast. I asked Richard Waitt to join them for the first part of the season. The engineers were John R. House, Jr., who was in charge, James W. Schoonmaker, Jr. (Bill), and Harold E. Fiebelman (Butch).

On October 14, 1971, the eight of us flew commercial to Honolulu, then rented a car to take us to Hickam AFB, where we checked into the BOQ. We did not have much of a wait this time. By the evening of the 15th we were on a MAC (Military Air Craft) C-141, with a refueling stop at Pago Pago, and crossed the International Dateline and went on to Christchurch, arriving on the morning of the October 17. Early on October 20, we were off to the ice by C-141, a five-hour flight at 37,000 feet altitude. The C-141 is, of course, a wheeled aircraft, but the McMurdo ice runway was still sufficient for it, so this aircraft was being used exclusively for the Christchurch to McMurdo run until ice conditions deteriorated in midseason. Temperature upon arrival was 5°F, a "heat wave" for this early in the field season.

I concluded my first letter to Karen, Scott, and Jill, written on the October 20 flight, with "it is good to be back." That was certainly true, although overseeing the field party meant more responsibilities, meetings with Navy and USARP staff and Air-Operations people, visits with the press, supervising the gathering of supplies, endless calculations of "weights and cubes" for the ever-changing configurations of the C-130 flights, making sure the motor toboggans were prepared sufficiently, and so on. Steve was to be our mechanic, Karl our cook, Walt our radioman, and I our weatherman, so each man organized and assembled all that their job required. But we all pitched in on whatever had to be done. All took snowcraft, radio, and weather schools. Then we had Jamesway-making school, for we were scheduled to erect a twenty-four-foot-long Jamesway (3,700 lbs) at our put-in site. The site I had selected was in the Ferguson Nunataks, northern Lassiter Coast. Putting together the Jamesway and the Preway stove inside, while the Herc waited, was to be the classic example of "some assembly required." The maximum waiting time for a running Herc to stay on the ground there was about ninety minutes, for Navy rules meant that shelter, warmth, and radio communications had to be established before they could leave. Fortunately, most times they gave us a lot of slack on that rule, for they were extremely eager to get back to the comforts and bar at McMurdo!

The USARP hired the private company of Holmes and Narver to bring in all supplies, manage warehouses for each supply type (food, motor toboggans, camping gear, gas), help the field parties pack and palletize gear, and get pallets to Williams field. McMurdo has a staff of about a thousand persons during the summer and about two hundred during the winter. Most of these are employed by Holmes and Narver. The winter is characterized by darkness, isolation, loneliness, and paranoia, with scary storms lashing the buildings and a sparse population that may not include somebody with whom you can become friends; a science job may allow you to become mentally immersed, but a menial job may not provide enough mental stimulation or satisfaction. Winter-over personnel are notoriously inept, and some display symptoms of mental illness. Either these individuals were that way when they applied for the job, or they got that way during the winter. We discovered that winter jobs that called for pulling supplies for summer field parties were rarely done and so we had to do them ourselves. So, day after day, we collected and boxed what we needed, only to have a protective warehouse manager resent our pulling things off his shelves. I had to have a series of heart-to-heart discussions with these managers and their superiors before things ran smoothly. On October 28, I spent the morning with a man and his forklift taking the last of our boxes of supplies to a terminal for weighing and storage—until a sudden southerly storm with about 40 or 50 mph winds and visibility of ten feet put a halt to this, too. Due to all these chores, we had little time to relax or sight-see, and all of us knew that there were terrific seals and penguins and ice cliffs that merited a visit. For all these reasons my next letter to my family did not happen until October 31, Halloween.

It was obvious that motor toboggans would continue to be our Achilles' heel, for they had not been serviced or shown care over the Antarctic winter. We needed six motor toboggans. Most that had been used in the previous field season were worn out or abused, and some had been left outside over the winter. Unlike most other field parties, we intended to put many miles on each. The winter-over crew had replaced three of the engines, so we did the other three, then rigged each for remote operation. Then I demonstrated the USGS remote-driving method as we all practiced on the ice shelf. During one such trip with five motor toboggans, on a nice, clear November 1 afternoon, Gene yelled that he saw a storm coming, about ten or fifteen miles to the south. We barely had time to turn the toboggans around and head toward McMurdo before the storm hit with 50 mph winds and a ground blizzard, and visibility reduced to ten feet. McMurdo disappeared. I kept the party together and steered for a half mile on a bearing to intercept the road, then led the way back north to McMurdo. Even though we traveled with our backs to the wind, our faces were solid sheets of ice when we finally reached the toboggan warehouse—a good lesson for all about the rapid onset and consequences of storms.

Getting us to the field was on the Navy's priority list, but only after they resupplied Pole Station, then a put-in of a French party, then six flights of fuel for Byrd Station. Pole Station must have a temperature above 50° below zero for a plane to go there, otherwise the hydraulic fluid in a C-130 turns to slush. By November 3, Pole had been resupplied and reopened, but the fuel for Byrd was not put in, partly because of bad weather and a bad fuel bladder, and partly because their skiway had not been dragged (a chain that bulldozers used for that had been buried by a storm). I declared a day of rest. We all walked over to tour Scott Base; one of the Kiwi geologists there had been a fellow student in Austin, Texas. For exercise and fun, the Kiwis had dog teams, with the huskies staked out adjacent to the buildings. We learned that anyone who wanted to pet the dogs not only had to visit every dog but to give each the same amount of time, starting with the lead dog. Over the next few days, others took trips along the sea ice and shore of Ross Island to see wildlife and climb rocks. But with Air-Ops declaring the possibility of flights to Lassiter, yet with Byrd still only partly operational, I spent days working with pilots to reconfigure the loads, including tearing down pallets to lighten them. I wrote Christmas cards, then persuaded one of the pilots take the bundle of cards to the Pole for postmarks. And I was able to get in a phone patch with Karen! I also found the time to send in an absentee ballot for George McGovern for President, although I knew his chances were bleak. By then I knew that Richard Nixon was a crook, but few other voters had figured that out!

On the evening of November 6, the first flight with the geologists (except Richard) took off. Our second Ferguson put-in plane followed two hours later. A little over three hours later, our flight descended into Byrd and the landing gear was lowered—but a ground blizzard and low clouds prevented us from landing. The pilot, Navy Commander John B. Dana, told me that he could have landed but it would likely have damaged the skis. We returned to McMurdo, then had to offload the planes, and only tumbled into bed twelve hours after this first call to the air field. The weather remained good at Ferguson but bad at Byrd. The next day, plans changed so that now we were to go straight to Ferguson, so we had to reconfigure the load once again. On November 8, at Gene's insistence, we had a party (straight vodka was the beverage!) to celebrate the Russian Revolution, one of Russia's biggest national holidays.

On November 13, just after midnight, we made another flight attempt, this time straight to Lassiter, a 6.5-hour trip at 23,000 feet. The pilot was John Dana once again. The co-pilot was Bruce Campbell, whom I had gotten to know in the 1970–1971 season, a good guy and kind of a hot dog. Bruce once flew through the ash at Mount Erebus, another time buzzed McMurdo, and once nearly hit an ice breaker on the Ross Sea. At an air show in Christchurch, he got permission to fly below three thousand feet but he never told anybody how much below—a mere fifty feet above the beach! Yet these two, like most other pilots I got to know, were true professionals and just loved to fly, especially Hercs. We had the five geologists and three huge pallets of gear, plus JATO bottles stacked in the tail, for a total weight of about 12,500 lbs. Two NSF officials joined us to witness this put-in of the remotest of the remote field parties. Before takeoff, above the whine of the engines, John reminded all of us that "when this aircraft hits the deck out there, you'll think it's breaking apart," as I noted in an article in *Summit* magazine. A second plane, with a load of 14,000 lbs, was about an hour behind. The weather, which looked "iffy" when we took off, improved as we flew. We passed over the Ellsworth Mountains, then as we flew over the Sweeney and Behrendt Mountains of the Orville Coast, we found that the weather was now deteriorating at Lassiter, true to form. It was partly cloudy, with dense clouds at several levels, and I came into the cockpit to identify the Ferguson Nunataks. Fortunately, there were only light clouds over the area where we wanted to land. I returned to the cabin and we fastened seat belts. We dragged the prospective "runway" once, just touching the hard snow surface with the skis to test snow thickness. On the next pass John put the plane on the ground and reversed the engines. I pointed out where the camp was to be and we taxied to it.

We were in a cluster of small jagged nunataks that rose about five hundred feet above the ice. It was about 1 pm local time and it was calm and 20°F. We freight-trained the pallets. Because the second plane was already on the way, we had time only to set up a Scott tent and establish radio communication and begin building the Jamesway, before our plane left. The second plane arrived an hour later and freight-trained its load. This plane brought with it a couple of Navy builders, and their expertise sped up the Jamesway construction. We gave them back the aluminum pallets from the first plane and, with the Jamesway about a third completed, the builders and plane departed in about an hour. We then finished putting up the Jamesway on our own, got the stove started, stored the food, fuel, and snowmobiles, had a meal, and got to bed about 3 am local time on November 14, exhausted and with the temperature at 4°.

A 20-mph wind followed as we slept, and we spent the next day in it, putting things away. Late in the day the wind attained our standard estimate of about 50+ mph. This wind lasted another day, exposing some errors in our Jamesway construction. A clogged fuel line in the Preway heater also had to be identified and fixed. We tightened things up and soon had a secure home. I followed the Wetmore Jamesway plan, with a dining table and the Preway stove in the middle, separated into two rooms by a tarpaulin. The eating and work room had, in addition to the dining table and Preway, tables on one side where we prepared food, and tables on the other side where we had the radios, aerial photos, maps, small library, and field notebooks. The other room, the sleeping room, held the cots, sleeping bags, and personal items.

During the storm, we learned from McMurdo that the topographic party was delayed, perhaps for weeks, although no reason was given.

While waiting for the storm to subside, we built a snow trench near the Jamesway as an emergency shelter should the Jamesway blow away. A snow trench starts with a linear pit at least six feet deep and three feet wide, cut in the hard snow using aluminum snow saws, which are about a foot long and resemble a hand wood saw except that the teeth are larger and not sharpened. The trench was at least a dozen feet long. Then, about two feet above the bottom of the trench, shelves at least three feet wide were cut on both sides, open to the top. Blocks taken from the trench and shelves were used to build a roof, with a gentle peak, over the trench and shelves. The shelves were where people would sleep, on their mattress pads and sleeping bags. The trench allowed people to stand up and move around. Cold air flowed into the trench. The entrance was a plywood door at the top of one end.

On November 16, the storm finally eased up, allowing us to explore the geology of some of the Ferguson Nunataks. The next afternoon, anxious to get on the trail, we headed south toward Mt. Coman, a prominent mountain east of the crest of the Antarctic Peninsula and twenty-five miles south-southwest of the Jamesway. The elevation given for Mt. Coman on the internet (www.peakbagger.com) is 8,720 feet, which sounds about right but, not knowing how the elevation was determined, perhaps should be regarded with skepticism.

The motor toboggan Gene and I shared was slow because of carburetor problems, and we consequently stopped for the night at 7 pm and set up our tents. The next morning, we replaced the carburetor and, in whiteout due to stratus clouds just above the surface, explored and identified the geology of the southern end of the Ferguson Nunataks. The weather deteriorated as we set out for Mt. Coman, and the clouds dropped to the ground. The temperature was 10°. We knew the compass direction to the mountain, so we steered in this direction for about two hours, visibility at one hundred yards. Then the clouds lifted a couple hundred feet and we saw a black rock mass in the distance, which we took to be Coman. We headed toward this for the next hour, until the cover of my toboggan's transmission started to come off, dumping transmission fluid all over the snow. We set up our tents at 6 pm and found that the vibration had sheared off three bolts holding the cover, another symptom of using decrepit toboggans. The air temperature was minus 9°. As during the previous Lassiter seasons, we had to jury-rig the motor toboggans. We had a spare motor toboggan at the Jamesway, but were not confident that it was much better than this one. Besides, before we left, we discovered that its transmission handle (a half-inch steel bar) had snapped off, presumably while it was being loaded into the C-130. Steel becomes brittle in constant cold, and motor toboggans are particularly vulnerable. Ever resourceful, the next morning Steve and Karl strapped the cover down with wire and we continued. We reached Mount Coman and explored its geology and that of the nearby nunataks, then continued south-southeast toward the Playfair Mountains, but within a couple of miles were hit with a raging ground blizzard and had to make camp. The snow kicked up by the wind created a beautiful sight when one looked west at the sun: a huge vertical circular ring, called a sun dog, around the sun and, cutting it, a second but horizontal circular ring!

The storm continued into November 19 and we were confined to our tents, as I documented in a letter to my family. Nonetheless, this was much better than McMurdo, and we knew it. All persons were in good spirits and working well together. Steve had suddenly become the most valuable member of the field party due to his proficiency with fixing internal combustion engines. Gene had warmed up and showed a good sense of humor despite his broken English. I wrote "Strange, a bunch of grown men steering decrepit old orange motor toboggans by ropes through zero-degree weather, eating meat-bar stew, sleeping in sleeping bags and Scott tents, and being slightly cold all day long except when in a sleeping bag. Yes, very weird!"

November 21, according to a letter I wrote on Thanksgiving to Karen, Scott, and Jill, was "one of the worst I have spent in Antarctica. So bad that it was comical." On that day we made it into the western Playfair Mountains, and as Gene and I drove to the base of a nunatak, we heard a clanking noise. We did the geology for an hour, then returned to investigate the FoxTrac. It became clear that the bearing on the rear axle of the toboggan, around which the track goes, had worn out. The other men showed up, and Steve, Karl, and I spent the rest of the afternoon on a cold day removing the sprocket from the axle. The bearing had burned up. We had no spare, and at 5 pm I declared the toboggan officially dead. We used the other toboggan to take two loaded Nansen sledges west to what ended up being Camp 29, about five miles west of Mt. Kane. I had wanted the camp to be four miles farther east, at the base of Mt. Kane. So, we continued east, flagging the trail. But while driving the remaining toboggan in a whiteout, I found myself going down a steep slope, with crevasses off to my left. We stopped and applied brakes to the Nansens by wrapping rope around the runners. A few hundred yards farther east, in the gloom, I found myself on blue ice and brought the toboggan to a stop. The front skis of the toboggan were a foot over the ledge of a precipice, a crevasse over twenty feet wide that was only partly bridged by snow. As we were wrenching the front of the toboggan away from the edge and turning it partly around, somebody yelled "look out" and we turned and watched the rear Nansen's runners sink into the snow bridge of yet another crevasse. The void (or hole in the bridge) showed as black, for the crevasse was deep. Fortunately, the bridges held long enough for us to get the toboggan and the first Nansen off the bridges, then unload the rear Nansen and pull it out of the deep snow of the snow bridge. Then we headed back west, following the flags for three miles, to establish Camp 29. While some of us set up the camp and made our daily radio contact, two others took the toboggan back to the dead toboggan and retrieved the other two Nansens. When they returned at 10 pm, by which time it was -9°, we had our meat-bar stew. As if the day were not bad enough, a radio message told us that the put-in of the topographic party was to be delayed for two weeks, if not canceled entirely. No reasons were given. A few days later, however, we learned that the cause of the delay was that one of the Hercs, presumably the first one, damaged a ski during our put-in, and it was not going to be repaired until early December. (At times when I read my old letters I wonder how we remained forever positive, enthusiastic, and unafraid in the light of events that could just as easily have been disastrous. Youth feels invincible, I suppose! I have also reflected on the matter-of-fact, another-day-at-the-office mindset revealed by the letters. All I can say is that we settled into a routine of daily work that seemed natural, as opposed to unusual.)

The next several days, including Thanksgiving (Thursday, November 23), were the most sunny and beautiful that we had seen so far. Ice crystals on the top of the snow sparkled like diamonds. Steve cooked the Thanksgiving dinner, consisting of two chickens we had thawed out, a pot-full of vegetables, accompanied by hot buttered rums! We all delighted in doing lots of geology and getting back late. In the sun, we scouted the crevasse field we had so dangerously encountered on of the 21st, and in a whiteout on the 24th we relayed our Nansens around it and eastward to Camp 30 at the western base of Mount Kane. We explored and documented the geology of that area. One day Karl saw a Snowy Petrel which, combined with the siting of one the previous day, brought us to a grand total of two for the season. Karl remarked, in the accent of W.C. Fields, "Ah yes, the Snowy Petrel, the most common bird on the Lassiter Coast." On a sunny November 25 (temperature about 10° to 15°), we moved north about fifteen miles to the Rivera Peaks and established Camp 31. To do so, we returned to the abandoned FoxTrac and stripped it of all parts we thought we might need, including the engine. It was a downhill trek out of the Playfairs, then a level ride across the five-mile-wide Swann Glacier, and we were able to transport all of us, plus the four Nansen sledges. Climbing up into the Rivera Peaks, however, required relaying Nansens. In the Rivera Peaks and adjacent Watson Peaks, we did the geology in two days in 20 mph winds and low visibility. We got our first glimpse of the Weddell Sea, with open leads in the ice, from the eastern end of the Rivera Peaks. At that point we were thirty-five miles southeast of the Jamesway, and always lurking in the back of our minds was the health of our lone motor toboggan, for that distance would have been a long way to man-haul all our equipment!

We need not have worried, for we returned to the Jamesway without incident on November 30. Part of that trip required traveling on a compass bearing in low-visibility clouds. The toboggan pulled all of us and the four loaded Nansen sledges, a weight of 3,500 to 4,000 pounds, not including the toboggan itself, in an eight-hour trip going at full power. We had lobster tails for dinner! There followed a period of waiting, as the Navy dithered and the weather did too, as reflected in daily messages that indicated that bureaucrats were changing their minds, back and forth, as to what to do with the topographic party. One option that the bureaucrats were considering was to drop the topographic party at the Ferguson Nunataks and combine their delivery with our resupply flight, which then would be nearly a month early. Then six hours later, the plan would change. In the interim we did day trips to document the geology at nearby nunataks.

Communicating by short-wave radio can be difficult, what with occasional blackouts or weak transmissions, commonly requiring that we and the recipient spell out words phonetically. Some messages had to be relayed through a US base, generally Palmer, Byrd, or Pole stations, or infrequently a British base, or even a US field party who could comprehend the message better. Walt became pretty good at this, even sitting outside with the generator thundering.

Late on December 4th and into the next morning, Walt and I alternated sending weather reports to McMurdo, while the other slept. The weather was pretty good on the 5th, except for a 10-mph wind. We spent most of the day trying to get the two motor toboggans working, for each had carburetor problems, until by early afternoon we were being asked to send hourly weather reports, which generally meant that an aircraft was in the air. So once we had fixed one of the toboggans, I marked the runway with orange flags. Steve got the second toboggan working after diagnosing the problem as a bad spark plug.

By mid-afternoon, we learned that two aircraft were heading our way with the long-lost topo party and the resupply items. But Karl, who had been tracking the weather, then noted that the barometric pressure had just dropped like a rock. Soon after, we were in a heavy snowfall and heavy cloud cover/whiteout, with the temperature at 8°F. The planes turned back.

It was dark and we fired up the Coleman lantern and had a dinner of tenderloins and French fries. Gene, who had been tinkering with the transistor radio, got the BBC and learned that we were on the eve of the Apollo 17 flight (the last one in which people walked on the moon). Gene remarked that "it is easier to get to the moon than it is to the Lassiter Coast!" I penned a letter to Karen and kids that noted that this had been a frustrating day for those reasons, here on the eve of my thirtieth birthday! Steve and Karl spent the following day trying to repair a clutch on one of the motor toboggans, which had been damaged when the winter-over McMurdo mechanic had used a ball-peen hammer to install the clutch belt "the easy way." Then we celebrated my birthday with whiskey sours, steak and lobster tails, and a cake that Steve put together using pancake flour with pudding frosting and three candles. That evening was followed by a day of 20 mph winds and ground blizzard, during which we stayed in.

On December 8, with only 95 percent clouds and a mild whiteout, our weather reports persuaded two aircraft to head our way, leaving McMurdo at 3 am our time. But the weather deteriorated, with clouds lowering and visibility dropping to five to ten miles; the first plane got overhead at about 9:30 am local, but a half hour was all the fuel he could spare so he had to turn home. Fifteen minutes later, huge holes in the clouds appeared and we were confident that at least the second plane, two hours behind, would make it. Not so, for the weather deteriorated by the time it got here. Then, with the plane gone, it became partly sunny!

Frustrated and depressed, and going stir crazy in the Jamesway, I proposed we head north, across the Meinardus Glacier, to do some geology. We loaded up and headed off at about 3 pm. Halfway across the glacier, the weather changed again: medium snowfall and 200 yards visibility, so we headed the rest of the way across on a compass bearing to a nunatak we saw on the other side. We set up the tents, exhilarated to be on the trail again. We were in the southern edge of an unnamed range west of the Dana Mountains. The weather was not much better, but we mapped the geology of nearby nunataks anyway. The bad weather broke off at about 11 am and we returned to the tents and packed up. Then one of the toboggans would not start until we replaced a carburetor. We headed eight miles to the east, just west of Mt. Barkow, where we established Camp 32. We spent December 9 and the next day doing geology. I got to climb Barkow, something Walt would have liked to do, but he stayed in camp sending weather reports to McMurdo every three hours to let them know the weather was good, while pretending we were still at the Jamesway. McMurdo did not seem too interested, radio communication was weak, and on the evening of the 10th the visibility deteriorated again.

At 2 am on December 12, I got up to do the late shift on the radio and learned that an aircraft or two would take off in two hours. We had less than a day left to finish the geology of this unnamed range. I woke Steve up and we packed up half the camp, asking the others to stay and finish the geology. We got going about 5 am, made the trip back to the Jamesway in about four hours, stopping along the trail each hour to crank up the generator and give the weather report. At 11 am, we learned that something needed to be fixed on the Herc, and they were not going to be able to leave until 5 pm local time. That, we surmised, was too long for the weather to hold and, sure enough, stratus at the surface moved in and we were down to two hundred yards visibility. Walt, Karl, and Gene arrived at about 7 pm, just in time for a steak dinner.

Much of our mapping on this part of the Lassiter Coast was in igneous plutons, whereas farther south, Latady sedimentary rocks dominated. One day about this time, Walt came upon a bunch of large quartz crystals that must have formed in a gas bubble (or geode) in a pluton, and he chucked a bunch of the crystals in a gunny sack. He hauled the sack home and, over dinner suggested that we all draw straws and reach into the bag, without looking, and take one crystal. I still have mine—a four-inch-wide and seven-inch-long hexagonal smoky-quartz crystal with a terminated end—on a bookshelf in my office.

The weather was mostly good on the 13th but the Navy made no attempt to fly. We therefore packed up again, hoping to head out into the Werner Mountains the next day, ten or fifteen miles to the east. Walt transmitted hourly weather reports throughout the night, and at 3:30 am on the 14th I spelled him and learned that it was cloudy in McMurdo, plus "night," so I did not expect much. I had difficulty getting out with the radio all night, due to a radio blackout, in which transmissions were weak, garbled, and rife with static on both ends. I could not get out at all at 6 am, 7 am, and 8 am. Then, while we were preparing breakfast at 9 am, as I was about to send another weather report, I heard a conversation between Byrd and McMurdo in which they were talking about two planes in the air to Lassiter. I then tried to reach Palmer Station, a backup we commonly used, but they had not heard the conversation between Byrd and McMurdo (radio blackouts are strange things). I told Palmer that we would tune the radio to the aircraft frequency, which requires changing the antenna, switching some dials on the radio, and tuning again. Walt was awake and was better at it than I was. And ten minutes later, I heard the voice of Bruce Campbell say "aircraft three-one-niner," one of the C-130s, and piloted by my favorite commander. That first (?) plane was only fifty miles away from us at that point. We figured that if anyone could get in, he could. There were solid clouds at about 2,000 feet above the surface. I asked Karl, Steve, and Gene to drive down to the airstrip two miles away and get snow off the orange panels we had laid out for the aircraft. Walt and I heard the aircraft and directed him in, but first he had to get in under the clouds by going to the west. After several passes, he spotted us and, with no time to waste at Ferguson International Airport, he put the aircraft down near the Jamesway! It was 11 am, December 14. It was our resupply flight, and out stepped Richard Waitt with two sacks of mail, then the load of two motor toboggans, lots of fuel drums of mogas and white gas, and some food and supplies, was freight-trained. Low on aviation fuel, Bruce could not wait around to take onboard the stuff we had for him (rocks, aluminum pallets), and after the ten minutes it took to offload, he was gone. When I thanked him over the radio on his way back to McMurdo, Bruce told me that he had had only enough fuel to make one more pass at landing. So finally, we had some luck, and a good pilot! The other aircraft, with the three topographers, either returned or never took off.

The priority during the season was to build Siple Station, on the northern side of the Ellsworth Mountains and about four hundred miles from Lassiter, so that people could winter over in it. Apparently, Bruce Campbell's Herc had been one of several that had been transporting materials to build Siple Station. The construction was behind schedule because of continuous bad weather, and another bad day there freed up his aircraft.

Despite the morning hour, we all felt the need to celebrate with Richard. Even though we were packed and ready to go, I declared a National Day of Celebration. We all had a few beers, and Gene pulled out a bottle of vodka. And we read mail. News of the outside world came via Newsweek magazines and Pravda newspapers. We agreed that we would open Christmas presents later, on or around Christmas. Walt and I continued to send weather reports, with me taking the late shift of the early morning of December 15. But there was no word of flights carrying topographic engineers. Richard regaled us with stories of McMurdo and the outside world. One of his interesting revelations was that the official word that the first put-in Herc had been damaged during the landing was false. Instead, the plane had glanced off a peak in the fog a few minutes after takeoff, and had jammed its ski into the fuselage. The crew worked on it while in flight and managed to get it lowered so that they could land at Byrd. Over the next day or two, engineers came out from McMurdo and repaired it sufficiently to enable them to fly it back to McMurdo, then to Christchurch for repair. Therefore, much of the field season for other remote field parties was curtailed or canceled. We felt lucky to be here!

The next day we headed ten miles east to the Werner Mountains, where we established Camp 33. We did the geology in lots of fog and the warmest temperatures of the season. During one of those days, when I collected the weather information at 7:30 am, I discovered a temperature of 28°F under clear skies. Two hours later we recorded 31°F, and doubtless it became warmer than that, although later fog dropped the temperature somewhat. Above-freezing temperatures are not good news. An interesting side-effect was that ski poles and shovels would fall over because they would absorb heat and melt the surrounding area. Also, any dark object lying on the snow surface would absorb heat and melt its way downward into the ice and snow. Thus, within a couple hours, wires, rocks, even ropes, etc. were at the bottom of a hole a couple of inches deep. Even an orange trail flag on a bamboo pole melted downward.

One day, in dense fog, after I had let Gene off at one end of a ridge, I drove the motor toboggan blindly down a steep slope toward a nunatak, and just when it came into view, I heard a clunk and a piece of pipe over a foot long came out from under the toboggan. I knew the problem was serious but I thought I would complete the objective, so I turned off the FoxTrac and hiked (I should have skied!) the last hundred yards to rock. When I was about a hundred feet from the rock, I realized that something seemed strange underfoot and I attempted to look more carefully through the medium whiteout at the surface. I muttered to myself, "For Christ's sake, I'm walking on a damned …" but before I could think "crevasse," I'd fallen into one! Fortunately, I stopped myself with my outstretched arms, waist deep and with feet dangling in air. It was a small crevasse, only about five feet wide, but it was deep and blue down below me. I climbed out and crossed the crevasse a hundred feet away where the bridge looked better, and did the geology. Three quarters of an hour later I returned to the motor toboggan, which I started and then followed my tracks back toward Gene, a couple of miles away. I visited a few more outcrops along the way, amazed that the toboggan ran at all. But it could not last forever, although fortunately the end came about the time I got Gene in sight, through the murk. Then the machine went clunk, clunk, clunk, and stopped. The axle on which the track moved had broken through the rest of the way. The only way to repair it would be with a spare rail and some welding equipment. We had neither, so I declared #125 officially dead, and Gene and I skied the few miles back to camp in the soft and slushy snow. We later cannibalized the motor toboggan, including its engine, and moved on. With that we were down to two motor toboggans, each to be shared by three geologists, although one remained at the Jamesway for the topographers, if they ever showed up.

Our way to the northeastern Werner Mountains was blocked by rock hills and mountains, so to get there to establish Camp 34 on December 20, we first went south onto the Douglas Glacier, then east down it, then northward back into the range. I led the way with one toboggan, the other following about two miles behind me. At one point on the glacier, steering as usual by ropes, I noticed in front of me a few low snow hills, each two or three feet high and ten to twenty feet long, known as haycocks. These are snow mounds that form like dunes near open holes where the snow bridge to a crevasse has broken through, and the turbulence caused by the hole on the wind currents results in snow piling up near the hole. So, I had a bad feeling that Gene, Steve, and I were entering a crevasse field. To my right were more haycocks and a gaping hole. As I looked back, I saw a hole open on a ten-foot-wide crevasse over which I had just passed, and I pointed it out to Gene and Steve, then shut down the motor toboggan. We were in a field of small crevasses, each one to five yards wide, most of which were bridged. The sun was out, so we could see them, and we took lots of pictures. We turned the motor toboggan around, passed back over the ten-foot crevasse and found a safer way before we turned north and, in an hour, established Camp 34.

The next day we traveled down a small unnamed tributary glacier that flowed east, to access large areas of bedrock at Mt. Fell and Mt. Hemmingsen, north of this glacier, and at Mt. Virdin, south of it. As I steered the toboggan east down the gentle glacier slope, I started seeing crevasses on my left. Seeing no way to elude them, I passed over three or four of them and just kept going. Fortunately, they were small, no more than five yards wide and bridged by snow. The glacier went through a rock narrow, with a steeper slope and a few more bridged crevasses. After we did the geology, we had to return the same way, and Steve passed over at least one hole through which the toboggan had broken earlier.

It was summer now, and the weather was (relatively) warm, commonly at or above 20°F. We stripped down to long-sleeved underwear tops as we hiked. Melting ice and snow left small pools of water on the rocks, welcome oases. Occasionally temperatures exceeded freezing. The warm weather led to more fog, but if we could reach bedrock before it arrived in the warm part of the day, we could follow our tracks back in the late afternoon. Fortunately, the motor toboggans did not bog down much in the wet snow, unlike what had happened during the 1970–1971 season. Then, during one radio transmission from Camp 34, we received the bad news that the topographic party had been canceled, at least its work on the Lassiter Coast, although its members had managed to do some topographic measurements at Byrd, Pole, and McMurdo stations. The priority had been the construction of Siple Station.

On December 26, we were at Camp 35 in the southeastern Werner Mountains and, with the geology of the range explored and documented, we were ready to return the roughly thirty-seven miles to the Jamesway. But weather pinned us down for the day. It had snowed all the previous night, resulting in three inches of fresh snow, and low visibility, whiteout, and a 15-mph wind marked the new day— Steve's twenty-sixth birthday! We had spent the previous days, including Christmas, productively, visiting several mountains and nunataks in whiteout, occasionally with patchy fog. But the visibility clanked down to nothing as we went to bed. Someone remarked that we had "a white Christmas with a vengeance."

We had consumed almost the last of our alcoholic beverages the night before, Christmas Eve. Therefore, all we had was one swig each from our last bottle of brandy. It was the dullest Christmas any of us could remember, except for the fact we were in this great place called Antarctica! Gene had three emergency flares, which he set off to liven things up! They resembled small Roman candles.

On December 27 we returned to the Jamesway in weather only slightly better than the day before, except that we had one-mile visibility when we set off at about 10 am. We were able to follow our old tracks for the first five miles, but then visibility dropped to a mere hundred yards and the whiteout was too dense to see tracks. We went on a bearing for the next four miles, when visibility improved to two or three miles and we crossed the Douglas Glacier easily. The trip back to the Jamesway required that we return to the abandoned FoxTrac to get the gear box, about the only thing that Karl and I had not stripped before. During the last several days, one of our working toboggans had had a badly growling transmission and we feared that it might fail at any time. Also, one of our Nansen sledges was battered by several misadventures of flipping it. We were a ragged mess. We spotted the abandoned toboggan and headed toward it. But my toboggan had a tough time plowing through the loose snow and my clutch was slipping, so Steve, Karl, and Richard headed down a hill with their toboggan to the abandoned toboggan. Then, with its gearbox in hand, they caught up with us. We had put a new drive belt on, tightened the chain, and rearranged our load so that they took the heavier Nansens. The visibility came and went as we traveled, and occasionally we steered on a compass bearing, but fortunately we were in familiar country. Eventually the clouds lifted a little and we saw the Jamesway, which we reached at 7:30 pm. We declared this to be Christmas Eve and gorged ourselves on lobster tails! On December 28, "Christmas," we opened presents, including a box my mom sent containing a tiny Christmas tree, tiny stockings, tiny candy bars, a small crèche, New Year's tooting horns, bells, and beads. Another box contained fruitcake and cheese. We wrote letters and took it easy in the Jamesway, even though it was a warm, clear day. I compiled our geologic map. The next day we packed up and worked on the new motor toboggan, and replaced one engine mount on another FoxTrac. And we celebrated Steve's birthday that we had largely ignored a few days before.

We left the Jamesway and headed northward for the Dana Mountains late on the morning of December 30, following repair of the chain sprocket of the "new" toboggan. We did some geology along the way, had to replace a drive belt on another of our toboggans, and completed about thirty-eight miles with the three remaining motor toboggans, then established Camp 36 that evening. The toboggans averaged 6 mph at a gas consumption of 5 mpg. We were near the northern edge (73° South) of the 1967 1:500,000-scale USGS Antarctica sketch map of Ellsworth Land (east part)— Palmer Land (south part)— and no sketch map had been published north of it, so all the ranges north of our map became the "Unknown Mountains." (It was not until 1979 that the USGS published the Antarctica sketch map of Palmer Land [north part], and it showed that the mountains north of the Mosby Glacier belonged to the Carey Range, and those north of the Careys belonged to the Wegener Range.) These ranges were no longer part of the Lassiter Coast but instead were in the southern Black Coast. Our aerial photos, however, extended north of the 1967 sketch map.

On New Year's Eve we put in a long field day, with Gene and me in parts of the Dana Mountains north of the 1967 sketch map. When we returned to the tents at 7:30 pm, it was -1°F. New Year's Eve is one of the biggest holidays in Russia—about like our Christmas—so we joined Gene in drinking all the contents of one of his bottles of vodka in celebration. On New Year's Day, we had another long field day.

On January 2, we moved to Camp 37, despite many carburetor problems due to icing from the new cold weather. After a glorious two to three weeks of summer, fall was coming. As we moved east to the new camp, the wind picked up from about 5 to about 20 mph, another sign of fall, and clouds (whiteout) moved in. We traveled down an unnamed small glacier south of Galan Ridge in a moderate whiteout. Just after I passed a steep place on the glacier surface, I crossed one crevasse, then another. As I looked around for more, I saw Gene gesturing wildly for me to stop. Realizing that he knew something that I did not, I cut the power. He had seen that we were breaking the bridges as we crossed them, with Steve and Richard following us closely and breaking bridges too. Karl and Walt were following next, and we waved them off before they reached the crevasse field. Though the crevasses were small at three to six feet across, we nonetheless turned the toboggan around and retreated. But in turning their FoxTrac around, Steve and Richard broke through a bridge, so that the rear of the toboggan dropped into the crevasse a foot or two, tilting the front of the toboggan twenty or thirty degrees up before it righted itself and pulled itself out. We traveled about a mile back, then south to establish Camp 37. The wind picked up that night, to about 40 mph the next day, although we were able to work all day in it. There were no clouds, however, so Gene and I also scouted out another (safer) way across the unnamed glacier to Galan Ridge.

On January 3, we could not get the generator started, no matter how hard we tried and how many four-letter words we used to encourage it! We even disassembled parts of it. Nor could we get out with a smaller, battery-powered radio that we rarely used, because its battery was dead. The Navy rule regarding radio communication is that you must call every day, and if you miss two days in a row, they come and get you—or so they threaten. Radio blackout, of course, does not count, for they can tell if a radio blackout is the cause for a field party missing radio communications. So, we decided that night that if we could not get it going on January 4, Walt and Steve would return to the Jamesway, where we had a spare generator. They took off on the morning of January 4 on a risky trip, for they were to travel alone nearly fifty miles with a decrepit motor toboggan. But we were nearly finished with the Dana Mountains, so we knew that we would follow them in two days. We gave them plenty of survival food and told them to sit tight in their tent if they could not make it. As it turned out, they returned to the Jamesway easily, re-established radio contact, then traveled north to finish up the geology in the unnamed range north of the Jamesway. There they established Camp 38, from where they explored the area around Mt. Heer.

On January 6, after several days in which we completed the mapping of the Dana Mountains, the remaining four of us headed back to the Jamesway. We left later than anticipated because of starting problems with one of the toboggans. The first twenty miles were the toughest part because we had some steep snow slopes to climb. It was fortunate that we made it up these grades, for then the fatigued superstructure that the toboggan track travels around broke in several places and our toboggan stopped dead in its tracks. I knew when I first saw it that it was all over, and that we would have to abandon it. We stripped it of anything we thought we might need and cached some of our fuel to lighten the load. With all of us on the one remaining toboggan, we completed the overall eleven-hour trip back to the Jamesway at 10:30 pm. There we found a note from Walt telling us they had made radio contact with McMurdo and where they were now, plus that they had recharged the battery-operated radio. We had a steak dinner and retired for the day at about 1 am, cold and exhausted.

On January 7, the stratus cloud cover dropped to the surface, with visibility about a mile, in light snowfall and whiteout. We tried to contact Walt and Steve using the battery radio during the morning and afternoon, with no luck, so we puttered around camp. In early evening, Karl and I went outside to do some chores and spotted a black dot in the middle of the Meinardus Glacier to the north. It was Walt and Steve, who arrived at 7 pm with the temperature at 9°F. Both front skis on their FoxTrac were broken due to sheared bolts (the skis had been weakened due to some drilling by the McMurdo winter-over mechanic) and were now held together by ropes. All of us were quite run down from our exertions in colder weather and trying to get as much done as the end of the season approached, and we agreed to take two days to recover. Gene had a cut on his thumb and I had bashed my bad knee. We all needed to be rested and recovered for our next, and last, traverse. Thus, we made the excuse to stuff ourselves with steak, chocolate bars, and the last of the Christmas food as we packed for our next traverse. Steve replaced the FoxTrac front skis. We inspected, repaired, and serviced as best we could the remaining two toboggans, both of which had many cracks in the metal. These were the two we had received on resupply, but they were only marginally better (if at all!) than the ones now long gone. By January 9 we were all in good shape and good spirits, and for exercise Steve and Karl skied up to our nearest nunatak. There Karl fell into a small crevasse up to his head. A not-too-concerned Steve took a picture of him.

January 10 was a beautiful clear, calm, warm day and we got under way at about 10:30 am. The sastrugi this time of the fall were higher and harder than before, and the surface was brutal. We towed five Nansen sledges, one with food, another with jerry cans of fuel, and three with camping gear. It was smooth going until about 4 pm, in the middle of the Haines Glacier. At that point one of the toboggans threw its drive chain. As with a bicycle or motorcycle, the chain transfers the power from the engine to the track. The force of the action tore off the chain guard, and sheared off some bolts too. The question was how the chain had become so loose as to be able to be thrown. It did not take long to figure it out, and the answer came from Gene, who owns only a bicycle and motorcycle but knew drive chains and in the last month had become handy with a FoxTrac. He noticed that the rear axle was loose because the bearing had burned out, the same reason that caused us to abandon our first toboggan. Fortunately, we were only about four miles from our third abandoned toboggan, and the rear axle was one of the parts we had not yet stripped. Karl, Steve, and I took the last toboggan and ran over to it, removing the axle as well as a few other items. That night and the next morning we put the ailing motor toboggan back together, then continued until about 12:30 pm on January 11. More misadventures followed. I got a rope tangled in the chain of my toboggan, breaking the chain and tearing out a sprocket that is used for tightening the chain. We repaired the chain and made our way to a previous gas cache, where we filled up. Then, at about 2:30 pm and only about five miles further on, the engine partly broke loose from its base when three of the four engine mounts sheared (the other toboggan had two bad engine mounts). Though this had been also a serious problem in the past, Holmes and Narver had never supplied us with spare engine mounts, despite our requests. All we had were those we had stripped from our three dead toboggans, and some of those we had already used up. We put on the two last remaining ones, then tied down both engines to their frames with wire and parachute cord. Given the large, end-of-season sastrugi that we were encountering, things did not look rosy. But at 4 pm we continued onward anyway. As it turned out, most of our motor toboggan problems were fortunately now behind us. But we certainly did not know that at the time, and our lack of confidence in the two remaining toboggans was stressful for all of us for the remainder of the season.

In an hour or so we reached a gas cache we had left on our previous trip, the Nansen containing two barrels of fuel. We picked up that Nansen sled and continued, and at 7 pm we had dropped down to the edge of the Mosby Glacier. We could see crevasses in this five-mile-wide glacier, and the aerial photos showed others. With binoculars, we scouted out a route, to the west and east of which were large crevasse fields, dotted with haycocks. Fortunately, the sun was out and visibility good. I led the way across and we had an easy time until we reached the far side, although along the way we had to skirt a crevasse at least fifty feet wide. On the far side, however, we concluded that to start up into this unnamed range we had to go through a field of small (three to ten feet wide) crevasses, for we saw no way around. We dropped parts of our loads before we went through what turned out to be the remainder of the crevasse field, although we realized this only when the other toboggan broke a hole in a bridge about two feet by four feet as it was coming to a stop. Several men who had taken off their skis fell up to their waists in crevasses, but once they put skis on, they were fine. We fooled around, probing bridges with ice axes and taking photos. The crevasses were everywhere, at least one every fifty feet or so. There was no easy way to do this, so Gene and I hooked ourselves to the Nansen sledge that bore the fuel barrels and headed across. I asked somebody to take home movies, for this seemed like the perfect time for an action shot of somebody going into a crevasse. I crossed at least a dozen crevasses in the quarter mile through the crevasse field, although I broke through only two bridges—but not enough to drop the toboggan or Nansen. Richard and Steve followed with part of their load and apparently picked a better path, for they broke through no bridges even though they crossed lots of crevasses. I returned by another route, breaking through bridges, to pick up our remaining two sleds, plus Karl and Walt. Steve and Richard followed my path (the blind leading the blind) with no ill effects. They picked up their last two sleds. Each team chose slightly different paths northward, and we all made it through, although my toboggan broke a hole in a bridge three by six feet wide. Gene planted flags, as we had done most of the way to this point. On the other side, we hooked up our three Nansens each and continued another three or four miles until 10:30 pm, where we established Camp 39, in "Unknown Mountains" (the Carey Range of the Black Coast). We were all cold, tired, and hungry, yet strangely happy (high on adrenaline) that we had successfully made it through the day, about forty-five miles. Along with our meat-bar stew we consumed another one of Gene's supply of bottles of vodka (of course, drinking it straight in the Russian tradition!) to help him celebrate his thirty-sixth birthday! We got to bed about 1 am.

On the morning of January 12, we arose late, at 9:30 am, to see fog moving up the Mosby, then overwhelming us. We had breakfast and waited for it to break off, but it thickened into stratus with light snowfall and a good whiteout. Conditions made it necessary to confine ourselves to camp. We threw a frisbee around for exercise and, for the first time in a long while we were able to write letters home. All of us were in high spirits, perhaps higher than before in the season. Not only were we in another unvisited mountain range, but this one had not yet been mapped. Plus, we were starting to anticipate the end of the season and the return to our families. But the next day, Friday 13th, Gene discovered a crack in a toboggan gas tank about two inches long and a quarter inch wide! Being too large to solder, we used soap to "heal" it, and later replaced the metal tank with a one-gallon polyurethane Jerry can that we had previously used to carry white gas for the Coleman stove. The other toboggan broke a chain, which we repaired quickly. Despite these little setbacks, we spent a warm day doing the geology in a heavy 1000-foot-ceiling cloud cover and light snowfall, in which visibility at times dropped to a hundred yards. Similar weather continued the next day, including a heavy whiteout despite visibility of several miles, and we did the nearby geology. Walt, atop a snowy part of a ridge between two rock outcrops on the ridge, fell through an overhanging snow cornice in the whiteout and slid down the 50° snow surface about three hundred feet. It happened so fast that he could not arrest much of the fall with his ice axe, although he slowed himself until he came to a crack in the ice, which jerked the ice axe out of his hands. Fortunately, he was alright.

January 15 and 16 saw heavy snowfall (four inches recorded on the 15th) and low visibility, so we stayed in camp and celebrated Richard's thirtieth birthday. McMurdo nagged us about when we could return, and they asked for weights and cubes for the flight. We suggested January 29 and gave them estimates of 14,000 lbs and 850 ft^3. We all had a debate about the likelihood and feasibility of manhauling back to the Jamesway, should the two decrepit toboggans fail at or near the same time. We calculated the weight we would need to manhaul back to survive the trip (food, tents, sleeping bags, stoves) and estimated less than 1000 lbs on two Nansens. I recommended that we hook up ropes to each of us and see what we could haul when on skis. The sled with the two fuel barrels alone was 1,000 lbs, not including the sled. We all got hooked up to it and found that we could move it with ease. We felt better and that issue died away.

On January 17, we headed north through the range, then east for a total of about fourteen miles, and established Camp 40. As we learned later, we were in the headwaters of the Fenton Glacier and just east of the Sverdrup Nunataks. The six to eight inches of light fluffy snow on the ground slowed our progress for the next three hours, then we set up the tents and did the nearby geology. Working with Steve that day, our toboggan broke a chain, which we had to fix, so we did not return until 10:30 pm, and found that the others had been about to send out a search party.

We got to bed about midnight, with the temperature cold but not recorded. The next day we moved camp again, doing geology along the way, into what we later came to learn was the southern Wegener Range; we established Camp 41 at 5:30 pm. Given the difficulty of trudging through about six inches of soft snow, we had to relay loads over the seven miles. A chili-flavored meat-bar stew tasted great!

On January 19, the temperature was 10° at 9 am, with clouds at about two thousand feet and a strong whiteout. The clouds dropped down to the surface in the afternoon, resulting in zero visibility, but by then we were doing the geology of a series of ridges. Our toboggan threw its chain four times, so Steve took a link out of it and we were fine for the rest of the day. We went down a steep slope to visit some rock outcrops and got into a crevasse field (only partly visible in the whiteout). As I steered the toboggan around to get out of it, I felt the toboggan tilt; Steve told me later that I had turned the toboggan so that it was parallel to, and on, the crevasse and I tore about ten feet of the snow bridge out before I got off the crevasse. We did more geology, then headed for the tents, spotting a track made by the other toboggan and following it back in low visibility. A message from McMurdo indicated that they wanted to get us on January 27. On January 20, it was 7° at 9 am, overcast with a 7 -mph wind. Steve and I worked on the geology east of camp, with carburetor problems all day due to freezing water in the carburetor; the throttle linkage also froze, forcing a ninety-minute delay. The wind picked up to 25 mph, then the clouds dropped to the surface, leaving us no visibility. At one point we went on a bearing but got all turned around, so instead followed our tracks all the way back to camp, arriving at 5 pm. The others arrived ten minutes later, about when we were ready to go look for them. On January 21, Gene, Walt, Steve, and I headed to the northwest to do the geology on the eastern part of the Peninsula Plateau, using one toboggan. It was cold fall day, but calm and with a light cloud cover. At the close of that day, I declared the end of the geologic mapping and proposed that we head back the following day, a trip of about fifty-five miles.

We left Camp 41 at 9:45 am of January 22. The weather was calm, partly cloudy, and cool. We went by way of Camp 39, where we had cached four boxes of rocks, arriving there after traveling thirteen miles in two hours. I next led the way across our famous small crevasse field on the northern edge of the Mosby Glacier. Gene, a trusting soul, drove while sitting on the seat of the other toboggan, following us. Neither toboggan broke any bridges. A cold front apparently rolled through as we followed our flags across the glacier, and it became cloudy and cold. On the southern side of the glacier, we crossed several large crevasses that we had not seen before. We had to relay loads up into the Dana Mountains, and we arrived very cold at the site of our last gas cache, near the abandoned toboggan, at about 7 pm. The whiteout was severe and we were all chilled, but we all wanted to make it to the Jamesway, twenty or twenty-five miles away. One person observed, as a point in favor of continuing, that our karma must have improved because this was the first traverse where we had not abandoned a motor toboggan. Our snow surface had been good all day, but here the sastrugi were big, and we could not see them in the whiteout. We crossed the Haines Glacier, then up a long grade. Richard tipped over the food sled on a sastrugi. Then throttle linkage problems, so from here we had to operate the throttle of our toboggan by hand, while sitting. We took turns with this cold chore, each driving the toboggan (with the white poly gas tank) until we ran the gallon tank dry. Then we would gas up and the next person drove, with the previous driver warmed by holding onto their rope attached to the last Nansen. Finally, with about two and a half hours to go, we spotted the Jamesway. Then, in addition to a bad whiteout, visibility dropped. The sastrugi were awful but somehow the toboggans held together. We crossed the Meinardus Glacier as the wind picked up, and we arrived at the Jamesway at 11:30 pm, all very chilled. We had traveled sixty-three miles in fifteen hours. Walt commented, "I don't think anybody suffered from heat exhaustion." All of us wore heavy gauntlet mittens, which go up to the elbows; these doubtless prevented frostbite. The temperature was 4° and the wind 8 to 10 mph. We prepared a steak dinner, and got to bed at 2 am.

On January 23, we were up at 9:30 to find a cold day, with a wind of 10 to 15 mph, partly cloudy. This was a day of rest, which we needed badly. We were like zombies all day, still exhausted. We talked with McMurdo, who told us to leave all FoxTracs here to keep the weight down on the pickup flight (they later changed their minds). We suspected that the rumored new Skidoo snowmobiles had arrived. We reminded them that we had 16 JATO bottles in camp, but they did not want them either. We spent the day packing rocks and other things for the trip, all of us observing that we were going to miss Lassiter. Gene presented each of us with a gift from Russia: notebooks, hand lenses, hammers, and a hat. I had one more geology day trip that I hoped to do, Schoofs Nunatak to the northwest, but I wanted a good day to do it. January 24 was not one of those good days, so I let McMurdo know we were ready to be picked up. We spent the day digging out the aluminum pallets, and putting gear and rock boxes on Nansens, ready for pulling to the plane. We prepared (driving up and down it) and marked the skiway. The weather improved and we let them know, although communications were poor. The temperature at midnight was -6°. On January 25, the weather remained good and we sent weather reports every three hours. At 7 pm, when I sent the weather, I received a message "please be advised that a plane is due into Lassiter at 0350 Zulu" (Greenwich time for local 11 pm). The message, however, was weak, so we could not be sure of it. We tried several radios, tried the aircraft frequency, etc. Pole Station relayed some messages, and after dinner we packed up most of what we had not done before. Strato-cumulus at 3,000 to 4,000 feet went from 20 percent to 90 percent but visibility was still good. Aircraft 917, piloted by John Dana, came in sight at their estimated time and landed shortly thereafter, the air temperature at 0°F. As the engines remained turning, we loaded it up in about an hour with about 14,000 lbs of gear, including ourselves. We attached two JATO bottles to each side of the fuselage, and off we went toward Byrd Station for refueling. At that point Richard learned from one of the crew that the Vietnam War was over, or at least the press was reporting a deal between combatants.

After a seven-hour flight, we came in sight of McMurdo at a little after midnight on January 27, McMurdo time. As we approached, the plane suddenly pulled up and we circled around. The crew scurried around the plane, concentrating on the hydraulics system for the nose ski. We learned later that the nose ski was dangling, apparently having been damaged on takeoff from Byrd. Therefore, John landed on the rear skis and came to almost a complete stop before he allowed the front end to drop down. Ours, we were soon to learn, was the last successful flight of #917. We were welcomed by a bunch of McMurdo officials, then we got our personal gear from outer Williams field to McMurdo. We arrived at our dormitory at about 3 am, and found a bag of mail! That took priority over sleeping, so we read our mail for a half hour. Next on the agenda was a bath and shower, our first in three months. We got to bed about 4 am and were up four hours later, refreshed despite only minimal sleep. We spent the day packing rocks for the ship, returning equipment to the various USARP warehouses, visiting with NSF officials, scheduling a haircut, preparing Gene's samples for a scheduled three-hour Herc flight to the Russian Vostok Station, and so forth. I got a ham operator to connect me with Karen, the jewel event of my day! She and the kids were fine, and were having a mild winter. Later, on a trip to outer Williams Field I saw our motor toboggans, which the pilots had decided to bring back to play with at outer Williams field. Unfortunately, when they did that, one broke down and had to be towed back to outer Willy; the last I saw of it, it was turned on its side. Lastly, USARP officials scheduled our trips back to Christchurch and the States.

On the morning of January 28, aircraft #917 crashed and burned while landing at South Pole Station, fortunately with no loss of life or serious injuries. Scuttlebutt plus conversations with aircraft officers at the bar revealed that again it was the nosewheel ski that dropped almost vertically on approach. But a patch of fog rolled onto the skiway after the aircraft was already committed to the landing. The pilot attempted to land with all the weight on the back end of the rear skis, but the momentary loss of visibility meant that the plane hit hard and unevenly.

The impact popped the struts loose from the rear skis, then the plane rocked forward, hitting the nosewheel ski. The wingtips, first one side then the other, hit the snow, tearing off both wings, and the engines burst into flames. The aircraft was the newest of the three Hercs being used this summer, although it was the least reliable and was way over its recommended airframe hours. It should have been retired years ago but, as with our FoxTracs, penny pinching by NSF now left them with only two Hercs, numbers #319 and #320. Of these, #319 was the grand-daddy (more airframe hours) of all C-130s in the world. Since the 1970–1971 season, when they had had five Hercs, the Navy had lost three Hercs and one Constellation to crashes, and had been forced to retire another Constellation due to old age. The planes of the 1972–1973 season were too old to do what they had to do, and all suffered many breakdowns, thus time lost for science.

There were several implications of this loss of an aircraft. First and most immediately, the continent was now virtually shut down as the Navy scrambled to get people out, with only two aircraft that had skis. Some people, including probably all Navy enlisted men, were to return by ship, which had just arrived at McMurdo. A future implication, we speculated, would be that the following year the scientific effort in Antarctica would be curtailed, with few or no remote field parties such as ours taken out there. Our next field season would almost certainly be postponed.

We spent the next few days finishing packing rocks and strapping the rock boxes, then putting them in large metal connex boxes destined for the ship. We shopped at the PX and wrote letters to our families. We had been scheduled to leave on January 31, but the plane (#319) was rerouted to Pole Station to medivac a guy with appendicitis, and then that same aircraft had problems with a strut in the landing gear that had to be fixed at the Pole. At breakfast on February 1, we learned that the aircraft, with the patient onboard, was on its way back. Twenty-four USARPs, including all of us except Gene, were to join this aircraft, which was to leave at 10:30 am. After an eight-and-a-half-hour flight we landed in Christchurch, went through customs, then were driven to the USARP warehouse (8 pm). After trading our USARP clothing for civvies, we took a cab to Warners Hotel. The next day, we confirmed plane reservations for our military flight back to Hickam AFB, then commercial flights on home. I called Karen, for we had planned a ten-day vacation in Maui.

It had been a trying field season but all of us felt good about it and how each person had performed and met the challenges and the weather. We had mapped about four thousand square miles of unexplored territory in the northern Lassiter Coast and southern Black Coast. In New Zealand, we said most of our goodbyes. Karl, Walt, and Steve decided to tour South Island, including trekking near Mt. Cook (New Zealand's highest) for a week or so, and I headed to North Island and Wellington to visit a bunch of active volcanoes and Kiwi geologists. Upon my return to Christchurch for the MAC flight, I had the opportunity to say my goodbye to Gene, who had just arrived from The Ice. In a day or so he would head back to Moscow the other way, through Sydney, Singapore, and Kuala Lumpur. I left Christchurch on the morning of February 13, crossed the International Date Line, and arrived at Hickam just before midnight of February 12. Karen arrived at Honolulu Airport on the afternoon of the 13th. It was a wonderful reunion, one both of us needed!

Brighton
DENVER
Aurora
Littleton
South Platte R.
25
Fairplay
Leadville
Aspen
Mt Elbert
4400
COLORADO
Great Sand Dunes National Park and Preserve
Walsenburg
Alamosa
San Luis
Conejos

4

Between Field Seasons, 1973–1977

It was great to return to family life! The 1968 song "Mony Mony", in which Tommy James and the Shondells shouted "it feels so good, yeah," typified this enthusiasm! Within hours, it seemed, Jill had grown from a baby to a walking infant, to a running child! Jill was the youngest in the neighborhood, and therefore a great hit. Karen treated her like a little doll, putting her in pretty dresses and strutting her around our part of Foothill Drive. One of Jill's many endearing practices (for she knew it would get a reaction) involved her security blanket, a small white plush baby blanket that was her favorite possession. She would hold one corner of it in her hand, then put the two middle fingers of that hand in her mouth to suck. When stressed, she would demand her "backey." To join Barney the dog, we acquired a cat or two as well as goldfish, hamsters, and the like.

Scott was always solicitous and protective of Jill. He attended the local school but did not excel. We were not concerned until, a year or so later, tests showed that he had learning disabilities (dyslexia). He saw words backwards, and reading was difficult. Of course, schoolwork depends largely on reading, so all his courses was affected. Unlike most kids, and certainly different from Karen and me when we were growing up, school was not fun for him. Nonetheless, he was brighter than anybody else in the family. He had a work ethic and kept at it and, as years went by, he was able to overcome many of the problems. But academics were never enjoyable for him, so he never went to college. Art skills, however, were his love and we encouraged his many successes in this. He was able to draw people and animals especially well, and by high school he could produce pictures of them using paints and watercolors of which anybody would be proud and would treasure. He had incredible taste because of his artistic talent. He also could repair everything, from digital things to anything mechanical, because he could figure out intuitively how it worked and what might go wrong, a clear sign of a fertile brain. Thus, friends and relatives all had lists of broken things for Scott to fix for free. Scott also became a wonderful cook, even self-published a recipe book in 2005. Scott has a sweet, beautiful, loving, and loyal personality. He ended up being tall and handsome. He was a terrific athlete but did not have a competitive spirit, so never took up any organized sports. When he was in high school, we often went downhill skiing together. The most enjoyable part of that was going as fast as we could down a ski run; when we did, he always beat me to the bottom! Peter and Arlene Dragalin, from across the street, had two boys who were about the same age as Scott, so Peter and I co-led a local Boy Scouts troop, and we insisted that its credo be to get the kids camping in the mountains once a month throughout the year, good weather or bad. Peter promised that he would always be there to lead the meetings, and I promised that I would always be there to lead the camping trips. Winter camping was like Antarctica, and always as fun for the kids as my adventures there had been! Scouting, of course, was another activity that enabled Scott and me to be together.

Scott turned into an intelligent, if not brilliant, caring adult, married well to Patricia (Patty) L. (Lynn) Osborne, a beautiful blonde who was equally intelligent and caring and had gotten a college education. They moved to New York City for their first jobs, eventually made their way to southern Maine and started a family, first Kathryn and then Chris. Scott worked with his Apple computer and started Scott Design LLC, which created advertising brochures, websites, and the like. After the kids left the nest, Patty opened a shop to sell fabrics and sewing materials, which was surprisingly successful. Wonderful parents and both successful independent businesspeople, Patty and Scott became two of my greatest heroes due to their morality, compassion for others, hard work, competence, and willingness to try new business ventures as recessions came and went. In 2020, when the COVID-19 pandemic overwhelmed the country (killing Patty's dad in Colorado), Scott became a baker, using their own kitchen to make pies, bread, cookies, muffins, and the like to sell to restaurants and at county fairs and country markets. As usual, this business was also successful.

But back to when Scott was a child. Living adjacent to the mountains meant that we spent a lot of time there. We hiked, had picnics, explored old mining towns, and skied downhill. I also joined the Optimist Club with Peter Dragalin, which required morning meetings once a week and various types of charitable work. One neighbor, Prebble Harres and his wife Pansy Napoli, who lived next to the Dragalins, had a large A-frame cabin at the ski resort of Breckenridge, in the mountains west of Denver. It had multiple bedrooms, and the neighbors, Karen, and I spent many weekends there. Preble, a big guy with a great sense of humor, had a jukebox, so we danced to rock-and-roll songs after skiing and dinner. In the morning, to get everybody up for breakfast and skiing, Preble would select "Amazing Grace" by the Royal Scots Dragoon Guards, the best instrumental version of this wonderful song, and play it at maximum decibels! Preble had paid for his cabin by being one of the last of the classic traveling salesmen, to ranchers on the Great Plains. He had a good occupation visiting farms and farm-related businesses in eastern Colorado, Nebraska, and Kansas, selling farm equipment, seed grains, etc.

Another family, which lived next to Prebble and Pansy, was that of Susan and Tom Fry, who had grown up in the Los Angeles area before moving to Colorado, where Tom drove a truck. Tom's upbringing including drag racing and fixing up hot rods on LA streets. They and their three kids, somewhat older than Scott, were hard working and ambitious, loved Mexico, and were fluent in Spanish, and soon they started a business bringing Mexican clay pots and figurines north to Denver to sell to homeowners. They became my ideal as entrepreneurs and business people, recognizing a need and moving quickly on it and using their contacts in Mexico, and their love for the people there, to capitalize on that need. They bought several old yellow school buses, tore the seats out, and used each bus to bring back their pots. Next, they found a run-down old motel in the southwestern outskirts of Denver, on a main north-south street that had been bypassed by the interstate to Colorado Springs, resulting in a neighborhood that lost its businesses and decayed. The motel was built around a courtyard, and soon they fixed it up by knocking out walls between bedroom units such that the rooms now held various types of pots on shelves. Within a few years, through advertising and the fact that the remodeled motel was tasteful and fun to visit, this became Denver's best place to buy Mexican clay pots and bric-a-brac. After a while, the Frys moved to a fancy new home in the mountains west of Littleton. We stayed in touch by Christmas cards and phone calls, but sadly Tom died in December 2020; Susan is still in the same house, and we occasionally call or write each other.

Lab work and writing reports on Antarctica and Utah began anew. In the summers of 1973 and 1974, I continued mapping in the Iron Springs mining district, focusing on where the iron deposits were. In 1974, I proposed to NSF to map the unexplored area adjacent on the south to the Lassiter Coast, called the Orville Coast. The NSF gave initial approval and encouragement, but the C-130 aircraft problems and other priorities delayed final approval. It was not until the months leading up to the 1977–1978 season that the Orville Coast project became a "go!" In the meantime, starting in July 1974, Ed Ruppel assigned me to oversee a project of mapping the Vernal 1:250,000-scale quadrangle, a Colorado Plateau 2-degree sheet like those that Paul Williams did, that spanned the Colorado-Utah border (one degree of longitude in each state), consisting of the eastern Uinta Mountains and basins north and south of it. The purpose was to study the framework of oil- and oil-shale-bearing lands in the Uinta basin, south of the mountains. (Oil shale is a fine-grained laminated brown or black shale so high in liquid and gaseous oil that you can smell it.) Energy resources had become a hot topic due to the events leading up to the 1970s "energy crisis." The region I was studying held the largest oil-shale deposits in the country, and the oil industry was involved in test after test in hopes of tapping the resource as an easy way to solve the crisis; it took more than a decade for the industry to realize that the waste products were an environmental nightmare, and therefore oil shale was never developed.

You might recall that the first part of the energy crisis was an oil crisis, which began in 1973 with a shortage in oil produced by the US, leading to greater dependence on foreign oil; then this was followed by a related phase of the oil crisis: the Arab oil embargo in response to the US decision to re-supply the Israel military after the Yom-Kippur war. The stock market crashed in 1973–1974, and the recession continued until 1975, in the Gerald R. Ford presidency (1974–1976). Experiencing the auto lines at gas pumps was a real shock for the American public.

The second part of the energy crisis was the Iranian revolution, including the hostage crisis in 1979–1980 and reduction in Iran's oil exports, late in the Jimmy Carter presidency, which ran between 1976 and 1980. Congress passed the Energy Security Act in 1980. With that background in mind, we return to my geological work.

My Vernal map area also contained the large Dinosaur National Monument, where paleontologists were still working Dinosaur Quarry. In 1974, Branch member Wallace R. Hansen joined me part time on the project, for he was an authority on the Uinta Mountains, among many other topics and places. Wally thought it would be a great idea for us to rent a river raft and boatman to take us through the rapids of the Canyon of Lodore, where the Green River cuts north to south through the entire Uinta Mountains, on its way to the Colorado River and then the Grand Canyon. We had some money in our budget, so we made a deal (at a reduced rate) to go in May, before the boatman's busy season. Newly hired Quaternary geologists Paul Carrara, who was also assigned part-time to the project, and Rich Madole, joined Wally and me. We stopped the boat at various places along the way to map the geology, the only way to see the rocks up close other than impossible hikes on foot. And we camped each night on the river bank. It was a great idea at the time, and we thought we were really putting one over on the USGS administration! Unfortunately, we endured five days of drenching rains, icy sleet, and local snowstorms, accompanied by hurricane winds! We mapped the canyon while witnessing firsthand nearly all the erosive processes that had made it!

> ***Wally*** *was born in 1920 in Salt Lake City, raised there, and received a B.S. in geology at the University of Utah in 1941. He started graduate work there too, but World War II intervened and, as an expert Utah skier, he volunteered to serve at a small military weather and rescue station on the Greenland ice cap. After his retirement from the USGS in 1985, he turned these adventures into a popular book. He was one of many veterans who was hired after*

the war by the USGS. There he met Ellen Edstrom, who had been hired as a USGS draftsperson at about the same time. Ellen, a beautiful Swede, was the love of Wally's life and they had four children and a happy and fulfilling life. Wally became a great field geologist, with a specialty in engineering geology. His first big project (1951–1955) was mapping the Flaming Gorge area of the northern side of the Uinta Mountains, where several dams were proposed. This he followed by mapping the area of Anchorage, Alaska in 1955–1956, which made him the obvious choice to analyze the great Alaska earthquake of 1964, as noted above. He next studied the Black Canyon of the Gunnison River, Colorado, in 1957–1965, where more dams had been proposed. The environmental movement came to the USGS in 1970 and, in response, Wally was put in charge of the giant Front Range Urban Corridor project, Colorado, where many dozens of maps, in the foothills and east of the Front Range and from virtually the Wyoming to New Mexico borders, portrayed the geology, water resources, gravel resources, engineering geology, slope stability (including landslides), and other practical aspects of this rapidly growing suburban and transportation corridor. This project became a model for others in the Washington, D.C. and San Francisco areas. One of the best and most prolific writers in the USGS, Wally published well over one hundred reports, many of them extensive final reports with maps (USGS Professional Papers). He received many awards, and was offered many speaking engagements and editorships. Quiet but hardly shy, he had the confidence, humor, extraordinary competence, and productivity that characterize many successful persons, and he was universally loved. Ellen died in 1996, and Wally died in 2017. That Ed put Wally on my project was an example of how the USGS mentored young people such as myself by assigning experienced and knowledgeable geologists to teach them. Consequently, I learned a great deal!

In ***Ed Ruppel's*** *fifth year (starting in July 1974) as Branch Chief, he chose me for a half-time administrative job as Assistant Branch Chief; I continued in that position for the next year too, working with his successor, Paul Williams. That year with Ed followed the year in which images of cars waiting in line for hours at gas pumps were common on the news. The Congress suddenly realized that geologists had a part to play in finding oil and gas. After many years in which the USGS could hire only to replace retirees, our Branch that year was suddenly advertising more than a dozen new geologist positions. Most other branches also added people. At our branch, Ed headed the process of announcing available positions, after which he was inundated with résumés. After a long period, helped by committees of Branch members, he interviewed the favored applicants and put the twelve résumés of the people he wanted in a pile on a table in his office. One day he received an unfavorable letter of reference about one of them*

> *and, after verifying the claim, picked up the résumé and put it in the reject pile, which was quite a bit higher. The remaining eleven became the mainstay of the Branch for a decade or two. Several more were added the next year by his successor Paul Williams. A few geologists from these two good hiring years lasted as long as 2022 before retiring. Most became close friends of mine. Those included Karl Kellogg, Paul Carrara, Ralph Shroba, Chet Wallace, Mel Kuntz, Rich Madole, Dave Lidke, and some others whom I will reference below.*

Within a few days of becoming Assistant Branch Chief, Ed handed me a couple of articles to read. They dealt with "groupthink," an affliction that can affect decisions made by groups, even presumably those made up of intelligent and thoughtful people. The classic example of groupthink is the Bay of Pigs invasion of Cuba during the John F. Kennedy administration, a decision made by "the brightest and the best" of that administration. A much earlier example was the dismissal of the notion that the huge naval base at Pearl Harbor, Honolulu was vulnerable to Japanese attack—this by all officials, including those in the military and in the Franklin D. Roosevelt administration. The articles noted that such lapses in judgment characteristically take place when individuals are particularly concerned about retaining the approval of fellow members of a work group. In other words, members of the group just go along with a group decision rather than raising doubts or even playing devil's advocate. Groupthink is a malady of management, and Ed recognized that as part of management I was now more prone to it. In 1995, it would be the major contributing factor in the USGS for the disastrous Reduction in Force (RIF), as described in Chapter 8.

An example of a cohort that tried its best to prevent groupthink was the Obama administration. Perhaps for that reason, the administration made few major bad decisions. In contrast, the Trump administration expressly embraced groupthink because the leader chose persons for his cabinet and staff who were sycophants, and the more groveling these persons did, the more he liked and praised them. To Trump, actual qualifications for a job were of no importance. Senator Ted Cruz, Governor Ron DeSantis, and Vice President Mike Pence were examples of these bootlickers. Cruz embraced Trump even though Trump insulted the attractiveness of his wife and accused his father of assisting in the assassination of President Kennedy. Pence attained his previous political offices by pretending to be a man of God but proved over four years to be a moral coward, although by the January 6, 2021, insurrection at the Capital he found a backbone and was widely praised for this even though it was a mandated part of his job. Therefore, most decisions made by the Trump administration made little sense; each day one wondered what bad thing would happen next. When COVID-19 swamped the world in early 2020, we had our answer: an administration with no leadership, a delay of seventy days before significant action, then halting and inconsistent moves and public addresses that produced more damage than help. Groupthink squared!

My main job as Assistant Branch Chief was to shepherd Branch manuscripts through the system, including by assigning peer reviews. But there were other semi-onerous jobs, including committee meetings, hiring, and supervising field assistants and technicians, filling in for Ed, etc. Mapping the Vernal 2-degree sheet continued until September 1977, amidst the administrative chores. I was mapping the many 7.5-minute quadrangles at a pace of nearly one a day. But we had some revenge, by picking some quadrangles for detailed mapping. Most of these were in Dinosaur National Monument, a beautiful park of mountains rising above the Green River and a famous quarry long mined for dinosaur fossils. But on October 1, the Orville Coast project started (next subchapter.)

In June 1975, Ed was happy to relinquish the job of Branch Chief and to return to doing science within the Branch. Paul Williams was chosen for the typical five-year tour for Branch Chiefs, until 1980, and I continued as Assistant Branch Chief under him during his first year. Paul was the best Branch Chief I ever knew: like most other Branch Chiefs, he believed deeply in quality science and productivity among his geologists, though his sense of humor and his quiet friendliness suggested to some that he was a pushover. But he had a sharp eye for competence and the backbone to deal with incompetence. He therefore removed or fired those who were not doing the taxpayer any good, and he was always on the lookout for talent. One of those that caught his eye was Bart Ekren, whom he brought in from the Nevada Test Site, then entering a new phase with Gary Dixon ready to take over (see below). Paul also hired a half dozen newly minted geologists, primarily Quaternary geologists. Paul strengthened the Branch mapping program, including publishing the last of the Colorado Plateau 1:250,000 sheets, the Flagstaff quadrangle.

With my term over as Assistant Branch Chief, I returned to the Vernal project full time, as well as continuing with Iron Springs maps and Antarctica reports. In the spring and summer of 1977, Paul Williams, other party members, and I prepared three reports on the Lassiter Coast area and a summary of the metallic mineral resources of Antarctica for a once-every-five-years Antarctic symposium. We presented our papers in August 1977 at the University of Wisconsin, Madison before an international group of more than two hundred scientists. Cam Craddock was the editor of the huge volume derived from these papers and published in 1982. Meanwhile, the main product of the Vernal project was the preliminary version (1978), final black-and-white version (1979), and final color version (1985) of the geologic map. Paul Carrara did the surficial map of the 2-degree quadrangle. Of the 7.5-minute quadrangles of Dinosaur

National Monument, Wally was the senior author of seven quadrangles and I the senior author of five, published between 1977 and 1982. Wally was the senior author in 1983 of a summary map at 1:50,000-scale of the Monument, as well as of several popular small books on the Uintas and on river rafting.

1977 – 1978
Orville Coast

5 The 1977–1978 Field Season on The Ice

Coming up fast was the next field season, this time to the Orville Coast, south of the Lassiter Coast. The Ronne Antarctic Research Expedition discovered the mountains of the Orville Coast during the 1947–1948 season and named them for Captain Howard T. Orville, US Navy, who coordinated the meteorological part of that expedition. I signed up Karl S. Kellogg and Paul E. Carrara from the branch, then asked Walt Vennum of Sonoma State University to be my second-in-charge again. This was Paul's first trip to the ice, but in addition to a strong field background he had a spectacular sense of humor. From résumés, I picked a University of Colorado grad student, J. Michael Boyles, to come along. I recognized that, in addition to a strong field background in geology, Mike had a talent for mechanical work, so I planned to put him in charge of the snowmobiles. Four USGS topographic engineers would join us, and fortunately George Schirmacher was available to lead the group. The others were Robin Schmutzler, David E. Reed, and Harry E. Zelfs.

> *For a foreign exchange scientist, I selected an invertebrate paleontologist from the British Antarctic Survey (BAS),* ***Dr. Michael R.A. (Robert Alexander) Thomson****, born in 1942, to join us. Paul Williams and I had gotten to know Mike over the years as the best paleontologist (yet a true field geologist) working in the Antarctic Peninsula, with a particularly dry sense of humor. Besides, the Antarctic Peninsula was British "territory" in Antarctica. Mike had earned a B.A. in geology at the University of Birmingham, with which the Earth Science Division of BAS had long been affiliated. With this degree in hand, he immediately "enlisted" for field work in British Antarctic territory, which would be applied toward a Ph.D. thesis. However, field seasons for young BAS persons then going to Antarctica required a tour of two and a half years (three austral field seasons and two intervening Antarctic winters). This he did from December 1963 to March 1966, out of Stonington and Adelaide Island bases (British), and mapping by dog sled along the east coast of Alexander Island, a huge island off the western coast of the southern Antarctic Peninsula. This east coast included a spectacular east-facing cliff, a north-trending fault scarp like ones in the Great Basin of the US. Mike specialized in Cretaceous sedimentary rocks, and his dissertation became a detailed study of*

the ammonite fossils. He received the Ph.D. from the University of Birmingham in 1969 and spent the rest of his career with BAS, where he spent at least fourteen more field seasons in the Antarctic Peninsula and Scotia Arc. In 1976, the Earth Science Division (led by Charles Swithinbank, from whom we will hear again, as with MRAT) of BAS moved to Cambridge, England, and Mike was one of three geology section heads then. He became Chief Geologist in 1984. He was an extraordinarily effective and highly regarded administrator and leader who was devoted to his staff and to colleague geologists and geophysicists whom he supervised, although when he joined us we never heard about this side of him. He had a quiet, modest nature and never related stories about the administrative side of his career while in the tents during our field season. I only learned how gifted he was from obituaries after he died in January 2020.

I asked ***Thomas S. Laudon****, a Professor of Geology at the University of Wisconsin, Oshkosh, to join us at Orville Coast. Tom had been teaching geology at Oshkosh since 1963 (and continued until 1999). He established, and for thirty-five years led, their summer geology field camp that visited the Canadian Rockies to as far north as the Yukon, where he regaled his students around the campfire with tall tales about cheap whiskey, cigars, and field geology. Born in 1932, Tom earned his B.S., M.S., and Ph.D. from the University of Wisconsin, Madison. His wife, Suzanne, put up with him, partly by excelling in her golf tournaments. Tom took ten trips to Antarctica, including as leader of a four-man expedition in November and December of 1965 into an area called eastern Ellsworth Land, just west of the Orville Coast, that we were planning to revisit. Therefore, it was only fair that I invite him to join us, an idea that delighted him. Pat Quilty, who accompanied Tom to eastern Ellsworth Land and later became Professor of Geology at Tasmania University, credits Tom with authoring or co-authoring seventeen reports. Tom could keep a person in stitches most of the time. I stayed in touch with Tom until his death in 2017.*

Here, it is appropriate to mention ***John C. Behrendt****, a friend of Tom's, whose life and Antarctic career had some remarkable parallels with Tom's. Born the same year as Tom, John also received his B.S., M.S., and Ph.D. at the University of Wisconsin, Madison, the latter degree in 1961. As noted in an earlier chapter, John's first trip to Antarctica, unfortunately, was spent wintering over as a junior geophysicist in 1956–1957 at Ellsworth Station, on the Filchner Ice Shelf along the Weddell Sea. In a much later memoir, John recorded the authoritarian antics of Captain Finn Ronne, whose last trip to The Ice included managing the base that season, providing good evidence of insanity while doing so! That did not dissuade John from making twelve more trips to Antarctica, doing geophysics during the International Geophysical Year and* *later for the USGS on the Ross Ice Shelf and in many other places. His office, in the Branch of Geophysics, was not far from mine in Denver, and we collaborated on some Antarctic mineral resources work. One outcrop (near Mt. Hirman) in the southeastern part of his then-unnamed namesake, which became the Behrendt Mountains of eastern Ellsworth Land, was visited during his Antarctic Peninsula Sno-Cat traverse of 1961–1962, and much of the geology in the range was mapped by Tom and his colleagues in 1965. Completing the mapping there was one target for the Orville Coast season.*

John Behrendt had many adventures, one of which became an Antarctic legend. It occurred in a Navy aircraft—more specifically a pre-Hercules aircraft, when ski-equipped equivalents of DC-3 and DC-6 aircraft were used. John was then twenty-eight years old, and leading his first Antarctic field party on November 22, 1960, flying out of Byrd Station. The aircraft, a Navy R4D8 (DC-3) was doing airborne geophysics, which involved towing a "bird" (magnetometer) as they were (unwittingly) approaching the Crary Mountains, when they encountered clouds. Both Art Ford and Tom were on the aircraft, so I heard many versions of the story. The only "visibility" as they went into the clouds was a radar altimeter, which soon recorded "targets" (mountains!) while a copilot was at the controls (the pilot was dozing in a seat near the navigator). John was in the pilot's seat guiding the science project. As the copilot pulled the nose up and climbed, the aircraft began to stall. Art and Tom could see rocks out the starboard (right) side, as the bird/magnetometer bounced along the snow or rock surface for seventeen seconds before it broke off. Then the outer edge of the starboard wing hit rock! The pilot jumped into John's seat and grabbed the controls, made a vertical bank to the port side, and pushed the nose down to pick up airspeed to get out of the stall. The radar altimeter showed the ground a mere two hundred feet below. Eventually, as endless minutes ticked away, the pilot (twenty-nine-year-old Marine Capt. Joe Walker) managed to right the plane and they circled and climbed to 13,500 feet. Later they learned that the aircraft had drifted off course by twenty miles while in clouds. Now to be fair, if they existed at all, air navigation charts were not controlled and mostly obsolete; in this case, charts showed no mountains in the area in which this crew encountered them. The pilot asked John if he wanted to continue, and John agreed. They opened the rear door, reeled in the magnetometer cable, and put a spare bird out. However, the enthusiasm for their research that day was strangely gone and soon all agreed to return to Byrd. Upon landing, they found a fist-sized dent in the wing tip.

As the years went by, John became the leading expert on the geophysics of Antarctica. I tell a little more of his story later.

For us, the season started on October 7, when we flew a commercial aircraft to Los Angeles, then Honolulu. We took a taxi next door to Hickam AFB, from which we were scheduled to leave the next day. We arrived in Christchurch by C-141 at 6:30 am on October 9. Having a few days free before our scheduled trip to The Ice, we rented a car and toured nearby towns and parks. At 1 pm on the 12th, a Starlifter took off for McMurdo, arriving at 6 pm. We were signed up for Snowcraft (October 20) and other courses. Because we were to be living in a Jamesway again, we took Jamesway and Preway school. The put-in spot I had designated was on the eastern side of Mt. Smart, in the southeastern Sweeney Mountains. In between my many meetings with USARP, Navy, and Holmes and Narver personnel, I started us all assembling and packing food (Paul), Nansens (Walt), field gear (Mike T.), bindings on downhill skis (Tom), snowmobiles (Mike B. and Karl), and radios (Walt). Fortunately, the radios had been upgraded. And far better, the FoxTracs had been ash-canned. We now had snowmobiles instead, specifically the Ski-Doo Alpine 640-ER, made by Bombardier Recreational Products in Canada; in fact, our party was to be the first significant Antarctic field tests of these machines. We checked out the best snowmobiles we could find, serviced them, and rigged them for remote driving. Then we spent hours field testing them on the sea ice, giving everyone experience with remote driving.

My journal for these days shows page after page of lists of things I had to do, and when, and I added check marks alongside each as I did them. One drain on my enthusiasm was new rules from Holmes and Narver that required us to fill out 8.5-inch by 12-inch paper forms, with six copies, that were to give weight, volume, and flight number, plus a ten-digit number, for every box we packed. Then they affixed red tags with this information to the boxes, and it had to match the paperwork. As was anticipated, things did not always match and it became my responsibility to fix it. I finalized "weights" and "cubes" for the flights, but now I had to do it dozens of times. Once we were able to get control of all these necessary chores, we had a little time to visit animals, send Christmas cards, hike to Scott Base, climb Observation Hill, etc. On October 21 we were ready to go, waiting on weather. Byrd Station was almost operational but its temperature then was -57° to -68°F., and Air Operations wanted it to be at least -54°. On October 25, McMurdo was racked by a storm that continued until October 28. The storm was designated "weather condition 1," which meant that the windchill is -100° or colder. Wind gusts of 100 mph were recorded, with a steady wind at about 40 mph. Temperatures ranged from 10° to -40°. Snowfall of about a foot was also part of the picture, leaving huge drifts all over McMurdo and knocking out the runway. As a result, nothing moved, least of all us! Once the weather brightened and the drifts were cleared, resupplying Byrd and Siple Stations had priority for the next several days.

In McMurdo and soon after we arrived in the field area, I received two or three letters from Karen. Then her correspondence stopped and I received no letters from her for the rest of the season. I continued to write to her and the kids. At the end of the season, when we returned to McMurdo, there was a five-page letter dated January 11 (postmarked January 17) that began "Dear Husband, Let's see. I really owe you a long letter." It was not signed. It stated, as an excuse, that she had not received my letters until January 10. The letter rambled on about a trip she and the kids had made to Florida and Disneyland, plus a few other things of no consequence. However, by the time I got this letter, I had figured it all out, although no friends or neighbors in Denver had ever written to me to help untangle the mystery. Of course, it was agonizing to go through this two-month reasoning process in which I had to conclude that she had been unfaithful. After I got back to Denver, friends told me that she had taken up with her hairdresser, Joe Cassio. I started divorce proceedings. The song "You are my Sunshine" by Ray Charles said it best!

On November 3, after dinner, seven geologists and two topographers got the word to go. Pickup trucks moved the pallets to the runway, where we had a whiteout, a temperature of -10°, and 15 mph winds. Jim Jaeger was the Aircraft Commander, with a second flight to be piloted by Bill Morgan. Flight 1 contained tents, whereas flight 2 held the Jamesway and two topographers. We were airborne by about 3 am McMurdo time. By the time we reached the Orville Coast, the clouds were at variable heights, from 500 to 5,000 feet above the surface, with five-mile visibility. Tom and I were called into the cockpit to look for and identify landmarks, but most were obscured by the clouds. We flew around for an agonizing hour before I saw something and asked Jim to swing the plane around 180°, for a last pass before our fuel levels would force us to abort the trip. When he did, the landmarks, namely peaks in the Sweeney Mountains that I recognized from the aerial photos, clicked in, bing, bing, bing, in the space of a few tens of seconds, with visibility from patches of sunlight passing between the clouds. I pointed to a camp location east of what I interpreted to be Mt. Smart and down we went. Low on fuel, Jim had no time to drag the skiway. It turned out to be a smooth landing because of four inches of new snow on the ground, but the snow required a JATO takeoff. Within an hour, we had freight-trained the load, attached JATO bottles, established radio contact, set up tents, and fired Coleman stoves. And off the plane went again! The aircraft altimeter gave us an elevation of 3,500 feet, and it was 6 pm on November 3, local time (11 am on November 4 in McMurdo). But immediately the weather closed in, with continuous cloud cover and low visibility by the time the second plane arrived, forcing it to turn back. With a half-mile visibility and whiteout, I was not even sure we were at what I had interpreted to be Mt. Smart, for I had seen it for less than a minute, and that uncertainty and queasiness continued through November 5, when a brief clearing enabled me to see a mountain several miles to the west that looked like Mt. Smart.

On November 5, in the poor visibility, ground fog, and low clouds that at times lifted a little, we built, for exercise and safety, two snow trenches near our tents. Early on November 6, the weather became clear and Walt began giving hourly weather at about 1 am. Then at 5:30, as the wind picked up, the generator quit due to a 20- to 30-mph ground blizzard. We tried a spare generator but it too quit. Then, after an hour of fumbling, we put a third generator in the snow trench and thankfully it started. We returned to transmitting our weather reports and learned that the second aircraft was in the air, being flown by Jim Jaeger. We saw a Snowy Petrel fly by, and we considered it a lucky sign! And sure enough, the ground blizzard stopped for a short time and the aircraft arrived at 11:30 am. George Schirmacher and Robin A. Schmutzler got out. Most of the load consisted of the Jamesway, which we set up that day. There was a third flight to come, containing most of the food, geology boxes, almost all the topo gear, and barrels of gas for the Preway, Ski-Doo, and Coleman stoves.

On November 7, while we waited for the third flight, we geologists explored and recorded the geology of nearby rocks. Mike Boyles and I worked together that day, on a trip south a few miles toward the Morgan Nunataks. Before we reached rocks, we found ourselves on a nice crevasse field, with haycocks; most crevasses were only two- to five-feet wide but some were more in the ten- to fifteen-foot range. For two days and part of the next day, there then followed a storm that I described in my journal as "horrendous." Snow sifted into the Jamesway until we tightened it up. It was a good introduction to the field season, for it seemed like we were in another world. But we had no telephones to answer or bills to pay, so no worries. Just basic survival! On November 10, the storm ended by late afternoon. We shoveled out and continued sorting out food and gear for our trips, sending a message to McMurdo to remind them that we could do nothing until the third flight arrived.

At this time, we also talked by radio with Rothera Station, the main UK base in Antarctica, which had been established in 1975 on the southern end of Adelaide Island after the skiway of their previous main Antarctic base, on the same island, became unusable. Adelaide Island was about fifty miles west of Stonington Island, the former base of Finn Ronne and his expedition. It was about 500 miles north of us and 250 miles south of Palmer Station, all bases being on the western side of the Antarctic Peninsula. Mike T. and I had agreed, well before the season started, that USARP would supply twenty barrels of JP4 for a Twin Otter, in the hope that the aircraft would take some of the geologists and topographers to isolated nunataks on the crest of the Antarctic Peninsula. The National Science Foundation (NSF) was aware of this, for I wrote it into the grant.

And of course, Rothera knew of it as well and was now following up with more concrete plans and dates. However, Rothera reported that they had had only six days of flying weather in four weeks so were way behind schedule.

On November 11, Mike B. and I mapped nearby nunataks, finding fossil tree trunks in the Latady Formation west of Mt. Smart. On another outcrop to the north that day, Mike T. found marine fossil ammonites, oysters, and clams. Like most paleontologists, Mike T. had the ability to spot fossils in rocks the rest of us walk over without seeing anything. By that day we were nearly out of gas for the snowmobiles, so future work was out. On November 12, fog settled in, but McMurdo asked for hourly weather, with a flight to lift off at 2:30 am our time on November 13. It never took off. The eleven of us had a good dinner, including four bottles of wine. I noticed the next day Paul smoking a joint, and I kiddingly admonished him on the risk he had taken getting his stash to the Orville Coast. He replied, "No problem, Pete, I mailed it to you!" He had intercepted the small box in the McMurdo mail!

I sent a radio message on November 14 to the USARP representative in McMurdo, Chris Shepherd, to tell him our situation once again. His excuse was that fuel was low at Pole Station and that Byrd Station was socked in. Fair enough, but on November 15 we ran out of Preway fuel oil, so we huddled around Coleman stoves. The following day, he told us that a plane would leave at 7 pm our time. We sent weather reports as usual, but the aircraft was delayed. By 4 am on November 17, our weather was perfect and we let McMurdo know. We also sent weather reports to Rothera, who wanted to send a Twin Otter from there to the Shackleton Mountains south of us. At 1:30 pm, we learned that, while on its way to us, the third plane had lost an engine and had had to return to the Pole. We were now down to eleven gallons of Coleman fuel. We speculated about whether this was all an evil experiment in disillusionment. At 8 pm, I asked Walt to contact the Pole for the hell of it, and he learned that the plane was being repaired, and would then return to McMurdo, and from there fly to Orville. So, they wanted weather reports starting at 9 pm. Our weather deteriorated (to clouds at 3,000 feet and a bad whiteout) but Walt sent reports from 9 pm to 2 am, when I took over. My first call, at 3 am (about 0°F) reported that the clouds had risen to 4,000 feet and that the pressure was climbing. I sat in the Jamesway, fully clothed, not wanting to waste Coleman fuel, and the weather continued to improve. When Robin took over the radio at 8 am, I went outside to shovel snow, just to get warm, and then I went back to sleep. Amazingly, a plane arrived at 1 pm on November 18, piloted by John Hutchins. We spent the afternoon unpacking, only to find that batteries to operate the electrotape (for measuring distances) and other topo gear had been left off the load, perhaps to lighten it. This meant the topographers were dead in the water! I sent Chris Shepherd an angry message, now requesting a fourth flight and, while they were at it, how about that JP4 that USARP and the British Antarctic Survey had agreed to?

The geology part of the field party, at least, could get started, and we spent the next day loading up our supplies. On November 20 at 11 am we headed south thirty-three miles to the Shelton Nunataks, where we did the geology and established Camp 1. My tentmate and teammate was Mike Boyles. The snow surface was good and we averaged a speed of 7 mph and gas consumption of 6 to 7 mpg with these new snowmobiles. The next day we moved another nine miles south, doing geology along the way. While walking to the first outcrop of the day, I fell into a small crevasse, going in over my head but landing on a small ledge. Mike threw me a rope and I climbed out. Clouds moved in as we continued and, as visibility was dwindling, at 8 pm we established Camp 2 in the Bean Peaks part of the southeastern Hauberg Mountains. We did the geology of the area on November 22, finding lots of marine invertebrate fossils. Then, in a fog on the 23rd, we moved ten miles to the east on a compass bearing to establish Camp 3. We called it "Camp Beautiful," for it afforded us spectacular views (when the fog lifted the next day) to the north and northeast of Mt. Dewe and Mt. Leek of the Haubergs.

Thanksgiving was November 24. With only one hundred yards' visibility, we had to wait until 2 pm before we headed out. Before I did, I got a message from Chris Shepherd giving lame excuses on why he had screwed up the batteries. He told us that other projects, including the resupply of Siple Station, had higher priority, and that a fourth flight might not be available until resupply. I fired off another angry message, asking that copies be sent to several higher-ups that I named, that the fourth flight "remains necessary to correct mistakes on your part" so that the engineers could begin their work. I reminded him that the fuel oil for the Preway heater was nearly out (a barrel lasts a week to ten days), and that I did not want those four men to go without heat or work until midseason. I also asked again for JP fuel for the U.K. Twin Otter. Having vented, we then turned to geology for the next several days.

We were soon near the southern end of the Orville Coast. About eighteen miles south of our tent camp was an isolated rock outcrop, Cape Zumberge, that marked the coastal edge of the subcontinent of West Antarctica. South and east of Cape Zumberge was permanent ice of the huge Ronne Ice Shelf. Cape Zumberge had been named after James H. Zumberge (1923–1992), a geologist and glaciologist who had been on several field parties to Antarctica, including as chief of a study of the Ross Ice Shelf during the International Geophysical Year (IGY) of 1957 to 1958. Larry Gould hired him to be Dean of the Earth Sciences School at the University of Arizona shortly thereafter, and together they turned the Geology Department there into a world class institution. One of James' hires was George H. Davis, my friend and University of Texas classmate who is mentioned at various places in this book because we stayed in touch and even worked together at times after grad school. George was not only a great hire but he likewise took up the baton and improved the department substantially, not only by his presence (he received many honorary doctorates, just like Larry) but by his hires and actions as professor, chairman, and provost. Zumberge, whom I never met but about whom I heard great things, went on to be president of several universities, including the University of Southern California (USC) from 1980 to 1991. When at USC, he wrote the foreword to Larry Gould's 1984 edition of *Cold.*

Getting to Cape Zumberge was a priority this season, but we knew that it would be a long and potentially dangerous trip, particularly if the snowmobile failed or the weather deteriorated. The weather had to be good, not least because we'd have to go on a compass bearing since Cape Zumberge does not form a promontory that can be seen from the north. The rock, if there were any (our aerial photos suggested that there was some, but some of it could have been a shadow), is coastline that faces south, so any rock had to be below the edge of a high south-facing ice cliff. I asked Walt and Mike Boyles to join me if conditions were right. We had only a few days in that tent camp, so we could not wait for conditions to be perfect. As I noted in a letter on November 27 to Karen and the kids, we woke up on the 26th to find the weather was clear, calm, and warm. But as we watched, fog moved in and soon engulfed the camp! Discouraged, we went back into a tent and ate breakfast. When we returned outside, we saw that the fog had miraculously burned off. We saddled up as fast as we could and were off to the south, going as fast as we could. After two hours on a compass bearing, not having seen anything but snow to the south, we climbed a slight snow hill and I stopped. The odometer indicated that we had gone only halfway but we were skeptical because we had been zipping along on a good snow surface. We sighted a white line (crevasse) far to the south and thought that might be our cliff. We took a five-minute break to eat candy bars, then proceeded. As we continued a few hundred yards, we crested the hill and its south slope became suddenly steeper and steeper. Then we saw huge gaping crevasses below us and about a hundred yards to the south. We braked to a halt and looked at what we had nearly gone into. Now we knew the odometer was wrong and that in fact we were at a crevassed bench above the coastal ice cliff. But the hill was broad and we saw no rock. We estimated that, if anything, our compass had taken us west of Zumberge. So, we turned east and, after a couple of miles we crossed another snow hill. And there below and in front of us was the rock and the entire crevassed ice cliff, probably four to five hundred feet high. We took pictures and motored over to it. Sure enough, the rock was about a hundred feet below the lip of the cliff. After dodging a crevasse field, we got above the rock at 2 pm and climbed down to the outcrop, which was about half a mile long.

The rock was a black slate loaded with ammonite fossils (the preserved remains of *Ammonoids*, a marine cephalopod with a shell coiled in a plane that became extinct 66 million years ago, with the dinosaurs)! We oohed and aahed and had a great time! We studied the rock and collected lots of fossils for Mike T., hauling them to the snowmobile in our backpacks. Although I did not make a note of it, I am sure I collected a sample to send to Jim Zumberge. Visiting this outcrop constituted an important geological discovery, for here the Latady Formation was quite different from how it appeared in the Haubergs. These fine-grained black rocks probably represented a deep-water, open-ocean environment, whereas the Latady rocks from which we had come represented a Jurassic sandy, near-shore, and on-shore coastal environment. At about 4 pm the wind started up and clouds were coming in, and we figured we should get back to camp. As we traveled back, the wind grew stronger, bringing a low ground blizzard. We drove as fast as possible, that is, about 10 mph. We pulled into camp at 7 pm, chilled but happy about the discovery. Very early the next day, a storm blasted the camp with 50 to 60 mph winds, and visibility so poor we could not see adjacent tents. We tightened the lines to the tents and piled more snow blocks on the outer flaps, had breakfast, and spent the day reading books (I was well into *The Hobbit*, appropriately enough). Paul prepared a shrimp meat-bar stew that was excellent.

The storm broke early in the morning on November 28, and it was sunny with a 10 mph wind the rest of the day. We dug out the sleds and snowmobiles. Ski-Doos have cowlings over the engines, unlike FoxTracs, and the snow had become packed under the cowlings, so it was tough to get the snowmobiles started until Mike B. cleaned the points. By then it was 4 pm, too late to move camp. When we made radio contact, we heard that a plane was scheduled, although George at the Jamesway had not responded. That was good news, plus Mike T. had his birthday today (35 years), so we celebrated! The next day at 7 am, we began sending hourly weather reports. George started communication at 10 am; the weather (partly cloudy, wind 10 mph) held, and the plane arrived at noon with 17,000 lbs of JP fuel (twenty barrels), a snowmobile, heating oil, mail, freshies, and at long last the batteries and other topo gear. Much relieved, we broke camp at 2 pm and moved about twenty miles to the northwest to establish, at 8 pm, Camp 4 in the northwestern Hauberg Mountains. We talked with George on the radio, and he was finally ready to set out.

November 30 was sunny with patches of fog. Because of my confidence in Mike B., we assigned ourselves the long snowmobile trips, and on this cold day we put in more than fifty miles on our snowmobile to visit scattered nunataks west of us, returning to camp at 8:30 pm. We had only five miles to go to the southeast to establish Camp 5, on the western side of Mt. Leek, and we attempted that on December 1. But we were stopped by fog moving west up the glaciers. Walt and I bumbled around in a hole between mountains in the fog, until we retreated out of the fog, just short of where I planned to go, and set up camp, then did the geology where it was clear. On December 2, Mike B. and I did another long trip, about forty miles, to a bunch of isolated nunataks to the north-northeast. That trip included reoccupying a station where topographic engineers from the 1969-1970 USGS season had reported fossils. We loaded up the Nansen sled with samples of invertebrates, and were back at camp to show them to Mike T. by 8 pm. December 3 "dawned" warm, calm, and sunny, but by 10 am we were partly socked in with low clouds. We were able to see enough to cross an unnamed tributary glacier and move about five miles to the southwest, near Mt. Dewe. Within two minutes of arriving, we were totally inundated by low clouds, with a temperature of 5°. But though we couldn't go out exploring, at least Mike T. had more fossils than he knew what to do with because all rocks in the Haubergs were sedimentary rocks of the Latady Formation!

Mike B. and I mapped a ridge near camp on December 4, where we also measured a thick, well-exposed section of Latady beds. At about 3 pm we came to several steep ice slopes that we would have to navigate with crampons. By then we each had at least 50 pounds of fossils in our packs and we were in a heavy whiteout with zero surface definition. We decided to be prudent—and had to return the way we came, with the heavy packs. Fortunately, the day's fossil haul was our best for the season so far, and among it was the oyster *Ostrea.*

It was calm, clear, and warm on December 5, and Tom and I went out to a ridge north of Mt. Dewe, whose peak is the crest of an anticline. While it was clear, we did a recon and set flags across crevasse fields on a steep, unnamed tributary glacier in preparation for the next day, when all of us would visit the Dewe area.

December 6 was a warm day and my 35th birthday. We successfully crossed the crevasses and mapped the rocks at and near Mt. Dewe. Karl spotted another Snowy Petrel, our good luck sign.

On the 7th, our objective was a trip of about fifty miles northwest to the Jamesway, and we needed to do the geology of a series of isolated nunataks along the way, so we were up at 6 am. We passed through the site of Camp 5, where we refueled with gas we had cached, then on to the Witte Nunataks. Unfortunately, the runners on one of the Nansen sleds collapsed, requiring two hours for one of the teams to fix it while the others did geology. Our snowmobile had carburetor problems (running, then dying, later diagnosed as a clogged fuel filter) but we pressed on as the wind picked up, with a ground blizzard, and we were lucky (it must have been the Snowy Petrel!) to be able to follow outbound tracks made by the topographers, and we arrived at the Jamesway at 6 pm. We warmed up next to the Preway stove, read our mail, and enjoyed something to eat besides meat-bar stew before fixing our snowmobile and the Nansen sledge, compiling the geology on maps, and preparing for the next traverse.

December 10 saw us heading northeast about twenty miles to Camp 7, just north of Mt. Jenkins, in the central Sweeney Mountains. This time Walt led the group in a cold, mild whiteout; he had gone part of the way when we were working out of the Jamesway a few weeks previously, so he supposedly knew where the crevasses were. Sure enough, he found them! He led us all through a large field of haycocks and crevasses, then onto crevassed blue ice. The snowmobiles and sleds broke through about a dozen crevasses before we reached the camp site. We ribbed him unmercifully as we set up the tents, then did geology until 7 pm.

Walt and I spent the next day mapping volcanic rocks near the camp, in a cold whiteout. The following day Paul and I worked together west of camp and north of Mt. Edward, on a long snowmobile trip that we would rate as delightful because it was a warm, clear day. We visited many small nunataks and found fossil wood in the sedimentary rocks, a sign that here the Latady had been deposited in a nonmarine environment, probably a flood plain. At our first outcrop, however, we looked back near where we had passed and noticed a nice (meaning dangerous!) crevasse field that we had fortunately somehow missed. We returned to camp about 7 pm, had dinner, then all went off to measure and study a good section of volcanic rocks that Walt had found two days before. We had to fight our way through a crevasse field, and when we returned at about midnight the day was not warm anymore.

On December 13 we moved to Camp 8, about twelve miles to the east (as the petrel flies) and just north of Potter Peaks. As we left Camp 7, a light wind started and soon it was at 20 mph, with a ground blizzard. We made it to the camp site as fast as we could across a small glacier and set up the tents. By then the wind was 30 mph, and we spent the rest of the day in our tents. The storm broke about 8 am the next day and, after digging out, were able to do geology until 7:30 pm. Before we left for field work, we got a message from Chris Shepherd saying that there would be no resupply flight this season, which did not exactly make my day because mogas (snowmobile gas) was low already, meaning an early end to the season.

On December 15 the visibility and whiteout came and went all day and it was cold, but we studied the nearby rocks. In the evening I sent another nasty message to Chris saying that "a decision to cancel our resupply flight would ensure that our season would be incomplete," that "it was never our intention to complete our work with the mogas now in camp," and that "the remaining unexplored parts of [the area] would have to be completed in a future field season under a new NSF grant at a cost of at least five flights, other major logistics, and much waste of manpower." Most of this Chris knew already, because I had already sent him our needs, including weights and cubes. I again requested that copies of my notes be sent to people at NSF above him and to Paul Williams above me.

We had one more day of geology to do out of this camp but it required a good weather day because the places that needed work were farther away. Instead, we had low visibility and severe whiteout, so we kept to our tents. The next day (December 17) was worse, so we stayed put. The following day winds came on, to about 50 mph, and we remained in camp. We read book after book, played cribbage and hearts, and talked. The storm raged, even intensified at "night," on the 18th. The tents flapped furiously in the wind, and drifts twenty feet long formed downwind from us. The snow pelted the side of the tent, like sand in a windstorm. The Coleman stove provided heat and we all read books or wrote letters. Then winds dropped to about 15 mph by 2 pm of the 19th. We made radio contact with the topographers and with Rothera Base, and it was clear that all we could do was hold tight. The storm picked up again on the 20th, with winds ranging between 30 and 50 mph. On December 21st, the winds were mostly gone, but we had about six inches of snow and low visibility. However, we received a message from McMurdo saying that our resupply flight was now scheduled for December 27 and that light snow was forecast for the next twelve to eighteen hours, then gradual clearing.

Two miles of visibility, light snow, clouds, and whiteout marked December 22, allowing us to do the geology of the areas we had been waiting to do, some requiring moderate snowmobile trips. The wind came back, picking up to about twenty mph, making for a cold day, but when we all got back to camp, at about 8 pm, the wind had stopped and it was almost warm. The next day we moved southeast another eleven miles to Camp 9, just north of Haggerty Peak, southeastern Sweeney Mountains. It was overcast with a light wind and we had to relay loads through significant snow, even up gentle rises. We had the tents set up by 3 pm, but because whiteout and low visibility returned as we set out to do geology, we had to travel on a compass bearing, and we had to return to camp by following our tracks.

December 24 was calm, clear, and warm, a beautiful time to do the geology of the rest of the Haggerty Peaks area, and we finished at 11:30 am. We packed camp and headed to the Jamesway, a straight shot thirty-five miles to the west. The trip to the Jamesway was made difficult by the deep snow, and we soon had clutch problems with one of the snowmobiles, requiring a one-hour delay to fix it. The other snowmobile also had track problems and could not get above a speed of 4 or 5 mph. At about 4 pm, clouds moved in and we started to lose visibility, but we kept at it. We arrived at the Jamesway at 6 pm, Christmas Eve. We found that wind-driven snow from one of the storms had both blocked the front door of the Jamesway and blown snow into our sleeping quarters, with a drift about five feet deep. This meant we had to dig ourselves into the Jamesway first, then dig and carry the drifted snow out of our bedroom before we started the Preway. That done, we celebrated with lobster tails and bottles of wine, getting to bed about 1 am!

At 6 am on Christmas Day, four cold topographers woke us when they stomped in, after a heroic all-night trip from the Wilkins Mountains, east of where we had come from. They cooked a meal and went to bed. We geologists got up on Christmas day at 10 am and started opening boxes that contained Christmas decorations. One from my parents contained a little Christmas tree, a miniature crèche, and a tiny train for under the tree. Decorations included a mobile with Santa Clauses and "Merry Christmas" figurines of Santa, reindeer, and other animals for thumbtacking to the walls of the Jamesway, red crepe Christmas bells, and eleven Christmas stockings with candy canes, candy, and tooters. We strung toilet paper from the rafters. There were presents for all, which we waited to open until the topographers got up. On our previous return to the Jamesway we had already eaten a fruitcake that my folks had sent. The day consisted of cleaning out the snow that had blown through everything, fixing snowmobiles and Nansen sleds, packing rocks and other items to be taken out by the resupply flight, then taking official photos of everybody. Christmas dinner consisted of steaks, scallops, and spaghetti, with bottles of wine! By the following day we had recovered sufficiently that we told stories, laughed together, and sung along with a tape recording I had of Willy Nelson songs. I compiled the geologic map from the aerial photos and our field notes, and we all packed up for the next traverse.

With no word on the resupply flight that had been scheduled for December 27, we decided to head out on our next traverse that day and let the topographers, who would remain in the Jamesway for a few more days, deal with the flight if it came. It was clear, almost calm, but cold, and we got going at 11 am, to Camp 10 in the northern Peterson Hills east of the Hauberg Mountains. We covered the sixty miles in seven and a half hours despite having to slog through fluffy snow much of the way.

The next day was clear with a 10-mph wind. Mike B. and I started mapping the geology to the east. By the second outcrop, the wind had picked up and we could see clouds moving toward us from the west. The third outcrop was far to the east, and by the time we arrived the wind was at about twenty mph and the clouds had moved in. We spotted another outcrop, about five miles to the south, and we headed for it and mapped it quickly. By now the wind was at twenty-five mph, with a whiteout and with visibility starting to decline, so we returned to outcrop three, then followed tracks west toward camp, directly into the wind. By about 3:30 pm, we realized that we were instead following (with difficulty) a track made by topographers when they headed to their former camp, which was about two miles south of our camp. By now visibility was a quarter mile, with ground blizzard, although occasionally we could see things a few miles away. We came upon the site of the topographers' camp and, after ninety minutes of looking, we found a trail we had spotted in the previous day's good weather that headed north to just east of our camp. Mike sat on the snowmobile seat to drive it, and I sat on the Nansen to look intently at the snow surface in the whiteout, as we slowly followed the trail at a speed of about one mph. We had walkie talkies and tried contacting Mike T. and Tom. Because they had a shorter traverse that day, we hoped they were back in camp, but we got no answer at four, five, and six pm. Finally, they picked up when we called at seven, when our visibility was fifty feet or less and wind was about forty mph, and we asked them to extend a line of flags north out from the tents. As we followed the tracks northward, we soon climbed up onto a higher area, a low snow plateau, where we had set the tents. Once on that plateau, we took a west bearing into the wind and headed toward where we thought camp might be. In these conditions, you have no sense of direction, and without tracks you can only rely on a compass bearing. I walked ahead of the snowmobile while Mike drove. As luck would have it, we spotted the most northerly flag that Mike T. and Tom had set for us, a quarter mile or so north of the tents. Once we reached it, at about 7:30 pm, we headed south along the line of flags and soon saw the tents. When we stumbled in the door to their tent, our beards and faces were sheets of ice. Walt, Karl, and Paul had not gotten back, and we tried to reach them by walkie talkies, with no luck. We were fine after hoosh and a few drinks of scotch, but all of us were worried about the missing team. By the time we hit the beds the others still hadn't returned.

When we got up on December 29, the wind had dropped to about ten mph and there was heavy snowfall, with visibility at about one hundred yards. I started the generator at 8:30 am to send weather reports to McMurdo and to talk to George and discovered that the resupply had been canceled due to aircraft problems, with another attempt scheduled in the evening. Rothera came on and said that, after January 1, they had a vital topographic survey they had to do. Therefore, flying with us, based out of the Jamesway, would have to be done within the next two days or never. Of course, their weather was not any better than ours.

As I was having this conversation with Rothera, who should poke his head into our tent but Karl! Unwittingly, they had camped only about 150 yards east of our tents, using their emergency Mead tent, and our generator had woken them up! To get there, like us they had followed the topographers' trail back to their former camp, getting there at 6 pm, just after we had been there. From there, they set out on a N. 20° W. bearing, and had eventually camped at about 7 pm. We had only just missed each other!

December 30 was calm but cloudy, with a visibility of a quarter mile. We dug out and got the snowmobiles working, then we all did the geology of the nunataks adjacent to camp. When we got back, we learned that the resupply flight had made it to us that evening. The next day, we had no visibility and stayed in camp. It was New Year's eve, and though I have no record of how we celebrated, I know we did! It was calm, warm, and mostly clear except for some low clouds on New Year's Day, and we moved eighteen miles northeast to Camp 11 in the southern Wilkins Mountains. Clouds on the ground rolled in as we pitched our tents, but occasionally we could see some nearby rock, so from four to eight pm we did geology, then followed our tracks back in no-visibility clouds.

January 2 brought us a warm, calm day, with high cirrus clouds, but we could see low clouds moving in. Mike B. and I went south, hurrying to reach nunataks about twelve miles to the south before the clouds surrounded us. Along the way, we crossed two small crevasse fields, then we had to navigate an unnamed, three-mile-wide, east-flowing, crevasse-ridden glacier that is a tributary to the Matthews Glacier. On this tributary glacier, we first passed over about a dozen crevasses three to twenty-five feet in width, where we saw bridges sag about four or five feet in the middle. But all the bridges held. Then the steering rope to the front ski broke, and we had to stop and fix it. That was followed by me falling on a bridge of a large crevasse when my ski tip caught on a sastrugi. We were able to stop the snowmobile by the kill cord. After that, we dodged other crevasses, then crossed a field of small crevasses on the far side, breaking through several bridges. The nunataks just beyond that were worth it all, for they were made up of black slate with some fossils, much like the rocks we had seen at Cape Zumberge. We returned to camp by an entirely different route, thankfully one free of crevasses.

The camp was totally socked in on January 3, but most of us had tracks to outcrops that needed more work. Mike and I returned to the black slate outcrops of late yesterday and measured the thickness of rocks there, then climbed five hundred feet vertically to the top of a ridge, where we found ourselves above the fog. Tom and Mike T. blundered north through the fog, came upon a nunatak by accident, and were able to study that.

The early morning of January 4 revealed a gentle east wind that was bringing fog up the glacier toward us. I wanted to move to Camp 12, about fifteen miles to the northeast in the northern Wilkins Mountain, so I proposed that we get an early start to try to get there before the fog rolled in, and that the others finish up the geology near camp, then follow our tracks northeast. Paul was ready to go, so I skipped breakfast and we hooked up the heaviest Nansen, with the fuel and rocks, and were off. We did some geology along the way and got to the site of Camp 12, where we dropped our load and headed off to do the geology. I scrambled up a ridge, while Paul returned to Camp 11 to help the others. I walked the ridge, and found lots of fossils. At about four pm, Paul returned after bringing a load to the new camp. I dropped him off on another ridge while I took the Ski-Doo to the north to visit some small nunataks, then returned to get him, then back by about 7 pm to set up the tents. The next day we had winds of about 30 mph and we stayed in camp.

We had a good day of doing geology on January 6, but January 7 was partly cloudy, cold (5° in the morning, 6° in the late afternoon), and with a 15 -mph wind, and although we wanted to move camp about twenty miles to the east, across the six-mile-wide Matthews Glacier, we dared not. Instead, we worked outcrops and nunataks we had missed in yesterday's clouds. At about 3 pm, the wind had died down and it was partly cloudy, but I could see clouds moving in from the north, so I went alone to scout a route across the Matthews Glacier. I encountered no crevasses and flagged the trail across, then returned at 6 pm in whiteout conditions, nearly hypothermic. I had lost feeling in most of my fingers, and it was painful to thaw them out during dinner. Bad weather the next day meant we stayed in camp. The weather was only a little better on the 9th but we decided to head to Camp 13 anyway, following the flags across the glacier and camping west of Mt. McKibben and south of McCaw Ridge. We set up camp quickly, then did some geology, but mostly we set trails to the largest ridges and nunataks, including Mt. McKibben, in case bad visibility returned. The sastrugi were large and the visibility had gone to a quarter mile when we returned, quite cold. Fall was approaching!

January 10 was a beautiful day, with a wind that died down as the day went on. It felt good to have a day like this. Mike B. and I intended to head east about twelve miles to visit the farthest east rock known, got most of the way to a ridge and took lots of pictures of the crevassed Hansen Inlet to the east, before the Ski-Doo started chugging slowly and strangely. We stopped and tipped it on its side and saw that we had a broken right driveshaft, something that would doom the snowmobile. We did the geology of the ridge, pointed the snowmobile back to the west and headed back, hoping that the left driveshaft would get us there. Camp 13 proved lucky, even though we had trouble climbing some sastrugi, and we made it back. During our radio contact that night with Pole Station, we learned that a BAS Twin Otter had arrived at the Jamesway two days before and had picked up Robin, then had gone on to Rothera for some topo work.

On January 11, it was cold and overcast. Mike B. stayed in camp to disassemble the snowmobile for parts, while Paul and I climbed thousand-foot McKibben and did the geology there. The next day, it was mostly cloudy when we broke camp and moved back west, now with two snowmobiles. We followed our tracks west for about eight miles, then headed south about five miles and established Camp 14 at noon. As the weather began to deteriorate, we all headed off to spend the afternoon doing geology, bumbling from nunatak to nunatak in low visibility and heavy whiteout.

January 13 was cold and overcast, with a slight wind. We were keen to get back across the Matthews Glacier to Camp 12, where we had cached fuel, rocks, and a Nansen, so we got up at 6 am and were on the road by 8:30. We intersected our trail from Camp 12 to 13, headed west in poor visibility, and made it the fifteen miles to the cache at about 10 am. We waxed the runners of the Nansens, for they had to carry a large load with only two Ski-Doos, then we loaded up and headed toward the Jamesway at a bearing of N. 75° W., with the wind picking up to 10 mph. At about 3 pm, with the wind at 25 mph, we quit and set up the tents, calling it Camp 15. Friday the 13 strikes back! We speculated that snowmobiles do not like to go back to the Jamesway. The wind increased during the "night." By 9 am, the wind had died to 10 mph and we packed up. We could now see Haggerty Peaks about ten miles due west, meaning that yesterday's compass bearing had gotten us off course, far to the north. The overnight wind had created large hard sastrugi—a bad thing—but it also had blown away the covering of soft snow and soft sastrugi—a good thing. In about an hour, as we passed south of Haggerty Peaks, the clouds thickened and the wind picked up. But the wind was at our backs. In three more hours, we reached the Jamesway, and there was a red Twin Otter parked alongside it! Gary Studd was the pilot, Les Starger was a geophysicist, and Gerry was the engineer. They had been at the Jamesway for three days, for the weather was bad everywhere else. So now we had a mob of fifteen, including Dan Hall, a USGS topographer who had been at Rothera! Fortunately, they brought some whiskey to share with us.

January 15 was Super Bowl Sunday, with the Denver Broncos playing the Dallas Cowboys. I worked on compiling our geologic map, and getting photos ready for our next traverse while the Broncos lost. The next day was overcast with the wind at twenty mph. Gary said that he had a window of time to fly a couple of us to do some geology, but we needed a good day for that. Instead, the storm picked up. So we prepared a new Nansen brought in on resupply, dug out aluminum pallets for transport back to McMurdo, and prepared for the next traverse. Paul Carrara erected a Scott tent so that he could take a bath, then heated water with a Coleman stove. He left the tent alone for five minutes to get his clean clothes, and the tent caught fire and burned up! He caught a lot of ribbing for that!

January 17 found us with a cloudy day and visibility that ranged from a quarter of a mile to fifteen miles, cold, and with a light wind. We gave up on the idea of doing any geology by aircraft, for at the first good weather BAS had many other things for the Twin Otter to do. Therefore, we packed up and pulled out at 3 pm, heading west, then north past Mt.

Ballard on a compass bearing and on to a bunch of isolated nunataks in the northwestern Sweeney Mountains, about twenty miles. The sketch map showed lots of nunataks, but we found that few of them existed. We reoccupied one of them that was, for it had been visited by George, who had noted copper stains. The camp had good views of the Sky-Hi Nunataks, about twenty miles to the west.

We did the geology of the scattered nunataks on the morning of the 18th, then packed up the camp and headed out toward Tollefson and Olander Nunataks, about fifty-five miles to the northwest and just past the crest of the southern Antarctic Peninsula. The weather was beautiful, clear, and calm. Before we left, we saw the Twin Otter about ten miles to the west, heading north to do topographic work. We could not see the nunataks we were aiming for, but we could see the Sky-Hi Nunataks along the way and a little to the west. After we passed these nunataks, about 2:30 pm, we started on a compass bearing of N. 40° W., and set up flags as we went. At about 5 pm, still with no Tollefson Nunatak in view, we climbed up a steep snow slope containing some crevasses off to the east. When the topography flattened at about 5:30, presumably on a plateau marking the crest of the peninsula, we saw Olander Nunatak dead ahead. We dropped off the presumed Continental Divide and encountered deep snow, went a little past Tollefson and set up Camp 17 at 7 pm. As we arrived at the camp site, we could see the Twin Otter taking off from an ice dome about 7 miles to the west. Paul's altimeter showed that we were 700 feet below (north of and on the Bellingshausen Sea side) the Continental Divide.

January 19 was another good day, and the barometric pressure was rising so we had hope for more good weather. Mike B. and I headed off to the northwest to pick up several isolated nunataks, namely Horner Nunatak that was shown on the sketch map to be twelve miles away, and Staack Nunatak shown as about twenty miles away. They were not as far as shown, and as we got to Horner we could see where the plane had landed and some footprints to the nunatak, presumably to establish a geoceiver station. Horner consisted of a precipitous cliff of Jurassic andesitic lava flows surrounded by a tricky steep ice slope (wind scoop) on the southern side and a snow cornice on top. Staack Nunatak was only three miles away and consisted of the same flows. The others studied Tollefson and Olander, plus some isolated smaller nunataks extending to the northeast. All were volcanic rocks that formed the axis of the Jurassic peninsula.

The good weather continued for a third day, so we followed our tracks back to the southeast, pulling our flags as we went, then swung south and went along the western side of the Sky-Hi Nunataks. There, as I steered the lead toboggan remotely, we got into a crevasse field off the northwestern end of the Nunataks, which prompted a few near misses. We made it to the southern end of the southern nunatak of the Sky-Hi Nunataks, where we established Camp 18. Snowcats of the Antarctic Peninsula Traverse (led by John Behrendt) of 1961–1962 had visited the northwestern end of the Nunataks as well as the same crevasse field just west of the bedrock there. Their track, with several 90-degree bends to get around crevasses, appears on the sketch map. We had tents up by 3 pm and were off to do the geology in a 5-to-10-mph wind. Mike and I went back to the west and northwest where we had run into the crevasses and, in addition to doing the geology, found a better path around the crevasses. According to Paul's altimeter, this camp is 1,600 feet below (south, on the Weddell Sea side) the Continental Divide.

The next day it was clear with a 10-mph wind, and I went with Karl and Walt to look at a possible copper deposit they had found in the northeastern part of the Sky-Hi Nunataks. I concluded that it resembled the porphyry-copper deposit we had mapped during our first field season, but this one was poorly exposed and consequently there was not much we could do to study it more.

Clouds built up and the winds picked up to 40 mph or so during the "night." The storm continued on January 22, with visibility between fifty feet and a quarter mile due to blowing snow, and so we stayed in our tents. It was Bing Crosby weather (warm and calm, but cloudy) on the morning of January 23, so we packed up the camp after waiting a while hoping for the weather to become clear. (It did not.) We headed southwest on a compass bearing toward the Merrick Mountains. We reached a cluster of three nunataks on the northern side of the range at three pm, dropped our loads, and did the geology for a couple of hours. The visibility then picked up, just enough for us to find our way, and we continued to just south of the Merrick Mountains to establish Camp 19. We had come about twenty-five miles, and now the visibility dropped to nothing. Tom and three colleagues had mapped part of the Merricks during November and December of 1965, so it was familiar territory to him.

On January 24, it was warm and calm, with a whiteout and low clouds (about five hundred feet above the surface). To do the geology, we moved north into the range but that required us to go higher, into the clouds, resulting in lower visibility. Mike, Paul, and I worked our way along the western side of the western ridge of the Merrick Mountains, capped by Mt. Matheson. They let me off on the ridge south of the mountain, and they continued north to explore the northern parts of the ridge. When one is in fog and it is calm, one can hear everything. A few minutes after they left, I heard Paul swearing, mumbling, talking animatedly. I learned later that they had got onto a steep, blue-ice slope in the mank, but had subsequently extricated themselves and continued north. I had a delightful day doing the geology of the ridge, including a first ascent of Mt. Matheson, a beautiful peak. The north side was a little tricky and I had a few self-arrests on the snow as I worked my way down. After that, the ridge consisted of narrow and steep rock outcrops with long snow drop-offs on either side. Mike and Paul later picked me up at a saddle north of Matheson, as we had agreed. We headed back south, with the weather now mostly cloudy but with decent visibility, allowing us to spot a nunatak just southwest of the range. Mike T. and Tom, also finishing up their day, joined us there.

January 25 was another similar day weatherwise, and we concentrated on the northeastern part of the range, including Mt. Berger. Paul and Mike dropped me off to walk the ridge that included Mt. Berger, and I had a pleasant day climbing it and mapping the geology. It was not as tough a mountain as Matheson, and clouds came and went. At one time, at or near the summit, I looked north to see Paul and Mike's Ski-Doo up against one of the nunataks in a complete whiteout. With no horizon, their rock outcrop and snowmobile seemed to be floating in white. As I worked my way down the snow slope on the northern side of Mt. Berger, my journal records that I "fell a few times but did not slide far" as I caught myself with my rock hammer. In a whiteout it is particularly tricky to work down a steep snow slope. I could not tell whether the snow slope below me was a hundred or a thousand feet to the bottom. I noted that I took "all sharp objects out of my pockets, put in pack, start down; as it is, it is closer to 100 feet." Mike and Paul came by soon and picked me up, and off we went to the northwest to visit several small nunataks. By then the visibility was about a hundred feet and, after we had visited the rocks, we headed off on a compass bearing until we intersected our track from the morning, then returned to camp.

January 26 was calm, clear, and cold, and we broke camp and headed southwest, traveling along the northwestern side of the Behrendt Mountains for about twenty miles, to the southern side of Mt. Trimpi and a place called Happy Valley, where we established Camp 20. It was to be our last camp of the season. Tom and his team had already mapped much of the Behrendt Mountains in 1965, but we thought we would collect fossils for Mike T. and study additional rock outcrops. Tom and I spent the afternoon using our rock hammers to break open concretions on one hill, for concretions commonly contain fossils. The two of us found almost nothing for all that whacking, but others in our group hit paydirt!

It was cool at "night" now, about -20°. We learned on the radio that the Twin Otter had finally left the Jamesway and our area, heading for the Russian Druzhnaya Station.

On January 27, it was cold but calm and clear. Mike B. and I tackled Mt. Neuner, north of Trimpi, from the north, but to get to rock we had to cross a bergschrund (a crevasse formed when snow and ice pull away, downslope, from rock). We roped up together and used crampons, then climbed to the mountain top once we got to rock. It had an overhanging cornice near the top, out onto which Mike walked before we realized what it was; fortunately, it did not collapse and send him down the slope! The sedimentary rocks on top were folded and overturned and we measured the thickness of the rock section, even though it was late and cold; we also found some of the best plant fossils we had seen all season. This is a common affliction for a field geologist—finding the best rocks late in the day when we are eager to head home!

A drop in barometric pressure suggested deteriorating weather on January 28, even though at breakfast time it was clear, calm, and almost warm. Considering our distance (about eighty miles) from the Jamesway and USARP's desire to come and get us, we packed the camp and started home at 11 am, heading south a dozen miles out of Happy Valley and past the last rock of the Behrendt Mountains, giving a wide berth to a large crevasse field south of Mt. Hirman. Then we headed southeast for about six miles toward the Quilty Nunataks before swinging east past the southern end of the Thomas Mountains (named for the newscaster Lowell Thomas) and heading northeast to the Jamesway, where George told us that the Navy was planning on two flights to pick us up.

From the Jamesway, Karl, Tom, and I did a trip to Mt. Ballard to the north to measure the sedimentary rock section there. Tom and I came at the mountain from the north, while Karl came from the south, and we met on the top. Some of the ridges were tricky, and a couple of times I started to slide down a steep snow slope toward rocks but was able to arrest myself with my ice axe.

Back at the Jamesway, the topographers had already packed up their gear and now we geologists started to pack ours, placing it on a Nansen, to be pulled by snowmobile to a waiting plane. We let McMurdo know that the weather was good and they could schedule pick-ups at any time. That "night," we started sending weather reports every two hours, but McMurdo did not seem to care. The weather held on the 30th and I compiled my geologic map, while George sent weather reports every four hours, including at 11 pm. At about 2 am on the 31st, with most of us asleep, George heard something and said "What's that?" We all woke up and heard an engine, then the plane buzzed the Jamesway. We slapped on our clothes and tuned the radio to the aircraft frequency to learn that they wanted to take only cargo. We started two Ski-Doos and they were moving with some of the cargo by the time the plane had landed and was heading back toward us. In less than an hour we had loaded onto the aircraft about 6,500 pounds of rock boxes, topo and geologic gear, and some of the camping gear. Jim Jaeger, the pilot, and Dave Srite, the co-pilot, told us that they had decided on the spur of the moment to make a run to Orville Coast from Bryan Coast to the northwest, where they had just pulled out a field party that was hunting meteorites. They made a JATO takeoff and were gone.

We went back to bed, slept late, and after breakfast went back to sorting and arranging the rest of the gear to take back, about 10,000 pounds. We let McMurdo know, and they told us that they would leave McMurdo at 10 pm our time and arrive Orville at 4 am on February 1. The weather remained good. After we had packed up everything, I went to Mt. Ballard alone to do some geology, returned at 7:30 pm, and had dinner with the others. Radio communication was impossible that evening and into the early morning of February 1, but nobody slept; instead, we played hearts and, sure enough, a Herc piloted by Commander Johnson arrived at 4 am. It took 6,500 lbs of Ski-Doos and 1,500 lbs of unused JATO bottles, but no passengers. A second plane, piloted by Jim Jaeger, arrived at 6 am. It took the rest, consisting of one Ski-Doo, three Nansens, personal gear, and eleven dirty men, and it was airborne before 7 am. We passed over the Pensacola Mountains and landed at Pole Station at 10:30 pm, our time, to refuel. Dave Reid stayed at the Pole and we were airborne once again soon after.

At McMurdo, we returned camping gear and Ski-Doos to warehouses, packed rocks, took a bath, took photos, and at the bar had a patched phone call to Karen to let her know we were out of the field. Over the next few days, in ones and twos, colleagues headed off to Christchurch. I visited and thanked various officials, returned first aid kits to the hospital, wrote letters, visited wildlife, and relaxed. I learned that about twenty staff members of Holmes and Narver, the subcontractor to N.S.F., had quit or been fired, partly due to their incompetence with regards to our field party. At 11 pm on February 8, I flew by Herc to Christchurch, arriving 7 am local time on the 9th. I traded cold-weather gear for civvies and checked into Warners Hotel, where to my surprise most of the Herc crew was too! I spent the next day visiting New Zealand geologists and some old USGS hands who were there, including Walt Seelig of Topographic Division, who had created a lot of the tricamera aerial photographs that we used and who had become a friend. The temperature in Christchurch was 95°! With those members of the Orville team still at Warners, we had a bunch of celebratory farewell drinks at the bar on the evening of the 10th. The rest of the trip was a blur. At 11:30 on February 11, I boarded a MAC (Military Air Command) flight on a C-141 that contained a Huey helicopter and fifty persons. After a stop at Pago Pago, we arrived at Hickam AFB at 4 am local time. From there we left almost immediately on another MAC flight to Point Mugu AFB in southern California, arriving at 2 pm. I took a bus to Los Angeles International Airport, had a quick flight that arrived at San Francisco at 8 pm, and from there flew to Denver, arriving just after midnight on the 12th.

Surrey
UTAH
NICOSIA
CYPRUS
Las Vegas
SYRIA
BEIRUT
DAMASCUS
Renton
Tacoma
OLYMPIA
Chehalis
Mt Rainier
Riffe Lake
Yakima River
Yakima
Mt St.Helens
Mt Adams
Lewis R.
RIYADH
Hofuf
QATAR
DOHA
Al-Kharj
Salwa
SAUDI
ARABIA
As Sulayyil

6

Between Field Seasons, 1978–1984

As with my trip from Christchurch to Denver, for a while my personal life after that was also a blur. As near as I can remember, we started the paperwork toward a divorce while putting our house on the market. I remained in the house until it was sold, whereas Karen and the kids moved into Joe Cassio's house in Wheat Ridge, a western suburb of Denver north of Lakewood. Other than a death in the family, a divorce is probably the worst thing that can happen to a person. A bitter divorce, which seems to go on forever, is particularly bad. At first, Karen seemed open to letting me spend time with the kids, generally every other weekend. But there was much wrangling over possessions and property, with letters to and from lawyers. The divorce took more than a year, becoming final in March of 1979. Scott was sixteen years old then, and Jill not yet eight. I bought a condominium in Lakewood. Though Karen got a job as an elementary school teacher, I still paid child support, for Karen had custody of the kids. Just before the final divorce, with the house sold and the furniture and belongings moved out, Karen and I met at the house to vacuum and clean the bare house before the new owners were to move in. She looked nice, for she had lost a little weight and had a pretty dress on. When we were nearly done, she asked, "Pete, would you like to get back together again?" I was unprepared for that, and totally dumbfounded. I tried to be nice as I said, "No, too much water has passed under the bridge for that."

After that incident, things became much uglier. Phone conversations were not necessarily pleasant. Visits with Jill were often sabotaged, so any plans I could make on Jill's weekend were frequently derailed. A seminal event came in the summer of 1980 when, after much finagling, negotiating, and planning, Jill and I were to take a week and go on a vacation together, to visit a national park. I picked her up on the appointed day and we started the trip, spending the first night in a motel on the way. In the morning, Jill asked me if she could call her mom, for her mom had asked that of her. By the time the call was over, Jill was hysterical and asking for her mommy. I could not calm her down. I had no idea what Karen had said to Jill, although it probably does not take a genius to figure it out. I agreed to take Jill home, and we returned that day. Things continued along those lines for the next few months. Jill tried her best to be a little diplomat. I refused to bad-mouth Karen or Joe in Jill's presence, but it was clear that this was not the case in the Cassio home. Brainwashing is an insidious thing, and Jill's demeanor indicated that it was taking effect. I concluded that she was complicit with the program, although that was not fair. Jill fared better at my parents' home on several acres near Newtown, Connecticut, and Jill spent a week or so with them every other summer. Within a year or two, the Cassios moved to the Orlando area of Florida, probably to separate me from my kids even more. Eventually the Cassios won the war and Jill and I grew apart, although we get in touch by phone on birthdays and holidays. We did not visit each other and I remember having no substantive conversation, other than about the weather or (eventually) her kids. Jill married a born-again (evangelical) high-school math teacher and wrestling coach, Andrew (Andy) Knutson. My wife, Dawna, and I, along with friends Paul Williams, Chet Wallace, and Mel Kuntz, attended their wedding in the Seattle area. It was not particularly pleasant, for Andy's evangelical family and church friends were in control of the shindig, over cries of "praise the lord" and "trophy wife," and they busied themselves pouring our bottles of wine and champagne down the sink unless we protected the bottles. Jill and Andy have three kids, Karsten, Callaway Jo, and Coen.

The one good thing that Joe Cassio did was that he got the family into water skiing in Florida. Jill, like Scott, is extremely well coordinated and athletic, and she ended up being one of the top ten women water skiers in the country and second in the world for about a decade. I paid for her college education, but part of it was supported by a water-ski scholarship. Water skiing became her profession, although she would have made more money clerking at WalMart. Scott was able to maintain greater objectivity and we remained close.

My condominium was conducive to having a lot of parties. I bought a jukebox and party toys, such as a foosball table and a pinball machine. It was like the line in the song, "today I'll buy no sorrows" in the song "Lookin' out my Back Door" by Credence Clearwater Revival. My best friends were the USGS geologists hired just after the energy crisis, most of them bachelors or divorcees (geologists have a hard time staying married). On Fridays after work, we would commonly stop at Lane's Tavern in Lakewood and have a few pitchers of beer, then go to somebody's house (commonly mine) and share a pizza, grill burgers on a back deck, or have a potluck meal. Rather than having the life of a carefree bachelor, however, I spent most weekends at my office compiling data or writing reports. I dated very little. Putting in sixty to eighty-hour work weeks was my main response to the divorce. In 1979, I was elected a Fellow of both The Explorers Club and the Geological Society of America. In 1986 the Board on Geographic Names named Rowley Massif in the Black Coast after me.

My best friend, from the time he had joined the branch in 1975, was ***Chester A. Wallace****, another of those types toward whom I tend to gravitate because they keep you laughing. Chet is about my age, had been an Assistant Professor at Georgia Southern University when we hired him, and is an expert in geochemistry, sedimentology, sedimentary stratigraphy, and structural geology, with a Ph.D. from California, Santa Barbara. He came from a farming family in western New York state, buckled down with his studies after an older brother died tragically and got a BA from Antioch College. He got married in Santa Barbara but, sadly, while I was in the Orville Coast, he and his wife, Britt, got a divorce. They have two kids, Tory and Chester, who were best friends with Jill and Scott. Fortunately, their divorce was not as brutal as mine, for Britt stayed friends with Chet after that. Chet is another mapper, and his main USGS project was mapping the Butte 2-degree sheet in Montana as well as working in the Belt Basin of Montana, where Late Precambrian sedimentary rocks are thousands of feet thick and contain strata-bound copper deposits. There he worked with Jack E. Harrison, another old hand and great Branch field geologist who had done a lot of work on Belt rocks and their mineral resources. Chet also mapped Wilderness areas in Montana. Chet's story will continue below.*

Another good friend who joined us at Lane's Tavern, also hired at the same time as Chet but to the Paleontology and Stratigraphy Branch in Denver, was ***Bruce R. Wardlaw****. As the name of his branch suggests, his specialties were paleontology, especially conodonts (tiny tooth-like fossils that are abundant and thus useful for identifying and correlating Paleozoic [an Era of time, about 540 to 250 million years old] sedimentary rocks), as well as the stratigraphy of the rocks themselves, and he had a Ph.D. from Case Western Reserve. Bruce had also gone through a divorce, although it was a few years previously, but his ex-wife was fighting Bruce over time he wanted to spend with their early-teen daughter. Bruce had just started dating a beautiful woman, Jeanne, who turned twenty-one while we were in our Lane's Tavern stage. Despite her youth, Jeanne was more mature than Bruce, and when they married within a year or so I felt she was the best thing that had ever happened to him. My name for her was "Twenty-one." Bruce was a tall handsome guy with a booming laugh and bushy mustache, another terrific field geologist who had a big ego to match (but the ego was justified). During the RIF, when the Branch of Paleontology and Stratigraphy was terminated, Bruce was given an administrative job, which required him and his wife to move from Denver to Reston. Tragically, he died from a fall at their home in 2017, while still employed at the USGS in Reston.*

Finally, another friend who joined us at Lane's was ***David W. Moore****, a Quaternary geologist who had joined the branch at the*

> *same time as Chet and Bruce. Dave was quiet and was fighting demons from his times in Viet Nam, where he had flown Huey helicopters in combat. His then wife was unfaithful, providing him another level of hell, and of course they divorced when he returned, after which he started grad school, earning a Ph.D. at the University of Illinois, then was hired by our branch and remarried.*

In March 1978, with the Vernal project terminated and most maps and reports written and in press, I proposed to my Branch Chief, Paul Williams, to continue my work in Utah, specifically by focusing mapping on the Marysvale volcanic field itself. My project was named the Richfield, Utah 2-degree project, and it was to be a collaborative effort with an identically named project, in the Central Mineral Resources Branch, led by Thomas A. Steven, that had just begun. Chuck Thorman was the chief of this branch at the time; later, I became much closer to Chuck, another great field geologist with degrees from the University of Washington, including courses taken from Hoover Mackin. As proposed to Paul, I would concentrate on the geologic framework (geologic mapping) of the area, whereas Tom would focus on the mineral resources in addition to mapping. Tom brought in a branch colleague, Charles G. (Skip) Cunningham, and together we three friends were the principals of the two projects, although dozens of others joined us in the effort at various times. The project area was large, for it included not only most of the Marysvale volcanic field but the rest of the Richfield map, which extended west almost to the Utah/Nevada border, made up of many other ranges and volcanic fields. Our investigation was probably the most innovative and productive part of my career, as explained in the next chapter. Officially, this project continued until September 1986. But unofficially, it continues to this day.

When Paul Williams stepped down as Branch Chief in 1980 (he returned to science, as a Branch geologist), Robert B. Raup, Jr., took over and served until 1985. Bob's career had been largely in administration and he came from a job in the USGS headquarters in Reston, where he was Office Chief of Regional Geology. He was one of the few good ones whose career was largely as a manager, and he kept the branch on the same steady, happy, and productive course as Paul and Ed had. For two years of Bob's term, Chet Wallace was his Assistant Branch Chief.

Somewhere in this time frame, computers came in. At first it started with word processing, presumably beginning in 1979. Somewhere after 1983, the USGS began equipping the secretarial pool, and a few years later also individual scientists, with word processing machines made by NBI, Inc. No longer did we take our manuscripts to secretaries for typing; now we did that ourselves. Branch typists were moved into other jobs. In 1981, IBM personal computers, using Microsoft MS-DOS floppy disks, began to be manufactured; Apple MacIntosh computers were available by 1983. The formal beginning of the Internet was in 1985. I think that by 1990 the USGS was using IBM-type computers.

In 1982, ***Paul Williams*** *volunteered for a geologist position with the USGS Mission in Saudi Arabia. Jean and Marjorie moved there to be with Paul, a stay that lasted ten years. Established in 1963, the Mission worked out of a large walled diplomatic compound in Jeddah in western Saudi Arabia. The point of the Mission was to diversify the Saudi economy away from oil and gas production. To do this, it funded the USGS to do 1:100,000- and 1:250,000-scale regional mapping, geophysics, and mineral exploration in and near the Precambrian Shield of western Saudi Arabia, as well as to train Saudi scientists. The compound included houses for employees, as well as chemistry labs, administrative offices, a motor pool, field services, and a group to process and publish reports. At this time the group consisted of twenty-five to thirty American employees and 150 to 200 local employees, mostly Saudis, but also many from African nations. After a year of quadrangle mapping, in 1983 Paul was called upon to fix a problem, and to do so he was appointed Mission Chief Geologist. The problem was 149 delinquent technical reports and maps that had been promised (contracted) by the USGS for the huge amount of financial support the Saudis sent to the USGS, but these reports had not been completed or submitted and the Saudis were finally complaining to Reston. Paul initiated emergency procedures and, after two years, the backlog had been reduced to less than a dozen. Jean, a thorough no-nonsense professional, found a job as manager of a golf course/tennis/athletic club, known as the Dunes Club, that was open to workers and diplomats of the US and other foreign countries.*

In 1987, Paul was appointed Chief of the entire USGS mission; his management skills by then had become obvious. As expected, Paul's tour resulted in many accomplishments. One was mining and developing a large deposit of phosphate rock (Cretaceous to Paleocene [145 to 56 million years] in age) discovered in the 1960s in the northern part of the Kingdom, mapping of which was partly done by Chet Wallace. Phosphate, essential to agriculture as a fertilizer, was one way the Saudi economy became diversified. Monitoring of young basaltic eruptions led to a seismic network with data telemetered to the Jeddah office. Establishing that several water wells in the holy city of Mecca were contaminated led to an environmental study of this and other areas. An aeromagnetic map of part of the Kingdom by excellent geophysicist Marty Kane led to the discovery of a new oil field not far from Riyadh, the capital of Saudi Arabia. Paul remained in the job until 1992, when he retired from the USGS and the family returned to Denver.

My favorite story about Paul's time in Saudi was one he wrote down just before he died in March 2020, about an incident when he was mapping a 15-minute quadrangle in November 1987 from a field camp not far from the Iraqi border. At times Paul had helicopter support, but it had recently been withdrawn when some Iraqi troops crossed into Saudi Arabia not far from camp. Returning to camp one evening, a storm rolled in with torrents of rain and a cold wind. His Land Rover became stuck in a flooded wadi and he had to wade waist-deep across it. He noticed a bright light in the distance and aimed for it, crossing at least two more flooded wadis as he hiked about five miles in about three hours. It was a camp of the indigenous Bedouin ("Bedu"), although not the typical black goat-hair tent but a low concrete building; the light, on a pole, was powered by a generator. When he pounded on the door, a boy of about thirteen opened it and beckoned Paul inside to a large low room, where a group of mostly young men was seated around a small charcoal fire. A man in his fifties stepped forward, took Paul's hand, and led him to a seat at the circle. No words were spoken and Paul was handed a cup of sweet tea, cakes made from sheep or camel milk, and some dates. They soon communicated, for Paul had a set speech in Arabic stating that he worked for the Saudi Arabian Deputy Minister of Mineral Resources, and he answered questions about his family and "other questions that I understood less well." The host's name was Ibrahim Ar-Rumah, which gave the name to Wadi ar-Rumah that had almost washed Paul away. At about 9:30 pm, Paul's host dismissed his sons with a gesture, stacked two foam mattresses on the floor, made a bed with blankets and a pillow, and left a small pan for "my convenience." Paul took off his soaked clothes and got into bed, exchanging good nights (Allah Isalmak) with the man, and fell asleep immediately. When he awoke briefly at midnight, the boy was kneeling at Paul's bedside to take care of him if needed. When Paul awakened at dawn, the boy was sleeping on the floor beside Paul; Paul covered him with a blanket. In the morning, he was gone.

The morning dawned bright. Paul put on his clothes, which must have been dried overnight by an electric heater. Ibrahim and his sons provided breakfast. Soon they heard a vehicle approaching, driven by Paul's camp boss and driver, who had rescued Paul's Land Rover and found Ibrahim's camp. Paul wrote that we parted with "smiling hand-shakes." Within a week, helicopter support returned, and then Paul's "first order of business" was to fly with his camp boss (a Saudi) to Ibrahim's camp, where they offered him a flight in the helicopter. Ibrahim jumped in the front seat beside the pilot and they spent several hours flying and visiting nearby places that Ibrahim knew. They returned to Ibrahim's camp and went inside for tea, where Paul noticed a woman seated across the room. It was Ibrahim's wife! Paul's camp boss "told me that her presence was an honor both to her and to me. And her face was not covered!" They had lunch of the standard cakes, including a bowl of warm camel's milk that was passed around, while Paul used his halting Arabic, helped by his camp boss. Paul concluded his story with these sentences: "Since then, I have thought a lot about the experience that began in a desert storm. It exemplifies the fundamental humanity bestowed by a desert Bedouin on a total stranger, honoring the age-old tradition of his people. It is something that nations should emulate." Paul had experienced the "immutable law of the desert—to assist and care for anyone in difficulty or distress."

Too soon after my divorce, I started dating Adrienne B. Anderson, an archeologist with the National Park Service in Lakewood. She had a Ph.D. and she got along well with Jill. Her dissertation was a study of the famous and beautiful Folsom points (arrowheads) from the archeological site near Folsom, northern New Mexico, where bison had been discovered in 1929 having been killed by arrows and spears containing these large, distinctive points, wielded by what was then thought to be the earliest inhabitants (at least nine thousand years ago) in the Americas. We married in late November 1981. In the Park Service, Adrienne managed archeological programs. Archeology seemed to me to be the closest science to geology, for both are field oriented and occasionally the fields overlap through collaborative efforts. And, of course, archeology is fascinating. Adrienne's sister lived in Tucson, Arizona, and it so happened that through them my parents became next-door neighbors when Dad retired. Mom and Dad bought a one-story home on about half an acre at the far eastern end of East Prince Drive in northeastern Tucson. They had a pool that they occasionally used. They bought a small four-wheel-drive vehicle that they used for exploring in the desert with other retired friends. It was an enjoyable retirement. Therefore, we visited Tucson commonly.

Once, when Jill was there visiting Mom and Dad, Jill and I went to visit Larry Gould, who lived about two miles away and had by then retired from his last bit of active teaching at the Geology Department at the University of Arizona. It was good to see Larry. We talked about Antarctica and mutual friends, including his good friend and my Texas classmate George Davis, who then was chairman in the department. When Larry saw Jill, he said "My, you're very pretty!"

Adrienne and I sold my condominium and her place and bought a mountain home on six acres of land in the Front Range, at 5894 Meadow Drive, near Evergreen. It was above and west of the town of Morrison, and that was our mailing address. It was in ponderosas and aspen trees adjacent to undeveloped mountain land owned by the City of Denver on one side and a big ranch on the other side. I loved it, for we now made enough money to make this dream of mountain living come true. Elk herds commonly passed through the property, especially during the fall rut. The sound of bugling elk is one of my favorite sounds, and we heard it often! There, I continued my routine of jogging five miles every other morning on roads near the house, an especially great way to start a morning because I commonly saw deer, elk, and other animals along the way!

Stateside lab work, data compilation, collaboration with others, and report writing about the Orville Coast season and previous seasons continued part-time during October 1977 through September 1978, and June 1979 through September 1983. The biggest push was on three reports on the Orville Coast and on Antarctic mineral resources, plus another one senior-authored by Tom Laudon, during the winter and spring of 1982. These were for the once-every-five-year Antarctic symposium, this year held at the University of Adelaide in South Australia (to coincide with the hundredth anniversary of the birth of Australia's greatest hero, Sir Douglas Mawson), where about two hundred scientists presented their findings. August of 1982 therefore was a working vacation in Australia, at which I presented and submitted the three papers. Adrienne and I spent nearly a month visiting many parts of the country. It was a wonderful vacation, in a beautiful country. We visited and snorkeled in the Great Barrier Reef, where one memory was that our swim was accompanied by a few ten-foot-long sea snakes, which we learned later are poisonous. At Adelaide we rented a car and drove north through opal fields and various national parks, many of them featuring adorable hopping marsupial animals. The largest of these are, of course, kangaroos, which we would occasionally see crossing the road. We saw wallabies, which look like small kangaroos but do not have the cute, deer-like face of kangaroos. One national park had koala bears, although these were becoming rare as forest habitats were being cut down for development. Eventually we ended up at Alice Springs, a relatively small town in the interior and a capital of sorts to aborigines. Over the years, these people had been treated worse than we Americans treated our Blacks and American Indians (if that is possible!). Alice Springs, in the middle of a desert, is relatively flat but a large rock, then called Ayers Rock, lay two hundred miles to the southwest and within a brief flight from Alice Springs. So, we flew out there and climbed this beautiful mountain, then circumnavigated it on the desert floor. It is a barren red sandstone monolith that is several miles in circumference and has a relief of more than 1,000 feet (with an elevation of 3,000 feet). Later, because of its significance as a sacred place for the aborigines, tourists were barred from climbing it, and it was renamed Uluru. It is now a UNESCO World Heritage Site, for the spectacular hike around it reveals petroglyphs and springs, accompanied by the typical small spiny trees and bushes. The symposium volume was published in 1983.

Eventually, and sadly, my marriage to Adrienne failed, and we divorced in April 1986. I am not sure why it failed, although we were both stubborn and fought a lot. Perhaps the reason this marriage, as with my first, did not last was because we were so commonly apart, including during my next Antarctic expedition. In the divorce settlement, I bought out Adrienne's share of the house on Meadow Drive.

1 Tom Steven and the Marysvale Volcanic Field

Serendipity has marked my life. When I joined the Branch of Central Regional Geology in 1970, I quickly introduced myself to two branch geologists who were mapping and writing breakthrough reports on the San Juan Mountains of southwestern Colorado, the largest volcanic field in the US. These were Thomas A. Steven and Peter W. Lipman, two unusually productive geologists who were doing the best study in the world of an extinct, deeply eroded Tertiary (68 to 3 million years old) volcanic field. In the San Juans, Tom and Pete had described some of the largest ash-flow tuffs and their source calderas on earth, and along the way they had discovered metallic mineral resources related to the calderas. Isotopic dates of igneous rocks were becoming more accurate, so the history of this volcanic field was the best done at the time. Other volcanic fields started to be studied, following the methods of Steven and Lipman. At the time of my hiring, the edges of perhaps the second largest volcanic field in the West, the Marysvale volcanic field, were being studied by these methods. These studies were: (1) Paul Williams (his 1967 dissertation) following distinctive ash-flow tuffs derived from the Great Basin, as recognized by Hoover Mackin, Earl Cook, and Paul, into the Marysvale field; (2) Paul Williams' reconnaissance mapping of the eastern edge of the field, as his Salina 2-degree sheet, about to be published (1971); and (3) John Anderson's and my dissertations and post-dissertation mapping on the southern and western flank of the volcanic field, plus adding the Iron Springs district, with our initial summary report submitted to the Geological Society of America (GSA, published in 1975). As in the San Juans, finding new mineral resources at Marysvale seemed likely, for the Marysvale volcanic field had been mined for many years for gold, alunite, silver-lead-zinc, and uranium. The first significant geologic study was done in the late 1930s by an excellent USGS geologist, Eugene Callaghan. Yet the concepts applied to its early study were now outdated, thanks to the newer work that Tom and Pete had done. The story of Marysvale can be traced through the research by Eugene, Tom, and others.

> ***Gene*** *(or Pat, as some called him)* ***Callaghan*** *(1904–1990) joined the USGS in 1928 as he was finishing up a Ph.D. from Columbia University (1931). His specialties were field geology and mineral resources, and by the late 1930s he*

had mapped enough of the mountains of the Marysvale field to publish a report on alunite, which is an aluminum potassium sulfate mineral that during World War I was mined south of the town of Marysvale for aluminum. The next year, he published a general report on the geology of the field, noting that the crest of the Tushar Mountains, in the heart of the field, consisted of gleaming white peaks (even in summer, when the snow is partly melted off) of rhyolite, including the highest, Mt. Belknap and Mt. Baldy. He also described beautifully exposed rhyolite ash-flow tuffs in the lower part of Clear Creek Canyon, where Interstate 70 crosses the northern Tushar Mountains. He named the ash flows the Joe Lott Tuff, after Joseph A. Lott, a pioneer born in 1855 who homesteaded 160 acres in Clear Creek Canyon and drove horse-drawn wagons of supplies and mail up a harrowing three thousand feet to the gold-mining town of Kimberly (at nine thousand feet elevation) in the Tushars. Gene postulated that those tuffs came from the gleaming white peaks (the word caldera had not yet been proposed). After World War II, Gene left the USGS to become a geology professor at Indiana University, then (from 1949 to 1957) the Director of the New Mexico Bureau of Mines and Mineral Resources. Gene mapped four 15-minute quadrangles (scale 1:62,500 scale, each about 235 square miles in size) and part of a fifth quadrangle making up the Maryvale field, but those findings were not published until the early 1960s, their compilation done by Raymond L. Parker, who made himself the second author, and later succeeded Dan Shawe as Chief of Central Mineral Resources after I joined the USGS. In 1965, Gene returned to Utah to become the Associate Director of the Utah Geological and Mineralogical Survey, as it was then called. In 1968, he became a professor and the first chairman of the newly organized Department of Geology and Geophysics at the University of Utah. He retired from these jobs in 1972, as his 1973 summary report on the mineral resources of the Marysvale field was in press by the Utah Survey, and then he consulted on mineral deposits. Gene had a soft-spoken, sweet, and gentle demeanor and cared deeply about teaching and his students, for which he was loved. We invited him to join us on field trips each year to the Marysvale area, led by Tom Steven, Skip Cunningham (see below), John Anderson, and me, with the summaries and road logs published later. On those trips, Gene stayed in the background, making only a few helpful comments even though we were changing his geologic contacts and interpretations in a big way! He could have been bitter or vengeful, a trait I have seen in other scientists when their ideas are questioned. Instead, Gene had the attitude of other great scientists, knowing that scientific conclusions are bound to change! The last I heard from Gene was in the late 1980s, after his beloved wife had died, when he called me about a cruise he was going to take to Antarctica.

***Tom Steven** was born in 1917 on a backwoods ranch near Grants Pass, Oregon, but the family soon moved with all belongings southward in a covered wagon pulled by four horses and driven by his tiny mother (while his dad, a "buckaroo," drove the cattle) to Crescent City, California. There, he grew up as a shy child who explored the mountains. Yet he was a good student, so he entered San Jose State College. On a whim, he took an introductory geology course, and this set the path for his career! He started graduate work at UCLA, but World War II intervened and he joined the USGS Strategic Minerals Program. He also did a stint (1945) in the Navy. Coincidentally, the USGS work included being an assistant to Hoover Mackin at Iron Springs, one of several of Tom's mapping studies during World War II. By 1950 he had completed his Ph.D. and begun his first major project, mapping and studying mineral deposits in the Creede and Summitville mining districts in the San Juan Mountains, with peaks that were among the highest in Colorado. By the 1960s he determined that the mineralization was controlled by calderas, which suggested undiscovered mineral deposits, and he began mapping the Durango 2-degree sheet (scale 1:250,000). He had many colleagues on his project, the most notable of whom was Pete Lipman, who had recently been hired. Pete, with a B.A. from Yale and a M.S. and Ph.D. from Stanford, became one of the best volcanologists and igneous petrologists in the world, even surpassing Tom, as Tom would willingly admit! Tom and Pete were great geologic mappers and extremely prolific writers, and their reports detailed features of the first large (up to 45-miles in diameter) calderas in the world, along with their mineral resources. Much of what we know of the caldera concept can be attributed to their reports. The division of labor in the San Juans was for Pete to map the general geology and Tom to map and zero in on the mineral resources, but there was a lot of overlap between their tasks. Once, during field work on a cold fall day, Tom came upon a herd of at least a hundred elk during a traverse and just kept walking through the herd, for he needed to get to an outcrop on the other side of them; individuals moved out of his way but nobody ran, and Tom likened the movement of the herd to brownian motion. The next chapter in Tom's life was the Marysvale project, detailed below. Tom continued at Marysvale until he retired from the USGS in 1985. Yet even after that, Marysvale reports as well as other studies—mostly in the Colorado Front Range—continued to be published. He died in 2013. One of the best writers in the USGS, Tom appeared spiritual in his wisdom, had a great love of nature and field work, exhibited much self-effacing humor, and showed a great deal of self-knowledge that he shared with others. If you look closely at the photo I have of him, you will see Dave Johnston laughing; almost certainly Tom just said something self-effacing, as I was to witness many times on field trips. He and his wonderful wife Grace had two children: Jim, who pre-deceased Tom, and Barbara, who is a gem.*

> *After retirement, Tom self-published two books of poems about his upbringing and his love of nature; some elicited laughs and others tears. He received many honors, but he was most proud of a legacy as a scientific leader who supported and mentored many others, including Pete L., Skip, and me. I thought of Tom in much the same way I thought of Hoover Mackin, and I co-authored a memorial to him.*

One of my favorite poems by Tom is "That coyote laughed at me," consisting of nine stanzas of which I give the last two. It details a howl of a nearby coyote as Tom was returning to camp on his mare, Sam, while he was mapping at Marysvale searching for mineral deposits. He was late getting back. Darkness was approaching, and he was tired and in no mood to listen to a coyote who seemed to be mocking him. But as Tom listened to the coyote, he thought differently and the subsequent poem expresses the joy we all found in field work.

But was his laugh at me? Or life?
Did he mock? Or celebrate?
We each were free in our own way
To do what we did best, and
Each had zest for what he did,
And neither needed envy. His hunt
That night was like my search—no guarantees
But tomorrow each would rise and go
And each would yip in joy

He laughed again, and I joined in.

When I joined the branch, Tom was finishing his studies and reports in the San Juans and was looking for his next project. I persuaded him to make it the Marysvale volcanic field. For I also was interested in continuing our work in Utah part-time, as Antarctica would allow, and there were more than enough mountains for us both! By 1974, Tom made his project focus the mapping of the Richfield 2-degree sheet, which lies west of Paul Williams' Salina 2-degree sheet and extends west almost to the Nevada border. The sheet contained most of the Marysvale volcanic field.

In 1972, bureaucrats deemed all USGS studies in Antarctica to be the purview of the Office of Central Mineral Resources, so I was transferred administratively to the Branch of Alaskan Mineral Resources, although my office did not move. My Iron Springs project had already been moved administratively into Central Mineral Resources. By early 1974, I had written most of my Antarctic reports about the 1969–1970, 1970–1971, and 1972–1973 field seasons, much of my Iron Springs field work was done, and the next Antarctic season had been delayed, so where was I to go next? Meanwhile, Tom's career subject generally was mineral resources, so some administrators thought that he was misplaced in the Office of Regional Geology. Somebody had a brainstorm, and the chiefs of the two branches traded Tom for me; back I came to Central Regional Geology and off Tom went to Central Mineral Resources. Tom moved offices to the other side of Building 25, but I stayed where I had been. Tom's Richfield 2-degree sheet was now to be funded by the Office of Mineral Resources, starting July 1974. My next project was to be a different 2-degree sheet, the Vernal 2-degree sheet, funded by the Office of Regional Geology. It began in July 1974, as described above. Yet I often wondered how any Branch Chief with a brain could have traded the great Tom Steven for the likes of me! As fate would have it, however, Iron Springs field checks, a short project with Pete Lipman concerning Utah geothermal resources, and informal collaboration with Tom kept me close to Tom's initial work on the Richfield project. Upon completion of the Vernal project, I proposed and began my own Richfield 2-degree project (housed in Regional Geology). But before that, the story of another outstanding geologist needs to be told.

> ***Charles (Skip) G. Cunningham*** *was born in 1940, and earned a bachelor's degree from Amherst College and a Ph.D. from Stanford University. He taught college for a year, then joined Central Mineral Resources Branch in 1974. Tom had just started the Richfield 2-degree project and he recognized Skip's potential, so he grabbed him in 1975 to do things that Tom or I could not. Specifically, Skip could apply new technologies and concepts, such as detailed chemistry, isotopes, and fluid inclusions to study ore-forming processes and exploration techniques for metallic and non-metallic mineral deposits. Skip remained with Tom's project until 1983, when he was asked to become Deputy Chief of the Office of Mineral Resources at USGS headquarters in Reston, Virginia. After four years of this administrative job, he began a decade-long project in which he applied the models he had discovered and developed at Marysvale to exploration of volcanic-rock-hosted and porphyry-copper mineral deposits elsewhere in the world, especially in the central Andes but also in Spain and China. Yet even then he returned for a brief time almost each summer to Marysvale—like the swallows to Capistrano—to learn and refine new models based on the areas of Marysvale he knew best. Like Tom, Skip was a prolific publisher and his discoveries in the Marysvale volcanic field continued past his retirement from the USGS in 2007. Skip died in 2017 after a long bout with Parkinson's disease had robbed him of most of his motor skills. He and his remarkable wife Cher have two bright daughters, Wendy and Betsy, the latter of whom became a geologist. Skip was a gentle and positive spirit with a ready smile, sense of humor, great modesty, and inner wisdom. He had a remarkable tolerance and patience toward others and made friends easily. On field trips or his later yearly Utah visits he was greeted with shouts of joy and*

recognition, as if he were a small-town football hero! Skip also enjoyed many honors and prizes. Whoever claimed that great scientists cannot be nice guys?

And one more name, from a few chapters back. John Anderson had continued beyond his dissertation work in the Markagunt Plateau once he was hired by Kent State University in the fall of 1965. He applied for and received NSF grants, enabling grad students to join him, as did I in the summers of 1968 and 1969 and the academic year of 1968–1969. Once my Richfield 2-degree project became official in 1978, John and his students had USGS support for the field work, which included mapping in the southern Tushar Mountains to join the mapping of Tom and Skip. As the mapping by John and his students continued through the 1970s and early 1980s, they discovered old (Miocene) and lithified (that is, now turned to rock) landslide breccia that, unlike easily identified, unlithified Quaternary landslides, was not confined to the base of fault scarps or of other high landforms, but instead coated parts of the top of the Markagunt Plateau. I will return to this story later.

The Marysvale volcanic field is about seventy-five miles in diameter, slopping over the western edge of the High Plateaus subprovince of the Colorado Plateau into the eastern part of the Great Basin. The volcanic rocks range in age from about 35 Ma (million years before present) to about 1 Ma. Of these, the rocks between about 35 Ma and 20 Ma are extremely voluminous and make up most of the field. These voluminous rocks are known as calc-alkaline, for they range in chemistry from andesite to low-silica rhyolite. They are underlain by intrusive rocks (solidified magma bodies known as stocks, plutons, or batholiths) that were the sources of the volcanic rocks and are locally exposed because of deep erosion into the volcanic pile. The heart of the Marysvale field contained the eruptive centers, most of which were stratovolcanoes (features like Mount St. Helens and Mount Rainier), although these volcanoes had been deeply eroded and their original landforms no longer exist. Ash-flow tuffs were known in the volcanic field, so their caldera sources must also be present, but only one caldera was known when we began work. That was Gene's, in the gleaming white peaks at the top of the Tushar Mountains, which Tom and Skip named the Mount Belknap caldera when they mapped its rims, about ten miles north to south and eight miles east to west. Tom and Skip concentrated on the eruptive centers, for these were the most likely sites for mineral deposits. Groundwater in aquifers above the hot magma of the plutons was turned into giant convection cells, much like what you see when you heat a pot of water to boiling on a stove. The heated hot water, known as hydrothermal fluid, that coursed through the overlying volcanic rocks led to alteration of the rocks and deposition of mineral deposits. Hydrothermal alteration breaks down the original minerals in the volcanic rocks into nondescript, fine-grained clay minerals that have distinctive and at times bright colors (mostly yellow, but also red, green, etc.) for which prospectors seek.

All the volcanic processes mentioned above, however, were pre-20-Ma ancient history. For the topography that we now see at and near Marysvale initially formed after 20 Ma and is continuing to form. It resulted in the Great Basin and its separation from the Colorado Plateau by large-scale faulting during the basin-range episode of deformation. The Great Basin-Colorado Plateau boundary is a large-displacement (of about 5000 to 15,000 feet) north-trending, high-angle, normal fault zone that lifted the Tushar Mountains (in the High Plateaus) on the east with respect to the Beaver basin (in the Great Basin) on the west. A normal fault is a fault in which one side moves vertically with respect to the other side, and most of them are high-angle, that is, they dip about 60° with respect to the horizontal ground surface. This fault zone, like a gazillion other northerly-trending normal fault zones in the area, may have started to form about twenty million years ago. But most displacement took place after about 10 Ma and is continuing today, resulting in the present topography. North-trending ranges alternate with north-trending basins in the Great Basin, which extends westward from central Utah to easternmost California. In the western part of the Colorado Plateau, the High Plateaus are a transition zone, where normal faults are less common and of less displacement than in the Great Basin. The main part of the Colorado Plateau, extending from east of the High Plateaus to about Grand Junction in western Colorado, consists of mostly flat-lying rocks that are little deformed. Synchronous with basin-range deformation, volcanic rocks were deposited between 20 Ma and 1 Ma; these rocks are of low volume and are referred to as bimodal, being either black basalt lava flows and cinder cones or white high-silica rhyolite (such as the Mount Belknap caldera) that erupted as lava flows, ash-flow tuffs, and central vents.

The volcanic field is named after Marysvale, a mining town just north of the center of the volcanic field at about 6,500 feet elevation. Marysvale was founded in the 1860s, when gold and silver were discovered just to the west in the Tushar Mountains. The town is along US Highway 89, which passes north and south in Sevier Valley, between the north-trending Tushar Mountains on the west and the north-trending Sevier Plateau on the east. The "big city" (almost eight thousand people now) in the field is Richfield, the county seat of Sevier County, in Sevier Valley about twenty-five miles north of Marysvale and on east-west Interstate Highway 70, which continues east to Denver. The town of Junction, where I stayed with the Piersons while doing my dissertation, is fifteen miles south of Marysvale, also in Sevier Valley. Sevier Valley is a typical north-trending downthrown basin, which broadens south of Marysvale and closes just north of Marysvale. Sevier Valley opens again less than ten miles north of Marysvale. The mass of hills in the middle, between the Tushar Mountains and Sevier Plateau on the west and east, respectively, consists of the Central Mining Area formed in the rocks within a caldera that I later found. Regardless, the north-flowing Sevier River, within Sevier Valley, maintained itself through the mass of hills and cut downward about a thousand feet to form Marysvale Canyon. Marysvale is now an attractive town of about four hundred people that is a popular tourist and hunting spot. With mountains rising all around, the setting of Marysvale is beautiful.

The Marysvale volcanic field has had a long history of mining but the field had not been subjected to a modern geologic study. Therefore, the framework that controls mineralization was largely unknown, and the prospect of finding new but buried deposits was substantial, or so we hoped. Many of the known mineral deposits were in the Tushar Mountains, but some were in the Central Mining Area and elsewhere. In 1888, gold was discovered in the Kimberly or Gold Mountain district, located high in the northwestern Tushar Mountains, to which Joe Lott and his horse teams took wagons. The district contained the most productive gold mines in the volcanic field. Mining ceased in 1907 but resumed in 1911 and continued intermittently until 1937. New drilling took place from 1981 to 1996 but it resulted in no production. Other smaller gold districts are present in the Tushar Mountains, including many earlier mines in Bullion Canyon southwest of Marysvale. The Deer Trail Mine, south of Bullion Canyon, was discovered in 1878 when several locals were hunting deer and found outcrops of lead and silver. Mining recovered these elements as well as gold, zinc, and copper in workings that were operated almost continuously until 1981, along a horizontal tunnel (adit) that eventually stretched two thousand feet. Another mining company bought the property in 2013, and new exploration began.

The Central Mining Area was mined for uranium in the 1950s. High-grade ore, including pitchblende, was mined from deep shafts and from adits off them. Local men did the mining at a time when the dangers of inhalation of uranium-bearing dust were not known, and most of them died from cancer over time. The Central Mining Area was one of the few hard-rock (underground) uranium mining districts in the Colorado Plateau. Elsewhere across the plateau the uranium boom was an exciting event, resulting in many small uranium pits that fueled the atomic weapons of the Cold War.

Mining districts attract all sorts of unsavory characters, and the reigning king of those in this part of nineteenth-century Utah was the outlaw Robert LeRoy Parker, alias "Butch Cassidy." He was born in 1866 in the town of Beaver, on the western flank of the Tushar Mountains, but was raised several miles south of Circleville, about thirty miles south of

Marysvale. If you knew where to look, his boyhood cabin was visible a few hundred yards west of Highway 89. But in 2017 it was commemorated as a history/bathroom roadstop, so it is no longer difficult to find. Of course, Butch became famous because of a popular 1969 movie "Butch Cassidy and the Sundance Kid" starring the late Paul Newman (as Butch) and Robert Redford. Now, both Circleville and Marysvale claim Butch as their own, for he is an asset in the tourist trade. Butch was said to have frequented Kimberly, although he caused no trouble there.

Big Rock Candy Mountain, in Marysvale Canyon, is a jagged, colorful, and picturesque mountain, mostly yellow with lesser tinges of orange, red, and white. It is a classic example of hydrothermally altered rock, in this case lava flows from a stratovolcano. At the base of Big Rock Candy Mountain is a motel, a former resort of some fame, with a café and gift shop. The café has its walls festooned with publicity photos of movie stars of the 1930s and 1940s, including John Wayne (if I remember correctly), all of whom came to the resort to hunt elk and mountain lions. (Hunting mountain lions may sound dangerous but it rarely is, for hound dogs chase them up trees, from which the "hunter" shoots them. The lions are invariably stuffed to demonstrate the great prowess of the human hunter!) The mountain's name came from a popular song "Big Rock Candy Mountain" that was claimed to have been written by Harry McClintock, a hobo who rode the rails during the Great Depression and published the song in 1928. Soon after the song became popular, locals erected a sign "Big Rock Candy Mountain" at the base of the yellow mountain, and a spring that issued from the yellow mountain about a hundred yards up a deep gulch became "Lemonade Springs." The song portrayed a hobo's idea of paradise, and locals adopted the two place names. Burl Ives later cleaned up the song's dirty lyrics, and his rendition became popular again, as a children's song, during the 1950s.

Each summer, starting in 1974, Tom, Skip, and later I set up USGS trailer homes in a small trailer park, Hoover's Truck Stop, several miles north of Marysvale and several miles south of Big Rock Candy Mountain, and spent the days mapping and doing related field research. John Anderson joined us for two seasons. Hoover's Truck Stop consisted of a gas station and café, whose specialty was chicken-fried steak. In those years, the cooks at the café were better than the ones at Big Rock Candy Mountain, and the locals and truckers knew it. The truck stop is on a river terrace just above the flood plain of the Sevier River and adjacent to US Highway 89. Back then, Ada Hoover owned and ran the café and gas station. She and her husband (deceased before we arrived) had been doing that since the Great Depression, when dancing and drinking were also part of the local entertainment. Ada's daughter Joan and Joan's husband Jim Anderson ran the trailer park. We became part of their family. At the time, Jim was an auto mechanic working in Richfield, about twenty miles to the north. In the 1950s, however, Jim had worked in the underground uranium mines, and cancer sadly took Jim in the 1990s. Tom, something of a mother hen, never went to bed until Skip and I returned, for Skip was working high in the Tushar Mountains, which reached an elevation of more than 12,000 feet, and some of the places I worked in the Sevier Plateau were logistically tricky, even though lower in elevation. Tom was hardly averse to strenuous field work and danger, for in his later years he relied on his mare Sam (Samantha) to get him from outcrop to outcrop. Tom had rigged a USGS pickup truck with a shell cover so it looked like a horse trailer, and Sam learned how to jump up into the bed; use of Sam, however, meant that Tom got back to camp earlier than Skip and me. Sam spent the night in a fenced pasture on the flood plain, with grass up to her waist, and she got fatter and fatter. In my later years I also acquired horses (my wife Dawna's idea), but not even in my most optimistic moments would I ever trust doing field work with any of them! Horses are amenable to the most stupidly scary things, done with the least provocation!

Skip and Tom were busy during the 1975 and 1976 summer field seasons unraveling the various volcanic rock units and their age and relationship to each other, known as working out the stratigraphy. They started in the highest part of the Tushar Mountains, where the gleaming white peaks were, and they soon mapped the Mount Belknap caldera. In addition to the Richfield 2-degree sheet, Tom, Skip, and (later) I did detailed mapping of the entire area by 7.5-minute (1:24,000 scale) quadrangles. Therefore, we all did detailed preliminary maps of the entire Marysvale field west of the Salina 2-degree sheet, to be released as black and white, author-linework USGS MF maps but intended to be assembled later as finished color 1:50,000 geologic maps. Reports were released as USGS Open-File Reports, for many concerned mineral resources and we wanted to get them to the public as soon as we could. In 1979, Skip, Tom, Harald Mehnert, and Chuck Naeser published two important USGS Bulletins that set out the geologic framework and isotopic ages of the heart of the volcanic field; that year I senior-authored a general summary with them. Field trips to show off our preliminary results were part of Tom's master plan to enlist as many collaborators as he could. Participants included not only Gene but others from the Utah Geological Survey (UGS), professors Lehi Hintze and Myron G. Best of Brigham Young University, geophysicist professors and students from the University of Utah, many persons from industry, and any others who were interested. Tom offered to have the USGS publish geologic maps through him, resulting in dozens of colored maps that filled in more western parts of the Richfield 2-degree sheet.

A significant part of the early effort by Tom and Skip was to enlist two persons in the Denver office who specialized in isotopic dating of igneous rocks. Isotopic dating determines the age of a rock based on the concentration of a particular radioactive isotope contained within it. The amount of the isotope in the rock is compared with the amount of the isotope's decayed products. From there the age can be determined using the known rate of decay of the isotope. These persons that Tom and Skip called upon were Harald Mehnert, probably the best in the business in K-Ar dating, and Chuck Naeser, who invented and practiced fission-track dating. Each method worked best on different types of rocks, so using both together broadened the number of absolute ages we could determine.

In 1975 to 1978, I published six 7.5-minute USGS Geologic Quadrangle Maps of the Iron Springs area and vicinity, plus a 15-minute quadrangle (four 7.5-minute quadrangles at 1:62,500 scale) in the Black Mountains.

But in 1975, the environmental movement brought the Marysvale project another opportunity. The technology was not yet economic for solar or wind energy, but geothermal energy seemed like a good possibility. The USGS received funding from the US Department of Energy (DOE) for exploration in areas where high-temperature rocks (magma) were near the surface. It had become known that young volcanoes (Pleistocene or younger, that is within one or two million years before the present time) was where we should look for geothermal energy. In the Marysvale field, volcanic rocks of the bimodal suite, namely basalts and rhyolites, were likely to be young. But basalt magma, although erupted at much higher temperature, comes from a great depth (the mantle), and its vents are thin dikes or small-diameter circular pipes that feed cinder cones; few basaltic vents are even a mile in diameter, so they were unlikely to hold their heat for more than a few tens of thousands of years. Most rhyolite eruptions, however, come from explosive volcanoes or calderas derived from sources in the upper Earth's crust that are melted by adjacent basaltic emplacement and form magma chambers (that will in time solidify to form granite plutons), so they are relatively shallow. Therefore, we looked for young rhyolite volcanoes, dated them to see if they were young enough, then encouraged industry to drill still-molten magma chambers.

Pete Lipman, his field assistant, and I spent several days in May 1975 sampling all the rhyolites I knew about, and the samples went to Harald Mehnert. When Harald started getting numbers within a month, the youngest ages, less than a million years before present (Pleistocene), were of volcanic domes on the crest of, and lava flows and ash-flow tuffs on the western flanks of, the central to northern Mineral Mountains, in the western part of the Marysvale field. A significant hot-spring complex, called Roosevelt hot springs, arose along the western range-front fault of the Mineral Mountains, not far from the lava flows. The springs had been the site of an eighteenth-century house of prostitution, now long gone. The lava flows were particularly interesting because they were obsidian, a black glass with no crystals in it, thus valued by ancient Indian Americans and their Fremont ancestors for the best available points (arrowheads) for arrows and spears. Those obsidian lava flows had been mined for points for thousands of years, and chemistry fingerprints of the obsidian showed that these points had been traded all over the west by these Indians. In early June, therefore, Pete and I were back for several days of mapping those young rhyolites. We stayed at the small (somewhat more than a thousand people) town of Milford, about a dozen miles west across the Escalante Desert (Valley) from the Mineral Mountains.

I have always liked Milford because it was, and in some ways still is, a mining town, with its few streets lined with small one-story homes, many with metal roofs. The town had one motel then, where we stayed, and one café. For many years, Milford was at the southern end of the Union Pacific Railroad line, which took ore from the mines back toward smelters in the Salt Lake City area. But by the time I worked out of Milford, the railroad continued south down the Escalante Valley, then west to Caliente, Nevada, and on to Las Vegas and southern California. A multigenerational Chinese family that came in with the railroad owned the single café, adjacent to the tracks. When the Chinese-American waitress or waiter passed through the swinging doors into or from the kitchen, one could see babies and toddlers playing on the floor and grandmothers and grandfathers directing the cooking. The café had two menus, one a café menu that was the same as café menus elsewhere in small towns in Utah and Nevada, consisting of chicken-fried steak, liver and onions, burgers and fries, beef "steaks" that were cut about a quarter of an inch thick and overdone (and over-priced), and not much else. The locals loved this fare because they were used to it. Fortunately, the other side of the menu at this café consisted of about twenty Chinese dishes, all of them delicious! Yet I never saw a local eat that food.

Phillips Petroleum Company was clued into the geothermal potential of the Mineral Mountains domes, but their geologists were secretive about their work. Also clued in were federal government agencies, who designated the area as the Roosevelt Known Geothermal Resource Area (KGRA). Framework geologic mapping in the KGRA was by two excellent geologists, Dennis L. Nielson and Bruce S. Sibbett, who by 1976 were part of a soft-money group, funded mostly by DOE, known as the University of Utah Research Institute (UURI), and they were mapping the area, with preliminary reports issued soon after and final reports published in 1980 and 1986. We worked closely with them, and our report on the Mineral Mountains rhyolites was published in 1978, with Pete the senior author. The Blundell geothermal power plant was commissioned in 1984 and expanded in 2007 by PacifiCorp Energy, the largest geothermal plant (38 megawatts) in Utah. Wells that tap the hot water go to depths of

2,500 to 6,500 feet. The water is heated by the underlying magma chamber that supplied the Mineral Mountains rhyolite domes, flows, and tuffs. I later compiled the geologic map of the KGRA, Mineral Mountains, and mountains to the east, known as the Beaver 1:100,000-scale quadrangle, as covered later.

My Vernal project ended at the close of September 1977, and my Marysvale project began when I returned from Antarctica in March 1978. The decade-plus Marysvale research, involving geologic mapping and related studies, with Tom and Skip was among the most rewarding of my career. At one time Tom added up our total two-project publications at more than 200 reports and maps. Parts of it have continued until the present with more publications, the latest being the study of our giant landslides (Chapter 11). I first "cleaned up" (did a better job of) my dissertation area in the southern Sevier Plateau, then continued mapping northward into new areas in the central and northern Sevier Plateau. An offshoot of our work on the young rhyolites, with me as senior author (1978), was the recognition of an east-west alignment of young rhyolites and of the range-front of the northern Black Mountains, making a "lineament" of east-trending features formed at the same time as north-trending basin-range faults; this was a germ of an idea that I would later enlarge upon under the name of "transverse zones," at least one of which (this one, in fact) spanned the entire Great Basin. Looking back on these busy years, I identified with the protagonist in "Rambunctious Boy", published by John Fogerty, my favorite songwriter (after Mozart).

On November 7, 1978, my friend from my first field season in Utah, Ray Lynn Pierson, then a Utah Highway Patrolman like his stepdad, Dean, was shot and killed by an escaped convict. Lynn, who had gotten good at spotting drivers who did not match their cars or who looked suspicious for some other reason, had pulled over a man along Utah Highway 20 at the eastern entrance to Bear Valley, where John Anderson had done his dissertation. As he approached the stopped pickup truck, the convict shot him in the heart through the back window of the truck. Lynn emptied his gun but was dead when he hit the ground. The man got away at the time but was captured the next day, confessed, and later was sentenced to life in prison. Lynn, born in 1949, had started on a good life after we had become friends when I rented the trailer home behind the Pierson's home in Junction. After graduating from high school, he had married Darlene Roberts and served in the Navy. They started a family of two boys and two girls and joined the Highway Patrol, where he served for five years until his death, living in their home in Panguitch. Lynn was the first person in the history of the Patrol to be killed in the line of duty, and his funeral was attended by the Governor and nearly all Highway Patrolmen in Utah. A strong woman with deep religious convictions, Darlene eventually remarried. In October 1979, I proposed to the US Board on Geographic Names (within the USGS in Reston) a name, Mount Pierson, in Lynn's honor, to be on the western scarp of the Sevier Plateau about a mile north of Mount Dutton. It was the second highest mountain (10,653 feet elevation) after Mount Dutton on the Plateau, and near where Lynn would drop me off for traverses off the scarp. It would have been visible from Highway 89 on virtually every day of his patrolman route. In July 1980, the name was approved. I had stayed in touch with Dean and Sheren by Christmas cards, and I contacted Darlene to let her know.

Kingston Canyon is a five-thousand-foot-deep east-west canyon that cuts through the north-trending southern Sevier Plateau. In the bottom of the canyon is the west-flowing East Fork of the Sevier River, popular with local fishermen for its rainbow trout and the reason for the canyon in the first place. In other words, as with the Grand Canyon, the river maintained itself (cut downward) as the Sevier Plateau was slowly lifted up along basin-range faults of the Sevier fault zone, which defines the western side of the range. In the bottom of the canyon, about halfway between the eastern and western ends, is a thousand-foot-high, sharp, small mountain called Phonolite Hill. The hill is a resistant white rhyolite vent, called a volcanic dome, that was formed exactly like the current dome in the bottom of the crater of Mount St. Helens. It resulted from sticky rhyolite magma that intruded upward in a vertical circular pipe and vented at the surface as short lava flows that piled up in a domal shape, outward from which was a shallow crater that ringed the vent. Because rhyolite is viscous, it does not create long lava flows of runny basalt that we see erupting at Hawaii, but instead results in spines or stubby flows of near-solid lava. The dome, of course, erupted after the canyon had been cut, but we did not know the age of either of them. I recognized these things when I mapped Phonolite Hill for my dissertation.

During the Fourth of July weekend in 1978, when all others except for Tom and me had returned to their families in Denver, Tom suggested we take a break from our mapping and give Phonolite Hill a second look. We had a delightful day in which we circumnavigated the mountain, about a mile in diameter, and climbed partway up its slope. Eroded pieces of the crater were exposed all around the base of the hill, and the crater contained garage-size rocks that had been spit out of the vent. We collected samples for isotopic dating by potassium-argon methods, which Harald Mehnert of the USGS in Denver could do for us. That would give us the age of the end of canyon cutting. On top of the Sevier Plateau, another (older) rhyolite flow was exposed. That would give us an age that predated both the uplift and the eastern tilting of the plateau by the Sevier fault zone. Six months later, when Harald gave us his numbers, we were amazed to see that the rhyolite on top had an age of about 7.5 million years, and the Phonolite Hill dome had an age of about 5 million years. That meant that basin-range deformation and uplift of at least six thousand feet to form the Sevier Plateau, plus concurrent canyon cutting, began here somewhat after 7.5 million years ago and continued for 2.5 million years during uplift and canyon cutting, after which the dome erupted. To my knowledge, this is the most tightly constrained span of years for uplift of a mountain range by basin-range faults in the entire Great Basin. We published our findings with the Geological Society of America Bulletin in 1981, but our conclusions were widely ignored. In 2014, the Utah Geological Survey ran the two ages again, using the more accurate and modern argon/argon dating method. The numbers were virtually the same and we repeated the same conclusion in the Panguitch 1:100,000-scale quadrangle that my colleague Bob Biek senior-authored in 2015. The American Association of State Geologists awarded Bob the Charles Mankin award for this map and report, the best by all the state geological surveys in the country for 2016. The award was not bestowed because of Kingston Canyon but for other things (including the world's largest terrestrial landslide, partly within the confines of the quadrangle, as covered in Chapter 11). Then, just to make sure the profession really knew about our discovery (nag, nag, nag), we published a short report on the age of canyon cutting in 2022.

When Paul Williams published the Salina 2-degree sheet in 1971, he defined a widely distributed ash-flow tuff that he called the tuff of Osiris, after a tiny ghost town east of the Sevier Plateau. Tom, Skip, John, and I found the tuff all over the place, for it was distinctive and of high volume. In our 1975 GSA report, John and I had described it and formally named it the Osiris Tuff, and Bob Fleck had dated it at 22.8 Ma. But where was its caldera source? As our mapping continued, we knew it had to be somewhere, and it had to be big! As I mapped northward in the Sevier Plateau, from my dissertation area, any caldera was nowhere to be seen. Yet on the western fault scarp of the central Sevier Plateau, erosion had exposed a large pluton that I assumed to be old (23 to 25 Ma), like most others exposed in the Tushar Mountains and Central Mining Area. But the pluton was similar in mineralogy to the Osiris Tuff, so we submitted samples of the pluton for isotopic dating in 1978.

On August 29, 1979, as vividly recalled in my field notes of the day, Skip returned to our Hoover's field camp after a break in Denver with his family, and that evening mentioned "by the way, Pete, I just got from Chuck Naeser a fission-track age of 21.5 million years old on that pluton on the scarp." Well, holy shit, that was far younger than I thought it was! The number meant that it postdated the Osiris Tuff and just might be the intracaldera pluton (commonly called the resurgent pluton) of the Osiris source caldera. Such a pluton is emplaced upward into the guts of a caldera shortly after the initial eruption (Osiris Tuff) and caldera collapse. Such a collapse consists of surface rocks that fall into the void in the crust created by the eruption of many cubic miles of subsurface magma that flash to tuff. As I tried to sleep that night, I recalled the similarity of the pluton with the Osiris Tuff and I remembered some screwy-looking outcrops not far from the pluton on the top of the plateau. These confusing features, which I could not explain at the time, just might be the caldera rim. I hardly slept that night, and early the next morning I was up on those screwy outcrops. Sure enough, they were the caldera rim, although for only about fifty feet, with trees and dirt covering up any other exposures. The next day less than a mile away I found more exposures, although of only a similar length. From there, the whole thing fell into place and I started outlining the shape of the southern edge of the caldera on the map. I had been working within the caldera for the last month or more and hadn't recognized it! The Monroe Peak caldera (which is what I named it) is the largest (at about twenty-five miles long east-west by twelve miles wide north-south) in the Marysvale volcanic field! But to determine that story took another year or more of mapping. I closed out that Marysvale field season in mid-October, 1979, happily mapping the caldera northward (it all now mostly made sense!), then bringing Skip and Tom into the mapping and interpretation. On the way back to Denver, I stopped in the Vernal area for one last day of traverses on that project. Now I could concentrate on mapping the Monroe Peak caldera! But on May 18, Mount St. Helens erupted, and clearly that took precedence. I tell that story in the next chapter.

Field work in Utah during the 1980 field season was hit or miss, as Mount St. Helens entered my life and field work on it became my highest priority. Yet finally, on July 19, 1980, with a presumed lull in the eruptions in Washington, I was back at Hoover's Truck Stop, visited Tom and Skip, then moved my trailer east to the tiny town (population four hundred) of Koosharem, in Grass Valley east of the Sevier Plateau. I parked the trailer at the small Koosharem Café and Motel, owned by Marian Hatch, a sweetheart. The town was just east of the eastern side of the Monroe Peak caldera, and I began mapping two 7.5-minute quadrangles that contained the eastern part of the caldera. At 7 pm on July 22, I heard that Mount St. Helens had erupted for the third time, and I checked in by phone with colleagues there and in Denver. I had invited Paul Williams to join me at Koosharem to map the caldera that erupted the tuff (Osiris Tuff) that he had discovered and named, and the next day Paul drove in from Denver. The following day he found a piece of the caldera wall! Then we found more parts of it; in fact, the wall was well exposed for miles in these two quadrangles, which were in his Salina sheet. On July 29, a field trip group with Tom, Skip, and Paul visited the wall and other features. Tom used his experience to explain outcrops that we couldn't quite fathom. On July 31, Paul returned to Denver while I continued mapping. But I felt Mount St. Helens calling, so I drove back to Denver on August 1 and, soon after, flew to Portland. Later, on September 13, I returned for more mapping, then back to Denver on September 30.

For 1981, my only field work in Utah was from June 29 to July 10. Temporarily, funds from my Mount St. Helens work enabled me to hire Allan M. Kaplan (see next chapter), a young associate, and in the year or so he was with me he obviously needed to see an ancient caldera. So, after a two-day tour of the Marysvale volcanic field and meeting Tom Steven, we spent a week mapping the Monroe Peak caldera out of Koosharem. In August 1981, I received the welcome news that I had been promoted to a GS-15, which is about the highest grade a scientist can get in the USGS. I knew that the Branch Promotion Committee had put me up for a promotion in the spring, but I had not expected it to happen this year. At the time, my bibliography listed sixty-seven published reports and geologic maps and ten abstracts. I am sure my fifteen minutes of fame at Mount St. Helens had helped.

By 1982, Tom, Skip, John, and I had finished most of the detailed mapping of the sixteen 7.5-minute quadrangles in the main part of the Marysvale field, and we had published them quickly as preliminary black and white maps. Skip compiled all those maps into a large (3.5 feet east-west by 3.3 feet north-south, 1:50,000 scale) wall map whose eastern line of longitude was the boundary between the Richfield and Salina 2-degree sheets; the western boundary was west of Beaver. The map included the western half of the Monroe Peak caldera. But the western margin of the caldera that I drew was put just west of the Sevier Plateau. That is because I had mapped the Sevier Plateau and Skip and Tom had mapped the hills in the Central Mining Area near Marysvale and had found no caldera! The map was submitted in early or mid-1982, to be a beautiful color map.

With the field work at Mount St. Helens having ended in August 1982 and Allan having gone back to school, I returned in early September to Koosharem to map quadrangles. In the third week of September, however, I was told to return to Denver by the end of the fiscal year because Congress had not passed a budget (financing our work). Skip and Tom headed back, but I elected to spend as long as I could in the field before October 1. I finished what I needed to do in the Koosharem areas and, with a week to go, moved west to work east and west of the Sevier fault scarp, including the Marysvale area, staying in Dean Pierson's old trailer in Junction. When I visited areas that Skip and Tom had already mapped and published in and near the Central Mining Area, I saw intracaldera tuffs that I had mapped at Monroe Peak, the high point on the western edge of the Sevier Plateau and within the caldera! In those days before I took I-70 back to Denver, I confirmed that the caldera margin was farther west than we had shown it, and I started to trace that margin. When I returned and broke the news to Skip, in the hope that we could pull the map out of editing and fix it, we learned that it was already too far along in the publication process. Our beautiful map was published in early to mid-1983 with that major inaccuracy. If we had been contractors building a new house or engineers designing a new car, we might have considered such an error to be a major sin. But science does not work that way. It does not wait for perfection to be achieved! None of us batted an eye at our mistake.

During the summer of 1983, I returned to the Marysvale area and fixed our mistake by remapping the Marysvale and Antelope Range 7.5-minute quadrangles. Tom put together a summary of all the calderas in the Marysvale volcanic field, including the corrected version of the Monroe Peak caldera, and published it in 1984. We found that the Central Mining District was entirely within, and therefore the mineral resources were partly controlled by, the Monroe Peak caldera. We published preliminary versions in 1985, then final versions of the two quadrangles in 1988 with the Utah Geological Survey (UGS). The maps had expanded texts because we had established new isotopic ages and more information on the timing of mineralization. The delay between 1985 and 1988 came because Helmut Doelling, the UGS geologist then in charge of their Geologic Mapping Division, let them sit on his desk and occasionally tinkered with them. In 1986, I wrote the Director of the UGS, Genevieve Atwood, to complain. She fixed the delay. Publication of the two new quadrangles is an example that science advances in part by fixing the mistakes of past workers, including me!

The year 1984 began with a trip to Saudi Arabia, working with my best friend Chet Wallace. Paul Williams remained Mission Chief Geologist there. One way to diversify the Saudi economy was to find gold, so our study area was the Murdama Basin in the northern part of the country because it contained the sites of King Solomon's mines, where gold had been recovered, according to the Bible. The Murdama Basin was a Precambrian marine basin behind (landward of) an offshore volcanic arc, so I could study the volcanic rocks and Chet could study his specialty, old (Precambrian) sedimentary rocks. I tell the story of our adventure, although it lasted a little less than a month (January 1984) in a separate subchapter below. It was written by Chet, with only a couple edits by me, and published in the 2016 USGS Retirees Newsletter. Although it was written for laughs, as many other remembrances by old geologists are in that newsletter, it is accurate. The robes that Saudi men wear brought to my humorous mind the enjoyable 1970 song "Lola", by The Kinks, a story about a drag queen who "walked like a woman and talked like a man." After field work, we published a scientific report on our Saudi studies.

The Utah field season of 1984 started early (late May) but was short because I had my fourth trip to Antarctica scheduled that fall. I also wanted to attend a field trip in early June by Grant C. Willis of the UGS, who had finished mapping the Salina 7.5-minute quadrangle, on the northeastern edge of the Marysvale volcanic field. The UGS had an enviable policy of scheduling field trips of their mapped areas when they finished the field work and had submitted their map and text for technical review. They advertised their field trips widely and they were attended not only by local industry geologists and geology professors and students, but also by the non-geologist public. Inasmuch as the non-geologist public was paying most of our salaries through taxes, this policy was a great method, politically, to get not only professional technical feedback but also views and concerns of the taxpayer. In part because of this, UGS is a successful and (mostly) well-funded geological survey.

> *I had known of **Grant Willis** before, for he had joined us on some Marysvale field trips. But what I found when I accompanied him for a day was a highly competent geologist about ten years younger than me. I was impressed with him, and I got to know him better. He had a bachelor's and a master's degree (supervised by Professor Lehi Hintze) from Brigham Young University. Rather than having a narrow specialty, he could map and study all rock types, all ages of rocks, and all kinds of structures. He was the best geologist in the Geologic Mapping Division of UGS, which was not well managed at the time. Regarding management, in 1995 he fortunately took over as head of that division, and he continued in that job until 2022. After Paul Williams, Grant was the best geologic manager by whom I was ever supervised. Both persons were similar in their calm, patient, extremely supportive demeanor and in being able to recognize and hire geologists who were talented and productive. But the part of the requirements of managers is also to seek money (seemingly always writing grant proposals!) from other agencies, and both Paul and Grant were able to do that well. Grant's wife Julie also then had a bachelor's and master's degree in geology and was (and continues to be) an excellent field geologist, but once they had a family, she stayed behind to raise the kids. When the children flew the nest, however, look out! In 2009, Julie B. Willis earned a Ph.D. in tectonics and seismology from the University of Utah, ended up being the Chairman of the Department of Geology at Brigham Young University—Idaho, in Rexburg, a department of well over one hundred geology students! After my retirement from the USGS in 2000, my consulting business worked extensively with the UGS in making geologic maps, and Grant ended up being my boss. In 2019, Grant received the Lehi Hintze Award for Outstanding Contributions to the Geology of Utah.*

The summer of 1984 involved office compilation and writing of three 15-minute quadrangles, to be published at 1:50,000 scale. One was north of Bryce Canyon National Park, and the other two were parts of John Anderson's and my dissertation areas. Inasmuch as most of these two quadrangles had been mapped by John Anderson, I made him senior author of both maps, even though his effort on compiling and writing text for the two maps was zilch. The first map was published in 1987, and John's two maps were published in 1990. Antarctica came next.

What was the value to the taxpayer of our work on the Marysvale field and the Richfield 2-degree sheet? The two hundred or so geologic maps and reports of the Marysvale field, the overall geologic map of the Richfield sheet itself (1990), and other maps we published that were authored by Lehi Hintze, Myron Best, and others put hundreds of metallic and nonmetallic mineral resources, as well as landslides and other geologic hazards in their geologic context. As just one example of a question that our maps answered, the huge Kimberly gold-silver mining district was now known to be adjacent to the north wall of the Mount Belknap caldera, so did that mean that a future mining geologist should look at other places along the wall for similar deposits? In this case, our geologic work showed that the answer was "no," for this district was tied to older (calc-alkaline) stocks, so that is where a mining geologist should look instead. I could go on, but you get the idea. Tom summarized in 1987 many dozens of mining districts and mineral resources throughout the 2-degree sheet; we now knew more about why they had formed. The taxpayer had been well served.

Photos arranged by chapter, with numbers on the left that are generally chronological. Most photos were taken by the author, except where noted, although some portraits were taken by unknown persons.

1) Arthur C. Rowley, "on the road" during consulting for R.T. Vanderbilt Company, November 1973.

2) Barbara J. DeWitt, at her marriage to Arthur C. Rowley, August 1941.

3) Christopher W. Rowley (on left) and Clark D. Rowley at Chris' Ph.D. commencement, May 1992.

4) Lawrence M. Gould, in Antarctica in 1957, from the dust cover of the 1984 edition of his book Cold.

5) John J. Anderson, in Cedar City, Utah, 1989; photo by Harry F. Filkorn.

6) J. Hoover Mackin, about 1960, from Utah Geological Association Guidebook *(2001), known as The Mackin Volume because it is dedicated to him.*

7) Paul L. Williams, about 2005: photo by Marjorie Edwards.

8) View east-southeast at the western fault scarp of the east-tilted Sevier Plateau, where Kingston Canyon cuts through the plateau. Phonolite Hill is in the middle of the canyon, visible above the southern end of Piute Reservoir; Awapa Plateau in the background.

9) Karen L. Ward, just before our marriage in 1969.

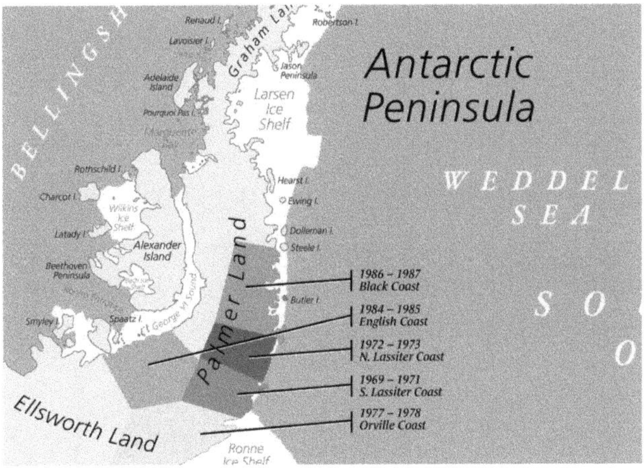

10) Map of the Antarctic Peninsula, showing in color the areas we mapped, with dates.

11) View south from McMurdo Station of the Royal Society Range of the Transantarctic Mountains, Antarctica, with Williams field (aircraft on the Ross Ice Shelf) in the foreground, November 1970.

12) Moving camp in the RARE Range, central Lassiter Coast, November 1970; Willis H. Nelson doing the remote driving of a Foxtrac motor toboggan, the author hanging on the rear Nansen sledge; photo by Richard L. Reynolds.

13) View northwest while walking a ridge that climbs to Mt. Laudon, Guettard Range, Lassiter Coast, November 1970; top of Mt. Laudon just off the view in the upper right, and Irvine Glacier to the left.

14) View south of Willis H. Nelson remote driving a Foxtrac to outcrops in the eastern RARE Range, with the Wetmore Glacier in the middle ground and the Latady Mountains in the background, December 1970.

15) View west of nunataks with hanging snow cornices in the southwest Playfair Mountains, with motor toboggan on the right,New Year's eve 1970.

16) View west of camp and Mt. Coman, on the peninsula plateau in the western part of the northern Lassiter Coast, November 1972.

17) View west at midnight of Walt Vennum on the summit of Mt. Kane, Playfair Mountains, January 1973; photo by Karl S. Kellogg.

18) Hercules C-130 pickup flight in whiteout at the Ferguson Nunataks, northern Lassiter Coast, February 1973; Jamesway hut at the far left.

19) Jill and Scott during a vacation at a dude ranch in Colorado, August 1975.

20) View west showing Michael A. Boyles sampling rocks at Cape Zumberge, with ice cliffs in the background that mark the south edge of the Orville Coast; the Ronne Ice Shelf is above and to the left of Mike, December 1977.

21) Adrienne B. Anderson, about 1980.

22) Thomas A. Steven (red vest), at timberline in the Summitville mining district, San Juan Mountains, Colorado, 1971; David A. Johnston, to the right of Tom; David then was field assistant to Peter W. Lipman, who took the picture; David would later die in the May 18, 1980 eruption of Mount St. Helens.

23) View south-southeast of the Central Mining District (uranium) just east of Marysvale; range in the left to right background is the Sevier Plateau.

24) The author as a U.S. Geological Survey Spokesman, June 1980; photo by Dwight R. ("Rocky") Crandell of TV news in Denver.

25) View southeast from Johnston Ridge of an ash flow (pyroclastic flow), during the eruption of August 7, 1980, traveling down the north slope of Mount St. Helens, Washington; photo by Peter W. Lipman and published by Rowley and others (1981, Fig. 293). (Back Cover Image)

26) View southeast from Johnston Ridge of the north flank of Mount St. Helens on October 18, 1980, while waiting out about two days of the October eruption, which was the last of the six major 1980 eruptions of pyroclastic flows; the white distal end of a new pyroclastic flow is just above the man's head and north of the "gun barrel;" ash and steam issue from a dome in the crater that erupted after the ash flow; photo published by Rowley and others, 1981, Fig. 287).

27) Chester A. Wallace in the Coliseum in Rome, January 1981.

28) The author at McMurdo Station, November 1984.

29) Storm at base camp south of Fitzgerald Bluffs, English Coast, December 30, 1984; photo by J. Michael O'Neill. (Cover Image)

30) Where did that crevasse come from? Janet W. Thomson climbing the western part of Mt. Rex, with snowmobile and Nansen sled to the right, January 1985; we never saw the bridged-over crevasse above Janet's head until the picture was taken.

31) Group photo at Mt. Peterson, January 1985; from left to right, J. Michael O'Neill, Janet W. Thomson, David J. Lidke, Karl S. Kellogg, Thomas S. Laudon, the author, and Walter R. Vennum.

32) Mary A. Siders on top of Mt. Evans, Colorado, about 1987.

33) Third British Antarctic Survey Twin Otter put-in flight at base tent camp at Kamenev Nunatak, Black Coast, December 9, 1986

34) View north of Mt. Jackson (highest mountain in the Antarctic Peninsula and 60 miles away), from camp west of Brennecke Nunatak, Black Coast, January 1987.

35) The field party at midseason, probably at Brennecke Nunatak, January 1987. Left to right, the author, Hein E. Wever, Anthony R. Simpson, Arthur B. Ford, Peter T. Marquis, and Brian C. Storey. (Cover Photo)

36) View north of Rowley Massif in distance (surmounted by Mt. Vennum, in clouds), 6 miles away across Odom Inlet of the Weddell Sea; author is taking notes in the Peck Range, Black Coast, December 1986; photo by Arthur B. Ford.

37) View northwest across Rainbow Canyon, south of the town of Caliente, Nevada; the canyon cuts through the Caliente caldera complex, so nearly all rocks are intracaldera ash-flow tuffs that filled the caldera and are thousands of feet thick.

38) Dawna, during archeological field work near Delamar ghost town, Nevada, about 2005.

39) From left to right, Gary L. Dixon, E. Bart Ekren, and the author, in Las Vegas 1999.

40) Andrew and Jill Knutson, in Tucson, Thanksgiving 1999.

41) Patty and Scott Rowley, in Denver, their wedding, September 1987.

42) View east of David B. Hacker (left) and Robert F. Biek at the "discovery site," at the sothern distal edge of the Markagunt gravity slide, Haycock Mountain, Markagunt Plateau, Utah, July 2012; the horizontal slide plane is at the level of the brim of Dave's hat.

43) View southwest of the 5000-foot scarp of the sothern Tushar Mountains, with the Markagunt Plateau on the left horizon; town of Circleville is just off the upper left of the view. Markagunt gravity slide makes up all areas in the view.

44) Dawna, with Oreo as a puppy, at our barn, April 2011.

2

Mount St. Helens

To work on an active volcano is a fantasy of almost any geologist who has spent time studying the deposits of extinct volcanoes—because this way you learn how volcanic rocks formed by first-hand observation without having to rely only on reading about volcanic activity. My opportunity came by unforeseen, wondrous luck: a phone call to my home on a Friday night—June 6, 1980—from Donal (Don) R. Mullineaux, a volcanologist in the USGS Branch of Engineering Geology, whom I greatly admired but did not know well. Don, then helping to manage USGS research on the Mount St. Helens disaster out of Vancouver, Washington, asked if I would take, ASAP, the temporary job of Information Scientist for Mount St. Helens, running USGS press conferences each morning in Vancouver, plus "other duties as assigned." The current (and first) Information Scientist, M.H. (Tim) Hait, Jr., who had been on the job since May 20, was getting burned out by the pressures of the job and the fifteen-hour work days, and he had asked to be relieved. Tim, a tall, good-looking geologist about ten years older than I was and with a Ph.D. from Pennsylvania State University, had been doing a great job. He was able to answer questions from dozens of reporters, few of whom had ever taken a college science course, and to explain simply and patiently the geology and meaning of eruptive events each day. He had a great desire to teach about geology in general, and Mount St. Helens in particular. He kept his comments to facts or reasonable scientific interpretations, and he avoided preaching, suggesting policy, or proposing restrictions to access. The job had been invented on May 20 because a full-time technical expert, familiar with the nuances and vocabulary of active volcanoes, was now needed more than a standard public-affairs person. Factual data about the volcano were needed to quell confusion, apprehension, fear, and rumor.

The eruption of Mount St. Helens on May 18, 1980, and its aftermath, ended up being the most important geologic news event that concerned USGS science in its history. Besides that, journalists had voted the story of Mount St. Helens the number one news story in the United States for 1980. For almost a year, Americans rushed to the TV whenever the news came on, for the video of an erupting volcano was compelling and exciting and, from day to day, nobody had much of an idea as to what it would do next. Mount St. Helens had been unpredictable and deadly, for the big and quite unexpected eruption of May 18 had killed fifty-seven persons. There were to be five more big eruptions, all in 1980, but none close in size to that first one, so the mountain held the attention of the country all year and beyond. Furthermore, the eruptions were visible from the population centers of Portland, Oregon (fifty miles to the southwest) and Seattle, Washington (a hundred miles to the north).

Many of Tim's comments in video in his press conferences were aired on local TV news and some on national TV news, and many of his oral comments in interviews were aired on radio. It was clear that he was going to be a tough act to follow. I had already planned a big weekend with the kids, but I was unsure of whether to say "yes" to Don's request for another reason: I was not particularly good at public speaking, and stage fright was a big reason. By then I had given oral scientific reports at many national and international meetings (plus their published abstracts), but such talks were the least enjoyable aspects of my job. Yet, I had also been in press conferences dealing with Antarctica. I asked Don if I could have a day or two to think about it, and he said "O.K., but make it quick."

So, what was the big deal about Mount St. Helens? For one thing, it was the only volcanic eruption in the twentieth century in the conterminous US except for Lassen Peak in northern California, which erupted from 1914 to 1917. But Lassen, a volcanic dome with only 2,000 feet of relief, erupted with little fanfare or press interest because the mountain and its eruptions, although as explosive as Mount St. Helens, were not big, killed no one, occurred far from population centers, and the eruptions coincided with World War I. In contrast, the events at Mount St. Helens began with an unexpected bang on March 20, 1980, that of an earthquake of 4.1-magnitude (Richter scale) at shallow depth just northwest of the summit, followed by aftershocks. The University of Washington (UW) Seismology Center, Seattle, which recorded the quake, was run by Professor Stephen D. Malone of the UW geology department, and he notified the USGS in Menlo Park. On March 21, as aftershocks continued, the US Forest Service (USFS) phoned the USGS for information, initiating USGS involvement. The UW and USGS collaborated and rushed monitoring devices and personnel to the mountain. The events made headline news. The next several days saw many more earthquakes, some at or greater than 4.0 magnitude. A gifted thirty-year-old USGS volcanologist from Menlo Park, David A. Johnston, who had only just finished his Ph.D. at the University of Washington (with the help of Stephen Malone) and had joined the USGS two years previously, was vacationing in the Seattle area and immediately went to the Center to help analyze the seismic data. On March 25, 22 quakes of greater than 4.0 (one was 4.7) were recorded in one eight-hour period, and Don Mullineaux arrived at the USFS headquarters of the Gifford Pinchot National Forest in Vancouver to provide technical assistance to the USFS. On March 26, the USFS, which administers US public land on and near Mount St. Helens, issued warnings, closed some roads, and restricted airspace. Initially, it was the Public Information Office of the USFS that issued information on Mount St. Helens. Press conferences were held in the USFS headquarters building, and later a local motel, where spokesmen for the USFS, State of Washington, local agencies, and public utilities provided general information. But since attention was focused on when the volcano would erupt, Don Mullineaux, and later Dwight R. (Rocky) Crandell and Robert L. Christiansen (all three of them USGS superstar volcanologists), became the focus of attention at the press conferences. The accelerating seismicity suggested that magma was moving upward in the vent. The USGS Public-Affairs Office was now issuing national news releases on the activity with increasing frequency.

Before 1980, Mount St. Helens was a beautiful, symmetrical stratovolcano, six miles across at its base and rising five thousand feet to a narrow, snow-covered summit above timberline and at an elevation of 9,677 feet. It was the fifth highest mountain in Washington (stratovolcano Mount Rainier, nearly fifty miles to the north, is the highest at 14,410 feet). People referred to Mount St. Helens as the "Fiji of America" because its classic form resembled fabled Mount Fiji of Japan. Mount St. Helens is forty-five miles north-northeast of Vancouver, which is across the Columbia River but a suburb of Portland. Beautiful, two-mile-long Spirit Lake and adjacent Spirit Lake lodge and Mount St. Helens lodge, nestled in a pine forest in the valley of the North Fork of the Toutle River, had been popular tourist spots on the northern side of the mountain. The end of the road to those lodges and other private cabins climbed partly up the northern side of the volcano to a viewpoint. Mount St. Helens, Mount Rainier, and Lassen Peak are part of the Cascade Range, a section of the Ring of Fire (volcanoes) that surrounds the Pacific Ocean basin.

Despite its beauty, Mount St. Helens is an extremely dangerous volcano. A ten-year study by two of the three mentioned superstars and among the country's best volcanologists, Rocky Crandell and Don Mullineaux, reached this conclusion. They found, based on some eyewitness accounts in the historical record and carbon-14 ages, that the volcano had erupted every hundred to three hundred years (the error bars on these types of evidence are large) since at least 1200 to 700 BC. The previous eruption had been in 1857, although no deposit from this had been recognized. They published an account of the dangers in 1975 in a remarkably prescient report in *Science*, with geochemist Meyer Rubin, in which they stated "an eruption [of Mount St. Helens] is likely within the next hundred years, possibly before the end of this century." Their final report in 1978 gave the evidence. These and other reports they and others authored were not only invaluable in helping USGS scientists to understand what was to follow, but gave the USGS an unusual amount of credibility with other government agencies and the public. Partly because of this background, the effort by the USGS, USFS, and others saved many lives, probably in the thousands, at Mount St. Helens. Several years later, the USGS predicted an eruption and saved thousands of lives at Mount Pinatubo in the Philippines. To my mind, these predictions and hazard studies marked the height of scientific and public respect for the USGS.

Ash eruptions began on March 27, 1980, only a week after the earthquakes began. A loud boom and a puff of ash around noon was followed two hours later by a 4.7-magnitude quake, which was the second strongest to date, and an ash plume rose seven thousand feet above the volcano, opening a crater. The events of this day became a huge magnet to the press, tourists, and every USGS scientist in specialties that might be of value to studying the mountain. Some scientists fanned out to make field observations, and others installed instruments to measure deformation on the volcano flank (tiltmeters and surveying by electronic distance-measuring devices), and composition of volcanic gases, heat, and seismicity. David Johnston and Tom Casadevall, whose specialty was volcanic gases, joined the field studies group on March 27. In an interview that day, David likened the volcano to dynamite with a lit fuse "but you don't know how long the fuse is." Until then, logging had been allowed in parts of the Pinchot National Forest surrounding the volcano. Over the next several days, about three hundred people in logging camps within a fifteen-mile radius were evacuated, the road to Spirit Lake and other roads were blockaded, and the citizens of Cougar to the southeast were advised to evacuate. Water was released from adjacent reservoirs. But longtime codger, widower, and owner of the Mount St. Helens Lodge, Harry Truman, refused to leave, so the press found him to be a popular interview subject. At one point he told Richard Waitt and Barry Voight that "geologists didn't know shit about" what was going to happen at Mount St. Helens. His bravado had the deleterious effect of giving backbone to many tourists and loggers ("give 'em hell, Harry," stated one newspaper, parroting the line voters once said to encourage President Harry Truman!), who doubted the scientists and insisted upon visiting the volcano close-up. The main then-perceived dangers discovered by USGS scientists were not in the ash but in cracks (crevasses, if you will) in a slumping ice field on the northern flank of the volcano as well as possible flooding and mudflows from melting ice. Soon, however, the USGS recognized evidence of other even more terrifying dangers.

Over the next several days, eruptions of ash and steam increased in intensity and volume, and volcanic ash fell over a wide swath of country and towns, carried by winds mostly to the east and southeast. Avalanches of snow, ice, and mud moved down the eastern flank. Seismicity included a 4.9-magnitude quake on March 28. The smell of sulfur dioxide in the gases indicated that magma was being tapped. USGS geologists collected samples of ash to see whether they were from old pulverized rock or from new magma. They were found to be from the former, indicating that for now these were phreatic eruptions (those triggered by subsurface water flashing to steam by hot rocks). Over the next week, however, the boring regularity of these relatively small eruptions led to a holiday atmosphere. Tourists and members of the press swarmed like locusts over, or evaded, roadblocks (easy to do because logging roads criss-crossed the area). Some sold t-shirts, small bottles of ash, and other souvenirs. Some "reporters" had a helicopter land them on the crater rim. On March 30, as many as seventy aircraft were flying around the volcano at the same time. Loggers and truckers demanded to be let back into restricted areas "to earn their living," and by April 1 most of them were indeed allowed back. Yet it was also on April 1 that the seismic signal included the first harmonic tremor, a likely sign of magma moving up the vent. A dozen major eruptions, including one plume that rose to an altitude of twenty-thousand feet, rocked the mountain on April 2. The National Guard was called in on April 4 to maintain roadblocks.

On April 7, UW seismologists confirmed that most of the earthquakes were originating beneath the northern flank of the volcano. On April 11, another 4.9-magnitude quake was recorded. Yet starting April 14, eruptive and seismic activity decreased, although another quake of 4.9 took place on April 19. Though by about April 20 eruptions had mostly ceased, by then scientists had found something far more ominous: the upper northern flank was bulging outward, indicating inflation due to upward emplacement of magma, as previously suggested by the seismicity. Observations were necessarily sporadic as spring weather produced overcast conditions, but on April 23, comparison of aerial photos taken over the past eleven days confirmed that "the bulge" had risen about 250 feet above the non-bulged flank of a year before. Surveying confirmed this analysis, while tiltmeters, which had to date given contradictory data, on April 27 provided more confirmation. More troublesome were abundant large fractures seen in the rock on April 28, once the snow had been melted off. The bulge was 320 feet high on April 29—alarming, in that this entire part of the upper flank looked unstable and could result in an avalanche of rock and ice if triggered by a quake, as Rocky Crandell warned that day. On April 30 Governor Dixie Lee Ray declared a "red zone" around the mountain at distances of three to eight miles from the peak. (Later in her career, Dixie Lee became a right-wing climate denier, as typified by her 1992 speech "Global warming and other environmental myths.")

On May 1, geologists calculated the bulge to be moving about two feet outward every twelve hours. That day, the USGS established an observation station, named Coldwater II, on top of a ridge that rose 1400 feet above the North Fork Toutle River, Spirit Lake, and the two lodges, and 5.7 miles north of the summit. It was just outside the red zone. On May 2, owners of some summer cabins on and near Spirit Lake stated that they would refuse to pay their property taxes because they had not been allowed to enter their property. By May 6, the bulge was growing at a rate of four to five feet per day, yet the volcano was quiet. After a lull of two weeks, the volcano resumed eruptions and seismicity on May 7, with an ash cloud rising to thirteen thousand feet, and on May 8 the harmonic tremor resumed. Most geologists felt that a gigantic landslide was due and that it might trigger an eruption. The landslide was considered the main danger, although nobody thought it would reach Coldwater II. Furthermore, most UW and USGS experts thought that any major eruption would be preceded by a seismic or eruptive warning. On May 14, rates of emission of SO_2 increased substantially, by then a sign (in retrospect, it was probably the "spike" that they were looking for, but they were expecting something more abrupt) that David Johnston and others knew was suggestive of an imminent eruption. On May 15, the outward movement of the bulge was five feet, yet tourists were still flocking to the mountain because the skies were blissfully clear and eruptions had ceased for several days.

On May 16, a Friday, workers at the Boy Scout and YMCA camps along Spirit Lake were allowed past the roadblocks to retrieve equipment. That inspired owners of private property to demand similar access, and some threatened to converge and go through the roadblocks. On Saturday, May 17, Dixie Ray relented and allowed fifty carloads of property owners to go to their property to get possessions, with a second caravan scheduled to leave at 10 am on May 18. Harry Glicken, David's twenty-two-year-old USGS field assistant, had been manning the Coldwater II observation site for the previous six days but had an appointment to visit with his Ph.D. advisor of the University of California, Santa Barbara, and would consequently be gone for several days. Don Swanson agreed to take over the observation post while Harry was gone, but a professor friend and his grad student from Germany wanted Don to show him around, so Don asked David to man the post, with the promise that he would relieve him Sunday morning. David said yes. Before he drove to Coldwater II on May 17, David worried that his readings of gases might have been contaminated or diluted by the air or groundwater, and realizing that gas samples taken directly from fumaroles in the crater would be more accurate, persuaded a helicopter pilot to take him to the crater edge, from where he descended about 150 feet, then returned rapidly (almost running down and back) with his samples. Harry was with him in the chopper but David forbade him to accompany him into the crater because of the dangers and because David was a marathon runner in excellent shape. Late that afternoon, David and Harry arrived at Coldwater II. Soon after, Carolyn Driedger, hydrologist with the USGS, Tacoma, and her field assistant, Mindy Brugman, a CalTech grad student, arrived with the intention of camping there overnight. David insisted "no," saying that this site was not safe and urging them to return to their motel. Fortunately, they agreed and soon left. Harry drove back to town in his truck at about 9 pm.

David spent the night in his USGS trailer. Harry Truman, of course, stayed at his lodge. Robert Kaseweter, a geochemist who had volunteered to be a volcano watcher and received permission to be in the red zone, and his fiancée, Beverly Wetherald, stayed in a cabin on Spirit Lake. These three were the only ones within the red zone. The rest were outside the red zone and considered themselves safe. These included Gerald Martin, a retired Navy radioman and volunteer volcano watcher for the Washington Department of Emergency Services, who had a camper on a ridgetop two miles north of Coldwater II, presumed to be well outside the red zone. Newlywed Reid Blackburn, a staff photographer for the Vancouver Columbian, on loan from the National Geographic, was camping at Coldwater I, outside the red zone three miles west-northwest of Coldwater II and 8.4 miles northwest of the volcano summit. Newlyweds John and Christy Killian were camping at Fawn Lake, about nine miles northwest of the summit of Mount St. Helens and beyond Coldwater I.

At 8:32 am on May 18, the peace of a beautiful Sunday was broken by a 5.1-magnitude earthquake, followed by many aftershocks and the cataclysmic eruption. The main earthquake immediately triggered a huge landslide made up of the broken and deformed rocks of the bulge, plus snow and ice, and the slide raced down the northern flank. This slide has been named the debris avalanche, and its emplacement and its deposits became the model for other huge volcanic landslides, including the world's biggest on land, which colleagues and I worked on in Utah, as covered in Chapter 11. Complete sequences of spectacular photographs, from several angles and locations, show the progress of the slide as well as small dark clouds that rose from below the base of the slide. As the debris avalanche moved down the flank, within seconds these dark clouds became a huge blast directed horizontally northward, known as the directed blast or lateral blast. Removing the upper flank of the volcano (the bulge) had in effect removed the rock cover that contained the magma, and the volatiles (gas bubbles) in the magma expanded rapidly as the caprock was removed and the lithostatic pressure dropped immediately, driving whatever remained of the cover rock outward as a spray of pulverized rock of mostly silt and sand size. The analogy here is of popping the cork on a bottle of champagne that has been shaken.

As pulsing hot lobes of blocks and fine rock, the debris avalanche hurtled the 3.5 miles down the flank of the volcano, at a speed of roughly 100 mph, into the valley of the North Fork of the Toutle River and onward, partly climbing and overtopping one low place in the ridge that contained Coldwater II. Most of the rock and mud of the debris avalanche, however, swung west and followed the river valley for about fourteen miles, taking about ten minutes to get that far. As it mixed with melted snow and ice from the volcano and with river water, it became a debris flow, then a mudflow, then a flood, thus causing significant damage downstream to bridges and homes all that day. Part of the debris avalanche also entered Spirit Lake, causing a huge wave that splashed onto the northern shore and raised the lakebed by about two hundred feet. The debris avalanche filled in the upper part of the valley of the North Fork to a thickness of at least 150 feet, with a hummocky upper surface. Later Barry Voight and Harry Glicken did most of the initial study of the debris avalanche, and Harry made similar landslides his life's work, hoping to awaken people to this hazard around the world. The Mount St. Helens debris avalanche was considered one of the world's largest at the time. Three side-by-side landslides that colleagues and I mapped and studied thirty years later, in Utah, were many times larger and about twenty million years older.

The directed blast was far deadlier than the debris avalanche. The blast began about twenty to thirty seconds after the 5.1 earthquake and in about eighty seconds destroyed a seismic station 3.5 miles north of the blast origin. This calculates to a speed of 220 to 250 mph for this debris-laden blast cloud, so it outran the debris avalanche. David Johnston, then outside his trailer taking measurements and relaying them by radio to Bob Christiansen in Vancouver when the quake struck, shouted into his radio "Vancouver, Vancouver, this is it" before the lateral blast overwhelmed him. (Harry Glicken was deeply shaken and thereafter blamed himself for the death of his friend. Sadly, Harry was one of thirty-nine persons killed later, in 1991, when an observation site was overrun by a pyroclastic flow at Mount Unzen, a volcanic dome in Japan, while Harry was a postdoctoral fellow at a Japanese university.) Gerald Martin saw what happen at Coldwater II and coolly radioed a description of the avalanche and blast, concluding "the camper and the car just over to the south of me [Coldwater II] are covered. It [the blast cloud] is going to get me, too" before he also was killed. The men's bodies were never found. Reid Blackburn also died, inside his Volvo. The Killians, like many others, were in the blowdown area where huge trees were torn from their roots and laid down like north-directed toothpicks, and they were killed; John's body was never found, while some skeletal remains of Christy were found a year later. The total death toll was fifty-seven persons, a terrible tragedy.

The directed-blast cloud spanned a northward arc of nearly 180° (twenty-three miles east to west) from the blast point and extended about eighteen miles to the north. Nearly all life in the inner six miles (north-south) was lost, and trees were stripped and snapped off or uprooted. Beyond that distance, trees were largely stripped and toppled. At the northern edge, trees were seared but left standing. The thickness of the directed-blast deposit was about three feet near the source and a half inch at its northern margin. Estimates of a temperature of the blast, based on melted plastic and other things, was about 680°F. Detailed written observations, with spectacular photos, by many persons on the ground and in the air, provided an accurate time frame of the events. Perhaps the best of these observations and photos were by geologists Keith and Dorothy Stoffel, who had chartered a plane and pilot in Yakima and were in the air above the volcano when the earthquake took place.

The debris avalanche took out the summit and northern flank of Mount St. Helens, and the directed blast cleared the new vent through the remainder of the flank and any landslides from the over-steepened inner walls of the new crater. The mountain lost a height of 1,300 feet. A gaping, enlarged crater faced northward, with higher walls on the southern, western, and eastern sides. Such a crater, with one side missing, was known before around the world and was called a sector collapse of a stratovolcano, and now we knew how it formed.

As the directed blast was continuing, and perhaps thirty seconds after it began, a vertical eruption column of ash of juvenile origin (magma) rose from the summit crater of the volcano. The UW recorded aftershocks, plus a quake of about 5-magnitude two minutes after the initial quake, then "strong seismic signals" were observed for another ten minutes, representing the upward movement of magma and its eruption. Within ten minutes after it began, the ash column had reached a height of fifteen miles, and wind blew it to the east-northeast. Daytime turned to darkness for as far as 120 miles downwind (at about 11 am, and sooner progressively closer). Yakima, about ninety miles away, was the first significant town to experience thick ash, but ash continued eastward through central to southeastern Washington, northern Idaho, southwestern Montana, and reached northwestern Wyoming by early evening. This is known as ash-fall or air-fall tuff, and it accumulated to thicknesses of eight inches immediately downwind of the vent and averaged 1.5 inches at Ritzville, Washington, about two hundred miles downwind. White to light-gray, it looks and falls like snow but does not melt. And because it is as heavy as rock or cement, it leads to the collapse of many roofs. The vertical ash column erupted steadily for nine hours, obscuring everything, and it was accompanied by strong seismicity.

The ash in parts of the eruption cloud collapsed from its weight onto the flanks of the volcano, from where it moved down the flanks driven by gravity at speeds of roughly 100 mph. Some of the masses of ash simply bubbled out or were blasted out (by the expanding gas in the bubbles) of the vent. These ash masses are pyroclastic flows, also known as ash flows, and their deposits are ash-flow tuff (or ignimbrites). These fast-moving masses consisted of juvenile ash (angular pieces of gas bubbles that had solidified and burst as the gas expanded in the lesser pressure of the atmosphere) and pumice chunks more than a foot long containing a small percentage of small crystals (phenocrysts) of minerals that had crystallized in the magma chamber. The masses of ash and pumice largely hugged the ground surface, like snow avalanches, as they moved down the northern volcano flank.

The emplacement temperature (when deposited) of the May 18 pyroclastic flows was about 780°F, with eruption temperature much greater. The flows were deposited on top of the blast deposits, which in turn had been deposited on top of the debris avalanche. (This layering of younger deposits on top of older deposits is called stratigraphy.) The ash flows consisted of lobes of ash and pumice that extended as far as almost 4.5 miles from the crater, almost to Spirit Lake and starting down the North Fork, with a maximum thickness of about 130 feet. Their top was a horizontal surface, forming what we call the "pumice plain." As they were deposited on top of stream water and small ponds dammed by the debris avalanche, the water flashed to steam and bored phreatic-explosion pits—most of which are about a hundred yards in diameter—upward through the pyroclastic flows.

By Monday, May 19, rescue operations and recovery of bodies began. Helicopters plucked out survivors, some of whom later died of their injuries. Some bodies were never found. These pilots performed remarkable feats of heroism and skill as they rescued about two hundred people. Towns started to clean up ash. Volcanic activity on Mount St. Helens slowed considerably. The ash column rose only occasionally to a thousand feet above the volcano and earthquakes were less frequent and smaller (less than magnitude 3). Yet now the fun-and-games atmosphere was over. Vendors still hawked t-shirts and trinkets, but the mood was serious and most citizens were scared, or at least respectful, of the volcano. Within two days, weather clouds cleared to reveal the mountain, now an amphitheater open to the north and with a relief of about 1800 feet. Trees had been flattened and gray ash several feet thick coated everything. President Jimmy Carter and Governors Dixie Ray and John Evans (Idaho) toured the devastated area on May 21. However, the quiet was soon interrupted on May 25, when the volcano erupted ash that reached an altitude of 24,000 feet in seven minutes, then on to an astonishing 45,000 feet. Harmonic tremor was noticed on the seismometers, along with swarms of small quakes. The eruption lasted five hours, and the wind direction drove most ash to the northwest, closing many airports. Other than ash, the eruption contributed several small pyroclastic flows that reached the pumice plain, and therefore we considered it one of the six significant eruptions of 1980. Largely because of poor weather, meaningful temperature measurements were not made. Governor Ray enlarged the red zone to twenty miles in all directions from the summit, with significant penalties for violating the ban. After this small eruption, things quieted down again until the second major eruption took place, on

June 12. That one involved me.

By Friday, June 6, when Don Mullineaux called me, I—like many others in the world—had come under the spell of Mount St. Helens. Many USGS and academic geologists, geophysicists, and hydrologists, whose specialties would be of use in understanding the volcano (but were motivated mainly by excitement), made reservations to Portland soon after the initial eruptions started (and only when they were on their way did they ask their Branch Chiefs for approval.) But I had been with the USGS for only ten years at this point, and did not think I had the seniority to presume I could go. In those ten years I had worked my way up the USGS grade system, to a GS-13 in 1975 (the average grade for a senior geologist) to GS-14 in 1978 (above average), but to me this did not seem to be enough. As I thought about Don's offer, it became obvious that Information Scientist was the only way I would get there. So, I called Don back that night and said "yes." I was on my way to Portland by noon on Sunday, then connected with Tim and some of the other geologists and seismologists. I knew some of the scientists (Richard Waitt, Pete Lipman), but I did not know most of the others. Many, like Pete Lipman, Rocky Crandell, and Don Mullineaux, were well established stars, and that list also included Robert L. Christiansen, Donald A. Swanson, James G. Moore, and Robert L. Schuster. Other geologists already on the mountain were more recent hires whose job was to study active volcanoes; these included Richard Waitt, Richard P. Hoblitt, Norman C. Banks, Thomas J. Casadevall, C. Dan Miller, and David A. Johnston. I would soon find all of those working on the mountain to be among the world's best in their specialties. Donald W. Peterson, a senior geologist, former Branch Chief, and senior administrator, was now heading the USGS effort (I had met him in 1970, during my return from Antarctica when Duff showed me the active Hawaiian flows and pit craters, for Don was then in charge of the Hawaii Volcano Observatory). His selection for Mount St. Helens was a good choice, for he had a Ph.D. from Stanford and was as good a volcanologist as any of his team. But he had chosen administration, a job that requires a different skill set from that of a geologist. Not all geologists succeed as administrators, but Don did, and most of the eventual success of what all of us did at Mount St. Helens can be attributed to him.

Tim had a few significant pointers on what to do and not do, but in keeping with the trust that USGS geologists and administrators had for each other, they were minimal. One, for example, was that I should answer all questions seriously, no matter how often they were asked, remembering that media people are typically struggling with strange words and complicated concepts. I was not quite sure exactly what Tim meant about "no matter how often asked," but I would soon understand completely! He also recommended that I add qualifiers to many answers, such as "I am not sure, but …," "the scientific evidence suggests … ," or

"I don't know at this time." Another bit of advice was not to recognize the notion of a "dumb question": if someone was asking, it meant they didn't know, so there was nothing "dumb" about it! A third point was to be aware of tricky situations, for yourself or for the USGS, in how you answer. This would include a real no-no such as predicting an eruption based on your gut feeling, recommending evacuations, criticizing another agency or a member of the media, and lots of others. And fourth, Tim suggested I double-check all facts and interpretations. He also had a few points to make about reporters and other media professionals. These included the fact that the Associated Press (AP) and United Press International (UPI) supplied stories to the entire country, so I should make sure they get the facts. Also, I should understand that any errors in the final, published newspaper or TV story might not be the fault of a reporter; for editors can shift meanings in a story and add unwarranted and inappropriate headlines. And finally, Tim reminded me that the 6 pm news will not wait on the USGS, so I needed to respect the media's deadlines.

The Federal Emergency Management Agency (FEMA) had rented the third floor, a large unfinished room, of a red-brick bank building in Vancouver, where desks were set up for eighteen FEMA information officers and secretaries as well as employees of the USFS, Army Corps of Engineers, Small Business Administration, State of Washington, USGS, and other agencies. Not only was FEMA recording and transmitting equipment added to make daily news reports, which radio stations could access by telephone, but FEMA teletype machines, word processors, and a photographic department also filled part of the room. The Information Scientist and several USGS public affairs persons sat at desks in an open area containing phones, each with the capacity for about a dozen incoming phone lines. On a scientific question, the public-affairs person would defer an incoming question to the Information Scientist. Answering such phones, mostly from print, radio, and TV reporters and news-wire services, but also from concerned citizens, professors/teachers, and geologists, took up most of our day. Whoever took the call had to understand that the conversation was probably being recorded at the other end, either for broadcast or to be quoted for a story on which the reporter was working, so one had to choose one's words carefully. I found such phone conversations to be one of the most stressful things I had to do, for I was always acutely aware of the implications of any unclear or unresearched answer to their questions, even their "dumb" questions. Some reporters came onto the third floor to do a live interview, notably the person designated as the Information Scientist (typically me), and we would go into an office or outside, with or without a video camera on. Televisions were sprinkled around the room, always on when news programs were on. Press conferences took place in a small auditorium on the second floor. In short, it was a bustling place where the Information Scientist spent most of his (and later, her) day.

The daily plan that Don Peterson had established for our team of about twenty to twenty-five USGS scientists and some others from UW or other universities was to start each day with an all-hands meeting at 7 am, in which each scientist gave a summary of the data he or she had collected from the previous day, with an analysis of what it meant, if anything. We scientists in the team rarely worked less than ten or twelve hours a day, and we never took any of a weekend off, partly because we were driven by the lives already lost. When someone became too exhausted by this unforgiving routine, they let Don know and he or they would find a temporary replacement while they returned home for a week or so. Commonly we had a bull session at the end of a day to analyze developments of that and previous days. Armed with new information after the morning meeting, scientists went off to do their day's work in their specialties, whereas the Information Scientist, with a few notes on a three-by-five-inch card, would head to the auditorium. The USGS Information Scientist led off each press conference at 9 a.m., after which the USFS, FEMA, and various offices of Washington State would give daily briefings on restrictions, weather, road openings and closings, and many other things. I watched Tim give his last press conference on Monday June 8. Video cameras were aimed at the podium from around the room, not only from the affiliates of the major TV news networks but also from foreign networks. Cameramen operated their video cameras while their reporters, poised to ask questions, sat in a chair in view of their camera.

After a more-or-less sleepless night in the Shilo Motel in Vancouver, then the 7 am daily USGS summary meeting, I took the lead at the 9 am press conference on June 9. Nobody, not even Don, had attempted to give me any coaching. Following the model of what Tim had done the previous day, I gave a briefing of the previous day's events, including seismicity, visual observations, gas studies, floods, deformation studies (tiltmeters, etc.), and other topics of interest. I tried to have something significant to say or to educate our audience in some way, rather than merely reciting data. As Tim had done, after about fifteen minutes of my monologue, I opened the floor for questions. That's where things got interesting! Almost always everybody wanted predictions, primarily concerning when the volcano would erupt next, and they wanted the likelihood on a scale of one to ten. This I could not give, although in my monologue I would interpret what the tiltmeter, gas studies, etc. might suggest in that regard. At my first news conference, I soon saw that most questions were in the realm of dumb, although I followed Tim's advice and never let the questioners suspect that I thought so. By now, most reporters—unless this was their first day—had been introduced to many simple geologic terms or they knew the implications of harmonic tremor or a spike in SO_2, but I still needed to keep geologic terms to a minimum unless I could define them quickly. The press conferences were informal: I do not remember wearing a tie, and I asked reporters to call me Pete.

Commonly a question would be asked that the reporters considered to be particularly insightful. After I answered it, a reporter from another affiliate would get up and ask the identical question. His cameraman would tape his question and my answer. Then the reporter from a third affiliate would ask the same question. I finally understood Tim's advice. Clearly, I should not tell reporters #2 and #3 that they should listen to my answer to reporter #1! This "question du jour" was apparently considered the best any of them could think of that day, and each network wanted their cameraman to show their reporter asking the question, then me answering it. I smiled at their game, and I tried to give a somewhat different answer to each, or to use a different explanatory sentence, which might clarify or embellish my answer to reporter #1. My answers had to be short, generally one or two sentences, so they could be shown that evening on TV. The last thing they wanted was a five-minute answer. Humor was welcome, but obviously had to be used sensitively.

I quickly learned never to watch the local or national news, especially when it concerned the volcano, for I knew that I was likely to be shown. Part of that was self-consciousness. But the main reason was that I did not want to catch a reporter or their editor butchering/misinterpreting what I had said, for I did not want to be irritated with that reporter the next day.

Although I was never too impressed with local TV reporters, at times I had a sit-down conversation (audio- and/or videotaped) or a telephone conversation (audiotaped) with a local or national print journalist who was doing a feature on the volcano. Their questions quickly made it clear whether these journalists were on the ball; the hard evidence came when or if they sent me a copy of their story. Some articles were excellent, and I came away with a huge respect for some journalists at some of the major newspapers and magazines. These people really knew how to write, and their human stories, including the tragedies of May 18, had a message that would commonly leave a lump in one's throat.

Many comments at our 7 am meetings or evening bull sessions expressed scientists' desires that a better understanding of explosive volcanoes and even volcano prediction could come out of all the destruction and sadness. I jotted down such comments in my field notebook, commonly attributing them to the person who had said them (the above sentiment, for examine, came from Pete Lipman, although many others said similar things). Toward this end, we threw every instrument we could think of at the mountain, hoping that we could somehow establish when Mount St. Helens would erupt again. As the song by The Who said, we were determined that we "won't get fooled again." We knew the volcano was not yet dead, but it remained "spooky" and unreadable. I tried to convey a sense of this desire in my introductory comments some days at press conferences. However, we never found a single secret key, and the next eruption surprised us once again. But overall, by the end of 1980 we knew a great deal more about stratovolcanoes and explosive eruptions than anybody had before, as reflected in the huge (844 pages and 62 individual reports) *Mount St. Helens volume* edited by Lipman and Mullineaux and published by the USGS at record speed in 1981. We had written the book on explosive eruptions! It is important to emphasize that all of us felt a sense of urgency to figure out this volcano before it killed anybody else, and to use that information to understand other volcanoes. As far as Mount St. Helens went, no serious injuries or deaths occurred after May 18 because all government agencies, including ours, did our jobs!

Friends in various cities later told me that the local affiliates fed some of their video of me, either from press conferences or one-on-one interviews, to the national news. Quite a few of my comments went national, but the only one that I remember was one on camera in which I was thrown a softball that I could easily have flubbed, when a journalist asked what was the main objective of that large team of USGS scientists. I replied that our main objective was to save lives. I did not have to remind the interviewer that we had already done that in spades, even though the fact that the big eruption came on a Sunday certainly helped. In fact, Governor Dixie Ray, in an appearance at the US Senate Appropriations Committee on June 10, had praised the "timely warnings" that she believed had saved "many, many lives."

Beginning with my first press conference, I found that my stage fright and apprehension of giving a short talk were minimal, and they certainly did not hinder what I was doing. The reason was primarily that most press conferences consisted of questions, so I did not have to memorize a speech. And I realized that anybody with some experience in volcanic rocks and with half a brain could do my job. Also, everything was so fast paced that I had no time to be scared. Just like skiing over crevasses in Antarctica, the job was just a job and quickly became almost routine. Routine, perhaps, but because of the frenetic activity and the unpredictability of the volcano, always exciting!

Many reporters' questions my first week, starting with Tim's Monday press conference, zeroed in on the upcoming Friday the 13th and the fact that that day coincided with a high tide. Tim and I quickly dismissed the Friday the 13th superstition. I believe I told them that I did not fear black cats either! The topic of earth tides, however, was hardly a dumb question given that the moon at high tide presumably must have a maximum pull on bedrock as well as on water, and consequently volcanic eruptions might happen more frequently during high tides. Some geologists had made that hypothesis, as reporters were aware. We had bounced the theory around several times in our 7 am meetings. I was not inclined to buy into it, nor were most others at those morning meetings, for the statistical evidence was suggestive but hardly compelling. After a couple of days of such questions, I remember saying with a smile and a little exasperation, "Look, if it erupts on Friday the 13th, I'll eat it!," thus coming awfully close to having one hell of a big meal!

Volcanic and seismic activity was low from June 5 to June 11. I joined three other scientists on a helicopter flight on June 10 to Spirit Lake, the ridge containing Coldwater II, and to several other stops to sample volcanic material. Thursday, June 12 likewise began with low activity, but by noon harmonic tremor began building, then increased in amplitude during the afternoon.

But I did not know that when I went to give a talk at the request of one of the County Commissioners of Clark County (the southwestern corner of Washington, which includes Vancouver, named for William Clark of Lewis and Clark fame). He and I went up there in his car and we found ourselves in this large auditorium at about 6 pm. The folding chairs were filled and aisles on the sides were standing room only. It was obvious that the folks were not there for a slide show or any form of entertainment. These people were scared! The airfall cloud of May 25 had passed over their heads and they had no idea what to expect next. I gave an update of what was happening, not knowing that harmonic tremor was building as I spoke. Remember that in those days there were no smart phones alerting us to events in real time. For the next two hours, I entertained questions, until they could ask no more. The commissioner and I headed back to the bank building. By then it was dark and raining, and soon we saw that something besides rain was landing on the windshield, and that it was muddy. Clearly it was windblown ash, and we could only conclude that we were in an eruption. We also smelled SO_2. The radio was no help, for we were between hourly news summaries. What we did not know was that harmonic tremor had peaked at a little after 7 pm, and an ash column that rose to 13,000 feet had followed five minutes later. Then tremor dropped off for two hours, returning to its maximum amplitude about 9 pm. National Weather Service radar in Seattle tracked the main eruption plume at 50,000 feet at 9:18 pm. Not knowing this, the commissioner left me at the front door to the bank building at about 9 pm and I took the elevator to the third floor. Nobody was there (yet!). As I entered the open area and saw my desk, I could see that all the buttons on the phone were flashing red. At that moment, before I picked up the phone, I was dead certain that an eruption was taking place. A few people soon showed up. Then they started streaming in. The prevailing winds blew the cloud west-southwest, over the area just north of Vancouver, with some ash (one-eighth of an inch overall) in Vancouver and a dusting in Portland and coastal Oregon and as far north as Seattle. The main eruption continued to at least midnight. It had just missed (or maybe it slopped a little into!) Friday the 13th.

We gave our first press conference at about 10 pm on Thursday and the second at midnight or 1 am. Of course, the darkness and cloud cover prevented any actual observations. But many data were telemetered in, so we had plenty to go on and to talk about. And those data indicated that the eruption was continuing as we spoke at the first press conference, if not the second. At about 2 am on the 13th, the National Weather Service reported another eruption cloud as high as 16,000 feet, indicating another eruptive pulse. We held another press conference at 3 am. Between these first three press conferences, one of the local TV crews asked if they could interview me outside. They had found a street light and they parked me under it, asked questions, and made a video of my response. It was one of my most vivid memories. I remember the ash falling like light snow on my head and shoulders, but I cannot remember any of their questions. At 4 am, the Weather Service measured a cloud that rose to 13,000 feet, and at 6 am harmonic tremor picked up. But by those two events, I had returned to my motel for some sleep, and we scheduled the next press conference at the standard 9 am, on the 13th. Rain continued most of the day, and a thick cloud cover prevented seeing much, although it was certain that pyroclastic flows had been deposited. The ash and rain made for slippery streets in Vancouver and Portland, and 15 mph speed limits were imposed. Many people donned face masks when outside. Small shallow earthquakes were abundant all day. No one could observe the crater area, but radar imagery suggested that a volcanic dome was filling the crater created by the ash eruptions. Late in the day, helicopter observations of the pumice plain showed that pyroclastic flows had formed two tongues that extended more than 4.5 miles from the crater, one of them virtually to Spirit Lake. This was the second largest eruption in terms of deposits. Later, we showed that the maximum thickness of the deposits was about fifteen feet. The emplacement temperature measured by Banks and Hoblitt was about 1100°F , hotter than the May 18 pyroclastic flows. Late in the afternoon I participated in a half-hour TV panel discussion in Portland with FEMA official Bob Stephens.

Saturday June 14 was overcast, with only a steam cloud visibly rising to 15,000 feet, but numerous small quakes and a decrease in harmonic tremor evident on seismograms. The weather cleared on Sunday and helicopter observation showed a volcanic dome about 700 feet in diameter and rising about 130 feet above the crater floor, resembling a muffin. Tiny spines or flows of viscous rhyolite oozed up in places through cracks in congealed parts of the dome, accompanied by small explosions of ash that shot to maximum heights of 700 feet. Volcanic domes, such as Lassen Peak and Mount Unzen, can be deadly, but this one seemed benign, and as years went by it was proved so. In subsequent eruptions in 1980, the dome was blown apart as pyroclastic flows shot out, then a new dome was erupted. The mountain, including its seismic signals, remained generally quiet for the next several days after that, although the dome had been growing about twenty vertical feet per day since Sunday. This scenario continued until June 22, when I had elected to "retire" as Information Scientist. Don Peterson had picked a replacement, Jane M. Buchanan-Banks, a geologist with the Hawaii Volcano Observatory. I believe Tim Hait returned for a day or two before Jane showed up and continued the Information Scientist tradition of doing an excellent job! Her husband, Norman S. Banks, was heading up the temperature studies of the various deposits on the volcano, as well as some studies of gasses. Norm, with a Ph.D. from the University of California, San Diego, was one of those I would consider an adrenalin junkie (my Antarctic colleague Walt Vennum is another example, because of his infatuation with mountain climbing), for he liked to take risks. I believe I heard that Norm once put his foot through the top crust of a lava flow in Hawaii, into magma, although his boot prevented most of the bad burns he would have gotten.

I flew back to Denver that afternoon. At my last press conference, I thanked the members of the press for their patience with me and I suggested that the government had done something right with its response to Mount St. Helens. I concluded my press conference with the prediction that what we had learned from Mount St. Helens would save lives in the future at other volcanoes. That certainly proved correct when we got to Mount Pinatubo. Several years later, with several other Information Scientists, I wrote a summary USGS open-file report of what the job of Information Scientist entailed in 1980, plus some do's and don'ts.

I never watched a video of my time at the microphone as Information Scientist. My two weeks were so hurried that now it seems like a nighttime dream, containing only a few short and vivid memories. After each long day, I grabbed dinner at the Shilo Motel with any colleague I could find returning from the field, then flopped into bed. However, months later I received a 35-mm slide from Rocky Crandell who had aimed his camera at the TV during an interview I gave on national TV news. About three months after my stints (I also appeared during the eruption of August 7) as Information Scientist, I happened upon Don Peterson in a hallway or an elevator at Vancouver. He asked whether he had ever told me about an incident with Joe Glutz (I forgot his name, if I ever knew it)—one (and the most influential) of the County Commissioners of Clark County. I said no, so Don told me the story. He said that Joe was the Commissioner who had been the most opposed to me becoming Information Scientist after Tim. The truth was that Joe had a great relationship with Tim and recognized that Tim was excellent, so he did not want anybody else in that job. A month or two after I had left on June 22, Don and Joe were in a meeting together, and in a dull moment Don teased Joe a little by asking him, "Well, how did Rowley work out?" Joe responded with a question of his own, "Where do you get these guys, anyway?"

Several days before I left Vancouver, I told Don of my interest in returning in the role of one of his scientists, and that my preference would be to work on the pyroclastic flows. Beforehand, I had asked Rick Hoblitt and anyone else there I thought might be interested in that topic if they wanted to do it. All said no. Rick's topic of study was the directed-blast deposits, so he spent a lot of time doing field work on the mountain, as we soon did too. Because of my extensive work on ash-flow tuffs in Utah and even in Antarctica, I considered these rock types to be the prize of all the topics about Mount St. Helens. But I needed help to do the study. Two USGS geologists, Norman S. MacLeod of Menlo Park and Mel A. Kuntz of Denver, expressed interest in joining me, and I secured funding from Don. Norm was a volcanologist, with a California at Santa Barbara Ph.D., an expert on the geology of the Pacific northwest, and a member of Western Mineral Resources Branch. Mel, of course, was a member of the same branch as I was, a former Carleton classmate, an igneous petrologist with a Stanford Ph.D., and was currently mapping Quaternary basalts on the Snake River Plain. Now we were all ready to have some fun!

And so it turned out that I did not spend long in Denver. My field notebook for 1980 shows that Mel and I flew to Portland on the afternoon of July 7, checked into the Shilo Motel again, then got over to the USGS bull session that evening. My field book entry for that day contained the words "science this time!" The next morning, we attended the 7:00 meeting, where we learned that the volcano was not doing much. While there, I got permission from Don Peterson to hire a physical science technician (a PST, the title used for an undergraduate geology major who would help us do lab work or assist in the field) to help us on the pyroclastic flows. I selected Allan M. Kaplan, a grad student who would join us soon. We quizzed some of the other field hands, notably Rick Hoblitt, about differences in lithology between the pyroclastic flows during the three eruptions so far. Donald A. Swanson, a Menlo Park volcanologist with a Johns Hopkins Ph.D. who had spent much of his career on the Columbia River basalts, was another field person we quizzed. A collegial atmosphere pervaded the entire USGS effort, with all persons volunteering helpful information no matter who was tackling which problem or topic.

Norm flew in to Portland in the morning. Vertical aerial photos of the volcano and surrounding affected areas had been flown regularly, and I got copies of sets of them taken after the three eruptions. By 1 pm, the three of us were on the standard form of transportation to the northern side of the mountain, a Bell Jet-Ranger helicopter, on contract to the USGS. It was Jet-Ranger #603, piloted by Bob Edwards, whom somebody described as a person "who doesn't give a shit"— considered a desirable trait because some of us wanted to get to places (if the pilot thought it possible) that a cautious pilot might not want to go. Most of the pilots were Vietnam veterans, who had seen plenty of things worse than a volcano and had more guts than any of us. Bob and the other pilots owned their helo and worked with their own colleague, a mechanic who also drove a light truck containing a tank of jet fuel, parked well outside the red zone. Once again, I had good rapport with pilots who just enjoyed the hell out of flying and clearly liked working with geologists who enjoyed the hell out of doing geology. We had multiple landings on the pumice plain that day, measuring sections of the pyroclastic flows of May 18, mostly in the walls of phreatic-explosion pits, and of the pyroclastic flows of June 12, and collecting samples for lab study. We took lots of photos. Like the other helo pilots, Bob monitored all events concerning the volcano by radio, especially seismicity and eruptions. When the volcano looked dead, he would shut down the helicopter's engine. When things were a little spooky, he would keep the rotor turning slowly. When things were scary, he would keep the rotor turning at almost full speed. As I remember, we experienced only the first of those three events that day. We got back to base at 6 pm.

That night we had dinner with Joseph G. Rosenbaum, a specialist in paleomagnetism who was taking on the job of assembling eyewitness accounts, collaborating with Richard Waitt. One eyewitness he interviewed was a baker who, with three others, had camped on the night of May 17 about thirteen miles north of the crater, unfortunately just inside what became the northern edge of the area of trees downed by the directed blast. His account was so amazing that I jotted notes (the story is also in Joe and Richard's report in the Lipman and Mullineaux volume, with many others that are similarly scary). At 8:32 am on the May 18, the baker first experienced a clear-air wind of 40 to 50 mph that he estimated to last from 4 to 15 seconds, followed by a cold, wet blast of ash and pumice chunks that was so dense that the campsite instantly became "night" and all the trees toppled, although he and his fellow campers could not see them coming down. They were buried by timber and ash, although two of them miraculously survived, apparently because they fell into a hole created by the root ball of a downed tree. At that point they could talk to each other but could see nothing. Within ten seconds, the blast became hot, at a temperature the baker estimated to be equivalent to sticking his head in a 300°F oven. Their hair began to sizzle as it singed. Pitch boiled out of trees and the heat caused minor burns for the next minute or so. Next there was a two- or three-minute respite in which they could see some blue sky, after which the ash-fall cloud from the pyroclastic flows passed over them and they could see nothing for a half-hour to an hour period, as if somebody were dumping a bag of ash over their heads.

After a day off, analyzing our notes, samples, and aerial photos, plus talking with Rick Hoblitt about some of the things he'd seen, we were back sampling from July 10 to 12 with Bob Edwards in helo #603 on the pumice plain. Most fumaroles, where small short-lived steam geysers had pushed through the pyroclastic flows, had left a red stain on the surface, but some fumaroles had left yellow sulfur crystals. Charred wood poked out of the flows. Bob let us off on the ridge containing the site of Coldwater II while he went to refuel, and we split up on our own traverses, studying mostly debris avalanche deposits, then met on the pumice plain for lunch. After lunch, we all headed off on our own traverses to map the deposits, plotting our stations on the aerial photos. Some traverses got almost to Spirit Lake. Our field notes recorded all sorts of wondrous things (such as the heat through our boots from the crusted-over top of the ash flow, the erosive power of the ash flows on the north flank, the fumaroles, the entire ambiance and noise of the inner crater, etc.) that we had never seen before in our careers. We dictated some of our observations into pocket tape recorders, as one might do with smart phones today. Bob parked his helo within eyesight and read a book.

The May18 pyroclastic flows were still hot but had a slight crust over them so we could walk on top of them most of the time, but occasionally we would sink in and the heat was painful though our blue jeans above our boots. Subsequently we bought gaiters to protect our ankles. Later in the summer, Mel hired another PST, Lisa McBroome, to help us with the lab analysis of the pyroclastic flows. Like Allan, she desperately wanted to see the flows in the field, so we flew her to Portland. When she showed up to get on the helicopter, she was wearing short shorts. I was shocked at her choice of clothing, for most of the volcanic deposits were warm to hot, and I asked her if she happened to have Levis or other pants with her, but she did not. The helo had no yellow protective Nomex pants to loan us. I knew that she was excited to get on the road, so I acquiesced and we lifted off. Sure enough, later in the day while we were working on the pumice plain, she broke through the surface and one leg sunk in, up to her thigh, and she could not get out and screamed. I was about thirty feet away from her when she did, and I rushed over and grabbed her arms and lifted her out vertically. With my walkie-talkie, I radioed in the helo, and it took her to the nearest hospital, which treated her burns. Fortunately, she did not need skin grafts. I always assumed that, being a petite woman, she sank in because of her small feet, as opposed to the others who had gunboats for feet. But more likely it was the god Pele who was responsible, seeing an opportunity to inflict maximum harm on the only one who was vulnerable!

The members of the USGS team had assigned informal names to various features of the mountain. The dome within the main crater had its own crater of ejecta (ash and rocks thrown out of the main crater before the dome erupted), which had formed during explosive eruptions that took place between or during mostly gentle eruptions of the small pasty rhyolite lava spires and flows that resulted in the dome. Surrounding the dome, therefore, was a ring-like low area where rain water accumulated as ponds, and this low area was surrounded by the higher, farther ring of the ejecta. The northern part of this somewhat higher (about 50 feet) ring, before one started down the open part of the northern flank, was called the "rampart." Below the rampart, a deep canyon had been cut in the northern flank by the pyroclastic flows moving rapidly down the flank; small streams of rain water followed each rainstorm. Pyroclastic flows are highly erosive, even though the older volcanic rocks of the stratovolcano were moderately resistant bedrock. But these flows gouged out some places in the canyon more than others, so the canyon did not have an even slope such as one finds in stream- and river-carved canyons. Instead, the flows had gouged out one place, leaving a cliff at its northern end, below which another area was gouged out. We called these "the stairsteps" (later to be named the Sasquatch Steps on the new topographic map), of which there were about a half dozen. The bottom of the stairsteps (canyon), where all pyroclastic flows dropped their deposits to create the pumice plain, we called "the gun barrel" because, when you were working there, you were looking up a deep canyon through which all the speeding pyroclastic flows were funneled. The gun barrel was a spooky place where you felt quite vulnerable.

July 13 saw us investigating the northern flank of the volcano. We looked for parts of the various pyroclastic flows, and the helo set down on the rampart, and from there we hiked down the stairsteps. On the rampart, we saw no erosion of unconsolidated deposits, such as would take place from collapse of big eruption columns (one theory for how pyroclastic flows form), so we figured that most of the pyroclastic flows had just bubbled out of the vent. The bedrock in the canyon, however, was scored by gently north-dipping fluting eroded by the ash flows. In the steps, we pounded in six-inch steel spikes so that the heads were flush with the bedrock, for we knew that more flows would be coming, and this might be a way to learn of the amounts of erosion into the bedrock (if erosion removed bedrock but not nails). The dome was then in a phase of almost continuous eruption, and steam and gases were visible at most times rising above the dome. Periodically we also saw small ash clouds rising from the dome. It was a noisy place, for there were also rock falls and small landslides coming off the inner walls of the amphitheater toward the dome. Earthquakes were common and would be heard and more commonly felt; the larger ones (magnitude 2 or 3) would buckle our knees. We wore yellow Nomex suits, presumably to protect us from heat, but mostly they were a source of amusement, for we knew that they would do nothing for us if there were an eruption. The only warning that could forecast an eruption was a sharp increase in seismicity, and the pilot was listening for this intently while he kept the rotor turning. On July 14 and 15 we used the helo to do more work on the rampart, the Spirit Lake area, and parts of the eastern flanks of the volcano and even the top of the volcano to see what we could find. Then we returned to our homes in Denver and Menlo Park. From there I was soon back at Hoover's Truck Stop in Marysvale to move my trailer to the small town of Koosharem and start mapping the eastern half of the Monroe Peak caldera.

Mount St. Helens erupted juvenile magma for the fourth time on July 22, as we learned on TV evening news. In terms of volumes of ash erupted, it was the third largest eruption. At least three huge ash columns were visible as far away as Tacoma, Washington in the afternoon to early evening. We soon learned that pyroclastic flows had been deposited. Banks and Hoblitt measured emplacement temperatures of the flows at about 1300°F , even hotter than before. We had more work to do! I checked in with colleagues in Vancouver, then with Mel and Norm. Aerial photos were taken when the ash clouds and weather cleared. Norm and I flew to Portland on August 3, and Mel came in the next day. Rick Hoblitt told us that three lobes had been deposited, the first in the stairsteps, another out on the pumice plain from the 7 pm eruption, and a third farther out onto the pumice plain at 7:30 pm. On the afternoon of the 4th we used the aerial photos and measured sections and sampled the new ash flows in most of the stairsteps, using Bob Edwards' Jet-Ranger. The youngest lobe of the pyroclastic flows from July 22 contained so-called breadcrust bombs, which are subspherical pumice chunks, most at or less than a foot long. The gas inside these bombs continues to expand after they have been erupted, so their outer solid surfaces split to look like the outer surface of some loaves of bread. Flights and our traverses continued on August 5th on the flanks of the volcano, the stairsteps, and the pumice plain. The ash flows of July 22 had eroded not only the sides of the stairsteps, as indicated by the metal stakes in the walls, but also some of the June 12 flows in the stairsteps and on the pumice plain. Most metal stakes that we had pounded into the bedrock walls in the stairsteps after each eruption were gone, so we recognized that we should measure erosion by each pyroclastic flow in feet rather than in few inches. On August 6 we flew on Bell Jet-Ranger #73N, piloted by Mike Holtzclaw, another good guy. We visited places on the eastern flank of the volcano, the stairsteps, and areas east of Spirit Lake.

On August 7, we made no excursions on the volcano. That ended up being a good move. In the morning, I replied to a phone call with Robert B. Raup, Jr., our branch chief at the time, about some administrative matters and meetings he thought I should attend concerning a possible new branch initiative. At noon, harmonic tremor began and then increased, and at 12:28 pm, a 2.3-magnitude earthquake took place and an eruption plume rose to 14,000 feet. Don Peterson called people out of the field, and spread the warning to the USFS, who requested that people evacuate the red zone. Seismograms at 3 pm recorded two quakes of similar magnitude, and at 3:30 pm a strange seismic signal that had been noticed just before the eruption of July 22. Volumes of SO_2 and CO_2 had decreased before the eruption of July 22 until immediately before the eruption, when volumes jumped up. During the morning and afternoon of August 7, volumes of SO_2 and CO_2 were similarly decreasing. Don called a press conference for 6 pm, and at 4 pm he enlisted me to take the lead as Information Scientist. At 4:20, as harmonic tremor and earthquakes increased to a high-level and continued for twenty minutes, pyroclastic flows erupted onto the pumice plain, and the local ABC-TV affiliate announced an eruption cloud to 20,000 feet, with ash falling in Vancouver. A team that included Pete Lipman and Rick Hoblitt was at the site of Coldwater II when this eruption took place, and they managed to take great photos of the ground-hugging cloud (see Figure 25), which we published in our report on pyroclastic flows in the Lipman and Mullineaux (1981) volume. Emplacement temperatures were about 1500°F, continuing the pattern of hotter magma. At 5:38, we again recorded high-level harmonic tremor, and radar mapped the eruption column at a whopping 44,000 feet. The winds were blowing to the north and northeast, so Yakima received only a slight dusting. I discussed all these things at the 6 pm press briefing, before opening it to questions. We set the next briefing for 10 pm.

Pulses in harmonic tremor and eruption plumes at 7:03, 7:10, and 7:30 pm announced a second eruption, with the ash column reaching 22,000 feet. Although it was dark outside by then, at our 10 pm press conference radar showed an eruption cloud at about 10,000 feet with low level harmonic tremor. I recapped the events of the day, then once again fielded questions, interrupted at 10:32 pm by a sharp increase in harmonic tremor and high-amplitude seismicity (although we did not know it until the press conference was over). Such seismicity, of course, was evidence of another eruption, the largest that day, with an ash column that rose to 36,000 feet, according to radar. That eruption deposited another series of pyroclastic-flow lobes on the pumice plain, as we discovered in the morning. Harmonic tremor and the eruption continued past 11:30 pm, when five quakes occurred, lasting until shortly before midnight when the ash column was measured 20,000 feet. As it turned out, the eruptions of August 7 marked the fourth largest of 1980, based on the volume of juvenile deposits.

Harmonic tremor was essentially over by the time of our 7 am all-hands meeting on August 8. As team members reviewed the seismic, gas, and other data, their preliminary analyses showed that each eruption had behaved a little differently, and we still had no silver bullet to forecast eruptions. A lot of UW and USGS scientists spent the day scratching their heads. For example, SO_2 and CO_2 levels did not jump back up immediately before the eruption. The closest previous eruption that it resembled was the one of June 12. Interestingly, this eruption on June 12 was the only one that coincided with high tides. Tim Hait took the 9 am press conference. Norm, Mel, and I spent the rest of the day on photo work and analysis of our field studies to date. A small dome filled the inner crater. The bull session at 7 pm yielded more questions than answers in forecasting future eruptions.

On August 9, we were back in the air, flying in Jet-Ranger #56Y, piloted by Mike Montgomery, another good pilot. This time we focused on sampling and measuring the new pyroclastic flows. We had two more days of similar field work. At the 7 pm bull session on the August 11, we discussed plans for the big USGS Professional Paper, concluding that the deadline for all our reports would be in January 1981. Pete Lipman and Don Mullineaux, who would edit the volume, emphasized the need to get this to the public while the 1980 events were still fresh in their minds. The USGS Director wanted the book to be published on the one-year anniversary of the start of eruptions, March 27, 1981. Despite the unusually quick production process, the volume turned out to be magnificent, of high quality, and with broad coverage of topics.

I flew back to Denver on August 12. Before I left, I sent breadcrust bombs to Carleton and Kent State geology departments and to several volcanologist friends as gifts. I still have a couple on the window sill of my office.

I returned to Portland, then drove to Vancouver on September 3, arriving in time for the 7 pm bull session. My field notebook records that I had a new title, until September 10, as Volcano Hazards Coordinator. These notes show meetings with Rocky Crandell concerning how to work with and notify officials in various agencies during future eruptions and emergencies. Collaboration with the USFS was especially important. Most of my duties involved discussions with various scientists as they came up with their best interpretations of the data in their topic of specialty, all to predict eruptions. I produced a written daily update to be forwarded to relevant agencies. When not doing that or attending meetings, I started writing our report on pyroclastic flows, and I got in one day of sampling and measuring sections of pyroclastic flows with Bob Edwards in helo #603. The dome was back, and growing slightly, but the volcano was relatively quiet. It had been a month since the last eruption. I returned to Denver on September 10.

Less than three weeks later, I was back in Vancouver in time for the 7 pm meeting on October 3. Though the volcano was quiet, its gas volumes were both up and down, for reasons we didn't yet understand. Much of Don Peterson's talk that evening was on deadlines for our reports. Reports were to be submitted to the various branches for their first review by November 14. Fortunately, our two reports—one a summary of all the pyroclastic flows, and the other on lab and chemical data showing differences between the ash flows—had been coming along well during the past month, partly due to Allan Kaplan's hard work. For the next several days, Norm and I continued field work with helicopters on the pyroclastic flows, while Mel stayed in Denver to supervise the lab work and to write his paper. Norm and I began focusing on dunes of ash known as surge deposits, mostly deposited during particularly high-speed pulses of the ash flows. This was a topic that we would look at more closely in the next year because exposures were good, but at this early date we had too many question marks to include it in our two nearly completed drafts. At our 7-pm bull session on October 7, our gas experts noted that levels of $SO_2$and CO_2 were falling, which by then we had been interpreting as being caused by a mass of magma clogging the vent, perhaps a suggestion of an upcoming eruption. On the afternoon of October 9, we noticed and relayed to Mike Montgomery intermittent low-level harmonic tremor, and we headed to the site of Coldwater II to wait it out. By about 6 pm, the tremor had become continuous, although still at a low level, and we flew back to Vancouver. It died out by about 3 am on October 10.

Nonetheless, October 10 was an eerie day, with various warning signs, none of them conclusive, plus high winds. We spent time waiting at Coldwater II, and the action started at 9:20 am with an eruption cloud to 20,000 feet, then ten minutes later an eruption cloud to 9,000 feet, then a half dozen other ash clouds, until about 11 am, when a good ash cloud was accompanied by harmonic tremor. When nothing else happened, we went back to sampling and studying the May 18 deposits south of Spirit Lake.

Gas volumes increased on October 13, but two days later Norm and I were back at work, flying with Mike Holtzclaw in helo #73N with Roy Bailey, another superstar volcanologist, and Don Swanson. We focused on surge deposits on the pumice plain by digging trenches in the deposits, which had already started to harden after a recent rain. On October 16 low clouds prevented Mike from getting airborne until early afternoon, when we managed to visit a large dune of May 18 surge on the eastern flank of the volcano before having to quit mid-afternoon to get Norm to the airport for a trip back to Menlo Park. At 11 pm, while having dinner at a restaurant in Portland with old friends, Sandra and Ed Metzger, neighbors from my days on Green Mountain, I was told that "it blew."

The eruption on October 16, I learned the next day, began with harmonic tremor and a shallow (mile deep) magnitude-3 quake at 7 pm; the tremor continued until 9:58 pm, when an eruption cloud rose to 46,000 feet. The eruption probably produced a pyroclastic flow, but fog obscured the volcano the next day, before the next eruption. Winds at 10 pm were toward the south, leaving a light dusting of ash. Seismicity increased over the course of the day on October 17, although without harmonic tremor. We had an 8 am meeting that day at which the seismologists suggested that another eruption was likely within hours. The word for those of us doing field work that day was to "be careful." Don Swanson noted some deformation in the inner crater according to shot lines, and that gases had decreased in volume the previous day. At 9:10 am, harmonic tremor began, and at 9:28 an eruption cloud rose to 50,000 feet, with a pyroclastic flow noted at 10:02 am by a spotter plane. At 1:00 pm Mike Holtzclaw and I, with a visiting volcanologist, lifted off in helo #73N, while Don Swanson and pilot Mike Montgomery lifted off in helo #56Y. We soon observed that the dome had been blown out and the pyroclastic flow, distinctively light gray, was in the amphitheater, stairsteps, and as far as 500 feet out from the base of the stairsteps, beyond the gun barrel. Norm Banks and Rick Hoblitt's temperature probes of the tuff revealed an emplacement temperature of at least 1560°F, another (and final) increase since May 18. We did many landings in the amphitheater, steps, and pumice plain to collect samples, then returned to Vancouver.

That night, at 9:12, another eruption took place, sending a cloud to 45,000 feet and ash fall blowing south-southeast on the wind. On October 18, the same visiting volcanologist and I were airborne by about 9 am in helo #73N, and soon confirmed that new, similarly sized pyroclastic flows had erupted the previous night, covering most of the ash flows from the previous morning's eruption. Between about 10:30 and 11:00 am, we were notified of bursts of seismic activity and we dropped down to Coldwater II to wait it out. When that activity subsided, we visited the amphitheater to collect and measure samples (most deposits were only a foot or two thick, so had cooled) of the new pyroclastic flow, then several stops on the steps to sample and to pound new, two-foot-long rebar stakes into the bedrock to test erosion.

Next, just after noon, we stopped at the gun barrel. As at the other stations that morning, our visits were short and Mike Montgomery kept the rotor going at almost full speed while we climbed out and made our way over to the front of the new lobe on the pumice plain only about a hundred feet from the Jet Ranger. Within a minute or two, I heard the radio crackle "it's erupting," prompting us to run at full speed back to the helo. While doing so, I looked up the gun barrel and saw an ash column shooting up into the sky! It took both of us about twenty seconds to get in the helo and close the doors, then another twelve seconds before we lifted off and were racing north at about 100 mph! The plume rose to a height of 20,000 to 25,000 feet, we estimated. The eruptions were photogenic and we took lots of pictures. We flew around above the crater during the eruption, watching the pyroclastic flow bubble out. We could not get too close, however, for the volcano was tossing big rocks into the air just below us.

The pyroclastic flow from this eruption barely made it beyond the rampart. Along with three other choppers, we set down at Coldwater II to wait for things to clear. We were radioed that another strong seismic signal took place at 2:18 pm, with increased activity and an eruption at 2:30 pm. The plume went to about 25,000 feet but this time there was no pyroclastic flow. Wind carried the ash to the southeast. Another burst of seismic activity took place at about 3:19 pm, but the new cloud went only 2000 feet above the crater, and seismic activity immediately died out. So we took off again and continued with our sampling and field work. We flew over the crater at 4:20 pm, noting that a new dome had formed, filling the inner crater. At 4:30 the helo radio told us that seismicity had increased, and other pulses of seismicity continued until almost 5 pm., although no ash plume was noted. We returned to Vancouver. This was to be the sixth and final eruption of 1980 (the events of the 16th through 18th being understood as a single long eruption), although for the next several years the dome would be periodically active, resulting in small lava flows/spines and small ash eruptions.

On October 19, Don Swanson, Don Peterson, and I spent most of the day in the amphitheater studying the dome. I wanted to give Allan Kaplan a chance to see the festivities, so he flew into Portland that day. We spent until October 23 doing more field work and collecting samples, visiting all the pyroclastic flows, surge deposits, dome, directed-blast deposits, and other deposits and features so that he could see first-hand what it was all about.

The weather was starting to get cold, with some light snow on the upper flanks of the volcano, so we mailed nine boxes of rocks to Mel, and flew back to Denver on October 24.

All told, 1980 had been quite a year! By far the biggest eruption had been on May 18, after which eruptions of magma (juvenile material) in general became progressively smaller and with longer intervals between them, until the last eruption on October 16 to 18. Along with the excitement of this news story, the year had seen much tragedy, with the final death toll at 57 persons, all on May 18. Most of the fatalities were due to suffocation by ash, although high temperatures killed some. The ash is chemically classified as rhyolite or dacite, rock types that are high in silica (quartz or SiO_2). Ash occurs as pointed, triangular-shaped microscopic fragments from the cooled surface of gas bubbles. As the magma breaks the surface (erupts), it freezes instantly to a solid state while bubbles of gas or steam in the magma continue to expand rapidly once outside the confining pressure of the magma chamber and vent. Ash, when ingested, destroys the linings of lungs, the gills of fish, and jet engines.

Yet the year also had its share of heroism and triumph, including search and rescue missions by helicopters that saved people. And the responsiveness and timely warnings by federal, state, and local government—all working closely together—likely saved thousands of lives, as Governor Dixie Lee Ray noted. The USGS had its finest hour because of its credibility, which started with the background reports by Rocky Crandell and Don Mullineaux and continued with the rapid work on the ground by these men and others, managed by Don Peterson. Without that credibility, government advice might have been ignored. Of course, luck was involved too, for the death toll would have been much worse if: (1) the 5.1-magnitude earthquake of May 18 had happened during the work week; and (2) the caravan of vehicles carrying homeowners whom the governor had allowed to gather up personal possessions, and which was scheduled to leave about an hour after what turned out to be the actual eruption, had happened before the eruption. Clearly the effort by the USGS is an example of the benefits of government-funded research directed at a scientific problem (when and what was the volcano going to erupt?), benefits of both immediate value to the public, but also more long-term value through knowledge and publications generated by the research.

The report edited by Pete Lipman and Don Mullineaux remains the best scientific study done to date on explosive volcanoes and their hazards. This huge book was put together by more than a hundred authors and co-authors, many of them colleagues from outside the USGS. Although intended for volcano experts rather than for the entertainment of the public, it is loaded with photographs, figures, and readable text that explain in detail all that happened in 1980 and how the results of each study serve humanity. The authors managed to meet their writing deadlines, the editors in our publication units dropped everything to get the document through production, and it was published when we said it would be— by the one-year anniversary of the first eruptions. It has saved lives, and it is now referenced when each new explosive volcano erupts.

The volume contains several articles in which David Johnston (who died on May 18) is a co-author, especially ones senior authored by Tom Casadevall, his close friend, who worked with David while they attempted to find gas signatures that would predict an eruption. The volume is dedicated to David, and this dedication page includes a brief biography and photo of him. The dedication was deeply felt, and his memory was on the mind of every one who worked at Mount St. Helens. The book's main objective therefore was to do what David's short life had done: to use everything at our disposal to predict eruptions of explosive volcanoes. Although we found no single answer, the profession now has things that might work the next time this or other such volcanoes erupt.

But Mel Kuntz, Norm MacLeod, Allan Kaplan, Lisa McBroome, and I were not yet done with Mount St. Helens. True, our two chapters in the big book had been put to bed in late 1980 and had been published in early 1981. But we had only just started to map the pyroclastic flows and write a report on the surge deposits. Allan, Norm, and I spent the first half of

August 1981 doing field work on the mountain, then mailing samples back to Denver for lab work, mostly by Allan. Things had changed when we got back to Vancouver. Except for steam and minor activity at the dome, the volcano was dead. There were no more 7 am or 7 pm meetings by the troops, let alone spending days in the FEMA bank building. There was none of the frenetic activity and excitement, and there were fewer scientists working non-stop, but there was still a lot of scientific clean-up to do, and monitoring continued. So, the work could be looked upon as routine, I suppose, because the mission was no longer to save lives. As I remember, the headquarters was the USFS regional office in Vancouver, and USGS information scientists and USFS information officers held press conferences out of that building. Congress had allocated funds for the USGS to build its own office building, but getting that done was nearly a year away. As before, we reached the mountain using contracted Jet-Rangers, but they no longer waited for us with rotors whirring while we did our studies. They would let us off on the mountain, commonly the rampart, and we walked down and met them at the end of the day on the pumice plain. Commonly we went the other way, climbing up the mountain. Traverses were therefore more leisurely and we had time to make observations and dig trenches in the pyroclastic flows, which were only slightly consolidated and had interiors that were still warm. Soon we concentrated on gathering data on the surge deposits on the flank of the volcano, which revealed much about how pyroclastic flows move; we published that report in 1985 with the Geological Society of America.

On May 18, 1982, Don Peterson dedicated a new USGS office building in Vancouver to monitor Mount St. Helens and other volcanoes in the Cascades, as well as to do other geology in the Pacific northwest. It was called the David A. Johnston Cascades Volcano Observatory in honor of our friend and colleague who had died on site. Allan and I worked out of that office during the first week of August doing final field checks on the pyroclastic flows and surges. Then, sadly for all of us, this good worker was off to continue his schooling. We submitted our two final reports for publication soon after. At least Allan got a lot of co-authorships out of helping us at MSH and Marysvale, and he learned a great deal.

As a postscript, my friend Robbie Gries sent me a beautiful biography of David Johnston that was authored by Melanie Holmes, a coincidental friend of David's sister Pat, and published in 2019. It describes an initially insecure, introverted boy who experienced death of friends and some family members early in his life, but overcame these traumas and his insecurities by adulthood through his intense drive to do something productive. His stage fright when giving talks was classic, even worse than mine (he fainted commonly), but his sense of humor helped him cope with that too. And becoming important and productive to society was something that he surely did! The story tells how this thirty-year-old was well on his way to becoming a world-class geologist, only to have his life cut short while doing exactly what he loved. For any young person today who is curious about some aspect of science yet feels inadequate in some way, this is a primer on how to soldier ahead and overcome these feelings. I can assure you that most of the authors in the volume *The 1980 Eruptions of Mount St. Helens, Washington*, including yours truly, identified with most of the feelings David had growing up. The story, by an author of remarkable perception and writing ability, is inspirational and positive despite the tragedy of a life cut short. Most of his and my USGS colleagues mentioned in this chapter are also discussed in Melanie's book, for they were supportive and loving friends to David. And, as you would expect, David's death and those of the 56 others motivated every one of us as we worked on this volcano to find something that would save lives by predicting the next eruption, or at least that of another volcano in the future. Part of this was therapeutic, and part of it was personal. David, whose life was aimed in this same direction to understand explosive volcanoes and save lives, would have agreed that we all succeeded, at least to a small extent. The seismic signals, including harmonic tremor, were promising, as were gas and deformation studies.

In her book, Melanie Holmes does an excellent job of explaining why authorities considered Coldwater II and other parts outside the red zone to be safe. First, the USGS and other government officials were under considerable pressure from loggers, business owners, and tourists to allow them onto the mountain, so a larger red zone was not politically feasible. But also knowledge about incidences of laterally directed blasts was poorly documented in volcanology at that time, although David and most other volcanologists there considered it a possibility. Holmes quotes Tom Casadevall (p. 111) as saying "Geology 101 teaches that volcanoes explode up!" Coldwater II was not overrun by the debris avalanche (the fast-moving landslide that resulted when the bulge collapsed). Instead, the ridge (now named Johnston Ridge) whose top was Coldwater II blocked the debris avalanche and deflected it westward down the Toutle Valley. That was what the USGS expected. But what they did not anticipate was the directed blast that killed David and most of the others, some specifically by falling trees or by suffocation. Holmes also pointed out that only Harry Truman and the couple in their cabin at Spirit Lake were within the red zone. She noted that the others who died were hardly foolhardy adventurers who skirted roadblocks and therefore might have deserved or at least tempted their fate (as some writers have said), but instead were people—some doing their jobs—who were outside the boundaries of the red zone during an apparent lull in eruptive activity and thought they were safe.

With that we shift, admittedly rather abruptly, to my next life adventure, as told by my best friend about our only collaborative study together, in Saudi Arabia, that he much later wrote for gags for a retirement newsletter (cited under his name in the Notes).

3

A Trip to Saudi Arabia

By Chester. A. Wallace, with corrections by P.D. Rowley

Pete Rowley and I were invited by Paul Williams (Chief Geologist, USGS Mission, Jeddah, Kingdom of Saudi Arabia) to visit the Mission during January 1984. Pete didn't have any tours to Antarctica or to Utah that winter, so he was able to go, and my ex-wife adjusted her schedule to take care of our kids. Dotty Kubicek, the Office of International Geology representative at the USGS in Denver, made all the flight and lodging arrangements, with layovers in London and Rome. In early January we flew from Denver to New York Kennedy and from Kennedy to London Heathrow, arriving at Heathrow at about 7 am London time. With forethought we purchased an imperial quart of Scotch at the Duty-Free Store at Kennedy. Our hotel was a Sheraton near Heathrow, not far from the Tube. The Hotel looked as if it had been designed by Soviet architects according to the Massive Cement School of Architecture. The building was ugly and the rooms were European small. We had to step into the hall to change our minds. With little sleep, we decided to explore London. Off we went on the Tube to visit Piccadilly Circus, Green Park, St. James's Park, Parliament Square, and Charring Cross, among other historic places. We found a place to eat dinner that we figured would not kill us (and remain within our per diem limits, which were skimpy for international travel), and after dark we returned to the Great Cement Monument for the night. I had no idea how we negotiated the Tube. Pete oversaw that.

On our return to the Sheraton, we decided to go through our suitcases to be sure that we did not have anything that might get us in trouble at Customs going into Saudi Arabia. That is when Pete discovered that his daughter, Jill, had carefully inserted dried elk turds between each layer of clothing in his suitcase. Had we been chosen for a detailed examination of our luggage at Heathrow, we would have been sent back to Kennedy on the next flight! I do not know how many of you have tried to flush dried elk turds down the crapper (Thos. Crapper invented the flush toilet in the U.K.), but the damned elk turds kept floating because they were dried, and it took an immense amount of water to finally dispose of Jill's gift. In the meantime, Pete and I also had to address the huge bottle of scotch that we bought, and we decided to try an experiment—drink the bottle of scotch and try to retain as much as possible in our bodies for the duration of our visit in dry Saudi Arabia. Pete vomited several times that night. I was OK, perhaps because I drank less scotch.

We managed to get to Heathrow the next day, went through customs and got to Saudia Airlines for our flight to Jeddah. Our experiment with scotch failed, and we got on the airplane dragging our hangovers with us. We flew Saudia Airlines because it was the National airline of the Kingdom, and we figured that we'd get less hassle flying Saudia than British Airways or Lufthansa. Wrong. Saudia sucked. Mostly we wanted to sleep anyhow.

On arrival in Jeddah, we went through the interminable process of Immigration Control and Customs. Long lines, pushing, jostling, and incredible slowness typified the entry process into Saudi Arabia. That done, we emerged into a large atrium at the airport that was hectic with people holding up signs for arriving passengers who were to meet sponsors, with people moving quickly to friends and relatives, and there was Paul Williams and Marty Kane, his chief geophysicist. We both breathed a sigh of relief. We quickly moved to Paul's vehicle in the parking lot with Paul shucking off all the contract "slaves" asking for tips for doing nothing (slavery was abolished in Saudi Arabia in 1962, but it was continued through penury contracts). Paul was a hard-ass regarding these folks, probably thanks to having lived in Jeddah for several years. Looking at Jeddah on our drive to a USGS compound (there were several USGS compounds) where we would be staying was exciting. What a culture warp! Most of the men walking around wore dresses! Pete and I needed several days to adapt to the time difference between Denver and Jeddah, but we managed. We met all the people in the Mission hierarchy that we needed to meet, and we did all the social events that were required of new arrivals.

Paul wanted us to go to the field, which is where we both wanted to be. First, there we were to examine the Murdama Group (Neoproterozoic [PDR edit: this is a geologic Era, from 1 billion to 540 million years ago]), a sedimentary unit that contained some volcanic rocks. The Murdama Group, located well north of Mecca, had been studied by Robert C. Greene (from Menlo Park originally). Bob, who was to show us where to go, had measured more than thirty miles of stratigraphic section through the Murdama Group, showing no faults and no folds. (Note: this is an inside joke from geologist authors to geologist readers, for in no place in the world has anybody else ever claimed that the thickness of a geologic formation or group is thirty miles!) Paul wanted an isotopic age on the Murdama Group, and Pete and I figured that intermediate and silicic volcanic rocks might give us the best chance of finding zircons. One of our tasks, then, would be to identify tuffs and lava flows. Then we were to try to determine relations between these igneous rocks and adjacent biblical gold mines, purportedly mined for King Solomon. We set off from Jeddah with a small caravan of vehicles. Bob knew where we were going, so Pete and I paid little attention to roads, towns, and desert tracks. Camp

Minders (a local group paid to set up camp logistics for USGS field parties) set up the tent at our first destination and Bob brought all the vittles that we needed. We looked at a lot of sills and dikes, and only one or two possible flows, and returned to camp. We carried on like this for several days, trying not to ridicule Bob's thirty-miles of measured section. Then Bob had to return to Jeddah, so we met a Skyvan at an unimproved airstrip, and Bob was gone. Skyvans were made by the Short Aircraft Company in Ireland and they were an ideal STOL (Short Takeoff and Landing) aircraft to supply camps and drilling crews at remote locations in the Arabian desert. Skyvans looked like a Volkswagen Microbus with short wings and had an airspeed of about 50 mph, or so it seemed. So, Pete and I were on our own to execute geologic research in a VERY foreign country where we had no handle on the language.

After Bob returned to Jeddah, Pete and I continued our field plans of doing field reconnaissance in the Murdama Group and examining old gold mines. We knew where we wanted to go and headed in that general direction on desert tracks. Pete's skills at navigating in the desert were remarkable, skills that transferred directly from his numerous mapping tours in Antarctica. We used 1:100,000-scale orthophotos to get to the places of interest using desert tracks, and Pete was never lost. I learned from Pete how to navigate in the desert, and these skills served me well for the next twenty-five years, mostly in Arabia. After working out of the first camp for several days, we (the Camp Minders, actually) moved camp about sixty miles from our first location. But we had to find the new location of our new camp, seemingly by magic, late that day after having spent the day working on the Murdama age problem. They had already set up our tent and put our gear in the tent at the new location. Camping in the desert was different from anything I had done at the USGS. My field area was in Montana where tall timber and streams were everywhere, and Pete had worked in Utah, where his desert was verdant compared to the desert in Arabia. At our Arabian camp there were virtually no plants from horizon to horizon. Camp was a mile or two along a desert track from the main highway. I am not sure why this place was selected, but the Camp Minders seemed to know this place, and we obliged.

We ate the remains of the food that Bob had left us while we planned our next few days of field work. We were off early each morning, onto the macadam road, and then onto desert tracks that Pete figured would get us to the places we wanted to be. Pete was covered by 1:100,000-scale maps in the passenger seat, and occasionally he would pop up to say "Go left (or right) here." Then he submerged himself under the pile of maps again. On our first day out from our new location, we stopped at a "bir" in the morning; a "bir" is a hand-dug water well. There were several birs in this area, and as we stared down one of these structures, which was dry, we realized that the snake skins that littered the bottom were cobra skins.

This discovery meant that cobras came to this place to molt and eat the small animals that searched for water. We hauled our butts to the safety of the Landcruiser and left.

For some reason, one day we stopped at a Bedouin camp. The men of the camp treated Pete and me as honored guests. We wanted to do field work, but they wanted to entertain us; we learned quickly not to stop at Bedouin camps. We were invited into the main tent where the veiled women brought tea for all and retreated behind vertical rugs that divided the main tent. The tea was heavily sugared. Coffee was offered, but their coffee was made of cardamom rather than coffee beans, and we declined the offer. One of the children was learning English at a weekly school for nomadic Bedouin so we tried to communicate with the male adults through this youngster using our Arabic-English dictionary. To celebrate the arrival of Americans the leader of this Bedouin tribe wanted to fire his shotgun into the air to let the neighboring camps know that he had important guests. He had some shotgun shells. I did not know the caliber of his shotgun, but it was an antique with a flared Damascus barrel. This elderly gentleman tried to cram a large shell into the breach of this antique shotgun. The shell bent. It became obvious that the shell was not going to fit into the shotgun, to our great relief. I am convinced that had he been able to fire that ancient shotgun, he would have been seriously injured or killed, which would have been a spectacular way to end our geologic careers in Arabia. The shotgun celebration was put away, and we continued our halting conversation with some of the young children. The language lessons were interesting and fun, and certainly less fearsome than the shotgun episode.

One day we looked up from an outcrop to see a huge dust cloud approaching from the south. This dust cloud was at least 1,000 ft. high. The cloud looked like something from a Hollywood set. We continued examining some outcrops for something or other, and the wind started to pick up. We agreed that we should get back to camp. The desert tracks are rough and our progress was slow, although our collective memory of how to return to camp was spot on. The sand storm descended on us before we got to the macadam highway, and visibility was only a few meters in front of our vehicle. The sand-storm started with airborne mud-balls. Mud-balls splatted onto the windscreen. Big, heavy, wet mud-balls. The mud-balls formed because high above the sandstorm was a rain storm. The rain drops fell through the sandstorm and turned into balls of mud as they accumulated sand, silt, and clays during the descent to the ground. We cleaned the windscreen with washer fluid although we could hear the godawful screeching noise as the sand scratched at the glass when wipers were on. Still, we made it to the macadam highway somehow, despite the near-zero visibility and howling wind.

Then, the question of when to cross onto the right lane of the macadam road became paramount. Tractor-trailer rigs did not slow down despite the sand storm. Timing became critical. When should we jump onto the highway and risk getting crushed by a huge truck? A few trucks emerged from the brown swirling cloud and disappeared into the cloud just as quickly, shaking our Landcruiser as they passed. We decided to count down from 10 to 0 and go. We did and survived.

Our next problem was to find the desert track that would take us to camp. We drove past the desert track that we thought we should take, so we pulled off the road, and turned around on the verge and headed back to the track. We could not be certain that this was the correct track, but it seemed right for some reason. I set the odometer, and we drove slowly for about one mile. We looked about in the swirling sand, but we could not see anything that looked like a camp, only a brown swirling cloud. Then I drove into a tent rope attached to a stake. Camp!

The Camp Minders had secured the tent edges with rocks to stop blowing sand from getting into the tent, and doubled the ropes and tent stakes. Inside the tent a light-brown mist of silt and clay slowly descended to the dirt floor, covering everything, including our sleeping bags, kitchen utensils, chairs, map cases—everything. The tent shuddered and shook as the wind buffeted the flimsy structure. I wondered whether the wind would blow down our tent and wreck us. We were stuck. The wind was hot. We managed dinner; albeit a gritty affair, and after that there was not much left to do but go to bed. The temperature must have been 90° F, and the only way to avoid the constant downward-floating silt and clay was to get into the sleeping bag, which made the 90° F even worse. Rivulets of sweat became tracks of mud running down our faces and neck. Light-brown dust continued to accumulate inside the tent. We were careful about water consumption, and we used a rope to attach ourselves to the tent when we went outside to relieve ourselves. The rope was a life line in case we could not see the tent in the sand storm. Cooking was difficult. The tent shook against the restraining ropes, and I wondered whether the tent might blow away. Imitating the essence of Pete's then-wife Adrienne (Pete got married a lot), I said to him: "YOU said that this was going to be a FUN trip! YOU said that we would have a good time! YOU said that we would have a cook and travel first class!" But NOOOOOOOOO." My imitation must have been decent because Pete laughed hard.

Two days. Sometime during the second night, the tent stopped shaking, and the howling stopped. Dawn was crisp and cool; visibility was limited only by the fine silt and clay that remained in the air, rather like the smoke we see in the western US during fire season. We stopped at the local fuel station to fill the fuel tanks on the Landcruiser. Fuel stations are stuck in the desert at improbable locations mainly to serve the nomadic Bedouins who need fuel for their Mercedes water trucks and single-axle, stake-body trucks; the Mercedes trucks haul bags of alfalfa pellets and water for the camels and goats. These stations also fueled the flotilla of white Toyota pick-up trucks that were the main form of transport for the Bedouin. The tanker trucks, stake-body trucks, and white pick-up trucks were provided to the Bedouin through the local tribal heads from the King. Traditionally the King has provided support to the Bedouin tribes because the tribes maintained an emergency food supply (camels and goats) should the Nation be cut off from normal routes of commerce. Moreover, the loyalty of the Bedouin was important to the royals because the Bedouin provided a stabilizing keel for the Kingdom, and Abdul Assiz, who founded the Kingdom, was from Bedouin stock. Besides, fuel stations were controlled by anonymous princes who profited handsomely from these investments.

At the fuel station, we needed a receipt to turn in to people in administration at the USGS Mission so we could get our money back. We had our handy Arabic-English dictionary to help secure the receipt. However, our pronunciation of the word for "receipt" baffled the attendant, who was hump-backed and couldn't hear well, and we couldn't understand him because he stuttered and had a facial defect that affected his speech. Eventually we arrived at a conclusion and we got the receipt that we needed, but the humor of trying to get the receipt from this person was not lost. Many years later, Richard Pryor's routine about the stuttering Chinese waiter brought back laughter about this isolated fuel station in the desert of Saudi Arabia. The fuel station also had non-alcoholic beer, although it took us less than a six-pack to conclude that it was not worth drinking.

After a few more days of work, we determined that we had enough time to sample several plutons on the return drive to Jeddah. We left camp to the Camp Minders and set off to Jeddah early in the morning. In the past I'd driven trucks in Chicago, I'd survived Paris roundabouts on my motorcycle, and I'd mastered traffic in Rome, but I was not mentally prepared to drive in Saudi Arabia. Driving on a two-lane highway there became a potentially fatal circus as a Mercedes sedan passed us as we were passing a slower truck. In fact, the Mercedes, which carried a family, blew by us on the verge (shoulder) to our left. This maneuver was surpassed (pun intended) when two sedans passed on each verge while we were passing a slower tractor-trailer. At this time in Arabia, when a sedan or sports car impaled itself into an oncoming truck or tractor-trailer, the wreckage was simply pushed to the side of the road where it remained, apparently forever. Usually, fire accompanied these impacts, so charred remains of the back part of the auto stuck out of the charred remains of the front of the truck. We saw a few wreckages in which the auto impacted a fuel hauler, and the wreckage was a melted mass. These autos contained entire families who perished along with the driver.

As we drove in the general direction of Jeddah, we stopped at a BRGM (Bureau de Recherches Géologique et Minières) Camp near Al Amar to contact any geologists who might be about, but the geologists were in the field or "resting" (we arrived at about 1:00 pm). Traditionally in France lunch is from 12:00 pm to 2: 00 p.m., so if geologists were in camp, they were not likely to appear. We moved on. As we approached our sampling targets, Pete took over as navigator. We had to collect about 100 lbs. of granodiorite to get enough zircon to provide a reliable age. I'd done this before in the Anaconda-Pintlar Wilderness, Montana, so I handled the hammer while Pete examined the results of my work to eliminate weathered and altered rocks. I do not recall how many samples we took as we returned to Jeddah. We got back late to our Jeddah compound, to the surprise of our house cook, but we were clapped out and needed sleep. For the next few days, we sorted samples, completed the forms needed by the Mission to process the samples, and completed descriptions of sampling sites and of the samples. We awaited the return of our Camp Minders because they had the samples that we collected early in this field trip. But for weeks the Camp Minders did not show, much to our frustration and Paul's anger.

After about a week we left again for the field, and this time we stayed in Paul Williams' camp near Hail, with helicopter support. Our task this time was to examine rocks that staff at the BRGM had mapped as Precambrian consolidated glacial till ("tillites") and affiliated deposits. This was a tough assignment. These matrix-supported conglomerates are difficult to classify unless one is lucky enough to find faceted cobbles, striated clasts, or drop-stones in laminated sediment that showed penetration of laminae (I spent considerable time examining tillites in California, Utah, Idaho, and Virginia in the USA with the USGS). Pete and I found all three elements of positive identification for tillites at various places in the basal Phanerozoic sequence [PDR edit: rocks that postdate 540 Ma] above Paul's beloved Neoproterozoic rocks.

Our helicopter pilot for this part of the trip was John Jensen, retired from the Royal Navy. The helicopter pilots were Brits or Americans, and they flew on contract for Special Flights, an arm of Saudia Airlines. John brought an extra dimension to our flights in that he was an expert on pictographs (hieroglyphics) of the Middle East. At some places we landed near sites that contained abundant pictographs and John deciphered the drawings for us, determined which tribe had done the drawings, and established when the drawings were made. After seeing some of the extensive drawing done on large, smooth rock faces, I found it amusing that the most recent drawings showed a pick-up truck with a camel in the back.

Paul Williams arrived at some point and we spent several days in the field with Paul as he squired us to various outcrops about which he wanted our thoughts. But soon we had to leave Paul's camp, so John volunteered to take us to the Hail airport to meet our Saudia flight back to Jeddah. John put his helicopter down next to the Boeing 737 near the departure lounge at the Hail airport, so theoretically we could simply get on the aircraft. Wrong. What followed was a tribute to panic and high disorder as we were intercepted by military with guns, with lots of wild gesticulation and shouts in Arabic, which we didn't understand, and lots of gun pointing. We were ushered into the terminal at gunpoint to go through security where we had to explain our rock hammers, knives, and other items that field geologists carry. Eventually we got onto the aircraft. By then, John was gone. Thanks John.

The errant Camp Minders finally arrived in Jeddah just before we left for Rome. Apparently, the Camp Minders had taken the opportunity to visit friends and relatives across Arabia on their return to Jeddah from our last camp. Inasmuch as they had many of our samples, we could not complete the paperwork needed to submit samples for laboratory analysis until they returned to Jeddah. Therefore, we spent the two days before we left Jeddah processing the massive volume of paperwork needed to submit samples for thin sections and isotopic ages. We completed the tasks needed to submit samples a few hours before we were to leave Jeddah.

Earlier that day we turned in the car that we had been using, so we needed transport to the airport. Our flight left at about 2 a.m. and we were to be taken to the airport by a driver who worked for the USGS Mission. Our fall-back position was that one of the great folks in Administration would take us to the airport. The driver did not arrive, and we contacted (awakened) our friend in Administration, who kindly took us to the airport.

Leaving Saudi Arabia was as difficult as getting into the country. All the information on our departure documents had to match the information on our entry documents. And, of course, we waited in line a lot, which is common in Arabia, at 2 a.m.

Pete and I were off to Rome for three days on our return to the USA. Because of our long day, I think that we slept during most of the flight to Rome. We arrived at the International Airport in Rome at about 6 a.m. local time, got through customs, and set out to find the Pensione that had been recommended by a friend. At the airport we had to fight off "independent" taxi drivers who all wanted to take us into Rome at extraordinary costs. I think that we took an "official" taxi from the airport, which deposited us, with our luggage, at the wrong place. We finally figured out where we should have been, and hauled our luggage to that place and checked in at the Pensione. Despite having made our reservations in advance at this place, the folks at the Pensione were not expecting us even though they had been alerted regarding our arrival time. Yep, we were in Italy.

I had not been in Rome for more than twenty years, so we needed a map to get around. The Pensione did not have any maps, so the map that I had grabbed at the airport would have to suffice. We decided not to waste any time sleeping, so we set off on foot to the Coliseum and Roman Forum. I discovered that many improvements (safety) had been made to the Coliseum and Forum in the years since I had been there, and it was all quite nice now. We self-toured the Coliseum and moved on to the Roman Forum, and later that day, to the Vatican. I have a picture of Pete standing in the Temple of the Vestal Virgins shading his eyes with his hand searching for a virgin in Rome. Apparently, he found none. But we saw wall-to-wall, cheek-to-jowl marble statues of them.

My memory fades about our time in Rome after that first day. I recall that Italy excavated bits and pieces of the Roman Empire and preserved them in various places in this great city, so fenced areas that contained remnants pop up in unusual places. We toured the Castelo d'Angelo, which was interesting with respect to the evolution of the papacy. Slowly my rudimentary Italian returned. At a restaurant, I reached back from my chair and plucked a blood orange from a tree, sharing wedges of it with Pete. It was the last day of January and Colorado was locked in snow and ice!

Our return flight was uneventful. Merciful sleep. First Kennedy and then Denver, and a huge sigh. Home at last. Kids. Dog. Normality, such as it was. An Open-File Report eventually followed.

Note: Wallace went to Saudi Arabia on a part-time basis in 1991, and continued with the USGS Mission in Jeddah until 2001, when the Mission was closed. From 1995 until 2001, Wallace was a contractor with the USGS in Arabia. He continued with the Saudi Geological Survey from 2001 until 2011 as a contractor. Rowley returned to Antarctica in 1984–1985 [PDR edit: see next chapter] and then began a dismal downward career spiral mapping in Nevada. He retired from the USGS in 2000. Both remain active in geology.

United States Department of the Interior
GEOLOGICAL SURVEY
BOX 25046 M.S.
DENVER FEDERAL CENTER
DENVER, COLORADO 80225
IN REPLY REFER TO:
SMOKEY
USA 20c
FEB 8 1985
U.S. GEOLOGICAL SURVEY
1984–1985
ENGLISH COAST GEOLOGY
Dr. Adrienne B. Anderson
5894 Meadow Drive
Morrison, Colorado
80465
Snow Hill I.
Robertson I.
Butler I.
George VI Sound
Palmer
Ronne
1984 – 1985
English Coast

7 The 1984–1985 Field Season on The Ice

I am not sure when I wrote the USGS proposal to NSF to explore the previously unvisited English Coast of Antarctica, but eventually the proposal was tentatively approved. Then it sat for several years for reasons that had to do with USARP logistics and long-term goals. The proposal was approved officially and its funding granted in the winter or spring of 1984, although I waited for a while before I made concrete plans because one could never be too sure about NSF. Eventually, I spent the summer and early fall staffing the field party.

The English Coast is north of eastern Ellsworth Land, which we visited in 1977–1978, meaning it is on the Bellingshausen Sea (southern Pacific Ocean) side rather than the Weddell Sea (Atlantic Ocean) side, as had been the case for previous field seasons. It was named for Captain Robert A.J. English, US Navy, who was captain of the ship *Bear of Oakland* during the First Byrd Antarctic Expedition. Walt Vennum (my second in charge) of Sonoma State, Karl Kellogg from the branch, and Tom Laudon of University of Wisconsin-Oshkosh, as previously, agreed to be part of the team. I picked two others from the branch. David J. Lidke had been hired as a PST to work with Chet Wallace in Montana, and had done some of our lab work on the 1980 pyroclastic flows. Then he earned a master's degree and moved up the ranks based on excellent mapping ability. He had a great sense of humor and good common sense, plus an exceptional work ethic. Ed Ruppel had hired J. Michael O'Neil, a structural geologist with a Ph.D. from the University of Colorado, during our hiring boom and he was mapping in Montana. I determined that he also needed to see Antarctica.

And finally, we needed a foreign exchange scientist, with a Brit being most logical because, after all, this was still British Antarctic Territory. And it was called the English Coast, no less! Art Ford had hired a woman on a field party to the Dufek intrusion in the Pensacola Range the previous year, and that seemed like a good idea this time. Janet W. (Wendy) Thomson, the wife of Mike Thomson and similarly a geologist with the British Antarctic Survey, has a B.S. in geology from London University and an M.S. in geology from Birmingham University. She had been working for BAS since 1964. Several of us knew her through Mike and enjoyed her sense of humor and work ethic. The Brits were even worse (in fact, a big fat zero) in allowing women to do field work on The Ice than we were, so we needed to right that wrong! Besides, as given in the title of the delightful song by Cyndi Lauper, "Girls Just Want to Have Fun", for Janet the English Coast sounded like the perfect place! Therefore, she accepted my offer for her to become the first British woman to do field work in Antarctica. She was tall, thin, and physically strong, and she was always at the head of the line when physical work was needed. Curiously, I later learned that Janet was born on the same day I was (December 6, 1942). One day during the season, when we were working together, we stopped the snowmobile at the base of a snow slope, took off our skis, and started hiking up to rock. I immediately fell into a small bergschrund, and she pulled me out. Then she fell into a small bergschrund and I pulled her out. We started laughing. That's the kind of fun we had!

It turned out that the English Coast was a more severe test, weatherwise, than any of my other Antarctic trips, and I am sure that all team members wondered at some time during the season what they, or Pete, had gotten them into. But all passed with flying colors!

After our Skyland conference in Virginia, we were ready to roll. Janet flew to Denver, stayed a few days with Adrienne and me, then the entire field party, except Walt and Tom, took the plane west together. We left Denver on the afternoon of November 17, later than during previous field seasons because USARP felt they could get people to Antarctica and the field more efficiently than before. After a four-hour layover in Los Angeles, we continued by commercial airlines to Honolulu. Next we flew all night by commercial airline to Auckland, New Zealand, then took a short Air New Zealand flight to Christchurch, arriving about 1 pm on November 19 (we had crossed the International Date Line and missed Sunday, November 18, entirely). We were met at the terminal by Margaret Lanyon and Walt Seelig, who gave us our room assignments at a motel about eight blocks from the city square. The next morning, we got fitted with our cold-weather clothing and were ready to fly to McMurdo, but the weather turned bad. So, we had two days of shopping and hiking in Christchurch. On November 22 we were up at 4 am, took a cab to the USARP clothing warehouse, put on the cold-weather gear, had breakfast, and loaded ourselves and our equipment on to a C-130 owned by the Royal New Zealand Air Force. We were among about twenty passengers in a shipment of cargo, mostly pallets of gas bottles and food. The 2400-mile trip took almost seven hours. The weather when we landed was calm and beautiful, with a temperature of about zero.

Our McMurdo quarters were once again a Jamesway. But as we began to assemble our gear for the trip, we noticed some penny-pinching by USARP, including limited amounts of the type of food we had enjoyed in the past, notably lobster tails, tenderloins, meat bars, and candy bars. The excuse was that field parties were more abundant this year than in the past, although clearly ours was the farthest out from McMurdo. Another decision was that this time we would not bring a Jamesway as a base camp; we would live only in Scott tents. Mike and Dave began work on our snowmobiles, of which we were to get only four, allowing us a spare of only one. We were allowed only four Scott tents (Janet was to have her own). Even radios were in short supply. We took the mandatory two-day snow-survival course. We field tested the snowmobiles. Tom Laudon arrived from Christchurch on November 27. Our days were filled with gathering and packing what we needed, and we were pretty much finished with that on December 5. On that day we were booked on a Huey helicopter for the Dry Valleys, fifty miles to the south, to collect samples of a well-known series of dikes for a geochemical and paleomagnetic study by Walt and Karl. This study had been built into our proposal. On our return at the end of the day, the pilot brought the helo along the edge of the annual ice about five to ten miles from McMurdo. There we saw several clusters of penguins, including Emperors, strolling along the edge, and seemingly gazing into space, and killer whales patrolling a few feet off the edge of the ice waiting for penguins. It was one of the most amazing sights of my Antarctic career!

I spent the next several days calculating weights and cubes for two flights, then waiting on weather in the English Coast. We had little to go on, mainly a satellite photo taken each morning. On December 10, Walt arrived from Christchurch. Finally, on December 11, we were ready to attempt the flight. We got up at 5:30 am to find that it had snowed about two inches overnight and it was overcast, creating a whiteout, with clouds near the ground. Outside looked like the inside of a milk bottle. But we hustled down to the airstrip, only to learn the flight had been delayed for a few hours. That gave us enough time to locate a drum of antifreeze for Karl's paleomagnetic core-drilling machine that had been left off the load, as well as for me to intercept a forklift that was about to put into our aircraft a pallet of gear intended for Pole Station! We loaded up (7,500 lbs of people, snowmobiles, Nansens, food, and fuel, plus USARP VIPs [Very Important People] who wanted to witness the longest and most spectacular put-in of the season). Then the C-130 taxied up and down the runway for two hours to pack down the soft snow. Once the plane had been refueled, the pilots asked all seven of us plus seven crew members (but not VIPs) to sit as far back in the plane as we could, to get the nose ski up, and then they attempted the takeoff. At about 3 pm, after the longest takeoff run I have ever been on, the Herc became airborne. We had 1,200 miles to go, with an estimated flight time of 5.5 hours. The second plane, with 12,000 pounds of equipment, plus more VIPs, was to take off from McMurdo once we were on the ground on the English Coast.

As I described in a letter of December 13 to Adrienne, in time the aircraft passed over the northern end of the Ellsworth Mountains. About

100 miles out from the English Coast, the pilot, Commander Dwight Fisher, asked me to come up to the cockpit to spot the landmarks for the landing site. There was a thin cloud cover over most of the area, but I saw Mt. Hassage and the Behrendt Mountains off to the east. The cloud cover prevented us from spotting anything to the south, especially Mt. Rex, the highest mountain in the area. But the northern front of the cloud cover ended about at Fitzgerald Buffs, "in a remarkable stroke of luck!" I spotted these bluffs, which on the photos have a steep, north-facing precipice, but from the south—our position—they are not nearly as distinctive. Once we were abreast of them, however, I confirmed them. I had wanted the camp to be just south of the Bluffs, but the clouds prevented that, so I suggested that the pilot land on any spot that looked good just south of the eastern Fitzgerald Bluffs. I returned to my seat and buckled up. The pilot made one pass to drag the surface to test it, then came back around and landed. It was a smooth and short landing, and "I knew right away that we had landed in soft snow, which I had feared!" It was 9 pm McMurdo time (local time of 4 am on December 11). The plane's altimeter said we were at 2,700 feet elevation. We were about seven miles east-southeast of the eastern end of Fitzgerald Bluffs. Dwight dropped the tailgate and we saw ground blizzard driven by about a 20-mph wind. We drove two Ski-Doos off the plane into the gale, then carried a generator and two radios well away from the aircraft's propwash, then started the generator. Meantime the plane freight-trained two pallets of our gear, then pushed out eight JATO bottles. By then we had the antenna up and the radios running, but we could not talk with anybody besides the pilot with either radio. We figured that the problem was in the antenna or antenna lead. Meanwhile four Nansen sledges and our hand-carry gear were offloaded. In accordance with NSF rules (three things required: radio contact, shelter, and heat), the Navy wanted to take us back, but I argued that they should leave us inasmuch as the second flight was due to depart McMurdo shortly. Dwight talked with USARP in McMurdo by radio and they finally relented. I asked them for another antenna and told them that we had enough food and fuel for two weeks. Then they were off!

This was not a great beginning to the season, but at least we were here. And, although we did not know it then, there was more of such weather and minor disasters to come, most the source of jokes now, because we survived them all! I am reminded of the pun "people are making apocalypse jokes like there's no tomorrow"—one of my favorites from author Vince Rozmiarek that is presently posted, at a rate of a new one each day, on a sign that can now be accessed by a Google search of the Indian Hills Community Center, at a town about five miles east of where Adrienne and I lived then in the mountains west of Morrison, Colorado.

We set up our tents in the wind, battened down the gear, and retired to a tent for our first communal meal. It was a meat bar stew that was light on the meat bars and heavy on mashed potato mix. We continued to try to contact or receive someone by radio, but after exhausting all combinations that did not work, we retired to our sleeping bags, exhausted, at 8 am local time. We slept until 4:30 pm local time on December 11 before getting up and making breakfast. The storm had abated but clouds had moved in. Our highest priority was the radios, and Walt, Karl, Mike, and I worked on them with no luck. Together, we also built a snow trench, for survival and to store gear. Finally, at about midnight, Karl and Mike isolated two failures that had occurred simultaneously in one radio (and one failure that affected the other radio): the antenna lead would not make contact with the antenna, and the coaxial cable inside one radio but near the lead had been damaged and was shorting out. But by now atmospheric conditions had become much worse and there was a lot of noise, so we still could not reach anybody. At least we knew we were over the hump, so we retired to the cook tent, which Tom and I shared, and prepared the evening hoosh. Before we could eat the meal, flight #2 came in over our heads! But by then the clouds were on the deck and the visibility was about 100 yards with a light snow. We started the generator and this time were able to reach both the aircraft and South Pole. We reported the weather and the C-130 left. As it did, we continued to talk with Pole. Then the generator stopped and could not be started, so we started the second one and resumed communication. In that com, we sent a message to McMurdo that when the second flight returned, we needed a new antenna, antenna lead, and another generator. Then back to our dinner, and finally our sleeping bags!

We slept like rocks until 3 pm. Conditions were the same, like being on the inside of a white beachball, and it was warm (as much as 30°F). We had breakfast, talked with folks at the Pole, and spent the day getting our gear off the pallets and organizing it in the snow trench. Mike and Dave found that the generator had seized up for some reason but they got it going again, only it did not look as if it would be reliable. I rigged the Ski-Doos and Nansens, ready to do our first traverse when the weather cleared. But the wind picked up and snowfall continued. On December 13, we got up at 8:30 am, finally off McMurdo time and onto local time! The storm continued, with winds of 20 to 30 mph and clouds. On December 14, the weather became good except for fog, but McMurdo was presumably busy with other things, and we dug out, enlarged the snow trench, fixed radios, and wrote letters. On the 15th, the weather was good but McMurdo remained uninterested, so we took off for our first field work, on Fitzgerald Bluffs. When we arrived at the outcrops an hour later, the wind had picked up to about 15 mph, with ground blizzard. Nonetheless, the snowmobiles and the remote steering system worked well and everybody was skiing like Olympians. We found that the rocks represented a Jurassic volcanic field and a Cretaceous pluton, and we were all energized to be whacking on rocks. The wind increased as a half hour went by, however, and something told me to cut the fun short. The clouds made for a heavy whiteout and the ground blizzard had blown away our outbound tracks, so we headed on a compass bearing in the general direction of camp and, about a half hour out from it, we spotted the yellow tents and made for them. Ten minutes after arriving, a major storm hit, with high winds, blowing snow, and 100 feet visibility. We had been lucky!

Most of our winds were southerlies, which were katabatic winds coming off the axis of the Antarctic Peninsula south of us. This storm let up during the morning of December 16, but clouds were on the surface and visibility was about zero, with a temperature just below freezing. The warm weather meant melting inside the tents and in our boots, and wet socks, insoles, boots, and other clothes hung from the peaks of all tents. The next several days were more of the same weather, but at about midnight on December 18 the clouds lifted and Karl began to radio weather reports every three hours. Walt spelled him in the morning and agreed to stay in camp to send weather reports every hour, starting at 11 am, while the rest of us did geology. At the time we were down to two gallons of Coleman fuel, so we restricted it to cooking and hoped for a plane! We had a good day doing geology, but the weather became cloudy by late morning and it started snowing. The next two days we had winds of about 50 mph and blowing snow.

On December 22, the storm abated during the day, and we began sending weather reports every three hours from midnight on. The radio had been blowing fuses because our generator had a bad habit of surging, and the last fuse failed that day, so we used a Ski-Doo fuse. But the barometric pressure was up and the temperature was down, so we hoped McMurdo would pay attention! Fortunately, at 3 am they radioed that a plane would be leaving in a half hour. At about 10 am on Sunday 23, with the weather holding, a Herc piloted by Perry Hill landed. He freight-trained most the load and told us that his altimeter said we were at 2960 feet elevation. He was low on fuel so could not spend time backloading the aluminum pallets. We watched as he tried to take off, a half dozen times, in the snow. At times the loose snow would wash over the nose of the plane and along the fuselage. It was agonizing to watch, and we thought that he might be stuck here with us, something that he and his crew would not be happy to do. Eventually, by taxiing around, he found a snow hill, powered up it, and was able to jump off the other end, enabling his nose ski to clear the ground rather than dragging him down! We radioed our thanks to him and his crew and our wishes for a merry Christmas, and spent the rest of the day reading letters from home, unpacking the load as the weather continued sunny and bright, and enjoying steaks and lobster tails for dinner!

On Christmas Eve, buoyed by the good weather and the arrival of our second flight Walt, Karl, and Dave did the geology in the Fitzgerald Bluffs area while the rest of us finished packing up for our first significant traverse. But by the end of the day, the clouds had socked us in. Mike cooked a turkey in a Coleman oven. The bad weather continued on Christmas Day, and we finished packing for our trip. The wind blew that night, followed by light snow in the morning of the 26th, but it started to clear by late morning and we were off to Camp 1 by mid-afternoon. We took three snowmobiles, each pulling two Nansen sledges. We traveled sixteen miles toward the east-southeast, arriving at 6 pm on the northern side of Schwartz Peak, a sharp spire that rose about four hundred feet above the ice. High south winds came up at about 9 pm, at times with gusts that we estimated were 70 or 80 mph, so strong that a person could barely stand up in them. As we were battening everything down, Tom noticed an eighteen-inch tear in the outer wall of Mike and Dave's tent. They took down the tent and moved in with Janet! We got to bed about 1 am. The storm abated by noon of the 27th, and Janet and Dave stayed behind to patch and sew the bad tent while the rest of us did the geology of Schwartz and its adjacent nunataks. The next day, in overcast but calm conditions, we had our first complete day of mapping. We spread out and tackled a series of named nunataks within fifteen miles of the camp, including Mt. Southern and Mt. Harry to the southwest and Marshall Nunatak to the east.

Hoping that the weather would hold and wanting to use the next day to head north on a second traverse, we packed up and headed back to our base camp, arriving at about 10 pm. But the weather did not hold, and December 29 was a day of clouds on the surface, no visibility, and light snowfall. That night the wind picked up and on the following day we measured winds at 30 mph, blowing the previous snowfall with it. I remarked in a letter to Adrienne the following day that everyone was bearing up well under conditions that could have been depressing to many people. Our sense of humor remained and we had lots of Christmas presents from the second flight to open and share, chores that we tackled together, and sufficient alcoholic beverages. On New Year's Eve, it was cloudy with a 10 to 15 mph wind but the visibility was sufficiently good for us to do the geology of the crest of Fitzgerald Bluff. Mike and I spent considerable time working on our snowmobile, which had starting problems, and eventually we traced it to a bad coil. We had fried chicken for dinner.

New Year's Day began with a 15-mph wind that increased during the day, but finally dropped to about 20 mph at about 5 pm. We were packed and anxious to start our second traverse and theorized that the warm daytime temperatures brought in low clouds and snowfall and that cooler "night" temperatures would also make our snow surfaces better. We had dinner and took off with three snowmobiles after 7 pm in about a 10-mph wind. The trip took us due north about five miles, just northeast of the eastern part of Fitzgerald Bluffs, where I went over a low snow hill and found myself in crevasses. Steering remotely, with Mike hanging on in back, I pulled hard right on the ropes, drove over several crevasses, but skirted the main part of the field. The following Ski-Doos got the idea and followed. We then swung left (west) for five miles past the northern front of Fitzgerald Bluffs to avoid several large crevasse fields north of us. The wind slackened and we had a sunny but cool (about 20°F) and pleasant evening, then headed northwest for about six miles, then swung almost due north. We refueled, and Mike took over the steering to travel about twenty-five miles to establish Camp 2 at the southern foot of Mt. Thornton. We were near the largest rock mass, in the center of an east-west string of rocky nunataks about thirty miles long, on the northern edge of the English Coast. We arrived at 4:30 am, got to bed at 6 am on January 2.

The weather held steady and we got up at noon, had breakfast, and headed off to do the geology at 3 pm. Janet, Dave, Mike, and I took one snowmobile west for about twelve miles, where we dropped off Mike and Janet at a nunatak, Mt. McCann, then continued another half dozen miles farther west, to Espencheid Nunatak, the westernmost of this cluster of nunataks. From here we looked north and saw annual ice of the Bellingshausen Sea, with big icebergs caught in it, about five hundred feet below us. We could see a small island beyond, Sims Island, with black rock that rose like a pinnacle about four hundred feet above the ice and had a dark ring around it, probably meltwater. Espencheid was surrounded by fifty-foot-wide crevasses, so we stopped the snowmobile and skied the last two hundred yards to rock. Here was something different: young (Tertiary) basalt lava flows that had erupted under water or ice, resulting in a deposit called hyaloclastite, a black and rust-brown glass shattered into sand- to boulder-size fragments. We did the geology, then returned to pick up Mike and Janet, who had found something called pillow basalt, where the basalt lava flows had erupted into water and created forms that looked like black pillows, stacked on each other. Meanwhile, Walt, Karl, and Tom had taken their snowmobile ten miles east of camp, to Mt. Benkert, and they had found more wondrous things! These included yellow airfall tuff from these basaltic volcanoes, with sweeping cross beds that probably were surge deposits like those we studied at Mount St. Helens. They also found waterfalls of meltwater and a rookery of Snowy Petrels. Walt fell into a crevasse over his head, and Karl and Tom had to dig him out! All rocks were part of a relatively young (Tertiary) basaltic province that continues west through Marie Byrd Land. We all returned to camp about 9 pm. As if that were not enough exercise for them, Walt, Karl, and Mike climbed Mt. Thornton that night after dinner, returning at about midnight. The next morning, we awoke to find ourselves in a storm, with 30- to 40-mph winds, which lasted all day. We had spaghetti for dinner (you will note that throughout our Antarctic trips the food menu is commonly mentioned because it was so important to keep us going, in addition to the camaraderie of having dinner together), and the storm continued the next night. It slackened at about 11 am on January 4, and we all attacked Mt. Thornton, finding yellow hyaloclastite, black pillow basalts, and ejecta of volcanic bombs and tuff. Snowy Petrels, which had nests in the rocks, flew around us, a beautiful and memorable experience!

January 5 was a gorgeous day, sunny with a light wind, and we dug out, had breakfast, and headed to Mt. Benkert to see the things Walt, Karl, and Tom had told us about. The peak rose to about 1,800 feet and we swarmed all over it like fleas on a dog, studying and photographing the tuffs and intertongued basalt flows and dikes. It was fun, for none of us had ever worked in these kinds of volcanic rocks before. This warm day on mostly black rocks produced cascades of water and blocks of ice crashing down the slopes. The Snowy Petrels above cackled like sea gulls, and were fun to watch: they come in like little white jet planes looking you over, their wing tips almost touching the snow surface we were on, then they shoot up and above you. They are the most beautiful bird I have ever seen!

We returned to Camp 2, had a long dinner, then packed up camp, and were on the road at midnight. I wanted to get the hell out of there for, as Tom had noted the previous night, our longest spell of good weather this season had been two days! We were on the trail by about midnight, in beautiful weather except for a 10-mph wind into which we had to drive as we headed south. After a couple of hours, as we neared Fitzgerald Bluffs, we could see ground blizzard pouring north over the Bluffs. We stopped to gas up, then continued, swinging east as we followed our tracks and the flags we had set out earlier. As we passed below the Bluffs, the ground blizzard was at about 20 mph, but we continued and soon were enveloped in a cloud with winds of 30 to 40 mph and visibility between fifty feet and a hundred yards. By some miracle, the Ski-Doos blasted along as fast as they could go, with no mechanical problems, and we made it back to our "base" at 6:30 am, by which time all of us were nearly hypothermic. We had been in ground blizzard for only two hours, but it seemed like days. The storm continued for two days, although winds dropped to a mere 20 mph, so we spent the days doing camp chores and readying ourselves for the next traverse.

That traverse seemed to come on January 8, when the first person up, at 7 am, noted that it was sunny with only a light breeze. That person woke the others, we had breakfast, and we were packed up at 10 am. Then wind came up at about 20 mph, with ground blizzard, after which it only got worse. We unpacked and set up the tents again.

On January 9 we tried this Chinese fire drill again after the wind slacked off to 15 mph in mid-afternoon. We packed the Nansens and headed out at 5:30 pm. We planned to visit a lot of nunataks to the south, along and south of the crest (axis) of the Antarctic Peninsula, which here trended about east-northeast but farther east trended northeast, then north. However, the sketch map and aerial photos showed crevasse fields due south of Camp 1. Therefore, we headed southeast for sixteen miles instead, past the southern side of Mount Harry, and up a steep snow slope past small crevasse fields to the north and south and into a light cloud cover. On that cloudy slope, we accidently tipped over several of the sledges, but after we got them upright and continued past the crevasses, we swung right and headed due south on a compass bearing. Visibility was about a quarter of a mile and a light snowfall came from the clouds. We continued up a gentle rise for about twenty miles, until I thought that we might be adjacent to several small isolated, unnamed nunataks that were shown on the sketch map. We established Camp 3 just before midnight, in a gentle east wind with a visibility of a hundred yards and whiteout, had dinner, and got to bed at about 2:30 am. The weather remained the same the next day, and not knowing where we were, we stayed put. All we could assume was that we were near the northwest- or west-northwest trending axial crest of the Antarctic Peninsula (our altimeter showed 4,200 feet). Wherever we were, it was lucky we stayed put, for the next day we would make the most important scientific discovery of the season!

The morning of January 11 was pretty much the same, but at about 2:30 pm one of us went outside and found that visibility had improved, although the whiteout and light wind remained. I suggested we get ready to leave, and as we started to pack up, one person saw a small nunatak off to the west, and another seemingly far beyond it, and we took compass bearings on both, in case visibility deteriorated. In the whiteout, there was no way to tell how far the nunataks were from us, but the nearest one looked close. So, while the others packed up, Walt, Dave, and I took one of the Ski-Doos and a Nansen, hoping to run over to it quickly, a distance we crudely estimated to be one or two miles. When we got to it, the odometer showed a little more than five miles, but the nunatak was quite small, about a hundred feet long and with an ice protuberance about thirty feet high. However, there were two tiny outcrops visible, one making up an area five by five feet, and the other ten by seven feet. The larger outcrop was a black sedimentary shale that I assumed to be the Latady Formation, and I saw plant material in the rock. I let Walt and Dave collect those fossils and record the geology, as I sat in the snow and tried to discern whether any photos showed such a nunatak. We looked at the other nunatak farther west and noted that it had rock exposures on it, and we also observed a two-hundred-foot precipice south of us, against which clouds to the south were banked.

When Walt and Dave were done, we walked back to the Ski-Doo, but thought better of leaving and, almost as an afterthought, walked the fifty feet away to see the smaller outcrop. I got on my hands and knees and looked at the south-dipping black shale and siltstone and noticed a tiny fragment of fossil leaf about an inch long. Then another. The others swarmed over the outcrop, and we soon were splitting the rocks along their bedding planes with the pick end of our rock hammers. Soon we had a whole collection of fossil leaves and stems. I bagged them and numbered them while the others continued searching and hammering, and they soon found far better samples. The leaves were several inches long, pointed at the stem and rounded at the end. After an hour, we loaded up the samples and took off to the west to see if we could get to the farther nunatak, but after three miles and apparently not significantly closer, we figured it was another twenty-five miles to the west, so we returned to the tents as the weather socked us in again.

We decided to stay at the camp that night and, over dinner, took out the fossils and passed them around. A bombshell moment occurred when Tom and Karl saw the leaves and pronounced them *Glossopteris*! This is a distinctive fossil flora of late Paleozoic, mostly Permian (a geologic Period, from 299 to 252 Ma) age. As such, it was far older than the Jurassic Latady Formation, meaning that here, near the crest of the peninsula, erosion had cut down through the Jurassic volcanic and sedimentary strata and exposed an older basement. But, even more fascinating, *Glossopteris* has been found in all the southern continents, including where John Anderson and Cam Craddock had reported in 1965 on its discovery in the Ellsworth Mountains. At that time, its presence was an important piece of evidence that all the southern continents had once been one supercontinent, Gondwanaland, and that they had later separated through "continental drift," now termed plate tectonics. After dinner, with the nunatak now invisible in the clouds, the others followed our tracks over to it and did their own collecting, virtually tearing apart all the bedrock. Janet proposed that we call the nunatak Erehwon, "nowhere" spelled backwards. We, especially Tom, described the discovery in later reports.

The whiteout and no-visibility conditions continued the next day. We took a Ski-Doo on a southeast, then southwest compass bearings to see if we could see anything or bump into more nunataks. The elevation was about the same, but we found nothing else of interest. Finally, on Sunday, January 13, Mike had gotten up early to find the day bright, with low clouds to the south and southeast but with good visibility in other directions. He noticed a nunatak to the southwest, but we still felt lost and could not tell its distance, so while Karl cooked breakfast, Mike and I took a snowmobile and headed toward it. After four miles, it did not appear to be any closer, so we concluded that it was Mt. Peterson. Then the clouds opened to the south and we spotted a mountain to the southeast, also of unknown distance. We drove toward it but it did not get noticeably closer, so we concluded that it was Mt. Rex. Then another east of it came into view, which must be Henkle Peak. By then we were engulfed by clouds but we took a compass bearing back to camp, arriving at 9:30 am, both of us hungry. Karl had saved us breakfast and the rest had packed the camp, and at least now we knew roughly where we were. We drove on a bearing southeast toward Henkle Peak, until the clouds lifted and we could make out the nunatak. We reached it at 2:30 pm, after a trip of about sixteen miles. Our altimeters showed an elevation of the base of the Peak at 3,350 feet, so we were south of the peninsula axis. It was not much of a peak, with a summit only about three hundred feet above its base made up of Jurassic lava flows. The weather was clear and above freezing; after doing the geology and climbing the peak, we cached two barrels of fuel there, for our anticipated trip to the Behrendt Mountains. Then we headed toward Mt. Rex, a beautiful peak about twenty-five miles to the south.

The Antarctic Peninsula Traverse of 1961–1962 had stopped on the southeastern side of Mt. Rex, so we were not the first to get there. At 8:30 pm, we established Camp 4, at an elevation of about 2500 feet, on the northern side of this mountain. It was clear and warm. After dinner, Walt climbed Mt. Rex, one of his many first ascents since he had begun working with us.

The day of January 14 continued clear and hot, and we got up at 7 am to take advantage of our good fortune, went in various directions to do the geology (Jurassic lava flows and ash-flow tuffs) of the base of the mountain and adjacent nunataks, agreeing to be back in camp by 3 pm. We needed to use the good weather to get to Mt. Peterson, about twenty-five miles to the northwest. We had the camp packed and were underway by 6 pm, but Mt. Peterson was on the other (northwest) side of the peninsula axis so we could not see it until we were almost there. We steered on a compass bearing. Visibility was unlimited but there was nothing to see except white (the continental divide) in front of us, and Mt. Rex and occasionally Henkle Peak retreating behind us. Finally, we topped the peninsula divide and spotted Mt. Peterson ahead. We arrived at 9 pm on the eastern side of Mt. Peterson and established Camp 5, and we had the tents up and dinner cooking in an hour. The elevation of camp was about 3,200 feet.

The good weather remained on January 15, our third good day in a row (a record!), and we did the geology. Peterson was a ridge with a relief of several hundred feet, with the "mountain" at one end of it. The landmark contained some of the most unusual geology of any we had seen before, mostly Jurassic volcanics at one end, in fault contact with sedimentary rocks of presumed Jurassic age at the other end. We scratched our heads a lot and debated the various interpretations. It was fun, and Dave had the final word. He approved the fault interpretation and stated that “otherwise we have to go for a magical explanation, which is the kind I generally like best.” As we laughed and argued, high, wispy cirrus clouds moved in, a warning of bad weather to come. Sure enough, zero visibility and 15 mph winds arrived the next day, but Tom and I went out, skiing in separate directions the short distance to rock, and we did the geology in more detail. There was much pleasure for me in doing the geology at a leisurely pace. It contrasted with this strange field season, during which we had to scramble to do the geology as fast as possible, then rot in the tents during the intervening bad weather. On January 17, the weather remained poor and we were able to do more detailed work at Peterson, for it was within walking distance. We were about forty miles south of our base camp, but there was a nunatak about fifteen miles north-northwest of Camp 5 at which I wanted to stop on the way back, but we could not see it.

On January 18, the clouds at the surface came in, then left, then came again, with visibility ranging from one to fifteen miles. At 9:30 am, we finally saw the nunatak I wanted to visit and got a bearing on it. We packed up and headed off, reaching it on that bearing at about 1 pm. We finished the geology and were ready to go at 4:30 pm. We had almost thirty miles to the north to get to our base camp, but a crevasse field was shown on the sketch map between six and eight miles south of camp, so we wanted to bear a little to the west. Visibility was about ten miles and the south wind was at about 10 mph, but we were in the middle of nowhere with nothing to see, so we headed a little west of north on a compass bearing for about twenty miles. During the last ten miles of that part of the trip, it snowed heavily and visibility dropped to a quarter mile. At about the twenty-mile mark, I figured we had cleared the crevasse field, so I changed the bearing to north-northeast. Before this traverse, Walt had laid two north-south miles of green flags west of camp and two north-south miles of red flags east of camp. We also had flagged the Henkle road heading southeast of camp. We continued in heavy snow on the north-northeast bearing for about 12.5 miles. Then Mike, who was driving our lead snowmobile and had the best eyesight of anybody, started waving his arms. He had hit the Henkle road! He took a hard left, and accelerated the Ski-Doo to as fast as it would go. We reached the base camp, about two miles away, at 8:30 pm, very much chilled. There were nine inches of new snow on the ground. We set up the tents, had dinner, and got to bed by about 1:30 am.

On January 19, the weather was lousy early in the day, so I let everybody sleep in to recover. And we needed to prepare for a trip to the Behrendt Mountains, so lousy weather was ok. We had a late breakfast, then the clouds and lack of visibility cleared off, so I suggested we do the northern and northwestern side of Fitzgerald Bluffs, where the north-facing cliffs were. We followed the road we had established north to Mt. Benkert, heading north from Camp 1, then west along the northern front of Fitzgerald Bluffs. There had been a bunch of avalanches of snow and rock that carried out lobes of material about a quarter mile north of the cliffs. We worked for several hours, until a storm came up suddenly and, from the cliffs, we lost sight of our snowmobiles. We made our way down the cliffs, found the Ski-Doos, and headed home. But the storm broke off after we had gone only five miles back east, so we had an enjoyable, relaxing, and beautiful trip the rest of the way, stopping occasionally to take photos of ice cliffs and crevasses. On January 20, we experienced heavy snowfall (two inches), above-freezing temperatures, near-zero visibility, and a south 15-mph wind, so we packed for our trip.

The poor weather had continued when the alarm went off at 7 am on January 21, so I went back to sleep until 8:30, when I looked out and noticed the weather had improved to partly cloudy and calm. I woke up the others, we had breakfast, and then the weather turned bad again. At 2 pm, with overcast conditions, one to three miles of visibility, and light snow, we took off on the Henkle road. We were carrying heavy loads (food and fuel) because it was to be a long traverse, so we dug out the fourth (spare) Ski-Doo, even though it had been having electrical problems. Despite that, we got bogged down in the six-inch light snow. We managed only about fifteen miles and were about three miles south of Mt. Harry when 30 mph winds and ground blizzard came on, and our visibility dropped to a quarter mile, the approximate distance between flags. Then we accidentally tipped over the fuel sledge while relaying up the same scarp with which we'd had trouble before, and then the winds increased and visibility went to nothing. We set up our tents there, calling it Camp 6, and got to bed after midnight, all of us totally worn out.

The next morning the ground blizzard continued until late morning, so we got up for breakfast and got under way at about 2 pm. South and east winds continued in our faces until we crossed the topographic divide, when a gentle east wind was the only obstacle. The electrical problems with the fourth snowmobile were too much to sort out, so we cached it near the divide. We went the rest of the way to Henkle Peak, arriving on the western side of the Peak at about 7:30 pm, establishing Camp 7, with an elevation of about 3600 feet. After dinner, Tom and I did more of the geology (sedimentary rocks of the Latady Formation) of Henkle Peak, getting back to camp at 1 am.

The weather was good in the early morning of January 23 but it deteriorated from then on, even though we left at 11 am. It was overcast the entire day, with a stiff north and northeast wind. Local patches of clouds destroyed the visibility from time to time. But for most of the trip we could see our destination, which was the Lyon Nunataks to the east-southeast, and when the clouds obscured it, we remained on a compass bearing. The Lyon Nunataks consist of a series of scattered low nunataks, the southwestern one of which had been visited by the Antarctic

Peninsula Traverse. We planted the last of our flags just before we arrived at camp, after a trip of about forty miles. Our gas consumption as we slogged through the soft snow was 2.5 mpg, about half what it normally was on good surfaces. This issue became a concern, for I had calculated the amount of fuel we needed based on higher miles per gallon, and now I had to cut some travel so we would have enough fuel to get back to Fitzgerald Bluffs. We arrived at 4:30 pm on the northern side of the same nunatak that the Peninsula Traverse had visited, set up Camp 8, then went and did the geology (mostly Jurassic ash-flow tuffs) until 9 pm, before returning for dinner and bed. A bolt that holds on a front ski broke on one snowmobile, but we scavenged another bolt and the ski held up. In addition to boxes full of rocks, we cached forty gallons of fuel at camp, which we hoped would be sufficient to get us back to our fuel cache at Henkle Peak.

The weather was good early in the morning of January 24, but when we got under way at noon it had become much worse. I wanted to get to the Henry Nunataks, about thirty miles to the southeast, so I headed on a compass bearing in that direction in four-inch-deep snow. We could see some small nunataks on our left (north) as we proceeded, under a heavy cloud cover and a 15 to 25 mph wind that was coming directly toward us. It was quite bitter! But we could not see anything in the direction of the Henry Nunataks except a snowy white hill on the horizon, so we headed for that. At about twenty-five miles we got to the top of that hill, which contained rock that we studied, and we enjoyed views both to the east at the Behrendt Mountains, and to the northeast at the Henry Nuntataks. We also were in about a 50-mph wind! We had gone too far south but, because the destination of this last traverse was the unexplored northern Behrendt Mountains, we continued another fifteen miles into the Behrendt Mountains. We had come about forty-five miles, arriving at roughly 7 pm in a 20-mph wind and low clouds that reduced visibility to about a mile at times. We called this Camp 9, at the northwestern foot of Mount Chandler. We got to bed around 11:30, with the temperature at freezing, disturbingly warm for this time in the season.

I got up at 7 am on January 25 in a snowfall of big wet flakes and a visibility of less than two hundred yards. I let the others sleep while I puttered around doing camp chores and worried, for I knew we were way out on a long, narrow limb. I estimated that we were 140 miles from Fitzgerald Bluffs, and USARP had been sending us messages saying that they wanted to pick us up there on February 1. I think that I had agreed to this, although we also had had an agreement a week before that our pickup date was to be February 5. As I waited for the others to get up, I put my worries in a letter to my wife. I began the letter, "there are some things we do in Antarctica that scare me, and this is one of them." Perhaps I should have used "I" instead of "we." The weather was my biggest worry, for I now had to wonder whether it would let us get back to Camp 1. Once the others got up and had breakfast, I felt better, although I did not voice my concerns to them. They doubtless had similar ones of their own. At about 2 pm, I needed some exercise, so Janet and I took a bearing and walked over to the nearest rock, which we could barely see through the clouds a quarter mile away, and climbed nine hundred-foot Mt. Chandler and did geology for the rest of the day. That was a delight!

January 26 brought a calm whiteout, with a morning temperature of about 30°F and good visibility. We all had a long, enjoyable, and productive day of doing the geology in the northern Behrendts. I also used the visibility to scout out a route back through the Henry Nunataks. Once in the tents again for dinner, we had a return of ground blizzard during the evening, and it continued all "night." The next day, a Sunday, was overcast and whited out, warm with a light snowfall and light wind but some visibility at times. As Dave and I were looking in the food box for meat bars or something to put in our backpacks for lunch, Dave handed me a couple of candy bars. I said "Thanks." He said, "You're welcome. My motto is to share, especially when we all have more than enough!" We packed up camp and, following my scouting of the previous day, we headed north on a bearing to the Henrys, then on a northwest bearing through the scattered nunataks of the Henrys in a north wind of 15 mph. We hop-scotched through the nunataks, each team running over to one nunatak when it popped into view, while the others went to another. The snow was deep and sticky, and we broke trail with difficulty, the engines going flat out yet producing not much above walking speed. With the geology done, we had the snowmobile driven by Janet, Tom, and Dave go first to break trail because it was carrying the lightest load. They traveled on the same northwest bearing as before because we could not see the Lyons Nunataks. One of the snowmobiles was running poorly, so we stopped and finally figured out that it needed a new fuel filter. With that in place, off we went again. By mid-afternoon, the surface became hard but with high, large sastrugi, and our speed consequently increased. By late afternoon we spotted the Lyons Nunataks and headed for them, but ten miles before we reached them the wind came up and a ground blizzard blocked our view. Mike and I took the lead and remained on a compass bearing, but soon the winds were too high (and in our face), and we had to stop at 8 pm and set up camp, which we called Camp 10, about five miles short of the Lyons Nunataks. Dave noted that on this trip, among others, the winds seemed to want to be in our face. Furthermore, when we pitched our tents, the winds were always coming through our front doors. He postulated, therefore, that the direction of the winds was "tent dependent," for which we had by now accumulated quite a bit of empirical data. We figured that this new theory could have a significant impact on the science of meteorology!

The ground blizzard continued until about 8 am on January 28, so we had breakfast, packed up, and at 11:30 am headed west in a light wind, partly cloudy conditions, and unlimited visibility. In a half hour we reached the site of Camp 8, where we had cached fuel and rocks. We had come forty-nine miles from Camp 9. We gassed up and left the empty fuel barrel, loaded the cached rocks, and headed off on a bearing, thankfully riding on a hard surface, even though it contained lots of sastrugi. About five miles west of the Lyons Nunataks, we intercepted the flagged Henkle road, which lifted our spirits. But the surface now was in soft snow, so our gas mileage went to hell; this became concerning. The trip droned on, following flags with no landmarks to see except for gently rolling ice topography. Henkle Peak was buried in clouds, so we could not see it, although we saw Mt. Rex a few times. We arrived at the Henkle fuel cache, a barrel with thirty gallons, at 5:30 pm, in heavy clouds with a visibility of about a hundred yards—obviously less than the quarter-mile distance between flags, so we made it only to about a mile west-northwest of the Peak before we had to quit, at 6:30, calling it Camp 11. We had come forty-five miles. We had another fifty miles to go, and two days in which to do it. We celebrated, hoping that this was not premature. Tom pulled out a bottle of Johnnie Walker black-label scotch that he had been hoarding, although he got almost halfway through it before he passed it around. Karl cooked our next-to-last steak meal, while Tom played a tape of Jerry Lee Lewis. After that he put on a tape of a blue-grass band called the Skillet Lickers. A line in one of their songs was "We need a lot more of Jesus and a lot less rock and roll." Amen to that, we all agreed!

January 29 began with five inches of heavy snowfall, prompting us to go back to sleep for another couple of hours until Janet woke us with the news that the visibility was becoming much better. We started packing camp while Karl cooked breakfast, with the weather calm with good visibility, although overcast. By the time we had finished breakfast, we had a 25-mph north wind and ground blizzard. We finished packing anyway, the wind subsided, then came back to about 15 mph. We got under way at 12:30 pm, but pulling our loads up the grade to the divide against the wind was difficult in the warm soft snow, which was sticking to our skis. Mike sat on the seat and drove our lead Ski-Doo, containing the lightest load, and broke trail. He did the driving because he had the best eyesight and was least likely to miss our trail flags. Occasionally we had to go on a bearing between flags. At what we had come to call the *Glossopteris* camp, the bearing of the road changed to nearly northerly, but here the north wind picked up, with ground blizzard and clouds on the surface. We missed seeing flags several times. Once when this happened, we started to pitch a tent to wait out the weather, only for there to be some slight improvement in the visibility that enabled us to see the next flag. Gas mileage was terrible, between 2 and 3 mpg, and we needed to be at 3 mpg to get us back to camp! Mike pressed on, spotting most flags, with Dave on one rope seeing a couple, and I on another rope giving the bearing. The wind picked up to 25 mph.

When we finally arrived at the abandoned Ski-Doo, about nine miles south of Mt. Harry, we were all a little hypothermic so I recommended we set up camp. It was 5:30 pm, and we had come thirty-one miles that day. We called it Camp 12, and we estimated that we had about twenty miles to go. We had ham steak and freeze-dried vegetables for dinner.

The wind blew all day on January 30, with winds from about 30 to 60 mph. Visibility was about thirty feet, and ground blizzard was heavy. The wind dropped a little in the late afternoon, then increased during the evening and "night." The next day, the conditions were the same when the alarm sounded at 7 am, so we went back to sleep until 9 am, by which time conditions had improved a little. We all sprang into action, dug out, had breakfast, and at 11 am the weather was calm with visibility of about ten miles, meaning we could see Mt. Harry! We got the fourth snowmobile going and were on our way by about 2 pm. Then up and over the topographic divide and on to our base camp, arriving about 4:30 pm. We immediately called Pole Station—and learned that our flight had been canceled! We dug out the pallets, JATO, and rock boxes. At 7 pm, south winds began, soon reaching a steady 20 to 25 mph and gusting to 30 mph. Nonetheless, we began sending hourly weather reports, all of us taking shifts. But during the "night" we could not reach Pole, for unknown reasons.

On February 1, katabatic south winds continued at about 30 mph, gusting higher. We found that the problem with the radio was again a broken antenna lead, and when we managed to contact Pole for the first time at 11 am, we learned that a plane was scheduled to arrive at about 3 pm. By 11 am, the winds had died and the barometric pressure was increasing, good signs, and we had told them that. We worked like mad digging out Ski-Doos and sleds, then packing gear and rock boxes on to them for quick transport to the aircraft. The weather at 3 pm was wind at 20 to 25 knots, with blowing snow, 90-percent clouds at 1000 feet above us, and a temperature of about 25°F . At 3 pm, a Herc indeed arrived overhead, stayed there for a while hoping for weather improvement, finally made a pass under the clouds and dragged the runway, then came around but hit hard and so pulled up and left us. It was almost 5 pm. The weather was about the same on February 2, but in the early morning of Sunday 3 there was steady improvement and we began sending weather reports every three hours. We had a heavy whiteout but good visibility and 10 mph winds, and we learned that a plane would leave McMurdo at 3 pm our time. The plane was on the ground at midnight, piloted by Ronald A. Stanton, in scattered clouds and a 10- to 15-mph wind. The snow surface was hard, so we thought that we would not need JATO. Consequently, we backloaded that as well as five aluminum aircraft pallets, snowmobiles, sledges, and smelly us—an estimated 12,000 lbs. We were all loaded by 1:30 am on February 4, local time. But the pilot could not get the heavy plane off the ground! The fix was to offload some JATO bottles and strap them to the fuselage, and we were airborne at 3 am. We refueled at Byrd Station, then arrived at McMurdo at 3 am of the February 5, McMurdo time.

At McMurdo we were all delighted to have heavenly hot showers, then real beds. And we were warm, for a change! The contrast between McMurdo (as ugly as the town is!) and what we had been doing just a few days before was, as you can imagine, truly astonishing. We had learned for ourselves why the English Coast was almost the last place explored in Antarctica! Yet despite an unexpected volume of snowfall and almost constant wind, we had explored an area of about five thousand square miles, about the size of Connecticut. And each of us realized that the field season taught us what we, and our colleagues, were made of! If we had come to the English Coast without much previous experience in Antarctica, it is likely that we would rarely have ventured out of our tents.

We spent the next couple of days returning stuff to USARP, getting rock boxes ready for shipment to the States, and so forth. We had a group meeting to clarify who was responsible for various topics for publication. Walt was scheduled to leave late on February 6 to get back to classes, and the rest of us left at almost midnight on the 8th. I sent a message to Kenneth A. Sargent, my Branch Chief, letting him know that all objectives had been met and that everybody was healthy and happy. And I sent a memo to the head of McMurdo USARP with an autopsy of the season. For once I could not complain about snowmobiles, and most of the rest was positive too! We were in Christchurch on the morning of February 9, and Adrienne met me at the Christchurch airport on the 11th. We did not leave New Zealand until March 4, when we departed Auckland. We spent most of our time on South Island, one of the most beautiful places in the world. It was a great vacation for both of us!

Qam
Meadow
4502
Springs
Qac
Tp
4580
QTf
38
Tli
Qac
29
14
41
5078
CREEK
82
36
5031
42
10
60

Between Field Seasons, 1985–1986

8

After I returned from The Ice in early 1985, Tom Steven and I understood that the Marysvale project was scheduled to end on October 1, 1986, so we had a lot to do. Getting manuscript maps and reports into the system took time, so I had no field season in the summer of 1985. I compiled and wrote up the two 7.5-minute geologic quadrangles in the Koosharem area, which showed the eastern side of the Monroe Peak caldera, and those were published in 1986. Tom senior-authored the Richfield 2-degree sheet, although it was not published until 1990. John Anderson and his students produced a spate of geologic maps, mostly for publication with the UGS. I completed work on the two 1:50,000-scale 15-minute maps, on which I made John the senior author because they showed mostly his dissertation area; these were published in 1990. I was also going through my second divorce, this time from Adrienne, which started in 1985 and dragged on and on. We signed the final papers in early April, 1986.

One bit of good news, which was announced when I was in Antarctica, was my appointment to the Science Advisory Committee, the most prestigious committee in the Geologic Division. It was a three-year term, starting in January 1985. The committee was made up of sixteen geologists, chaired then by David B. Stewart, a geologist (former Branch Chief) with Eastern Regional Geology in Reston. It was an elite bunch, the most stellar of whom was Dave, a living legend with many honors, whose wisdom and self-deprecating sense of humor were equally well known. It was fun to get to know him a little, for he was a career (1955–1995; died in 2015) igneous petrologist, with a B.A., M.A., and Ph.D. from Harvard, who had done mapping in Maine and bordering parts of Canada, high-temperature and high- pressure experimental studies of minerals, and the geochemistry of high-level radioactive waste. It seemed as if he could do damned near everything in geology! In 1981, Dave was the runner-up to Dallas Peck when the National Academy of Sciences picked our Director. The purpose of the Science Advisory Committee was to advise the Chief Geologist of Geologic Division on scientific problems and opportunities, and we met twice a year and in the interim prepared reports on various issues of the day. The Chief Geologist at the time was Robert M. Hamilton, who appointed me. Dave referred to our role as being the conscience of the Division, and we operated with independence and integrity even though at times we gave opinions that were not necessarily those of the Chief Geologist. The Water Resources Division had a nearly identical committee, and we met once a year to talk about mutual business. The star on that committee was Isaac (Ike) J. Winograd, who had a major role in recognizing the movement of groundwater by fracture flow (that is, flow along fractures/joints created by, and generally parallel to, fault zones) in the Death Valley regional groundwater flow system. This came from his long-term studies of the Nevada Test Site and Yucca Mountain, resulting in an important USGS Professional Paper in 1975. When I ended up at the Nevada Test Site in 1995, just after the RIF, that was a report that was the basis for much of our thinking.

Ken Sargent had been chosen as our Branch Chief in 1985 when Bob Raup's term had expired. Although Ken stayed on for only 1985 and 1986, moving to a higher-level administrative job in Reston; he was another steady hand for the Branch. He had a career of mapping volcanic rocks at the Nevada Test Site, among other duties, and was a friend of Bart Ekren, as well as a close friend and mentor of Gary L. Dixon, both of whom are described in Chapter 9. Glen A. Izett replaced Ken in 1986, another good branch chief whose research career had included studies of the ash from the Yellowstone caldera.

In late 1985, I started writing a big report on the Monroe Peak caldera, intended to be one of the summary reports for the Marysvale project. I compiled the map of the entire caldera complex at 1:50,000, and was nearly done with a hundred-manuscript-page text by the early spring of 1986. Then one spring morning, as I was happily polishing it, Ken Sargent strolled into my office with a different duty, referred to in the USGS by the cliché "other duties as assigned." He had just gotten a decision from the USGS Director that the USGS was going to propose a new line item in their next budget. It was to be for geologic mapping of the entire United States. The major benefactors of the funds for this proposal were the State geological surveys, as pass-through money from the USGS. The beauty of this idea was that each state's members in the US Senate and House of Representatives supported the act, even though few congressmen had ever heard of geologic mapping before then. The USGS had always done geologic mapping but they buried the exercise among other budget items in the belief that members of congress were too stupid to comprehend the value of mapping. In fact, a common question we received from these and other non-geologists was the refrain that many geologic maps had been done in previous years of the same areas, so why did geologists want to do another map? The answer was that map scales (detailed maps vs. reconnaissance maps) differ, and that concepts, based on work elsewhere, change with time; because science builds on the work of predecessors, new maps are more accurate. Nonetheless, it looked as if others were going to have to do a hell of a sales job. But I was happy to do my part!

The new job that Ken had in mind was for me was to compile the geologic mapping needs of all the states. The report was to be used as justification for the budget initiative. I was told to drop everything and was given a secretary with a word processor, to whom I would send my writeups on the geology and needs of all states, including Alaska and Hawaii. So away I went, with much help from other USGS geologists. My secretary's worst day of the week was Monday, when she would find a stack of handwritten and typed first drafts on her desk—my weekend's work. I would of course darken her office door other times of the week as well. I had two co-compilers: Fred Miller from the Reston office and Dave Miller from the Menlo Park office. Our job was to articulate, primarily to congressional staffers who would push the legislation, the practical economic benefits of geologic mapping in terms of mineral deposits, hazards, water resources, etc. Others from various branches around the country funneled reports to me on their areas, and I would put them into a coherent text. I worked weeks and weekends for months, finishing a draft by late spring. That gave me a short field season in Utah, which included visits to eastern Nevada, where I was contemplating proposing another project, this one in the Caliente caldera complex. When I returned to Denver in June, I had revisions to do on the mapping-needs report following technical reviews, before leaving for Antarctica. The report was tidied up and soon published as an 82-page USGS Open-File report. A lot of administrators in Reston and elsewhere worked hard at selling the program, and I always considered that my part was small. In fact, I learned in 2020 that my two co-authors, Fred and Dave, together had written most of the drafts and final National Geologic Mapping Act in late 1986 or early 1987 when they were doing an administrative tour together in the Regional Geology office in Reston. The funding was approved in 1987 and in all subsequent years to the present, for a wildly successful program, the National Cooperative Geologic Mapping Program, that funded State geological surveys, geology undergrad and grad students, and the USGS. I never got back to my report on the Monroe Peak caldera, although I still plan to do so. As proposed in the John Fogerty song, "Rock and Roll Girls", one of my favorites that had been released then, "life is just a rodeo and the trick is to ride until you make it to the bell." Then my 1986–1987 Antarctic field season arrived, to my relief!

As my Marysvale project wound down just before I left for Antarctica, I proposed to map in detail and study one of the largest calderas in the country, the Caliente caldera complex. It was to the west of Iron Springs, in Nevada, and I knew many of the ash-flow tuffs it erupted, plus it bounded the Delamar gold district, a big one. Most mapping in Nevada, because of the huge size of the state, had been done at reconnaissance scale. The geological map of Nevada, at 1:500,000 scale, had been published in 1978 by John H. (Jack) Stewart (1928–2013), a great geologist of the Branch of Western Mineral Resources (Menlo Park). Leading up to that, county maps of Nevada, at 1:250,000 scale, had been done by USGS geologists and published by the Nevada Bureau of Mines & Geology (NBMG). In Utah at the same time, Professor Lehi F. Hintze (1921–2014) of Brigham Young University, was working on the Utah state geological map at 1:500,000, published in 1980. Many of the Utah counties that contributed to that map had been done by geologists of the Utah Geological Survey. Both Jack and Lehi were geologic superstars, yet universally loved. Jack had a Ph.D. from Stanford, whereas Lehi had a Ph.D. from Columbia. Now I figured that it was time to do some detailed mapping in Nevada.

By the spring of 1986, I was dating a tall, beautiful geologist, Mary A. Siders, who was with the Geologic Mapping Division of the Utah Geological Survey. She was fifteen years younger than I was, but that did not stop me! When she was hired by Genevieve Atwood at the UGS a few years before, Genevieve had assigned her to map several quadrangles just west of the Iron Springs district. Another geologist hired by Genevieve, Michael A. Shubat with the Mineral Resources Division of UGS, was working with Mary by mapping his own quadrangle just to the east that contained a small silver district. We all got to know each other when I attended UGS field trips, and I led some of my own to show them my rocks and faults. Both Mike and Mary were good field geologists, with excellent senses of humor. For a short time, Mary had been a singer and banjo player in a bluegrass band. They recorded one album while she was there and she sang the lead on one track, the best on the album. She had a beautiful voice.

The two quadrangles that Mary was asked to study and map were boring, with sparse good exposures and no societal importance. Between one of these quadrangles and Mike's quadrangle was, however, an interesting one that I wanted to do, and I proposed that Mary, Mike, Quaternary geologist Gary E. Christenson of UGS, and I map it together and publish it with the USGS. It had practical significance because it contained a small geothermal district controlled by a huge basin-range fault near the small town of Newcastle. Hot water heated by the geothermal gradient was able to rise rapidly from depth along the fractured rocks of the fault zone, and shallow (about 200 feet) wells sunk near the fault tapped the hot water. This water was directed toward a half dozen huge greenhouses, which were then able to grow tomatoes and flowers cheaply all year long. The UGS had already done a study of the geothermal district, but nobody had mapped the quadrangle. The field work started in the summer of 1986 and finished in the summer of 1987. The Newcastle quadrangle was published by the USGS in 1990, with Mary the lead author.

Mary and I had Antarctica in common: she had done her Master's thesis for Ohio State University by mapping a piece of the Transantarctic Mountains. My field notes show that after a GSA field trip in Flagstaff, Arizona, Mary and I took a few days off in May 1986 and drove to the south rim of the Grand Canyon. There we hiked to the bottom in one day (9.5 miles), camped at the Bright Angel Campground at the bottom, and hiked out the next day. This was an overly ambitious way to see the Grand Canyon, but we were young … .

We were seeing each other regularly in the summer and we did some mapping together in and near Iron Springs. By the fall, we were talking marriage, with a plan that perhaps she could pursue a Ph.D. at the University of Colorado. She quit the UGS in Salt Lake City and we hauled all her belongings over to my place in Morrison. She lived there while I went to The Ice and she applied to the University of Colorado to pursue a study in hydrology in the Dept. of Geology. In the meantime, she found a job as a clerk at an Eddie Bauer clothing/camping store east of Morrison. We got engaged before I left for The Ice.

BRITISH ANTARCTIC SURVEY 1986–1987 U.S. GEOLOGICAL SURVEY
BLACK COAST GEOLOGY
SOUTH POLE
King George I.
Robertson I.
Air Mail
JOHN BISCOE ANTARCTIC SEASON 1986-1987 FALKLAND IS.
BRITISH ANTARCTIC TERRITORY
Ms Mary A. Sidus
5894 Meadow Drive
Morrison, Colorado
80465
U.S.A.
Ice Shelf
Steele I.
Butler I.
1986 – 1987
Black Coast

9 The 1986–1987 Field Season on The Ice

By the spring of 1986, NSF had accepted my USGS proposal for one last trip to The Ice, this time to finish mapping the unvisited part of the Black Coast, north of the Lassiter Coast. And we finally had a 1:500,000-scale sketch map of the Black Coast, for the USGS had published it in 1979. The grant proposal was for this to be a cooperative effort by the USGS and British Antarctica Survey, made up of two geologists from each agency. We would use BAS logistics to reach British "Antarctic Territory." This method is to fly from the U.K. down the Atlantic Ocean to the Falkland Islands, and from there cross the Drake Passage by ship (the *John Biscoe*) to their northern base (Damoy) on the Peninsula, then fly the rest of the way south to Rothera Base, the main BAS base.

The Black Coast was named for Richard Black, the Base Commander at East Base, one of two bases built during the US Antarctic Service Expedition of 1939–1941. Richard Byrd led the expedition, so this is commonly referred to as Byrd's Third Antarctic Expedition. Unlike the previous two (1928 to 1930, with Larry Gould; and 1933 to 1935), which were privately funded, this was the first to be funded largely by the US, under the Presidency of Franklin D. Roosevelt. East Base, abandoned but now a tourist spot for cruise ships, is on Stonington Island, about fifty miles east across Marguerite Bay from Rothera Base, adjacent to the western coast of the peninsula. East Base was later used by Finn Ronne during his expedition of 1947–1948. The largest mountain in the Antarctic Peninsula is beautiful Mount Jackson, rising to a sharp spire at an elevation of 10,446 feet, in the northern Black Coast. The mountain was discovered by field parties led by Richard Black. They named it after Andrew Jackson, seventh president of the US.

The BAS workhorse aircraft is a DeHavilland Canada DHC-6 Twin Otter, developed in 1964 to be powered by twin turboprops, and currently produced by Viking Air in British Columbia. Although not comparable in size and carrying capacity to a Herc, it is a reliable and powerful utility STOL (Short Takeoff and Landing) aircraft with a long range, although with a slow cruising speed of about 125 mph. Pilots bring the planes south each year through North and South America, then equip them with skis once in Antarctica. Some of you may have flown in one, especially if you go to exotic places, small towns, or the so-called Third World, for the aircraft can be configured to carry as many as twenty passengers. The British Antarctic Survey painted the planes red. Each Twin Otter flight could bring in at least 2,000 lbs (e.g., two Ski-Doos plus more) for our mapping, and we also planned on close air support to reach over a dozen isolated nunataks in two places: along the coast where large crevasse fields prohibited reaching them by snowmobiles, and small nunataks on the crest of the Antarctic Peninsula, far away from our main study area.

I chose Art Ford to be my colleague on this trip. By this time, he was the world's expert on the geology of Antarctica, having seen many parts of it during twelve previous field seasons, but he had not been to The Ice this way before. On the BAS side, the leader was to be Bryan C. Storey, a permanent BAS geologist with a Ph.D., a structural geologist, and a veteran of five previous trips to The Ice. I was to be co-leader with Bryan. He was to be joined by Dutchman Hein E. Wever, a temporary (two to three-year appointment) geologist with BAS, specializing in igneous rocks, and a rookie to Antarctica. The fun started on Sunday November 16, 1986, when Art flew from San Francisco to Denver, then he and I flew from Denver to Kennedy Airport in New York, then to Heathrow Airport, London, arriving at 7 am of November 17. Then we took a bus to Cambridge and a cab to BAS headquarters near Cambridge. There we found ourselves in the middle of a retirement party for Charles W.M. Swithinbank, exiting as Chief of the Earth Sciences Division of BAS, a position that Mike Thomson was to assume.

Charles (1926–2014) was a living legend, a British glaciologist, geologist, and pioneer of early studies of sea ice, ice shelves, and glaciers. After two years with the Royal Navy, he received the equivalent of a bachelor's degree from Oxford University (1946), then a Ph.D. (actually, Doctor of Philosophy) from the same university (1955). He spent the early part of his career studying Arctic sea ice and its effects on shipping, for the Scott Polar Research Institute in Cambridge. In the 1950s and later, he was on several well-publicized SnoCat traverses in Antarctica, including during the Geophysical Year. He joined BAS in 1974 to become head of the Earth Sciences Division, a position he held until his retirement party that we joined. He was much decorated, by Queen Elizabeth II and others, for his noteworthy career. Art had met him and worked with him before and had learned British camping methods from him. I found him to be charming and delightful. Charles told us that we were the first non-BAS scientists to work a full season with BAS in the peninsula. We of course also met with Bryan, Hein, the Thomsons, and Bob Pankhurst, who is a celebrated isotope expert. We stayed with the Thomsons that night.

On Tuesday, we took a tour of BAS, including getting our cold-weather clothing, then a tour of Cambridge University, including Kings College chapel (a choir was practicing). I visited with Stuart Agrell, Emeritus

Professor at Cambridge, with whom I had been working on an interesting breccia dike he had found a decade earlier on the eastern edge of the Sevier Plateau, a dike that contained small sapphires, though not of gem quality. Stuart was a tall, thin, spry, animated, and delightfully friendly person, a classic professor of petrology and petrography, the latter the microscopic examination of rocks in thin section. Janet and Mike had a party in our honor that night, in their beautiful refurbished old farmhouse on a third of an acre on the edge of town and about a ten-minute drive to BAS. Always a treat to visit with limeys, even though they talk funny (the latter the words of my friend Paul Carrara)!

On November 19 we began the day at BAS with meetings on geologic work that BAS was doing. We learned that each of us could receive and send a one-hundred-word radiogram a month, transmitted to us in the field by short-wave radio and with the text mailed to and from Cambridge and the States. Mid-afternoon we caught a bus to Brize-Norton RAF (Royal Air Force) Base. By 8 pm local time, the four of us were in the air on an eighteen-hour flight in a jet, a Lockheed TriStar 500, with a half-way refueling stop at Ascension Island, an isolated basaltic volcanic island and RAF station just south of the Equator in the mid-Atlantic. Ascension, like Iceland, is part of the mid-Atlantic ridge, which runs north-south down the middle of the Atlantic Ocean. The ridge, now volcanically active, formed by upwelling basaltic magma that rises, then spreads west and east to drive the two opposed giant conveyer belts of oceanic crust east and west in opposite directions away from the ridge. Therefore, the Atlantic Ocean grows wider each day, and the continental plates west and east of it (North and South America to the west, Europe and Africa to the east) farther apart each day.

The temperature was 73°F when we landed at Ascension at 4:30 am local time. After two hours on the ground and a couple of beers, we were airborne for Port Stanley, Falkland Islands. We arrived at Mount Pleasant RAF Base, which had replaced the airport that had been bombed out during the Falkland War. We landed in a snow squall with a temperature of 45°F , at 11 am local time on November 20. We took a bus to Stanley, arriving at 1 pm. Our quarters were on board the *John Biscoe*. We shopped at, and hiked around, the capital of Stanley, a pretty, well kept, and colorful small town.

The Falkland Islands are a cluster of islands making up about 4,700 square miles and located three hundred miles east of the southern end of Argentina, at latitude 52° south. The population is now about 3,400, most of whom live in Stanley. The population is made up of British citizens, descendants of settlers who came there in the 1840s, mostly from Great Britain, when the islands became a crown colony. The businesses are fishing, sheep farming, and tourism. There are large bird populations on the islands, including penguins. The landscape of hills and low mountains is barren, with much tundra and no trees. The climate is maritime, meaning cold, windy, and rainy with temperatures that range from 12° to 70°F . Snow flurries are possible any day of the year. We arrived only four years after the Falkland War between Argentina and Great Britain, and we were warned about ordinance still on the ground, should we want to do day hikes. The war began in April 1982, when the Argentine military invaded the Falklands, South Georgia, and the South Sandwich Islands. It had been a long-simmering territorial dispute between the two countries, but its eruption had been fomented as a distraction from widespread economic hardship in Argentina by a military dictatorship that had been in power since 1976. After an amphibious assault by Great Britain, then under Prime Minister Margaret Thatcher, the Argentinians surrendered in June. Sadly, it was at the expense of the deaths of 650 Argentines, 255 Brits, and 3 Falkland Islanders. This fiasco led to the collapse of the Argentine dictatorship.

The RRS (Royal Research Ship) *John Biscoe*, commissioned in 1956, was a supply and research vessel 210 feet long and 40 feet wide, with a crew of thirty-three led by Captain Malcolm Phelps. It was used to transport people and supplies to the BAS bases on the Antarctic Peninsula. Owned by BAS, with Stanley its home port, it was retired from service in 1991. The *Biscoe* fueled up in the morning of November 21, and we were under way at 4:30 pm, crossing the Drake Passage. Malcolm, a pleasant man, told us that we were only the fourth and fifth non-BAS people to travel on the ship. The first had been Paul Williams, with his banjo, who had visited a bunch of BAS bases in 1964¬–1965 and had done as much geology as he could along the way. The next day we crossed the Antarctic Convergence, an approximate and imaginary circle around Antarctica where cold, nutrient-laden water from Antarctica meets and descends beneath warmer water from the north. I worked on my Christmas cards. The waves and swells were moderate, although at dinner we learned to hold onto a table (bolted to the floor) with one hand and eat with the other. Dinner in the Captain's dining room, in fact, was quite formal: coats and ties were required (sweaters for breakfast and lunch), white tablecloths, wine, etc. During the night, the ship slowed as it encountered icebergs, and on November 23, a Sunday, we had moderate swells and saw more bergs. On the 24th, it was cloudy and 35°F and off and on all day we were in pack ice (new, relatively thin ice that you either go around or through), with a few seals and penguins visible on the ice flows. Various sea birds flew about, mostly petrels and gulls of several species, waiting for garbage.

Late in the day, we passed several islands, and the next day we went into Admiralty Bay of northern King George Island to pick up a field party there, but wind-blown brash ice (ice and chunks of bergs crushed by waves) prevented us from sending a rubber dingy to get them. Southern

King George Island contains many bases tucked into one bay and run by countries (Russia, China, Poland, India, Uruguay) who want a presence at the Antarctic Treaty table and found this to be the cheapest place to go logistically. We continued past Livingston Island, with peaks that rise to 5,800 feet, then an active volcano at Deception Island. On the morning of November 26, as the sun came out, we went through beautiful Gerlache Strait, with sheer walls of ice and rock that rise several thousand feet out of the sea, and by 9 am we had docked at the British Demoy Base, where there is a rookery of Gentoo Penguins. It is on Wiencke Island, not far from Port Lockroy, a former BAS base that is now a historic site and a destination for cruise ships, where tourists can post letters home with Antarctica stamps and postmarks. Surrounding mountains rise to about nine thousand feet. From there we hiked up to the landing strip to a waiting Twin Otter, and we were off by 11 am. The flight of about 250 miles was incredibly beautiful, in mostly sunny weather, with sharp peaks of the Antarctic Peninsula and its western islands below us. We arrived at Rothera Base at 1 pm.

Rothera Base, at the southeastern end of Adelaide Island, was established in 1975 to replace old Adelaide Base, whose skiway (runway) had deteriorated. Rothera is perhaps the most beautiful base on the continent. Art and I were transfixed by the setting, which was gorgeous in all directions, thanks to its mountains, glaciers, and ice cliffs. The base consists of a cluster of buildings on a rocky spit. The largest building is a two-story, green one with bedrooms, the base commander's office, a library, medical room, and bathrooms on the first floor. The second floor consists of a kitchen, dining room, lounge/bar, food- storage area, and weather bureau. Other buildings include garages, laboratories, a power plant, and warehouses. When we visited, a resident skua, George, sat outside the front door of the main building, waiting for a handout. At the time, the base had about thirty people, but it typically swelled to about fifty or sixty in the summer as field parties came and went. Fourteen people can winter over. The base commander while we were there was John Hall. It was a congenial group, with a lot of laughter. At the bar, I had my first taste of Talisker, a particularly smoky single-malt scotch whiskey. It immediately became my favorite scotch, and now I always have a bottle on hand. Rothera had an admirable tradition of everybody sharing the menial jobs ("gash duties") on a rotation basis, such as washing dishes, gathering snow blocks for water, preparing the dining hall, emptying toilets, collect trash ("gash" in limey vocabulary), etc. My turn came on December 4. And on Fridays, everybody pitched in to scrub down the base.

Crabeater seals (smaller than the Weddell seals at McMurdo, and somewhat furry) lounged on sea ice near the base. Every so often a group of Adelie penguins would stroll in, wobbling side to side as they walked, wings held out to their side (at other times, Adelies would move by getting on their bellies, then "swimming" along in the snow). The base had three dog teams, and most persons used the teams for fun, exercise, and Antarctic tradition. We spent the next few days packing in poor weather.

We met two General Assistants (GAs), Anthony (Tony) R. Simpson and Peter T. Marquis, who would join us in the field. Tony had wintered over and was in his second year, while Peter was beginning a two-year tour. Their training was primarily in mountaineering, and they possessed little if any scientific background. Peter had an appealing sense of humor but Tony was humorless. Peter was reasonable, but Tony considered his job to be that of mother hen, schooling all of us bozos on doing everything "by the book." After a while, he became a real pain in the ass, but we tolerated him. Besides, they had done most of the work to get us ready! We also attended radio school, first aid training, and snowcraft school. Art and I, who would work together, configured our Ski-Doo for remote steering.

One day, during an Adelie's visit to the base, he/she happened to be heading in the direction of the dog teams (three strings of husky-type dogs staked down outside, with about a dozen dogs in each string). Antarctic penguins evolved never having to fear land animals, so they occasionally will wander inside the length of the chain a dog is on, and be killed and eaten in the process. Seeing that the penguin was heading toward the dogs, two base persons tried to scare it off. But it was determined to head that way, so it went between the legs of the potential saviors and right into one of the lines of dogs. But all the dogs missed the "cheeky little bastard," and it even picked a fight with one dog. So, it got through scot free and continued its trek.

On December 3, the weather improved southeast of Rothera, and field parties started getting airlifted. Rothera had three Twin Otters available, piloted by Richard Hasler, chief pilot, and Steve and Brendan (I neglected to get their last names)— all good guys as pilots always are! The planes were being sent to some places we knew well: a geophysical crew to the English Coast, and geologists or glaciologists to Mount Smart (our Jamesway on the Orville Coast), as well as runs to Fossil Bluff station. This last put-in spot is on eastern Alexander Island, a 270-mile-long island off the western coast of the southern Antarctic Peninsula and northeast of the English Coast. Alexander Island is separated from the peninsula by linear, north-trending George VI Sound, a twenty- to thirty-mile-wide body of ice-covered water that is effectively a rift valley, in other words a graben dropped down by normal faults, with the relief of the land on either side of it being as much as seven thousand feet. This is an example of east-west crustal extension (pulling apart), such as in the Great Basin, with basalts (mostly hyaloclastites in Antarctica) accompanying the faulting, also as in the Great Basin. The northeastern tip of Alexander Island is about eighty miles south of Rothera. Unfortunately, the weather was not good on the Black Coast, so we stayed in Rothera. On December 6, I was invited to join a Twin Otter as it made some runs of one field party and of food, fuel, and equipment to bases and caches on Eklund Island (south of Alexander Island) and elsewhere, with a re-fueling stop at Fossil Bluff Station. It was a spectacular trip and a nice birthday present. The next day, Art was treated to a trip to the English Coast! The temperatures now were close to 45°F , which made it difficult for a loaded Twin Otter to become airborne in the sticky snow.

The weather on the Ronne Ice Shelf and eastern Lassiter and Black Coasts began to improve on December 8. We had six Twin-Otter loads to make us functional, and the put-in site (Camp 1) was to be on the southern side of Kamenev Nunatak, a name I had proposed to honor Gene Kamenev, my exchange scientist from the 1972–1973 field party (actually, I proposed the names, but the US Board on Geographic Names assigned specific landmarks). In fact, the area we were to explore this season had lots of names from many of our field parties, including Rowley Massif. Kamenev Nunatak is almost twenty-five miles south of Mt. Jackson, on the eastern edge of the peninsula plateau (formally named the Dyer Plateau in this area), which is the highest part of the Antarctic Peninsula, from which glaciers flow east and west. Kamenev Nunatak is about eighty miles north of the Mosby Glacier, which we (including Gene) crossed during our 1972-1973 field season, from the northern Lassiter Coast into the southern Black Coast.

Four flights left on December 9 during excellent, calm, clear ("dingle," the limey word for bright) weather, at one point above 40°F , with Bryan and Tony first to be put in. The temperature was down to 20°F when Art and I arrived at 5:30 pm. On our put-in, we had great views of Jackson, which towers above other peaks, but it was not visible from Kamenev Nunatak. Steve, who piloted this fourth load, reported that the plane's altimeter said that the elevation of the put-in camp was 4,260 feet. There were about eight inches of light snow on the ground, and we set up our tents and had skraj (limey for dinner), a stew of freeze-dried beef and vegetables. Then to gonk (limey for sleep) at 11:30. "All of this seemed so familiar," I wrote to Mary on December 16. We were all up at about 7 am on December 10, with the temperature about 10°F and a 5- to 10-mph west wind. We did the geology of Kamenev Nunataks (Jurassic ash-flow tuffs) and of scattered nunataks to the north (a large pluton). The last two flights, which included Hein and Peter, arrived while we were on the outcrop.

The BAS field rations were not terribly dissimilar from ours, although with less choice by the party chiefs. They had twenty-man-day food boxes set up ahead of time. Instead of meat bars, the staple was one-meal packets of dried beef, plus lesser one-meal packets of dried mutton. Freeze-dried vegetables (peas, mashed potatoes, mixed vegetables, sliced onions) were the other staple, also intended for the evening stew. British "biscuits"— in effect a large cracker that was slightly sweet—were also abundant in the food box, meant to replicate bread, and they were accompanied by cans of marmalade, jam, and of course Marmite (yeast extract to which most limeys, Kiwis, and Aussies are addicted; Aussies, however, have their own: Vegemite). Breakfasts consisted of oatmeal, muesli, and bacon. The rest of the goods in the food box was much like what we had on our field parties, although BAS favored tea over coffee, as one would expect. But as much as I searched the bottom of the food box, I could find no steaks or lobster tails!

The BAS had a safety system of Ski-Doo travel that they used, based on two snowmobiles for each two-man team: each snowmobile pulled a Nansen sledge, but the lead (or first) Nansen was connected by a hundred-foot rope to the second snowmobile, in case one Ski-Doo were swallowed by a crevasse. The two men, of course, had to sit and drive each snowmobile, wearing crash helmets and roped to the snowmobile. The General Assistant (GA) in this system made all decisions on logistics, including how we got there and in what weather we were allowed to travel. Each geologist was required to rope up with a GA to get to most outcrops, plus many other "rules." Not only was this a goofy way to travel, it was bureaucracy gone amok, with the scientist assumed to be a klutz and his scientific field program to be run by a young non-scientist who cared little for priorities and the time the conditions allowed to do them. But as we were accustomed, Art and I worked together, and this meant that our rules were relaxed: we drove our single Ski-Doo remotely while skiing, as in all our previous field seasons. We had no damned crash helmets either. The next day, we visited nunataks twenty miles to the northwest, with Mt. Jackson looming not far to the north. This was followed by several days of fog, keeping us in camp except for some hikes over to Kamenev to do a little more detail. I collected some samples of his nunatak to send to Gene, along with photos I had taken previously.

On Sunday December 14, a light westerly wind kept the fog to the east and we packed and headed southeast and then east, down the Condor Peninsula, which is between the Rankin Glacier to the north and the Gruening Glacier to the south. The day was clear, but as we went the soft snow became as much as a foot deep. We traveled about thirty miles, generally dropping lower and passing Boyer Spur (named for Steve), then crossing the Kellogg Glacier (named for Karl), but on uphill climbs we had to relay loads in the soft snow. Camp 2, on the central Condor Peninsula, afforded pleasant views of Schirmacher Massif (named for George) to the north, across the Rankin Glacier. From camp, a snow slope concealed any view of Rowley Massif, just east of George's namesake (both are about ten miles long north-south and six to eight miles wide east-west), but we could see Mt. Vennum (named for Walt) rising from the western side of Rowley Massif! The next day we visited the rocks farther east in stratus about a thousand feet above the surface; fog moved in during the day, but we were able to follow our tracks back to camp. We were fog-bound all day on the 16th, such that Hein and Tony's tent was barely visible. The same continued the next day until 3:30 pm, when we finally did some geology. On December 18, the fog cleared at about noon, and we did geology after that. Fog and light snow followed the next day, so that day was a bust too! Clearly open ocean to the east was supplying moisture that expressed itself as snowfall and fog from storm systems, despite common katabatic west winds.

On December 20, fog and snowfall did not clear until 2 pm, so Art and I tried to reach Mt. Showers, the eastern rocks of the Peninsula about ten miles away. We became socked in about a third of the way there, but we were able to get bearings on eastern parts of the Showers cluster and laid flags to it, from where we did the geology. But we could see no more rock to the east, so we returned. The following day, it was clear in the morning and we attempted to return by the flagged route, but the fog came in soon after and we again retreated to our tent. By mid-afternoon, some clearing allowed us to do the geology of a nearby nunatak. The other two teams had similar success (or lack of it) in going other directions from camp, but at least together we picked away at the geology. High winds, ground blizzard, and low visibility characterized December 22 and 23, so we stayed in camp. The 24th was clear for a while in the morning but we had problems with one of our snowmobiles; though we replaced a fuel filter, we still had air bubbles in the gas line. Finally, after clearing up those problems, we found that the soft snow was too deep to get to Showers. High west winds and ground blizzard began at 3 pm and continued all night. By 6 am on Christmas Day, the winds broke off and we agreed that it was best to get out of this mess. The winds had the fortunate effect of hardening the surface; otherwise, it would have been difficult going west, uphill. Nonetheless, we still had to relay loads west, back towards the peninsula plateau, in soft snow and significant cloud cover. Exhausted, we eventually made it about twenty miles before we established Camp 3, about ten miles north of Malva Bluff (named for Tony), and east of Kellogg Glacier. Although we were retracing our steps, we had not yet done the geology of this area. We had that opportunity the next day, when we had high clouds, temperature just above freezing, and light winds, and we all did geology. Art and I tried to get to Malva Bluff but found that it is in a deep hole, and soon we were on a steep blue-ice slope to the south in the whiteout, followed quickly by a 30-mph wind, so we retreated and did the geology of a ridge north of the Bluff instead, then continued out of the hole to a course that took us west to nunataks along the eastern side of Kellogg Glacier, before returning to Camp 3.

The morning of December 27 started with a light rain, moderate clouds, and fog coming and going, and four of us headed "home" to Kamenev Nunatak, a trip of more than twenty miles. Along the way, however, we were to do the geology of nunataks south of Kamenev, while Bryan and Pete would take a more northerly string of nunataks. Hein and Tony stayed at Camp 3 to finish nunataks to the north. Our first six miles was in a whiteout with decent visibility, but then fog and near-zero visibility settled in, so we went on a westerly compass bearing and made it to several nunataks, then set our compass bearing to the northwest and bumbled onward, accidentally flipping a sled on a side hill. For the next half dozen miles, we continued in the fog, then it lifted and we saw Bryan and Peter and their snowmobile train a couple of miles north of us, and then Kamenev Nunatak. We stopped at several more outcrops along the way, reached old Camp 1, and had our tent up by 8 pm. The next morning, we had snowfall and 15- to 20-mph winds during the morning and we did not get going until about noon, heading south-southeast to Boyer Spur, fifteen miles away. The winds continued all day, but we did the geology of many nunataks, including Boyer Spur. We took pictures of Kellogg Glacier for Karl, and we collected a small bottle of glacier ice for a souvenir for Karl. For some strange reason, I anticipated that it was unlikely it would still be ice when I handed it to him.

High winds came up during the night at Camp 1 and continued during the 29th and 30th. As I described in a letter on the 30th to Mary, we experienced a constant swishing sound of snow from the ground blizzard hitting the side of the tent and the wind howled against the ropes. At times the tent shook fiercely, booming like a bowstring. We spent the afternoon reading books while remaining in our sleeping bags, with our soft personal-gear bags as pillows. The primus was on for warmth, plus it heated water for coffee, but we tried to avoid drinking much to keep from going outside. I stated in the letter, "Oh well, at least we have no mosquitos here!" Hein and Tony, of course, were still in Camp 3, probably in the same wind. That day I had Bryan send a telex by radio to Rothera, and from there hard copy to Mary by the apparently circuitous route of aircraft, ship, and more aircraft to Cambridge and on to the States. The copy I now retain told her that we were put in at Kamenev on the 9th, had just returned from the Condor Peninsula, wished her a belated merry Christmas and a happy New Year, requested football scores, and asked her to relay the message to Art's wife Carole and to my branch chief, Glenn Izett. It concluded with "these limeys are not as bad as I thought they would be!" Ribbing people whom I like has always been my nature, but unfortunately those people are generally smarter than I am, so they always seem to get me back more than I give!

The wind stopped at 3:30 am on New Year's Eve, so we rose at 6:30 to get an early start on the trail. We could see fog lurking to the east but a steady 10-mph west wind kept it at bay, and we headed off in a heavy cloud cover and no surface definition to explore nunataks to the west. Bryan and Peter headed east to Mount Whiting (named for Ron). Moderately heavy snowfall, with visibility of less than a half mile, arrived in the afternoon of our traverse, and we went on a compass bearing to reach our last outcrop. Then, rather than following our tracks back to camp, we went on a compass bearing south for 5.5 kilometers (our odometers were in metric, British style), then east for 6 km, and bingo, we hit camp dead on at 6 pm! We went over to Bryan and Peter's tent for dinner and drinks that night, with the entertainment being the short-wave radio for our daily schedule with Rothera. As recorded in my letter of January 6, "Everybody in the field, plus Rothera, came up about 11:30 pm and we all talked and yelled and sang our way through midnight! Lots of drunken limeys on the airwaves then! Disgusting, these people."

New Year's Day was calm with high clouds—finally some good weather! We headed toward the southern ridge of Boyer Spur. To get there, we had to travel down the Kellogg Glacier. That proved to be challenging, as we fought our way through fields of blue ice and crevasses, taking one route around one crevasse field only to find ourselves in another, then another. After much time spent horsing around, we got to the southern Boyer Spur and did the geology, including collecting a granite boulder for a doorstop for Steve, then we were faced with wandering back through the crevasses on the glacier. At one point, going up a steep grade, the Ski-Doo started sputtering, so we had to stop to replace a spark plug before continuing. We made it to a last outcrop east of the glacier, picked up a doorstop for Karl too, and arrived back in camp at 8:30 pm. Hein and Tony were there, finally.

January 2 was clear but Tony wanted to service the snowmobiles rather than head off on our next traverse, so we all did what he wanted, but Art and I also found time to run off to do the geology of a nunatak to the west. The next day, as we might have predicted, we had low visibility in light- to moderate-snowfall and we stayed in camp, fixing the brakes on one of our snowmobiles but doing nothing else productive. This weather continued until 9 am on January 4, then in low clouds but with some visibility we started a traverse to attempt to reach Waitt Peaks (named for Richard) about sixteen miles to the northeast, across the Rankin Glacier. With Hein and Tony leading, we went about fourteen miles due north, but thick fog settled in and we were forced to stop and set up Camp 4. Bryan and Peter had gone off west of us, intending to meet us at Waitt Peaks, but they obviously were also beset and, as we later learned, were tented near Musson Nunatak. We were stuck in that camp in the headwaters of the Rankin Glacier with low visibility, for January 5, 6, and 7. Hoar frost precipitated on the tent ropes from the moist air in the fog that moved northwest up the glacier. Thick clouds were above us, and in the whiteout we appeared to be floating in milk. The four of us talked among ourselves and with Bryan and Peter on the radio, and we agreed that if the weather magically broke off in fine style and we had surface definition, we would be able to cross the heavily crevassed Rankin Glacier to Waitt Peaks. But if the mank (another limey word) let up just slightly but the whiteout continued, we would run back to Kamenev camp, then do our next traverse to the many previously unexplored areas to the south and west, where fog was less likely. We were aware that the fog was coming off ice-free parts of the Weddell Sea and working its way up glaciers, and we feared being locked in by fog until the end of the season. We also agreed that the three parties (or at least Art and I) would go our separate ways on our next traverse. Art and I were the prime advocates of this change, for we had seen too many opportunities for travel and mapping killed by Tony and his insistence on caution and so-called "safety first." The season was coming to an end and we could not squander any more opportunities. We also learned that our resupply flight was scheduled soon (they were asking for weather reports) at Brennecke Nunataks, about forty miles south of Kamenev camp.

On January 8, visibility improved to about a mile but the fog drifted in and out and the heavy clouds continued above. Tony vacillated as to what we should do, although Art and I made plenty of "helpful suggestions" that it was time to retreat south. This was a heavy blow for me because the only way to travel to the Schirmacher and Rowley massifs was to come in from the north. Finally, we forced the issue by packing the tent; they did the same, and the four of us blundered off in the fog back to Kamenev, arriving in slightly clearing weather. Bryan and Peter had one more day to finish up the Musson Nunatak geology. Once back at Kamenev, Art and I knew we had to take advantage of any break in the weather, so we packed for the next traverse. But when we were ready to go at 4 pm, the fog had settled back in. Tony vacillated again, so we headed off on a westerly compass bearing, where a nunatak was shown on the sketch map. Hein and Tony followed, but three miles out we flipped a Nansen on a huge sastrugi that was invisible in the whiteout. That requires unpacking the sled, turning it upright, then repacking, amidst much swearing. While we dithered, Hein and Tony finally made up their minds and told us they would go south to Owen Peak, across the Gruening Glacier and about fifteen miles south of where we were then, and they took off. Seeing their last sled disappear into the fog was a great feeling! We mounted up and traveled on a compass bearing, hitting our nunatak in twelve miles as the fog lifted a bit, then doing the geology there. We then headed southwest, climbing upward and running out past the cloud cover. The surface was hard and beautiful and we blasted along with a 10- to 15-mph wind at our backs. It was a beautiful evening and our landmarks were visible as we made the fifteen miles and set up our tent (Camp 5 for us) at 10:30 pm about two miles north of Lokey Peak, a nice sharp nunatak in the western Guthridge Nunataks of the Gutenko Mountains, on the peninsula plateau. We had covered about forty-five miles that day and were quite chilled. Art's pocket altimeter said that we were 1,600 feet higher in elevation than Kamenev, making our elevation about 6,000 feet. Guy Guthridge, a long-time employee of NSF's Polar Programs, who at times worked out of McMurdo and Christchurch, was the namesake here. Bill Lokey, who managed the Berg Field Center in McMurdo for years in the early 1970s, was the other. In fact, I proposed his name, for he had been very helpful, and Art bumps into him occasionally, for Bill now lives near Carole's home town in northern California. They were some of the many good guys at NSF, and the next day we collected mantle-size rocks for both. By 2018 Guy was the volunteer editor for The Antarctican Society newsletter, and he wrote a one-page note for the newsletter about the day he received his rock, and quoted my entire letter that accompanied it about what it was (Jurassic ash-flow tuff) and what the day was like (cold and windy) when Art and I grabbed it; his rock and my letter "occupies a bookshelf" in his home.

We got a late start on January 9. We called Rothera at 8 am because we had missed the previous night's radio schedule. We learned that Hein and Tony were beset by fog and had not yet made it to Owen Peak, and Bryan and Peter were still at Musson. But we had few clouds, good visibility, and only a light east wind. But by the time we had done radio work and packed our sled for geology work, the wind was at 15 mph and fog had hit the nunataks to the northeast, where I'd wanted to go. Instead, we sped off to the nearest rock, Lokey Peak, reaching it in fog and wind. I dropped Art off at the northern end, while I worked south along its ridge. When I got back to Art ninety minutes later, we could see Walcott Peak, six miles to the west and above the fog and ground blizzard, so off we went; when it became invisible, we traveled by a compass bearing. We did the geology of it and several nunataks west of it, and by then we were west of the fog. With good visibility, we decided to set up our tent; it had been our most productive day so far in the Black Coast. On the negative side, each of us alternated falling into bergschrunds at every nunatak we visited, although fortunately not any deeper than our hips. But the weather was too good to pass up, so we took off for another five miles to the west to Mt. Jukkola, a pluton amongst continuous Jurassic volcanic rocks. Then many more rock stations, as the wind picked up, with ground blizzard, hitting us in the face as we returned east. With visibility at about a half mile, we had to go on compass bearings, although Lokey Peak became visible from time to time. The trip back seemed endless and we missed camp, passing just to the south of it even though I must have crossed our morning's tracks. Just as we were having visions of spending the night in our emergency pup tent, Lokey Peak popped into view for a minute. We swung around and got our bearings then found our tracks, partly drifted over. The tent at 8 pm was a beautiful sight! We had traveled more than forty miles. Art crawled into the tent and lay on his back on his cot for fifteen minutes, totally zonked! After a couple of scotches, however, he revived nicely. We called Rothera, had dinner, and retired late, with ground blizzard lashing the tent.

When we poked our heads out of the tent at 6:30 am on January 10, fog had surrounded us, so we went back to sleep. By about noon, it broke off and we headed northeast to some nunataks whose geology was shown on a BAS geologic map. A couple of empty boxes protruded from the snow, so clearly a plane had visited it before. But the rock type was incorrect as mapped, meaning that the person who had visited these rocks could not have been a geologist. We ambled on, but an east wind picked up to 10- to 15-mph and continued all day. We next headed to some small nunataks about six miles to the west, where we found ourselves on the top of a south-facing snow scarp as a ground blizzard started. Trying to find two rock outcrops shown on the sketch map and presumably invisible because they were below that scarp, we came upon a hole about thirty feet across, obviously a collapsed part of a bridge over a crevasse, about 150 feet ahead of me. As I noted in a letter to Mary on January 14, "the thing looked like an elephant trap out there where you don't expect it, but not being an elephant, I went around." Looking for the outcrops, we found a way south off the scarp, the edge of which was crevassed. Art let go of the rope from the rear Nansen and skied down the scarp, while I put rope brakes on the runners and headed south. We found and headed for the rocks in a mild whiteout (high clouds above) and finally reached them after about three miles. "By then the ground blizzard was severe, so that was short geologizing. Then six miles almost due south to a tiny nunatak, another endless trip!" That nunatak had a flagpole at it, although the red flag was mostly torn off, and there were two empty drums for avgas (aviation gas). So, the nunatak had been visited but the pilot had not recorded the rock type because it was on no BAS map. That nunatak was another short visit for us! This time, however, we could see camp, fortunately only four miles to the east, and we made for it, getting back at 9 pm, very cold.

On January 11, we had a thick cloud cover and the same infernal east wind, at about 10 mph. We packed camp and were ready to go about noon. Doing geology along the way, we traveled about twenty-eight miles southwest, establishing our Camp 6 (at 6 pm) in the central Blanchard Nunataks. Art estimated that our elevation was about 5,000 feet. The next day, January 12, the weather was perfect—warm, calm, with few clouds. Too perfect. By the time we had gotten to our first nunatak, about five miles to the west, fog had come in from the west. We bumbled from one nunatak to another, mostly in fog but spotting nunataks when the mank momentarily lifted, and got back to camp about 8 pm. The next day, we had fog and hoarfrost with a half mile visibility. We explored a ridge just west of camp, flagging our trail on the way over. After we did, the fog broke, although we had a thick cloud cover above. We therefore took the opportunity to go east, to nunataks that were on the sketch map but that we could not see because, as we learned, the rock was on a north-facing scarp and invisible from the west. We went on a compass bearing and, at the correct distance in whiteout, bumbled upon a sharp hill with a north-facing overhanging snow cornice and crevasses; rock was below, a faulted pluton. The sun came out for an hour in the late afternoon and we took pictures of these features, with snow crystals in the air—beautiful and surrealistic!

We were back at camp by 7:30 pm, almost on time for our radio schedule. As I gassed up the Ski-Doo, threw a tarp over it, and labeled and packed rocks outside, Art went into the tent to start radio communications, learning that the Broncos were going to the Super Bowl! But then he yelled that I had an air letter, a written message from Mary that they would read over the air in ten minutes. So we huddled next to the radio for my first communication from Denver. It was, of course, a Christmas/New Year greeting in which Mary had enlisted comments not only from Jill, Scott, Mom, and Dad but also from the Chet Wallace family, following a Christmas get-together of all of them in Denver. But what was the main message that they had concocted? Mostly terrible, terrible abuse directed at me! I have the fax of the message that apparently was sent to Cambridge, then somehow to Rothera. It contained such things as "Christmas isn't the same without you! It's better!" Of course, the worst part of it was that it was a radio transmission, so all the other BAS field parties, and perhaps others as well, heard it. Each BAS field party was given letters; the one for Art and me was "w," which when given phonetically in all radio transmissions became "whiskey," certainly very appropriate because we were drinking a lot of it. Bryan/Peter had "uniform," and Hein/Tony had "victor," certainly less exciting. The point here is that the number of parties in the field was large. So lots of people heard this message of awful insults!

We had planned on moving camp to the south on the morning of January 14, but instead we had 20-mph east winds and ground blizzard, with near-zero visibility. I had time to write letters to Mary and to Chet and his family. The one to Chet was composed during cocktail hour, as the wind was getting even worse, in which I described that wind-driven snow was lashing the tent, and the primus also made some noise, and I concluded my paragraph with "hardships are aplenty here in Antarctica!" I also described my disappointment at not getting to Rowley Massif because "I had planned, when I got to Rowley Massif, to fill up 15 to 20 boxes of rocks for shipment back to the States, for I knew that many persons, probably including yourselves, would want to purchase some. Alas, my hopes for fortune have been dashed again." In amongst other news as to what we were doing I had to get back to the main point, namely the air letter: "not only did she [Mary] not find enough abuse within her own mind and the small minds of my children and parents, but also she had to seek out the teeny, tiny minds of you." And I predicted that, meanwhile, all the other field parties up to, and perhaps beyond, parties u, v, and w, were now "trying to reconcile the wonderful impression I made when I met them, with what they heard last night over the air waves!!" Of course, I admitted that those wonderful impressions were "dispensed in the bar at Rothera."

The wind intensified to an estimated 50- or 60-mph on January 15, with fierce ground blizzard and visibility of fifty to a hundred feet. That continued through the following day, although it had lessened by the evening. On January 17, the wind died to an almost constant 15- to 20-mph, although we experienced several lulls. On January 18, an east wind was light, clouds were east of us, and it was mostly sunny. We dug out and were on the trail by 12:30 pm. We visited many outcrops along the way, but I flipped a sled on sastrugi, tearing holes in the outer wall of our Scott tent that was strapped on top. The surface was hard and the sastrugi were huge, with some three feet high, and I had to sit on the seat of the Ski-Doo to keep everything upright, getting a sore butt and shoulders. What with all the visits to do the geology, we went about thirty miles, although our straight-line distance was only seventeen miles due south of Camp 6. At that point, I flipped a Nansen again, at about 10 pm. Art skied up to me and asked me to pose for a picture, so I put one foot up on the overturned sled to look like a big game hunter with his trophy, a disgusted look on my face. We had had enough, and we established Camp 7 in that very spot! The Ski-Doo was overheating and we could not figure out how to fix it, although we suspected our heavy loads and the strain of navigating slowly over the sastrugi were to blame. A bad spark plug was possible too. We had dinner about 12:30 am. Despite the significant geology we had done, the last entry for the day in my field notebook was "overall, a very bad day." We were at 4,300 feet elevation in several unnamed nunataks about a dozen miles north of Bergen Nunataks.

January 19, a Monday, brought clear weather and a light wind at 7 am. When we made radio contact at 8 am, we learned that the resupply flight was about to leave. Then the wind picked up, getting stronger and stronger. It was another lay-up day. As we stayed in the tent in what seemed to be Mach 1 winds, Art resorted to using a piss bottle rather than venturing outside, mumbling "pretty good [employment]..., [we have] two days' work then five days off ... one of those cushy government jobs!" We did another radio com at noon, and learned that the resupply flight had landed about 1.5 miles east of the northern one of the Brennecke Nunataks, about twenty-five miles east of our camp. Brian and Peter were working in the Holmes Hills about ten miles northeast of that landing site, and they moved their camp to the resupply put-in site. Hein and Tony were still in the Owen Peak area about twenty-five miles north-northeast of the landing site. On the January 20 the wind picked up to an estimated 30 to 50 mph, with visibility essentially zero due to blowing snow.

The wind at Camp 7 died down by 6 am the following day, although it was still overcast. Art took one snowmobile to the nearest outcrop while I changed spark plugs on the bad snowmobile. One spark plug was badly fouled and the other was not in great shape either. I replaced both, and Art returned. We packed camp and headed south at 12:30 pm, soon stopping to go in opposite directions to do the geology of two nearby nunataks. We started east at 3 pm toward Brennecke, and it soon became a steady climb up toward the top of the peninsula plateau. With no landmarks, we went on a compass bearing, and the last pitch was quite steep, such that the Ski-Doos overheated because of their heavy load of rocks. Just before reaching the top, a Snowy Petrel came in low to look us over, first Art and then me, before it headed east. On top we stopped to let the snowmobiles cool off and we gassed up; the elevation that Art recorded was more than six thousand feet. As we got into the western Brennecke Nunataks, our hard surface with low sastrugi became eight inches of soft snow, and the Ski-Doos started heating up again. The eastern Brennecke Nunataks were on the eastern edge of the plateau, and we could see that the resupply cache and Bryan and Peter's camp were about three hundred feet lower on the eastern slope. With snow getting thicker and thicker eastward, we decided to camp about three miles west of the landing site, above the break in slope. A main reason was that the route south toward the sites of our next traverses, the DuToit and Wilson Mountains, would have to head up the three-hundred-foot grade and pass through our camp. Furthermore, south of the cache and Bryan and Peter's camp was a significant crevasse field. Therefore, we dropped our loads but took one snowmobile down to visit with Bryan and Peter and, more importantly, to get our mail—the first of the season! We reached their camp at 9 pm, in thick snow, visited awhile, then returned west to our camp, had the tent up at 11 pm, and then dinner. We had come about forty miles all told. Camp 8 was at 5,600 feet according to Art's altimeter. But the most important part of the day was the incredible pleasure of reading mail, such that we did not get to bed until about 1 am. In the meantime, Hein and Tony arrived and camped near Bryan and Peter. Then another plane came in with mogas (for the snowmobiles) and avgas. After all the peace and quiet, Bryan and Peter must have wondered if they were at Grand Central Station!

January 22 was mostly clear with scattered low clouds and a 10-mph west wind. Bryan/Peter and Hein/Tony headed south, then southeast for the DuToit Mountains, and we planned to follow their tracks the following day. But first we had to fix fouled plugs on Art's snowmobile and prepare for that traverse. It was warm and we worked in shirtsleeves on Art's snowmobile, sorted rocks for shipment to Rothera, and gathered food and fuel for the next traverse. From our tent we could see Mt. Jackson far to the north, a beautiful sight! The next day was marked by low clouds rising from at the ice surface to a thousand feet. By the time we headed off at about 9:30 am, the clouds were on the surface, and we had to follow our colleagues' tracks in low visibility conditions and deep snow that had had at least a day of warm weather to harden it a little. At about twelve miles, Hein and Tony's tracks took off to the north and we continued in Bryan and Peter's tracks until the clouds broke and we saw our first nunatak of the day. Our odometer showed thirty miles but we still did not know where we were, although our mileage indicated that we were farther than where Bryan said he would be. We dropped our loads and continued with one snowmobile on the tracks for three more miles, and there was their camp. They were off doing geology, so we went into their tent and found a USGS sketch map on which they had plotted their location, in the southern DuToit Mountains. So now we knew where we were, but we wanted to work the nunataks well to the north. We left them a note to that effect, returned to our load, and headed north.

The clouds came down to the surface again and we continued a compass bearing for about six miles, when the sky cleared partially and we could see a ridge, in the north-central DuToit Mountains. We dropped our loads and established Camp 9. We had come forty-three miles, and the elevation was 4,500 feet. We could see a lower basin (a small unnamed tributary glacier) to the north, surrounded by ridges, but given our heavy loads and the soft snow, we dared not drop into it. Instead, we took one snowmobile and flagged a route down the basin axis so that we could get at those ridges in bad weather. When we returned, it was 5 pm and we set up the tent and had stew.

The weather was high cirrus and calm in the morning of Saturday January 24, but soon fog moved up the slope and had us socked in by the time we got going at 9:30, sitting down on the seats while driving our two snowmobiles. We followed our trail over some sharp passes, requiring rope brakes on the Nansens we were towing, and we dropped down into the basin. Art mapped the ridges on the east while I went west and worked my way to the northern end. When the fog came in, I went on a compass bearing as best as I could, and when the clouds cleared for a while, I made for rocks by eye.

It was a warm day and thoroughly enjoyable and productive for both of us. We worked rapidly because the geology was simple, mostly a batholith of fresh, coarse-grained granite. I reached the northern end of the basin by early afternoon, at the intrusive contact of the batholith with the Latady Formation. I crossed a bergschrund and scaled a peak with great views to the west toward a small unnamed series of hills where we supposed Hein and Tony must be mapping, with the heavily crevassed Beaumont Glacier beyond them and separating them from the Brennecke Nunataks. I followed my tracks halfway back to the south and spent two hours in the fog in mid-afternoon collecting samples for paleomagnetism, a time-consuming process.

I knew that Art was not far away, for I could hear his Ski-Doo. Working on a calm day in fog is an eerie sensation, for sounds seem to be magnified and the direction from which they are coming is hard to discern. Eventually the fog broke and I spotted Art with my binoculars at the northern end of his eastern side, so I finished my mapping and drove over to him. The basin plus the ridges and mountains on either side made a massif, extending north-south about eleven miles and east-west about six miles. The massif had no name yet, so we decided that we ought to name it the Peck Range for Dallas L. Peck (1929–2005), our then and long-time (1981–1993) Director (the eleventh) of the USGS, and one of the few Directors of whom we thought highly. We grabbed a couple of samples from "his" range for a doorstop in his home. We returned in the mank, pulling our flags as we went, and arrived back at 6 pm, having put thirty-two miles on my odometer. We remembered Dallas's leadership methods over drinks and dinner. Before he was Director, Dallas' main research was mapping quadrangles in the Sierra Nevada batholith and studying plutonic rocks there, so the granite here was appropriate to name in his honor, and this batholith could aptly be named the Peck batholith. Dallas had an admirable trait when he traveled for meetings from Reston to the outlying centers: during breaks or before or after meetings, he would stop unannounced in the offices of his geologists. This was always a high honor and a great morale booster, for he knew you and your project even though you had never seen him before, and he would ask you specifically how it was going and whether you needed any help. The people he visited were not necessarily his friends or geologists with high GS grades, for the one visit I remember was ten to fifteen years after I joined the USGS. Neither Art or I ever experienced that from other Directors, let alone many people above our Branch Chiefs.

When our field season was over, as I did after each field season, I turned in Dallas' name, plus those of Bryan, Hein, Peter, Tony, and pilots and others who had helped us in the 1986–1987 season, to the US Board on Geographic Names. The name for Dallas was the only one for which I suggested a location. Curiously, the Board assigned the northern and highest spire on the Peck Range the name Mount Wever, after Hein. That peak was probably the one I climbed on the 24th to get great views to the west. Several years later, after names were approved (1989), I received a personal letter from Dallas, who told me that his rock was on his living-room mantle! Dallas was succeeded as director by Gordon Eaton, whom I will discuss below, and Dallas retired from the USGS in 1995, just as things in the USGS were going to hell.

Superbowl Sunday, January 25, was another day of clouds at the surface and low visibility, plus higher clouds that created a solid whiteout. Nonetheless we managed to drive to all the nearby nunataks that we could see. We did the geology of a series of ridges until 2 pm, by which time we had completed that task. On the last nunatak, I almost tumbled off a ledge in the whiteout. Back in our tent, we were able to tune into the Superbowl on Armed Forces Radio using our little transistor radio, only to hear the Broncos lose to the New York Giants. I don't want to talk about it!

On top of clouds at the surface, winds of 25- to 30-mph arrived on Monday. We stayed in our tents and I used the time to send an air letter to Mary. I told her of our travels and noted that we "still tolerate these limeys." After I sent it, Joe, the radio man in Rothera, said "Oh, by the way, I have one for you!" Thus, I got one from Mary, and the day became much better! Mary and the cats were doing fine.

The wind died at 9:30 am on January 27, and we packed the camp and were ready to leave by 1 pm. Despite the severe whiteout, the visibility was one or two miles. So we headed due east, doing the geology of five nunataks as the clouds dropped down to the surface. By 3:30 pm, as we crossed the western headwaters of the Bohnecke Glacier separating the DuToit Mountains from the Wilson Mountains, the visibility was largely gone and we went on a southeast compass bearing for four miles as the surface of fluffy snow became thicker, then a northeast bearing for another four miles in total whiteout and with no visibility. When my snowmobile bogged down at 5 pm in fluffy snow that was about eighteen inches thick, we called it quits on what appeared to be a sidehill sloping west, calling it Camp 10 in the southwestern Wilson Mountains. We had come twenty-five miles but had little idea where we were! Camp elevation was 4,200 feet.

The Wilson Mountains were named after John Tuzo Wilson (1908–1993), a Canadian geophysicist and professor at the University of Toronto, who visited Antarctica once in 1958. But visiting the ice was not what led to a range being named after him. Instead, this range was part of a series of mountains and peaks in the Black Coast that were named for geologists who were instrumental in discovering the important concept of continental drift, renamed plate tectonics in the early 1960s. (In contrast, the Lassiter Coast contained a series of mountains that were named after the seventeenth- and eighteenth-century scientists who founded geology, as noted above.) Though he had a Princeton doctorate, Tuzo Wilson at first disparaged the idea of continental drift in many of his early publications, but by 1960 new geophysical evidence from the new science of marine geology had identified a mechanism by which not only continents moved around but also sea floors moved and carried continents. In 1961, Wilson published his most important work, in which he defined a new type of fault called a transform fault. These existed as east-west structures primarily in the ocean basins; he put these transform faults in the context of movement of the sea floors. (In my 1998 paper, I defined transverse faults and made these east-west faults on the continents analogous to transform faults on the sea floor.)

The 1960s were exciting times for students in geology, and Wilson's paper and many others were required reading in my Texas structural geology and tectonics courses as we debated the pros and cons of continental drift. By the early to middle 1960s, the marine-geophysical evidence became so overwhelming that few doubters of drift remained. Wikipedia defines plate tectonics as the idea that rigid outer layers of the Earth (the crust and part of the upper mantle), namely the lithosphere, are broken into numerous pieces, or plates, that move independently over the weaker underlying asthenosphere.

Before 1960 there were plenty of doubters. In fact, drift advocates were regarded with laughter, as when Alfred L. Wegener (1880–1930), a German researcher who made his name in meteorology, first proposed that all the continents were once a supercontinent, which when translated from German to the Greek was named Pangaea ("all the earth"). The Wegener Range, in the southern Black Coast, is just north of the Fenton Glacier, in turn north of the Mosby Glacier, both of which we crossed during the 1972–1973 season when we were the first to map in the Carey and Wegener Ranges. In fact, the Wegener Range might have been a worthy tribute if one considered merely Wegener's work in meteorology, for as part of that research he led four early expeditions to Greenland. But the real reason for the name is that Wegener was the first to develop a comprehensive theory (others had previously recognized some aspects of the jigsaw puzzle) that pointed out that the western coast of Africa looked like the eastern coast of South America, and it is more obvious when one matches the continental shelves (which Wegener tried his best to do). From there, Wegener took on the other continents and matched them correctly based on geography. Though he published a book on this in 1912, it went largely unnoticed by English speakers because its first edition was in German! But when he produced his third edition (1922), he translated it into English; a more elaborate version was published in 1924 as *The Origin of Continents and Oceans*. That was when the laughter really began, even though he already had an inkling of the likely mechanisms at work in the theory he was proposing. Later, as the geology of the matched continents began to be compared (Du Toit), and then finally in the 1950s with the rise of the field of paleomagnetism (S.K. Runcorn and others), geologists began to understand the wisdom in Wegener's jigsaw puzzle and the evidence put an end to the laughter.

Some of the other geological giants are commemorated in the Black Coast. The Du Toit Mountains are named for Alexander L. du

Toit (1878–1948), a South African geologist who mapped much of that country early in his career and published the results. In 1923 he received a grant from the Carnegie Institution of Washington to visit continents adjacent to Africa to compare their geology. In 1937 he published his book *Our Wandering Continents* that presented the geologic evidence of previous geologic connections between continents, although he departed somewhat from Wegener by proposing that Pangaea could be broken into two supercontinents, Laurasia (North America, Europe, and Asia) to the north and Gondwana (or Gondwanaland, for the southern continents) to the south. The Runcorn Glacier, a tributary to the Beaumont Glacier north of the Du Toit Mountains, was named after S.K. Runcorn. The Holmes Hills lie west of the Runcorn Glacier and are named for Arthur Holmes (1890–1965), a British/Scottish geologist who was chairman of the departments of geology at Durham University (1924–1943) and the University of Edinburgh (1943–1956). He was the author of a popular introductory textbook titled *Principles of Physical Geology*, first published in 1944 and containing a chapter on continental drift. He was an early proponent of Wegener's ideas and suggested that mantle convection was the driving force behind the movement of the plates in opposite directions from the Mid-Atlantic Ridge. This convection idea was correct, and Hess and Dietz (see below) explicitly documented it. Holmes also did the first isotopic dating, on uranium-lead, in 1911. The Runcorn Glacier is west of the Hess Mountains, an eastern prominence of which is Dietz Bluff. Harry H. Hess (1906–1969) was a professor of geology at Princeton University from 1948 to 1969, and in 1962 published undeniable evidence of sea-floor spreading; this report was a reprint of one he had published in 1960 that researchers had largely overlooked because he had had it published by the Office of Naval Research—not a particularly prominent outlet. Robert S. Dietz (1914¬–1995) was a marine geologist who worked for the US Coast and Geodetic Survey and other government agencies during his career and published an important paper in 1961 that independently reported much the same thing as Hess, although because Hess had presented his preprint a year earlier, he is given slightly more credit for this discovery; thus, their two Antarctic landmarks are adjacent to one another. Dietz coined the term sea-floor spreading. Both Hess and Dietz proposed that new oceanic crust formed by upwelling at the mid-ocean ridges. Dietz also did pioneering work on impact craters, including being the first person to recognize that the Sudbury basin in Ontario was such a crater.

A few months after I returned to the States from the Black Coast and sent a rock to Bob Dietz, I received a small package from him that contained a two by one-inch meteorite in a blue zippered pouch, that I keep in the top drawer of the desk at which I typed this book. It is from Meteor Crater, Arizona, a state to which he had retired, and it contained a photocopy of a page about this impact crater and a personal note in red ink thanking me for the rock and informing me that he had accompanied Byrd on his third Antarctic Expedition (I assume he was at West Base, Little America). I had read the reports and books by all these giants of geology in Bob Boyer's Tectonics course in grad school. Of course, Bob Dietz is not the only one for whom I collected samples from their named features. The last page of my field notebook shows that I also sent rocks to Tuzo Wilson, Dallas Peck, Steve Boyer, Karl Kellogg, Walt Vennum, George Schirmacher, Guy Guthridge, Bill Lokey, and Gene Kamenev.

January 28 gave us poor visibility, heavy clouds above, an east wind at 10 mph, and moderately heavy snowfall. We stayed in camp. I took a walk for about three hours on a compass bearing of north-northeast, looking for rock; by counting my paces I estimated that I had hiked about a mile out. I found nothing, but at least I got some exercise. However, my return was harder than I expected it to be because my tracks were largely snowed in and the whiteout made them doubly hard to see. That made it awfully nice to see that yellow tent appear in the murk. In addition, I was back in time for snacks (generally sardines, canned cheese, and marmite on a biscuit) and our small ration of scotch.

At 7 am on January 29 we could see some nunataks to the west, but by 8:30 the visibility was gone and we had light snowfall. We stayed in, once again. It was a calm whiteout, with some stratus in the morning of January 30, but we had enough holes in the clouds to see some nunataks, and after breakfast we headed off to explore their geology. By comparing our aerial photos with the outcrops that we could see, we found we were still in the western headwaters of the Bohnecke Glacier, with a ten-mile-long, northwest-trending ridge to our east and northeast (but hidden in clouds), between the eastern and western headwaters of Bohnecke. With fog wafting up from the glaciers in a light south wind, at times above us but mostly at the surface, we attacked only the rock that we could see. Our first outcrop was east up into a pass that crosses that ridge, then south at the northern end of another northwest-trending ridge on the eastern side of the Defant Glacier (west of the Bohnecke Glacier). The snow at the surface was more than a foot thick and we bumbled through it—the fog and the whiteout—and returned to camp at about 4 pm as clouds settled in. While we had put in twenty-five miles on our snowmobile, ours was a less than productive day.

Saturday, January 31, to our total astonishment was clear (dingle)! We were on the trail by 9 am heading toward the northwestern end of the ten-mile ridge east of camp. From a pass there, we went northeast down a small unnamed tributary glacier draining northward into the Beaumont Glacier, then we shifted eastward to the northern end of the ridge, where we did the geology. The soft snow beneath us was now nowhere near as deep, and we consequently made good time. We continued eastward into the northern end of the Wilson Mountains, exploring one outcrop after another. There was, however, a steep north-facing precipice, where an unnamed, heavily crevassed, tributary glacier drained into the Beaumont Glacier south of Dietz Bluff, and I dared not take the Ski-Doo northward to get at the rocks at the northern tip of the Wilson Mountains. We got back to Camp at 8 pm, having put more than fifty miles on our snowmobile.

February 1 began with beautiful weather. This time we went south, through deep snow, around the southeastern end of the ten-mile ridge. There and south of there, we did the geology, then headed north along the eastern side of the ridge. The snow depth was much better on this eastern side but thick fog blew in from the north. We were able to do some geology and, when the fog lifted briefly, I flagged a trail up to the pass through the ridge that we had visited on January 30, in anticipation of our return. Shifting winds blew the fog back to the north and we headed north for a dense cluster of nunataks east of where we had been at the end of the previous day. But the fog blew back in and we spent mid-afternoon to early evening bumping from nunatak to nunatak. Our visits to each rock outcrop were quick, generally five minutes to a half hour, for we were short of fuel by then on this traverse. Furthermore, it was late in the season and we had lots still to do. By 7 pm, we could no longer work because of the fog and started back, fifteen miles if we could find the flags to the pass. We returned on a south compass bearing, with the fog so thick that surface definition was poor. Despite that, we luckily found our tracks, then the flagged trail, and we went west, up and over the pass. There was no fog on the west side of the ridge, and we reached camp at 8:30 pm.

On the morning of Monday, February 2, the stratus was both at the surface and at height, so we had no visibility as well as a whiteout. It looked like another work week of bad weather, following a weekend of frantic mapping. We talked with Rothera during our scheduled radio check-in, and learned that they were ready to provide air support to those nunataks far away or inaccessible by snowmobile. I also received a radio letter from Mary, which made my day, although she told me that my four-wheel-drive car, an International Harvester Scout that I had left her for the winter, was no longer working and so she was shopping for a Subaru. She also warned me that a gift package to Art and me was on the way from our USGS friends ("expect a care package from the guys … know what I mean?"). We also learned from the radio that Bryan/Peter and Hein/Tony were now back at Brennecke Nunataks. Tony said he had cached two jerry cans of mogas for our return. The previous day I had heard a clanking in the track of my snowmobile, so I suggested to Art that the bogie wheels of the track might need grease and, because the day looked grim for travel, that we service the Ski-Doos. In the middle of doing so, the weather improved. So after we'd finished work on the snowmobiles, we packed up and headed west. Yet as we proceeded out of the wind shadow of the long ridge, the wind picked up and visibility dropped to zero, obscuring where we wanted to go. After a couple of miles, we quit and set up the tent.

On February 3, we had stratus at the surface, snowfall, and winds at 15 to 20 mph out of the south, whereas Tony told us on the radio that they had 30 mph winds out of the northeast. Our winds picked up later. The next three days brought more of the same. On February 6, however, with fresh snowfall up to our knees, the weather appeared to be breaking, and we dug out, then cut a trail through the snow and planted flags for about three miles to the west, west of which the surface was harder. We learned on the radio that a plane was at Fossil Bluff poised to visit us. As the weekend started, sure enough, it was calm, with unlimited visibility, 50 percent clouds, and moderate surface definition. We got up at 5 am, missed the scheduled radio call because we had packed the radio and everything else, and headed west at 8 am. By 10 am we had only reached the end of those flags, for the Nansens flipped over several times in the soft snow, requiring us to untie the load, unpack the sled, turn it upright, and put everything back together. On several other occasions, the Nansen tipped 40 or 50 degrees off vertical but I saw it happening and stopped it before the machine went over; then we returned to it, dug out the runners, and pushed it upright.

After the last flag, the snow depth improved significantly to about two inches and we blasted along, going uphill toward the peninsula plateau. At about ten miles from Brennicke we passed within three miles of the nunatak where Tony had stashed the mogas, but I figured that we could make it back with what we had, so we pressed on. Art went on ahead, cutting a trail through the snow because he had a lighter Nansen, and now we had significant sastrugi. Predictably, I flipped my Nansen on one of those. At 4 pm, however, we reached the Brennecke fuel cache and our colleagues' tents, having come about fifty miles. The guys were out working the geology. We were expecting the Twin Otter to be there already, given the good weather, and drank two of our precious beers in disgust when we realized they weren't. We later learned that the weather at Fossil Bluff had been bad, preventing a takeoff.

Art and I were not long disappointed, for February 8 was still the weekend (bad weather during the week, good weather on weekends!). And, sure enough, it was sunny and calm when Richard Hasler and his co-pilot Martyn R. Bailey left Fossil Bluff and arrived in Bravo Charley with a load of avgas and some resupply items. Martyn was a rookie to Antarctica this season. Except for Peter and Martyn, we all climbed on board and took off at 10 am, heading for rock exposures along the coast that were inaccessible by snowmobiles due to crevasses. This trip, as with the subsequent ones, was spectacular in its scenery and we shot roll after roll of film (in those days, all cameras used film). Our first landing was northeast of Dietz Bluff, about thirty-five miles to the east-northeast of Brennecke and rising above the sea ice of western Hilton Inlet. That sea ice was badly crevassed but Richard found a crevasse-free spot and put the plane down. Unfortunately, atop that ice was a foot of newly fallen snow making pedestrian travel and airplane takeoffs difficult. We were about a mile and a half from the Bluff, a sharp promontory. Bryan, Art, and I roped up and skied over to one spot on the Bluff, while Tony and Hein went to another. It was shirtsleeve weather and we sweated plenty as Bryan and I took turns breaking trail through the snow. We visited the outcrop, made up of faulted plutonic rock (quartz diorite), for a half an hour to collect samples, including one for Bob Dietz. Then it took us another half hour to get back to the plane. Hein and Tony walked and used snowshoes while on belay (Tony's "safety first" method), and it took them an extra hour to return. Then plane trouble ensued because of the soft snow. We could not get off the ground until our eighth attempt—when we all sat in the back of the plane to force the nose ski up; by then it was 4:30 pm. From the plane, the Hess Mountains looked like they were of the same rock type.

Our next stop was on the southern side of Cape Knowles, about thirty miles to the northeast, at the eastern end of the Condor Peninsula. An aeromagnetic survey here had disclosed a positive magnetic anomaly, perhaps a pluton. Here we again landed on sea ice, but only two hundred yards from rock. We roped up and walked to the outcrop, all falling in little crevasses on the way over and back. It was another plutonic rock, this time a gabbro (with the same composition as basalt). We took off at 7:30 pm, flying south about twenty miles across Hilton Inlet to a nunatak about five miles east of Hjort Massif, in effect the northeastern end of the Wilson Mountains. More plutonic rock. We took off at 9:30 pm and were soon back at camp, where Peter and Martyn had made curry stew for us. We got to bed about midnight.

February 9 was cloudless and almost calm. This time Art and Tony stayed behind. Our intent was to explore Rowley Massif, about sixty miles to the northeast. We flew around it and Schirmacher Massif to the west, looking for contact points between granite and shale that we could map. Then we set down just east of Rowley Massif, on a crevasse-free part of sea ice on Odom Inlet. Again, deep snow covered the sea ice. As with the outcrops of the previous day, nobody had been on these rocks before. It was less than a half mile to rock and we just walked over to it in the deep snow, not roped together because we did not have Tony to nag us. It was a black, medium-grained gabbro and I collected some paleomagnetic samples for Karl Kellogg and a large sample for Walt's mantle but forgot to collect any samples for myself (or, if I did, I lost them).

It was another day of shirtsleeve weather, and the skis of the Twin

Otter stuck to the snow. To unstick them, we stood on boxes and pushed up and down on the plane's wing, and at other times we had to dig out the skis. On our first attempt, Martyn gave it a try but he hit a bump and the plane started to lurch to the port side; Richard reached up and cut the power. Finally, after five attempts, the last of them when we sat in the rear of the aircraft, we became airborne.

Our next stop was Schirmacher Massif. We landed in soft snow on the Cline Glacier, which separates Rowley and Schirmacher Massifs. From there we had a similar walk and saw the same rock type. Also similarly, it took us five attempts to get off the ground, which we finally managed at 4:15 pm. The last stop of the day was the eastern side of Hjort Massif, where the previous day we thought we had seen sedimentary rocks. Here we landed on a hard surface of white and blue sea ice. We spent two hours on the outcrops of sedimentary rocks and lava flows that we had never seen before on the Black Coast. We arrived back at Brennecke at 9 pm. This time Tony and Art had cooked a meal from fresh food that Richard and Martyn had secretly brought in: ham, sausage, broccoli, and mashed potatoes. What a welcome treat and a perfect cap to a great day!

We were surprised to wake on February 10 to see more great weather. Our mission this day was to explore the widely spaced and isolated nunataks on the peninsula plateau west of the northern Lassiter Coast and southern Black Coast. Hein and Peter remained behind in camp. We were airborne by 9 am, heading southwest. We expected hard sastrugi surfaces and, sure enough, we had them, and consequently we popped a couple of rivets and put a few dings in the skis, but otherwise it was another great day. We passed over the Journal Nunataks at about forty miles out, then the Jensen Nunataks a further forty miles (both apparently visited before), then south thirty miles to Mt. Vang, which had an elevation of about five thousand feet according to Art's altimeter. Here Martyn made a beautiful landing on a difficult surface. I complimented him and he gave us a big smile and asked that his picture be taken, for this was his first Antarctic open-field landing! We next landed at Gunn Peaks, about ten miles to the east. Bryan and Tony walked over to one nunatak, and Art and I to another. Our nunatak yielded one of our most exciting discoveries of the season: Jurassic bedded chert and limestone interbedded with airfall and ash-flow tuff. We collected gobs of the limestone, thinking that there might be microfossils in it. Next, we landed at Galkin Nunatak, about a dozen miles farther east. Then Toth Nunataks, about twenty-two miles farther east. Art's altimeter where we landed showed us at 6,800 feet, and the peak at Toth itself rose another 500 feet. Then, in the general direction of home, we went forty miles to the northwest and landed at Jensen Nunataks, where the sastrugi were large and hard. All of these were west of Lassiter Coast. Then, a dozen miles northeast, to an unnamed nunatak west of the Black Coast, where Art recorded an elevation of 4,900 feet. Next, we flew about forty miles to the northeast to the Bergen Nunataks, which Art and I had almost reached on our traverse along the peninsula plateau. All these nunataks, except at Bergen Nunataks, were high-silica volcanics, mostly ash-flow tuffs, of the Jurassic magmatic arc. We were back at 8 pm to find that Peter had cooked us another good stew!

Sometime in this span, the care package about which Mary had warned me arrived at Cambridge. In the past, in keeping with my being the brunt of practical jokes whenever I went to Antarctica, I had received a foot-long box containing cheap paperbacks of limericks and jokes, a girlie magazine or two, and gag gifts of various sorts bought, no doubt, from a gag-gift store I commonly frequented (for Christmas presents for the kids) in the Cherry Creek area of Denver. But this time Mike Thomson in Cambridge had not forwarded the package. Instead, he sent a message that was read aloud during radio-com time on February 10 for all the field parties to hear. I quote it in its entirety, for it illustrates what at times I find myself up against. "A parcel has arrived from quote The Guys at the USGS unquote. Cannot get this to you now, so it will have to wait your return. However, knowing the company you keep, it may be no bad thing. Hope you are still managing to find some rocks in all that mank. Any fossils yet? Regards, Mike."

Incredibly, the weather held on Wednesday February 11, and on this day Tony and I stayed in. We did the geology of a series of the Brennecke Nunataks. The plane with the others went north to look at rocks considered to be old ("basement") by a BAS geologist in the 1970s, near Mount Jackson. They had a good day but concluded that the rocks were within large fault zones (mylonite), so their deformed appearance was the result of younger faults. They returned about 8 pm to a stew that Tony had put together. On the morning of the 12th, with our fuel nearly gone, and with the weather showing signs of deterioration (partly cloudy and falling barometric pressure), Richard proposed returning to Fossil Bluff, backloading rocks, then picking up drums of fuel, then returning. But he had trouble getting off in the snow until we removed eight boxes from the plane's cargo hold. It was about an hour each way for him, so we used the wait to take group portraits. The plane returned with fuel, we loaded up, and all of us except for Tony and Art were in the air by 2:30 pm en route to the Sverdrup Nunataks about thirty miles to the south. We had gotten close to these nunataks, which are just west of the Wegener Range, during the 1972–1973 season. This time, we landed and did the geology: like the Brennecke Nunataks, they were on the eastern edge of the peninsula plateau, and the Jurassic volcanic rocks of both nunatak clusters were similar. Within an hour or so on the outcrop, however, Tony radioed to tell us that stratus was building at the surface. We hurried home, barely able to see the camp in light snowfall and stratus on and near the surface, but the plane had the camp on radar and the surface definition was not too bad and fortunately we landed just before 5 pm. Within a half hour, the visibility went to zilch.

Friday, February 13, we spent in the clouds and light snowfall at Brennecke, with a red Twin Otter at the other end of our line of tents. The four geologists spent much of the day talking about what we had found this season, and its meaning with respect to other parts of peninsula geology. The Jurassic magmatic arc that made up the spine of the peninsula intertongued with sedimentary rocks to the east that make up the Latady Formation. These sediments formed in what is called a back-arc basin, a marine basin formed by rifting (faulting). The sedimentary rocks at Hjort Massif may be the oldest of these sediments, pre-dating Latady. All these rocks were folded and thrust faulted, forming mylonite shear zones, and were then intruded by Cretaceous plutons, which dominated in areal distribution. I am ashamed to say that I described this geology in a letter to Mary, just one day before Valentine's Day. But getting home was on my mind as well, and I told her that, too! In fact, the lyrics in one of the greatest rock and roll songs of all time, "Louie Louie" by the Kingsmen, included "a fine little girl, she waitin' for me."

Breaking our month-long trend of good weather on the weekends, Valentine's Day saw high winds, with gusts to 40 to 50-mph, all morning and afternoon. It was generally clear during this windy period but stratus came in by late afternoon. The winds died on February 15, but instead we had visibility of about a quarter mile due to clouds at the surface and light snowfall; winds returned at 10 pm, holding at a steady 20 mph. Continued winds with gusts to about 50 mph characterized the following day. On the 17th the winds died at about 6 am and Richard decided to get all of us to Fossil Bluff base. The field season was over! Unfortunately, Fossil Bluff weather was worse but we got underway anyway. Richard calculated that four flights were needed to take all our equipment and rocks. Bryan, Tony, and some rocks went out first, at about 11 am, and they landed at Fossil Bluff in a ceiling of 500 to 1,000 feet. The second load included Art, Hein, and me, leaving at 2 pm and arriving at 3 pm. They decided to send Tony back to Brennecke on the return because they needed somebody to send weather reports if they could not get in all four loads that day. And indeed, they got the third load out before the weather worsened, but not the fourth. Tony and Peter were the lucky ones who remained!

At Fossil Bluff, we set up tents because the base was too small to accommodate beds for all of us plus five others who had been pulled out of the field a few days previously. This number would grow over the next few days. All of us needed to get to Rothera, a ninety-minute flight away, but the weather was overcast at Fossil Bluff and worse at Rothera! Fossil Bluff is a BAS refueling stop built in 1961 and is occupied for the most part during the summer by two persons for radio communication (to transmit weather reports), although it houses four persons comfortably. It sits at the base of Giza Peak, bare rock that rises about two thousand feet. It consists of a main building, painted red and with a pitched roof, and several smaller buildings, namely a garage, food storage hut, meteorology hut, generator hut, and a one-room emergency hut in case the main building catches fire. The porch of the main building has a beautiful view of George VI Sound to the east.

The main building consists of a single room, about the size of a standard living room. I here give an overly expansive description of this room because after a while it seemed as if I had spent a major part of my life there, even though I now remember nothing! The northern wall has a counter where we wash dishes, with a large (3-feet on each side) plastic container of water with a spigot. We generally melt blocks of ice for water, but during our current temperatures we tapped a huge melt pool about a hundred feet south of the hut. Above and to the left of the counter is a small library of paperback books (I had soon read a couple dozen of them!). South of the counter is a large wood- and coal-burning stove which has a built-in basin that can hold the equivalent of six gallons of ice blocks for water or for warming the water, with a spigot at its bottom. The southern wall holds four built-in bunks. The western wall has a counter where we did our cooking and meal preparation, with shelves and cupboards above it for food. The eastern wall contains a partly enclosed cubby hole where the radio gear resides, including a desk and stool for the radioman. A table and six chairs occupy the center of the room. The front door, off the northern end of the eastern wall, consists of two doors to provide an air lock between them. Between the two doors is a tiny anteroom (vestibule) used as a food pantry. Another pantry with shelves is on the exterior southern side of the hut; it is used for camping gear and holds a chemical toilet, which is emptied each day.

For amusement and exercise, on February 18 David I.M. MacDonald, a permanent BAS employee and stratigrapher, took us out to look at the Fossil Bluff Formation, a sedimentary unit exposed in the nearby cliffs. Dave, a Scotsman with a brogue, Ph.D., and lively sense of humor, was the geologist whom Art and I had come to know when we all sat at the captain's mess on the *Biscoe*. His initials—DIMM—amused small minds like ours, but he was one smart cookie!

Weatherwise, our ceiling outside remained at about 1,000 to 2,000 feet, with temperatures about freezing on this day and many others to follow.

On February 19, the stratus clouds were at the surface for most of the day and our field party did the cooking for the base that night. My treat for the day was an air letter from Mary, who told me she had bought a used 4x4 Subaru to get her to and from the mountains on snowy days.

On the 20th, the clouds lifted at Fossil Bluff but the weather was poor at Brennecke, so Richard attempted to make it to Rothera at 11 am. Not long after taking off, Rothera became socked in and Richard had to return, experiencing icing conditions when he came back through a thick cloud layer above.

February 21 saw improving conditions at Brennecke and other parts to the south, but not to the north. Therefore, Richard picked up a BAS glaciological field party and brought them back to Fossil Bluff. Then he took another field party to Siple Station, far to the south. Steve and another Twin Otter, which had been configured to do aeromagnetic surveying, brought a load of stuff from Siple to Fossil Bluff, then returned to Siple. So now we had two Twin Otter crews eating steaks and lobster tails at the US base! On February 22, the weather was bad at Siple and Rothera, and twelve people remained at Fossil Bluff. The weather was bad everywhere the next day. On the 24th, the weather was good in all places apparently, and the remaining Twin Otter at Rothera departed, although it did not come here. The two aircraft at Siple were not ready to leave because it was too cold (-30°F). On February 25, Fossil Bluff had high clouds but good surface definition, but Rothera had clouds at the surface and 10- to 20-mph winds. Late in the afternoon, however, both aircraft from Siple arrived, although Richard's plane went back to get more stuff, then returned at 11:30 pm. There were now twenty-one people at Fossil Bluff! On February 26 we had light snowfall and clouds as low as 1500 feet above the surface, and my notebook recorded a long conversation with Jeff Renner of the aeromag field party on what they had found this season and the geological implications of those finds. Clearly things must have been dull if I spent the day talking geophysics! The next day Rothera had high winds, whereas we had heavy snowfall, with a temperature around freezing. I read books.

On February 28, things were the same except we had rain instead of snow. The twenty-one people were in seven tents outside, plus one sleeping in the garage, another in the food store, and another in the emergency hut. Only four slept in the main hut. I noted that this is a "zoo, with animals." On Sunday March 1, the pilots dithered, being particularly cautious, and then when they were finally ready to go in the evening, they found a flat tire on the aircraft. The weather most of the day was a light wind and total cloud cover with a ceiling of 1,000 to 3,000 feet. The *Bransfield* had been at Rothera for a week, waiting for many of us (Bryan, Hein, Dave MacDonald, Art, and me) and ready to head to Port Stanley. We five had been scheduled on a flight from Stanley to Brize-Norton on March 10. I voiced my frustrations in my notebook and in a letter to Mary. Steve, the permanent base commander at Fossil Bluff, and Steve, the pilot, baked bread and told funny stories about snowcats, Ski-Doos, and caterpillars they had known, while most others were reading, writing end-of-season reports, and burning trash. By then it was almost pitch dark at midnight and we could see stars!

At 7 am on Monday, March 2, we had clouds at 3,000 to 4,000 feet and we learned that Rothera had scattered clouds and no wind, so we scrambled. At 8 am, the five of us, plus two glaciologists, the pilot, and co-pilot lifted off. We were on the ground at Rothera by 10 am. We spent the day sorting, numbering, boxing, and banding rocks, and assembling our gear. At 11 pm, I was surprised to learn that I could call Mary (we were two time zones east of New York), for I had not known that before. I called. I have no record of what either of us said, but I can imagine that it was mighty emotional and nice! The following day, I wrote an air letter to Mary and puttered around the base. In the air letter, I expanded on some of the stuff in the phone call, and I worried about Tony and Peter ("Peter and Tony and one load of gear, however, stay in the clouds of Brennecke, as a remote weather station"). I concluded with "We are fine and are still managing to put up with all these limeys but, then, we are very tolerant men. Call Carole and Glen. I am nuts about you." I found a record with a Mozart clarinet concerto and played it. It was warm (35°F) and sunny. The seals were sunning themselves on the rocky beach and skuas and seagulls were squawking in the brash ice that fills the bay, as I noted in a postcard to Mary.

On March 4, the RRS *Bransfield* was in at 1:30 pm. We finished packing, then boarded. The skipper was John Cole. The *Bransfield*, an ice-strengthened cargo ship launched in 1970, is about 325 feet long, has a crew of 37, and a maximum speed of 13 knots. She continued as BAS's primary cargo ship for twenty-nine years, was sold in 1999, and scrapped the next year. We were under way at 2:30 pm. It was partly to totally cloudy but visibility was five miles, and we watched icebergs and ice-clad peaks until they were no longer visible. Soon we were in the open sea, with no more views of land. We were in effect making a bee-line for Stanley to catch our scheduled flight. In the wardroom we played cards until early morning with the captain, Dave MacDonald, and others. By mid-morning of March 5, we had gotten beyond most of the icebergs and the seas were calm, but we had some fog and, in the afternoon, scattered bergs. Bryan, Hein, Art, and I talked geology all afternoon, summarizing what we had seen to get an early start on the reports we needed to write. By 9 pm we were off the South Shetland Islands, in swells, and we started across the Drake Passage. On March 6, we had five miles of visibility and encountered large swells but no heavy seas. Still, few people made it to breakfast. I did gash duties (emptied trash, did dishes, etc.) that day, wrote Patty and Scott, and started a short report for *Antarctic Journal of the United States*, a journal in which, over the years, I have published a lot of short, preliminary reports (not much more than abstracts), especially just after field seasons. On March 7, we passed the Antarctic Convergence shortly after midnight, and during the morning the swells were heavy, the winds were at about 20- to 30-mph, and we had good waves, termed a Force 10 gale. But the winds were at our back, so it wasn't nearly as bad as it sounds.

On March 8, however, the winds were at about 50 mph with gusts to 65, which put them at hurricane-force winds (when above 63 knots). The crew was concerned about getting the boat crosswise in these swells. In fact, it was a Force 12 gale (out of 17). We made it to Stanley harbor early in the morning, where the waves were choppy but not bad. We learned, however, that an Argentine ship was in trouble (sending distress calls) an estimated hundred miles northwest of us. But soon we were dragging our anchor in the choppy seas, and it made sense to head to the outer bay (Port William). Here we spent the day going around in a rectangular route, waiting for the weather to improve. I finished up a draft of my paper for *AJUS*, typed it, and gave photocopies to the three others for coauthor review. On March 9, the winds were a little better but we still stayed in the outer bay, going round and round. It was overcast, with good visibility. Finally, at 3 pm, we approached Stanley, anchored about a thousand feet offshore, and took a launch ashore at 3:45 pm. I gave Mary a phone call, shopped for gifts, and shot up a roll of film, then returned to the ship at 5:15. We had dinner on board, then by 7 pm it was calm enough to get to a dock. Ashore, we found a bar (there were three in Stanley) and had a few beers with a Falkland Islands councilman named Lewis Clifton, then visited the Victory and Globe Bars until they closed at 10 pm, then returned to the ship.

On March 10, as scheduled, we were up at 9 am and took a minibus to the Mt. Pleasant Airport, arriving at 11 am. We lifted off at about 2:30 pm on a RAF Tristar 500, a large jet with ten seats across and thirty-two rows. It was a seven-hour flight to Ascension Island, arriving at 12:30 am of the 11th. After a crew change, we took off again at 2 pm. It was a nine-hour flight to Gütersloh, Germany. From there my notes are sparse, although I recorded that we later got to Brize-Norton RAF Base in the UK, then successfully cleared customs. We again stayed with the Thomsons, whose hospitality was again extraordinary. On March 12, they once more hooked us up with Stuart Agrell. Then we had drinks with Tom and Sue Laudon, who were visiting, and dinner with the Swithinbanks. Then, on another Friday the 13th, we cleaned up details, mostly with Janet, concerning who does what with rock samples, who gets isotopic ages, proposes topographic names, who publishes what, etc. On the 14th, Art and I were up at 7 am and Janet and Mike drove us to a bus station in Cambridge, from where we caught a ride to Heathrow Airport. Though we said goodbye to Mike and Janet, we knew it would only be a short time before we'd see each other again, at the next Antarctic convention, in Cambridge, as noted in the next chapter. Near Heathrow, Art and I made reservations at Hotel Isis at 11:30 am, then went into London to shop. The next day, we boarded a jet at Heathrow at about 2 pm, arrived in New York Kennedy Airport at 5:30 pm, then to Denver at about 9 pm. As in the song “Louie Louie”, I had gotten home again, and Mary met me at the airport.

CHAPTER 7

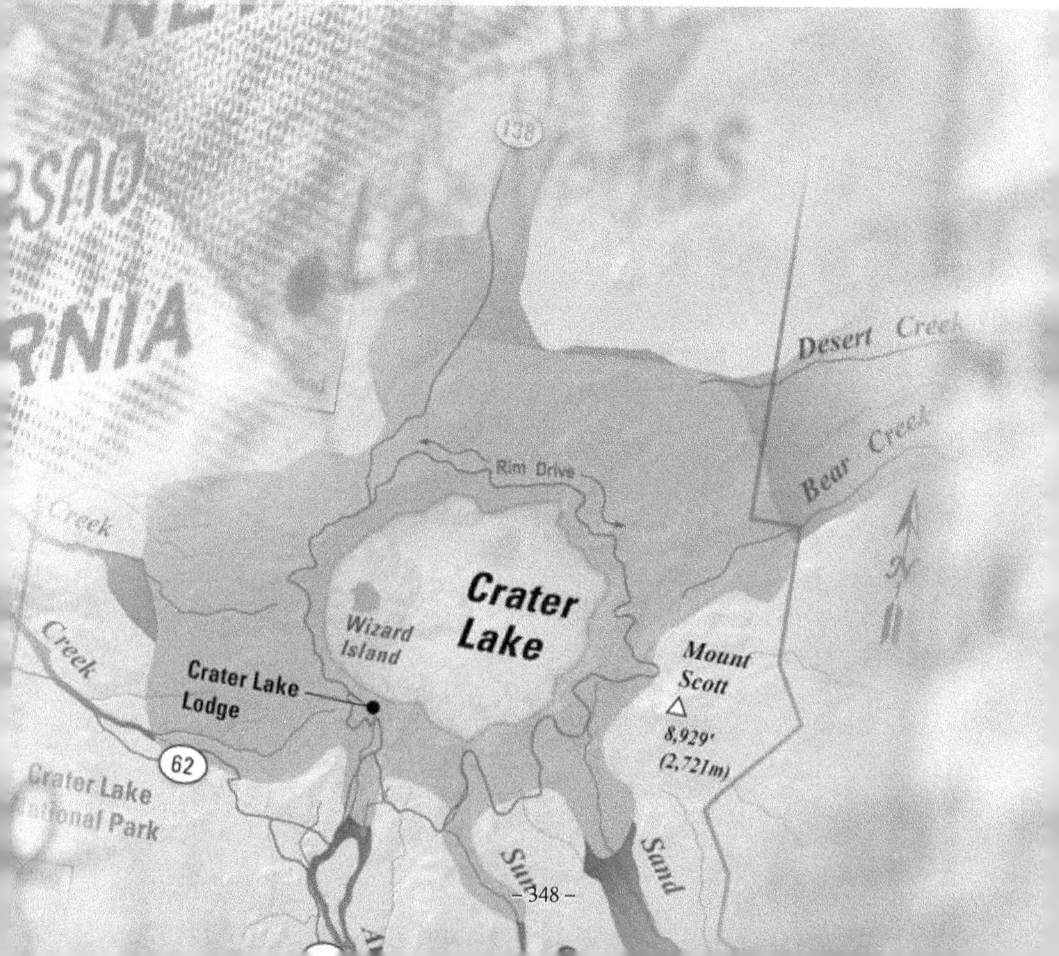
138
Desert Creek
Rim Drive
Bear Creek
Creek
Crater Lake
Wizard Island
Creek
Crater Lake Lodge
Mount Scott
8,929′ (2,721m)
62
Crater Lake National Park
Sand

The spring of 1987 was idyllic. Sometime during this time period, we learned that Mary had been accepted to a Ph.D. program at the University of Colorado, Boulder. She was to start school in the fall, specializing in hydrology. At work, I completed several Antarctic reports for the next international Antarctic conference, as mentioned above to be held during August 1987 in Cambridge, UK. Mike Thomson was the senior editor of the conference volume, and Janet was the third editor. By mid-May, Mary and I were headed out for field work. We stopped off first to spend time with my

CALDERAS IN NEVADA AND UTAH

parents in Tucson, then westward to finish mapping the Newcastle quadrangle in Utah, then westward again to start our new project mapping one of the world's largest caldera complexes, the Caliente caldera complex, named because its western end is near Caliente, Nevada. And while there, we got married!

The Carter and Reagan presidencies (1977–1981, then 1981–1989) were generally stable, optimistic, and without drama. The Carter years had returned the nation to stability and relative prosperity after the madness of Watergate and the corruption of Richard Nixon, but Carter's term included the energy crisis, a short recession due to that, and the Iran hostage crisis. The Reagan years had a recession of its own, a foolish tax cut ("trickle-down economics" that has never worked) that increased the national debt, an unfortunate rollback of Great Society liberalism and of federal protective legislation, yet the cold war ended as the Soviet Union dissolved.

For his contributions to my sense of enjoyment during these times, especially with his positive view of science, I must mention cartoonist Gary Larson. Starting in January 1, 1980, cartoons of his strip *The Far Side* were distributed to 1,900 newspapers. A digital search on the online encyclopedia Wikipedia reveals that, just before he became famous in 1980, his day job while distributing his early cartoons was as a cruelty investigator for the Humane Society. His empathy toward animals, his high morals, and his love of science were obvious from his cartoons, and if one looked carefully, one could see his strong environmentalism. Copies went on workplace bulletin boards and were widely distributed. The cartoons were particularly good at self-effacing humor and at recognizing the stupidity of the things we all do, many of them shown by animals whom he personified. When we no longer received our daily dose of his cartoons on January 1, 1995, most of us went into shock. A private person, Gary seemed to withdraw both his cartoons and himself personally at the same time. His past cartoons remained around for less than five years in desk calendars, greetings cards, and anthology books. Then he and his cartoons were gone, except that his cartoon-anthology books can still be bought and a daily desk calendar returned in 2023. He had only retired, but with that a small part of many of us had died. Yet, this amazingly creative man's writings should be taught in graduate-school writing courses, alongside Ernest Hemingway and Maya Angelou. Many of us claimed our humor to be "Gary Larson type humor."

Many of Larson's cartoons related to what we scientists did. Inasmuch as most scientists are professional writers, one cartoon that resonated with me showed a bunch of early (1787, no doubt) Americans, with wigs, standing around a table starting work on the *Constitution of the US*, while another man of their group was sitting at this table with a parchment in front of him and poised with a quill pen, asking "So, then . . . would that be 'us the people' or 'we the people'?" Another, which portrays scientists' eureka moments, shows a frazzled and exasperated scientist, clearly Einstein, at a blackboard that is covered in formulas ($E=mc^7$, $E=mc^8$), the last one at the bottom being $E=mc^{10}$. A cleaning lady with a feather duster in one hand has just rearranged and tidied up the things on his desk in front of the blackboard, and as he looks back at her she is saying to herself ,"Now that desk looks better, everything's squared away, yessir, squaaaaared away." And for the times in which we, as scientists, cannot for the life of us explain what our data mean, there is the cartoon of the thin owner of a bakery standing behind a glass case loaded with doughnuts talking to his employee in front of the glass case with a broom in his hand. The employee is horrendously fat, and the owner is saying "Well shoot! I just can't figure it out. I'm moving over five hundred doughnuts a day, but I'm still just barely squeakin' by!" Along the same lines is a cartoon showing two cavemen crouching by a fire, holding and cooking legs of meat but grimacing from the pain of the flames hitting their fingers. A third man, standing behind them, sees another man, wearing glasses and at a fire a little in the distance, who is cooking his meat with a stick through it. The third man says to his grimacing companions "Hey! Look what Zog do!" And for those who are incredulous and baffled by the data they are looking at, there is the cartoon of two pioneers with rifles in their hands crouched behind a barrel between covered wagons, watching a rain of arrows, some aflame, that are coming down above them and lighting their wagons on fire. One is saying to the other "Hey! They're lighting their arrows! Can they DO that?" And, of course, there is the cartoon that puts a damper on that common denominator that motivates all scientists. It shows a detective and a policeman standing in the door of a room in which four cats are dead, two slouched over onto their extended arms, one on its back with four feet in the air, and another slouched in a corner. There is a blackboard on which are drawn various formulas and a circle shown with a line of radius. A microscope is on a table. The detective is saying to the sarge, "Notice all the computations, theoretical scribblings, and lab equipment, Norm . . . Yes, curiosity killed these cats!" Of course, there is also the takedown of many intellectuals, such as those with fancy degrees, in which a chubby pre-teen boy is at the front door of a building labeled "Midvale School for the Gifted." He is pushing hard against the door, which has a prominent sign on it that says "PULL."

When Mary and I gave oral talks on our geology at scientific meetings, we would commonly include a slide of one of Gary's cartoons just to get the talk off on the right foot. One of Mary's favorites was the view from the back of a classroom, with one student with a raised hand. The professor in the front is alongside a picture of a cow on the blackboard, with an arrow pointing to the animal stating "COW." The professor is addressing the student with the raised hand, "Yes…I believe there's a question there in the back." Mary might begin her talk by showing this on the screen and saying, "I will start my talk by being as elementary as I can about my field area." My favorite slide for the start of a talk was like hers but tended to insult the audience more. It is also a view from the back of a class of a professor at a blackboard, with a pointer, showing a picture of a person and a duck, with arrows pointing to both that says "TOP" and "BOTTOM." To the right of those is a duck upside down and it has an X through it and is labeled "WRONG." The cartoon caption reads "People who don't know which end is up!"

And finally, pertaining to the song that provided some inspiration for the 1986–1987 field season (previous chapter), Gary even had a cartoon for that. It shows two elephants, one sitting down in front of a piano and the other standing to teach him and stating, "Smash your left hand down here three times, then twice up in this area, then three times right about here…That's 'Louie Louie.'"

Optimism about the future, consisting of marriage, Gary Larson humor, and a fascinating new scientific problem, was soon to come up against reality.

BARCO

1

A new large project, supervised by Branch Chief Glenn Izett, was to start at the new fiscal year (October 1, 1987). It was the Basin and Range-Colorado Plateau project, or BARCO, to be led by Robert B. Scott, an excellent and productive geologist a few years older than me. It was to consist of mapping 7.5-minute quadrangles in SW Utah, SE Nevada, and NW Arizona, to understand this part of the Great Basin and its transition zone to the Colorado Plateau. The Arizona part of the project was to be run out of the Flagstaff office of the USGS, primarily by George Billingsley. The Utah part was a southward extension from the Marysvale project by Branch colleagues Florian Maldonado, Ed Sable, and Dave Moore (Dave is the Vietnam veteran introduced in Chapter 4 who once flew Huey helicopters in combat). That part allowed continued support for John Anderson and his students at Kent State University. It included a re-examination of the large old landslide spread over the crest of the plateau that John and his students had mapped. They found that the landslide was Miocene and covered about two hundred square miles, but questions about how and why it had formed had yielded unresolved and competing hypotheses.

The Nevada part of BARCO was to be the largest effort. My Caliente project, which had started a year earlier, was folded into BARCO. Bob Scott was to concentrate on the Kane Springs Wash caldera complex about fifteen miles south of the Caliente caldera complex. From those calderas, we would extend our 7.5-minute quadrangle mapping outward. William R. (Ric) Page, who had been brought along by Gary Dixon and ended up being an excellent geologic mapper, was to concentrate on areas of Paleozoic rocks, mainly thick deformed limestone and dolomite sedimentary rocks that formed the great carbonate aquifer of Nevada. R. Ernest (Ernie) Anderson also joined us part time due to his knowledge of faults and Nevada geology. Although he was partly a product of the tutelage and mentorship of Bart Ekren, he had by then become a famous and productive geologist in the Branch of Earthquakes, Tectonics, and Risk (more on Ernie below). We had three Quaternary geologists, WC Swadley (WC—without periods, because the letters stood for no words—was the name that Swadley was given by his parents, and his younger brother named him "Dub" for "Double U," so thereafter all others called him Dub), Ralph R. Shroba, and Van S. Williams, who would work closely with each of us. We also collaborated with Myron G. Best, professor at BYU, and Gary J. Axen, professor at UCLA, who had mapped in Nevada near the BARCO area. Isotopic dating, isotope work, and paleomagnetic studies meant that we also worked with Harald H. Mehnert, Lawrence W. Snee, C. Sherman Gromme, Joseph G. Rosenbaum, Daniel M. Unruh, James R. Budahn, and Mark R. Hudson of other USGS branches. It was quite a team!

Much of the Nevada part of BARCO had been published in 1977 as the 1:250,000-scale geologic map of the Cenozoic rocks of Lincoln County, Nevada, by E. Bart Ekren, Gary L. Dixon, and others. This map defined the Caliente caldera complex but it was shown on the map as a mere ten inches long, and as being entirely all in Nevada (in short, running forty miles east-west and twenty miles north-south). These two remarkable geologists were to become colleagues and close friends in this new chapter in my life.

> ***Bart Ekren*** *was born in 1923 in South Dakota and raised in Kensal, North Dakota, where his parents owned the grocery store, hardware store, and town theatre. He began college at the University of North Dakota in chemistry and physics. But World War II came along and he enlisted. He served in the US Army from 1942 to 1946, in infantry combat duty with the 75th Infantry Division, supporting General Patton's drive from the lowlands across the Rhine River into the Ardennes Forest and Germany; he reached the rank of Sergeant. His service included combat in the Battle of the Bulge (December 1944), where he spent thirty days during a cold winter in a fox hole, seemingly just yards from the German lines. He rarely talked about his harrowing war experiences, except to friends, but one story of which he was proud was when he captured some SS officers in the cellar of a damaged farmhouse, then delivered them safely to an Army prison camp, this at a time when prisoners on both sides were routinely executed, especially SS officers. He returned to college, to do a degree in geology and mathematics this time, and married Doris Mullis in 1948. They started their family of four children and, after graduation (1949), Bart worked as a geophysicist with Texaco in Louisiana. He joined the USGS in 1951 for a year, returned to the oil business, then back to the USGS for good in 1953 during the uranium boom, out of the Grand Junction, Colorado office. His early research was mapping uranium deposits on the Colorado Plateau. He later mentored Paul Williams, who began his USGS career in 1957 in the same office, doing similar research. The next year this office closed and both moved to Building 25 in the Denver Federal Center.*

Nevada research, resulting in Bart's deep love for this state, began in 1962 when he mapped 7.5-minute quadrangles at the Nevada Test Site to establish the geologic framework needed to site more than eight hundred underground nuclear tests during the Cold War. The USGS hired Ernie Anderson in 1963 to assist Bart, amid the desire of the Atomic Energy Commission (AEC) to move test areas farther northward from their facility at Mercury as city leaders in Las Vegas (sixty miles to the southeast) became increasingly nervous about shaking buildings. Over the next few years, Bart, Ernie, and colleagues mapped northward, then into the adjacent Nellis Air Force Bombing and Gunnery Range, a huge area that was closed to the public for aircraft bombing and shooting. Attempted coordination with Air Force personnel to get on the Gunnery Range went too slowly and did not seem amenable to "getting the job done," in Bart's mind. Therefore, Bart and his colleagues resorted to cat-and-mouse tactics with border guards as they moved eastward from range to range (always watching for incoming aircraft doing target practice). Eventually the military came around and cooperated, and Bart and colleagues determined that the area was not suitable for nuclear tests. Instead, they chose calderas on the northern side of the NTS, such as on Pahute Mesa, where thick sections of intracaldera ash-flow tuff were exposed. After several years, Ernie left to begin his own structural study of the Lake Mead area east of Las Vegas. Once, as Bart was discussing Ernie's progress on this study, he said "Ernie, this work will one day make you famous." Those words, which meant a lot coming from his mentor, came true: among other things, Ernie discovered detachment faults (normal faults that are surprisingly low-angle, some almost horizontal, yet commonly with large displacements).

As the AEC expanded its focus, in 1966 Bart became Project Supervisor to coordinate all geologic, hydrologic, and geophysical studies; he led field parties of ten to thirty geologists, always out-mapping them all. From there he became Program Coordinator for all work on the NTS and Gunnery Range, as well as a five thousand square-mile area in central Nevada where a nuclear test was detonated. His advice in meetings in Washington, D.C. with officials of the USGS and AEC (later renamed the US Department of Energy [DOE]) led to much of the present configuration of the NTS and the Yucca Mountain nuclear storage facility. Because many tests were sited in thick intracaldera tuffs, Bart became an expert at mapping calderas, which are particularly complex. Few geologists have discovered more calderas, and much of our knowledge about them came from his NTS mapping. He extended his mapping into the Walker Lane, an enigmatic, northerly-trending structural and mineral belt just to the west, along the Nevada-California border. He also did county mapping (1:250,000 scale) as part of the USGS effort to produce a geologic map of Nevada, which John H. (Jack) Stewart authored. Bart's county map led to a report on east-west "lineaments" (my work drew on many of his ideas, but in 1998 I changed the term to "transverse zones"), that formed at the same time as basin-range extension, that were at least two hundred miles long and were an additional focus for mineral resources. His county map and report were published in 1976 and 1977, respectively. Jack Stewart, with colleagues, published a report on other east-west structures in Nevada in 1977, and I published my first report on them in Utah in 1978. Bart had great intelligence, leadership qualities, work ethic, common sense, and modesty, along with a dry, self-effacing wit. Many colleagues revered him because of his supportive nature. Gary Dixon was mentored for a longer period than Ernie, and when Bart left NTS work in 1977, Gary had been sufficiently trained to take over the mapping and management duties at NTS.

Bart's exit from NTS and related work was because Paul Williams, by then Branch Chief of Central Regional Geology in Denver, saw an opportunity to upgrade his Branch and offered Bart an opportunity to map the Owyhee geothermal area in southwestern Idaho. By then we had moved into a new office complex several miles northwest of the Federal Center, but Bart developed allergies to the carpeting and found that the windows could not be opened for relief. However, Chuck Thorman, at the time Branch Chief of Central Mineral Resources on the Federal Center, recognized Bart's talents and was happy to offer Bart a position in his Branch in 1979, and he went back to Building 25. Bart's new CMR project was not far away, in the Challis 2-degree sheet in central Idaho, which included the Challis volcanic field. Bart discovered two major gold deposits and a major caldera and continued, with some interruptions, with the project until he retired in 1985. He also concluded that some large volcanic sheets, long interpreted as lava flows, were high-temperature ash-flow tuffs that were derived from such great depths in the crust that they did not produce typical calderas. Chuck remembers visiting Bart at his Idaho field camp, and in particular Bart's dog "Clean," so named because it licked each person's meal dishes to perfection, so that they needed no further washing—or so Bart told everybody! Bart's interruptions to the Challis work included mapping three(!) 1-degree sheets (1:100,000 scale) in the Precambrian shield of Saudi Arabia. Post-retirement stories of Bart will continue below.

2 Caliente

Nevada is a starkly beautiful place of alternating north-trending mountain ranges and valleys. Outside of the metro centers of Las Vegas and Reno, it is sparsely populated, significantly adding to its appeal. In fact, most people I know who have spent significant time in rural Nevada love the state, although most of us can do without Las Vegas. All our Nevada mapping was to be in Lincoln County, larger than either Vermont or New Hampshire but with a fraction of the population (only five thousand people). The three largest towns in the county are Pioche in the north, Caliente twenty miles to the south, and Panaca halfway in between, each with not much more than a thousand people.

The attractive town of Caliente is at 4,300 feet elevation and lies about twenty-five miles west of the Utah border in the northwestern corner of the Caliente caldera complex. Named for a series of hot (caliente) springs on the northern side of town, it was founded in 1901 as a ranching community to supply hay and beef to the active mining towns of Delamar, about twenty miles to the southwest, and of Pioche. In 1905, the Union Pacific Railroad was completed from Milford, Utah, to Caliente not only to supply the mining districts but primarily to push from there southward through beautiful, two-thousand-foot-deep Rainbow Canyon, then on to Las Vegas and Los Angeles. The population was 5,000 in 1910, as Caliente became a railroad company town that supplied not only water for the steam engines from several cold springs but also mechanical and other services. Along the main street on the northern side of town one can see a dozen or so identical row houses, originally built for supervisors working for the railroad. At one time the population reached five thousand. In 1923 a railroad depot of Spanish Mission architecture was built in the center of town, a beautiful white two-story building (now housing city administrative offices and a library). But steam engines were replaced by diesel locomotives in the 1940s, and the population rapidly declined.

In the caldera complex, the magma bodies (intrusions) that supplied the tuff eruptions resulted in two small gold districts within the caldera, one large gold district (Delamar) along the western edge of the caldera, and another small gold district along the northwestern edge. The Delamar gold district was the largest gold district in Nevada during its brief heyday, 1892 to 1909. The historical archeology of the district was told in a delightful University of Nevada, Las Vegas Masters thesis in Archeology by

Dawna E. Ferris, whom we will meet later! As she told it, in 1891, two prospectors (brothers) carrying a monkey wrench because they had no hammer, found gold in an arroyo now named Monkey Wrench Wash. A mining camp sprang up, named Ferguson after the prospectors. In 1894, Joseph DeLamar bought the most important mines and renamed the camp Delamar, at the western base of what became the Delamar Mountains, a relatively low range. The town of Delamar, its population of three thousand by 1897, sprang up southwest of the main pit, with a quarter-mile-long, east-west main street that is now a long-dead ghost town of stone foundations and partial buildings. Although at an elevation of about six thousand feet, it was a hot dusty townsite, absent of trees and even sagebrush, with its main form of vegetation prickly low blackbush. A huge stamp mill (where ore was crushed by pounding) was built, run by steam, and consequently the Delamar Mountains (with a high point of 7,700 feet) were partly denuded of trees, mostly pinion or juniper, chopped down for fuel. Water was pumped west across the Delamar Mountains from Meadow Valley Wash, the perennial stream in Rainbow Canyon twelve miles to the east. But this water, initially distributed by wagon, was too valuable to be used in the mining or milling of the gold ore. The gold was locked in a tightly cemented Precambrian sandstone known as quartzite, which therefore was mined and milled dry, resulting in clouds of dust that workers inhaled, causing silicosis and death, in some instances within a month or two. The dust was so deadly because quartzite resembles glass that, when crushed, becomes jagged shards, much like volcanic ash. Delamar consequently became known as "The Widow Maker." At first, all supplies for the mines and citizens had to be transported by mule train from the railhead in Milford, 150 miles to the northeast. In 1902, DeLamar sold out and the new owners installed a new stamp mill in 1903, although most mines ceased operation by 1909.

Lumber was a valuable commodity in this area where the pinyon-juniper trees were too small to be cut into boards. Therefore, virtually all the wood frames of the Delamar buildings, which had been hauled from Pioche when mining declined there, in turn were shipped in wagons, as mining largely ceased there by about 1910, and erected again at the ranching town of Alamo, twenty miles to the southwest. From 1929 to 1934, during the Great Depression, a revival of mining took place and the post office and school were reopened. Much of the ghost town, mine headframes, and some stone mine buildings survive. The ghost town has remained mostly intact because it was far enough off the beaten track, about fifteen miles south of the main (Ely to Las Vegas) highway, US 93, on a dirt road along the north-trending Delamar Valley.

The gold resources were my reason for the project., which was to map the Caliente caldera complex and surrounding areas. A caldera complex denotes where more than one significant tuff has erupted, each leaving a hole (caldera) in its wake. The only other geologist assigned to my project was L. David Nealey, a go-to-hell geochemist and igneous petrologist. He was not involved in the geologic mapping but was the brains that would interpret my field work on these igneous rocks. Dave, with a master's degree from the University of Northern Arizona and a Ph.D. from the University of New Mexico, had spent much of his early career working out of the Flagstaff office on the San Francisco volcanic field, but then transferred to Denver.

> *Another colleague and friend is* ***Laurence W. Snee****, an* $^{40}Ar/^{39}Ar$ *dating specialist according to his main duty, but a complete geologist with a long history of field studies. A few years younger than me, he quit undergraduate studies at the University of Rochester to join the US Marine Corps and serve in Vietnam. In his thirteen months there, during his 1966 to 1969 enlistment, he was on thirty-four combat missions, twenty-seven of them long-range reconnaissance patrols deep within enemy-controlled territory. Here he saw much combat and received a Bronze Star for valor. One of the members of his unit, a black private who had just joined, threw himself on a grenade during a fire fight while on a patrol in which Larry was not participating. This private died while saving others; he was posthumously awarded the Congressional Medal of Honor. In 2018 Larry was one of a group who joined in the launching of a new destroyer named to honor that man, the* USS Ralph Johnson. *Following his tour, Larry returned to finish his bachelor's degree, after transferring to Florida State University to be near his folks. He majored in geology, biology, and history and participated in the NSF research vessel* Eltanin's *fiftieth cruise to do marine geology in Antarctic waters. Then he set out for a master's and Ph.D. in geology at Ohio State University, brought in by John Sutter, as noted above. He received support from the USGS in Denver before and after finishing the Ph.D. in 1983. From 1983 to 1986 he was Assistant Professor of geology at Oregon State University, then joined the USGS when he was hired by Chuck Thorman, then Chief of the Branch of Central Mineral Resources. Later Larry joined the Branch of Isotope Geology, where he was the creator and director of the Central Region Argon Geochronology Lab. He and Dave Nealey accompanied me on collecting trips to Caliente. Unfortunately, most of Larry's Argon ages were never published in final form because the project was stopped abruptly by the Reduction in Force (RIF) in 1995, as covered in the next chapter. In 2022, Larry and I began putting together his dates and seven of my quadrangles, for publication by the Nevada Bureau of Mines & Geology, under Director James E. Faulds; the first two of these, the Caliente NW and Chokecherry Mountain quadrangles, were published in 2023.*

In May 1987, after visiting my parents and doing field checks on the Newcastle quadrangle, Mary and I drove west to Caliente to start the new project, before heading to the Fifth International Symposium on Antarctic Earth Sciences in Cambridge, England. Along the way, we detoured to Pioche, Nevada, the county seat, and got a marriage license. On June 2, we took the day off and returned to Pioche and got married, with Judge Sarah K. Getker presiding, and her clerk as witness, at the "million-dollar courthouse." Pioche is an attractive small town, nestled among abandoned mine head frames and buildings. The camp was founded in 1864 with the opening of a silver mine but soon was abandoned due to Indian raids, until it was re-established in 1868. Francois Pioche, a San Francisco financier, bought the town in 1869 and sunk more mines. By the early 1870s, the town had grown to be one of the most important silver mining districts in the US, and a wild one at that! The population in 1872 was six thousand people, and the town contained 72 saloons, 32 houses of prostitution, 2 theaters, 2 breweries, 2 gravity-fed water systems, 2 fire companies, and a livery stable housing 300 horses. Legendary Boot Hill cemetery contained 72 men killed by violent means, primarily by gunfights—this before the first natural death! The "million-dollar courthouse" (due to cost overruns, even then), included a jail in the rear of the building, and was built in 1871. Ore was shipped south to a stamp mill at Bullionville (now totally gone), almost halfway to Panaca. During World War II, additional lead-zinc deposits were mined.

Our honeymoon, of course, was during July in the UK, where we rented a car and toured England, Wales, and Scotland before the Antarctic symposium at Cambridge University. It included a geologic field trip sponsored by the symposium to see old rocks in the Scottish Highlands. We visited lots of castles, saw Stonehenge, London, and Edinburgh, drank lots of great British ale (the world's best beer), ate delicious pub food, and had a relaxing vacation, some of it among old friends we had both known for years. The symposium volume, of which Mike was the senior editor but on which Janet Thomson did most of the work (as Mike states in the book's introduction), was published in 1991.

I was not to see Janet or Mike again, although we continued to communicate through Christmas cards and e-mails. Mike retired from BAS in 2002, and Janet in 2003. After more than fourteen trips to Antarctica beyond his initial one, and more than one hundred publications, Mike in retirement became a carpenter when he and Janet moved to an old house (Stone House) in a remote area (Yorkshire) and remodeled it, becoming involved in the local community; they helped renovate Gayle Mill, an historic water-powered sawmill, and, to help fund the mill, Mike produced items on his lathe for sale, including percussion instruments for local children. Sadly, Mike died on January 19, 2020, from a brain tumor that had been only discovered ten days before. In the Antarctic Peninsula, Mike has Thomson Rock named after him, while Janet has Thomson Summit (a peak in the Behrendt Mountains, after I submitted her name following our 1984–1985 season) and Thomson Glacier in the peninsula named after her. Before retirement, Janet was Head of the Mapping and Geophysical Information Center at BAS, received the Fuchs Medal from BAS in 2001, and received the Polar Medal in 2003. The golden years for BAS geology, in terms of publications and scientific discoveries, were the years when Charles Swithinbank and the Thomsons ran the Earth Sciences Division, after which funding was cut and the Survey became more of a bureaucracy.

Upon our return to the US, Mary and I settled in at our home on Meadow Drive. Built into a hill facing east toward an undeveloped low mountain owned by the city of Denver, its main level was a living room with picture windows on three sides that looked across a meadow (of Meadow Drive) at the mountain. It isolated us from the rat race. Mary started classes at Boulder, about an hour's drive away. We had two cats, one a big, affectionate, long-haired, unkempt, neutered tom cat named Scruffy that the Williams had asked us to take when they headed to Saudi Arabia.

After Cambridge, in early November I returned to the Caliente area and mapped for a month. In mid-November, my field notes record a phone conversation in which Mary told me she was snowed in, with twenty-two inches of snow on the ground and it still coming down. I bought a slightly used dark blue 1986 Toyota 4Runner that always got us home in winter, and I also used it for field work because government field vehicles were undependable. Sadly, the car long outlasted our marriage.

In mid-June, 1988, a forest fire started in Wyoming and soon moved into Yellowstone National Park, and several dozen other fires followed, caused by lightning and humans. At first, the fires were left alone, but by July winds accelerated them and they combined, so thousands of fire fighters took up the fight. The fires eventually claimed an extraordinary 1.2 million acres, of which nearly 800,000 acres were in the park—36 percent of the park! Old Faithful was threatened and some buildings there burned. Only rain and eventually snow on September 11 extinguished most of it. Mary's advisor, Professor Donald D. Runnells, suggested this might be an ideal Ph.D. dissertation topic, and they received funding to study what the effect of the fire was on the groundwater. Had it become polluted by the ash from the fire? There were many water wells and monitoring wells that could be sampled for water chemistry, and for the next year or two, Mary (I joined her once or twice) sampled the wells and started analyzing the results. The conclusions would turn out to be what science refers to as "negative results," for the pollution was minimal, but it was a terrific study nonetheless, whose value was significant and whose report would be read and cited by others whose areas have been devastated by fire.

The next several years were happy, as we had both embarked on all-consuming ventures. Mary was engrossed in her schoolwork. Bob

Scott, others, and I hauled USGS trailers out to Youngs RV trailer park on the southern side of Caliente, and generally had both spring and fall field seasons because the summers were hot. We cooked our own meals and socialized sparingly, for we were exhausted when we got back each night, but the camaraderie was good and we did not go to bed before the others had returned. Our field areas were remote, on dirt roads where we rarely saw a single rancher or traveler all day, yet few of us had the patience to enlist field assistants. The owners of the trailer park were Chad and Brenda Young, descendants of Brigham Young, one of the founders of the Mormon Church. At the end of each field season, we only had to drag the trailers next door, to the Bureau of Land Management compound, for storage. In late March of 1988, I visited my folks in Tucson after Dad had a stroke. Fortunately, his brain was intact and he only had paralysis on one side of his body, necessitating a walker to get around. Mom was a good nurse. Even though their travels abroad stopped, they had a nice place to relax, and they had made many friends. They had a good retirement, and the stroke seemed to slow them down only a little.

Mapping calderas is exciting and it requires an experienced geologist. The Great Basin has many dozens of calderas, and the one I was studying was one of the largest. Even though Mount St. Helens erupted ash-flow tuffs six times over six months of 1980, none was big enough to leave a caldera. Big ash-flow eruptions, however, may be many cubic miles in volume. The magma is high in volatiles (water and gases) that rapidly drive the eruption of volcanic ash. The ash is mobile because it is riding on a frictionless layer of hot gas, so outflow distances may reach up to a hundred miles in all directions. Speeds are in the range of 100 mph. When such a huge volume of magma is erupted instantly from a depth of perhaps a mile, the rock above it will collapse into the hole left by the magma. This subsidence results in a subcircular depression, a caldera. The instantaneous subsidence, along a steep circular fault called the structural wall of the caldera, commonly is more than a mile vertically, like a piston in an engine. A subvertical structural wall a mile high will not remain long, but upper parts will slough off (landslide) into the caldera. These rock masses from the rim, which may be hundreds of yards if not a mile long, will fall violently into the hole, commonly brecciating (breaking apart) in the process, a rock type known as caldera megabreccia. Smaller ash-flow eruptions will follow, some getting out of the caldera but most filling the caldera and burying the megabreccia layers. The whole process, from eruption to extinction of each large (initial) outflow tuff unit (they are distinctive, and we give them names), is fast as far as geologic time goes, perhaps a hundred thousand years or less (there is a joke that when a person tells you that a million years is a short period of time, you can be assured that that person is a geologist!).

Years passed quickly, with early spring and late fall field seasons in and adjacent to the Caliente caldera complex. The rest of the time I spent doing lab work on the rocks and writing up the results for publication as both geologic maps and as reports. Then in 1992 my world came crashing down. At about the time that Mary's dissertation was finished in 1992 and she was awaiting her Ph.D., she asked for a divorce, and she soon moved out. The procedure took a while, but was amicable. The divorce was finalized in April 1993. It broke my heart. The 1988 song by Poison, "Every Rose Has its Thorn", seemed particularly meaningful. My response was to bear down on my science, probably what had caused the divorce in the first place! I worked long hours and weekends, and publications were quick to follow. My field notebooks show long field seasons in 1993 and 1994, mostly in the Caliente area. But the fall 1994 field season was to be my last one there, and the end of the project.

The fall of 1994 coincided with a political catastrophe that affected the entire USGS. This was an after-effect of the mid-term elections of 1994, two years after Democrat Bill Clinton was first elected President. In these elections, on Tuesday November 8, the Republicans won majorities in the House of Representatives (with a swing of 54 members from Democrat to Republican) and Senate. The Minority Whip of the former House, Newt Gingrich, became the Speaker of the new House, and Bob Dole became Majority Leader of the Senate. Gingrich was an ultra-right conservative with an ego as tall as Mount Everest and an agenda that was bad news for any thinking human. Gingrich had pushed a bunch of right-wing ideas that he called Contract with America, and these caught the public's attention, then secured their vote. Gingrich and his cohorts had made a concerted effort to gain the vote of the religious right, specifically evangelicals, "born-agains," and others who believe in the literal interpretation of the Bible. Most of these persons are poorly educated and have no use for science. This election resulted in the start of willful damage by Republicans to the concept of democracy in America, and Americans swallowed lies by members of the party. As P.T. Barnum once said, "No man ever lost money by underestimating the intelligence of the American public."

For me, as I was finishing up my field work in an area south of Caliente, that election of only a few days away was far from mind. Nor could I have guessed its results. Besides, just before that election, two noteworthy events, one an animal encounter and the other a eureka moment, happened in quick succession. On November 1, 1994, I was mapping volcanic rocks in the Caliente caldera complex on a traverse in a western part of a Wilderness Study Area about five miles east of where I had parked my Toyota 4Runner. The weather had quickly turned cloudy, windy, and cold; a cold front made its way through as I was returning west, into the wind, at the end of the day. As the juniper and pinion trees waved in the wind. I worked my way out of a canyon, heading to the car. I stopped to sample an interesting outcrop. As I did, I happened to look back at the deer trail I had been following. There, on the trail about twenty-five feet away, was a beautiful adult mountain lion. I do not know how long he (or she) had been following me. I stood up, with my rock hammer in hand, and admired it, for I had never seen one in the wild before (or since). I knew that all I had to do was wave an arm and it would be gone like a ghost, so I wanted to take in the view of this beautiful animal as long as I could. After a couple of minutes, I did wave an arm above my head. But the lion did not move! How curious, I thought. I raised both arms and yelled. Nothing! Then it occurred to me that it had been stalking me. And, worse, I had only a rock hammer and Swiss army knife, and the cougar was doubtless better at climbing pinyons or junipers than I was. So, for my next trick, I moved toward the animal purposely, all the while yelling and waving my arms. That worked, and it moved into the brush out of sight. I hefted my pack and headed west about twenty feet. But I looked back, and there it was again, at about the same distance away. I did several iterations of approaching the lion, but each time it reappeared as I continued west toward the truck. This method clearly was not working. It made no menacing moves or sounds, just watched me. I glanced at my feet, and spied several baseball-size rocks. I picked one up, wound up, and heaved it as hard as I could, and it passed over the animal's head. The second rock, however, was effective: it hit the lion in the middle of the chest. It quietly retreated and I never saw it again, though I looked back over my shoulder many times as I retraced my steps the long five miles to the truck.

That night it snowed about a foot. I spent part of the day trying to get into parts of the field area in four-wheel drive, but the wet, heavy snow made such progress too difficult. I waited another day, then winterized my trailer, and before 7 am of Saturday November 5, I had started for home.

The drive from Caliente to Denver eats up a twelve-hour day. Whether from Caliente or southwest Utah, I have made this beautiful, lonely trip more than a hundred times. The last time, in August 2021, for example, I drove from New Harmony, Utah, to Grand Junction, Colorado, to visit my son, Scott, and his wife, Patty; then I stayed focused by playing albums by Bruce Springsteen (*Letter to You*) and John Fogerty (*The Long Road Home*) to pass the time. The trip from Richfield, Utah, to Grand Junction via I-70 crosses the Colorado Plateau, through buttes and mesas in Paul Williams country and George Davis country, my name for their classic geologic work. There is not much traffic, mostly long-distance truckers bringing their goods from L.A. to Denver, slowly grinding up the long grades with their flashers on, as well as occasional sedans with Colorado license plates anticipating their return home. The vistas at each hilltop are spectacular. Heading east from Salina, Utah, the rocks are tan and gray with a few black coal beds making up the Cretaceous time interval; sparse vegetation consists of rabbit brush and sage. Then the climb up onto the giant bulge known as the San Rafael Swell, made up of reds and yellows of the Jurassic, notably old friend Navajo Sandstone, amidst pinyon-juniper. Talus spills down the bare cliffs. The eastward descent off the Swell is abrupt, with the rocks dipping eastward (a monocline), then flattening as one reaches the town of Green River and the bridge over the actual Green River. Here the rocks are back within the Cretaceous, mostly grays making up thousands of feet of Mancos Shale, much of it containing the clay bentonite, which swells due to moisture, so the interstate roadway is deformed into a roller coaster in miniature. The Book Cliffs rise just north of Green River; laccolith mountains of the Henrys, Abajos, and La Sals (Larry Gould's dissertation area), are in the distance to the south. The shale and rabbit brush continue east to the Colorado border, past exits such as "Lone Tree (No Services)." Then to Grand Junction and the Rocky Mountains beyond.

By early afternoon of November 5, I was crossing the Swell, past the road sign that states "Denver 382 miles" (only six more hours to go!). Though the multicolored cliffed mesas and buttes are beautiful, they are also quite familiar, and as I barreled along past patches of snow on the Interstate at 75 mph on a bright fall day, with no music radio stations to tune to, my mind wandered to geologic mysteries that I had not understood. The one I happened to focus on that day was how the Great Basin that I had just left at Richfield had so many east-west fault zones, including what I named the Blue-Ribbon lineament that Interstate-70 followed in east-west Clear Creek Canyon just before the highway swings north to Richfield. I had published on that fault in 1978 but was still unsure how it could have formed at the same age (in the Miocene at least twenty million years ago and continuing today) as the much more common, north-south basin-range normal faults that define all the alternating north-trending basins (valleys) and ranges that span the Great Basin. Colleagues Bart Ekren and Gary Dixon had also published on other Great Basin east-west "lineaments" and they provided insights on which I expanded. I took out a pad and pencil and started writing.

As was well known, north-south basin-range faults form when the crust is pulled apart east-west. But how can you produce faults at 90 degrees to these dominant structures by the same east-west pull? Earlier, some geologists had suggested in published reports that the stress field had alternated quickly (then back again) between an orientation of east-west to form north-south faults and an orientation of north-south to form east-west faults, but that made no sense. Yet transform faults in the ocean basins, all of which trend east-west even though the stress field (that produce north-south oceanic ridges such as those that underlie Iceland) is also east-west. Transform faults, as J. Tuzo Wilson (of the Wilson

Mountains that I had visited in 1987) recognized, form parallel to the stress field because they represent boundaries between blocks north and south of them that are pulled (or pushed in the case of subduction zones) at different rates and amounts.

I jotted down ideas for a report the rest of the way back to Denver. In the next few weeks, I consulted Bart, Gary, and other colleagues and more carefully scanned the limited literature on east-west faults. Because these east-west structures were transverse (90 degrees) to the dominant basin-range faults, I followed several authors in naming them transverse zones. I considered them analogous to transform faults in the ocean basins; Bart and Gary had previously concluded much the same thing but had not developed the idea.

A year later, the USGS put together a volume on volcanic fields on the Colorado Plateau and I submitted a paper on the Marysvale volcanic field that enabled me to bring forward my transverse-zone ideas, including on the Blue-Ribbon "lineament" south of Richfield; my co-authors, including Tom Steven and Skip Cunningham, were fine with these concepts inasmuch as some transverse zones controlled some of the features in the Marysvale field. But this manuscript languished for two years because an author of another report in the volume could not meet his deadlines, and unfortunately for us, the editors of the volume waited patiently rather than going ahead without the late author. This ended up being a bit of good luck, for the wait allowed me to refine my ideas, which—I realized later—were rather half-baked at the time. Another volume, by the Geological Society of America, was proposed for articles on similar structures to mine, by James G. Faulds of the Nevada Bureau of Mines & Geology. I submitted an entirely new report, focusing on the origin of the Great Basin and how transverse zones related to this origin. It was published in 1998, the same year that the half-baked report finally was.

The broader significance of these ideas on transverse zones was my initial recognition that the Caliente caldera complex was a new type of caldera, which Ernie Anderson and I called a syntectonic caldera. As our mapping progressed through three published 7.5-minute quadrangles and seven others that we completed but were yet to publish, we recognized that the caldera was unusually elongated east-west (fifty miles east-west, six of which were in Utah; and twenty-two miles north-south). This was because it was bounded on its northern and southern sides by large east-west transverse zones (mostly faults) that were active (moving) as eruptions were going on. In fact, it had multiple eruptions (about ten, each the product of many ash flows) of huge ash-flow tuff units over a ten-million-year span of time (25 million years ago to 15 million years ago). So the caldera lengthened east-west through time, and the structural caldera walls were east-west faults. Some of the gold deposits were along these transverse zones that bounded the various calderas of the complex.

But by the time I was figuring this out, the project was terminated by the USGS Reduction in Force (RIF), as discussed in the next chapter. It was only in 2022, nearly thirty years after the RIF, that I returned to those seven quadrangles.

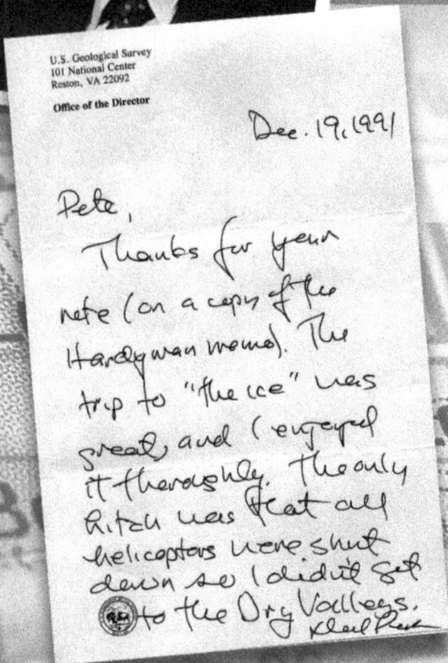

U.S. Geological Survey
101 National Center
Reston, VA 22092

Office of the Director

Dec. 19, 1991

Pete,

Thanks for your note (on a copy of the Hardyman memo). The trip to "the ice" was great, and I enjoyed it thoroughly. The only hitch was that all helicopters were shut down so I didn't get to the Dry Valleys.

CHAPTER 8

The repercussions of the November 1994 mid-term elections unfolded rapidly over the winter and spring. Few of the promises in Gingrich's Contract with America ever became law, for a Democrat president held veto power. But other disasters took place because Republicans controlled budgets and both houses of Congress. These disasters included elimination of the Office of Technology Assessment and the US Bureau of Mines, under the false guise of budget savings. And for us, the most important disaster in this time frame was when Gingrich urged elimination of

THE RISE AND FALL OF THE USGS

the USGS. This brought on rapid-motion events that took the USGS from (as in the classic 1971 song "American Pie" by Don McLean) a "broncin' buck" to its near-death. This was the 1995 USGS Reduction-in-Force (the RIF), when the government fired six hundred geologists from the Geologic Division.

By late 1993, Dallas Peck had left his job as USGS Director to return to doing science and finishing some of his reports, which he did until he officially retired in 1995. Gordon P. Eaton became Director in March of 1994, just in time to have stepped into this political bear trap. Before he had gone into administration, Dallas had achieved a certain amount of greatness in his science. The same could be said for all previous Directors, as well as for Gordon Eaton. After Gordon Eaton, however, all Directors except for Marcia McNutt (appointed by Barack Obama) were, at least as far as I could tell, merely career paper pushers. They never achieved any significant importance in their science, even though all had a Ph.D.

> ***Gordon P. Eaton*** *(1929-2022), known to all as Gordie, received a Ph.D. in geology and geophysics from the California Institute of Technology. He was a geophysicist with the USGS from 1967 to 1981, and for the latter part of this time his office was not far from mine in an office building complex in Lakewood, a few miles northwest of the Federal Center, and we bumped into each other from time to time. I always considered him a likeable colleague and great geophysicist, the reason for the latter being that he had become a good geologist before he called himself a geophysicist.*

> *Coincidentally, one of his reports provided geophysical support for my later ideas on transverse zones, and I talked with him about it and cited his report. But Gordie reached for higher goals beyond being an excellent scientist. He became President of Iowa State University, then Director of Lamont-Doherty Earth Observatory of Columbia University, then finally returned to become Director of the USGS. Now, dealing with the egomaniac Newt Gingrich, Gordie made what I think was the worst decision of his career.*

Although most of the details and reasons for what Gordie did are unknown to me because I had no firsthand knowledge of his thinking, I did have firsthand knowledge of the results of his actions. Perhaps it was not all his fault: funding cuts by Gingrich and the Republicans were to go into effect on October 1, 1995. And Gordie had some help from Benjamin A. Morgan III, Chief Geologist, who ignored a 1988 internal analysis that forecast funding problems from overstaffing. And Gordie had accomplices: his main one came when he appointed P. Patrick Leahy, a hydrologist (including Chief Hydrologist) with Water Resources Division (WRD) with little knowledge of the Geologic Division (GD) ethic or any GD personnel, to follow Ben Morgan and become Chief Geologist. Patrick seemed hostile to some geologic specialties of which he had little knowledge. To apply the song by Bob Dylan, we now had Tweedle Dee and Tweedle Dum, both to my mind lacking in judgment, who initiated the Reduction in Force (RIF). From the rumor mill, which was extremely active in the spring of 1995, Gordie had apparently agreed to fire six hundred permanent scientists, all of them geologists, in return for Gingrich not axing the entire agency. It is noteworthy that all the cuts were in Geologic Division, a fact that we probably owe to Patrick. The historical record of the RIF of 1995 is slim. One report still available on the internet at the time of this writing, from the September 1, 1994, issue of *Science News*, claimed that the number was ">500 employees (about one quarter of the staff of Geologic Division)," but for those who lived through this, the number we remember was six hundred. Was this decision by the USGS, and the other decisions to follow, examples of groupthink? Probably, inasmuch as I never saw evidence that Gordie was authoritarian, so he probably did not act entirely on his own and likely had consensus from underlings.

The rumors of the impending reduction in personnel during early 1995 preoccupied all geologists of GD, as management held the secret very closely as to how many and who would be fired. Virtually no science got done during that time. Instead, we had sleepless nights and excited talk in the halls each day as we all wondered whether our USGS careers were now over, and whether we should start applying for other jobs. And in fact, many worked on their résumés and looked for jobs outside the USGS. By that time our Branch Chief was John C. (Jack) Reed, a long-time member of the branch who agreed to do the dirty work; partly in his defense, he retired from the USGS shortly after. To be what they mistakenly considered "fair," Gordie, Patrick, and other administrators chose geologists from specialties or branches in the division that they considered now to be obsolete, such as Antarctic research, metamorphic and igneous petrology, the Paleontology and Stratigraphy Branch, and several other specialties. They also decided against allowing persons to be funded by multiple projects, which hit hard at such branches as Isotope Geology, where geologists and chemists did isotopic ages and other studies for multiple projects. Committees in Reston made the initial cuts, amid extremely bitter wrangling. In most cases, however, they fired people they disliked or did not know, and they saved their friends. All of us could have lived with a decision to fire based on incompetence or low productivity, but their method was to cut indiscriminately excellent, good, bad, or indifferent geologists based on their flawed formula. The entire exercise destroyed the trust—long built up—in the organization. In effect, it was the end of the USGS as we knew it.

In the spring, after months of waiting, a list appeared unannounced on a glass-enclosed bulletin board, in each of the major centers, listing those who were to be fired at the end of the fiscal year. For me, this was and remains "the day the music died" for the USGS, to quote a line from Don McLean's "American Pie". Names were not given on that list. Instead, the list displayed that person's title, GS grade, specialty, and perhaps project. Digesting and interpreting who that person was ranged between being an easy and a difficult task. As I remember, within another month or so, those who would be fired were given notice. Emotions ran high during the interim, ranging from tears to thankful relief. By being so incredibly subjective, it was clear that the decisions to fire were utterly unfair. Virtually no person who lived through the RIF, even if they survived, had much use for the agency after that. Obviously, those who did not survive felt worse. The survivors soon had a sole goal: to stay with the USGS until they qualified for their pension, then get the hell out.

Among the persons in our branch to be fired were Chet Wallace, Mel Kuntz, and Karl Kellogg. I survived but was saved and reassigned by Doug Morton, Gary Dixon, John Sutter, and Larry Snee to a management position with the Las Vegas project with Gary Dixon. Then, many of those who remained campaigned for those who had been fired, unfortunately with few good results. However, Bruce Bryant, a longtime branch member and close colleague and friend of Jack Reed, showed a high level of courage when he marched into Jack's office to show his indignation, then announced that from that moment he was retired, so that should free up a slot for one of the others. Apparently, that is how Karl remained! The lobbying effort by branch members somehow saved

Mel, although it took a while and he came back after the fiscal year with a different type of appointment (to be paid only by outside funding) when Gary Dixon and I snapped him up, so he lost his pension. Within the Office of Regional Geology, more than a hundred projects were reduced to seventeen big projects. Doug Morton's southern California study, our Las Vegas project, and our (shortly thereafter) Nevada Test Site project were three of these seventeen. BARCO, of course, was terminated in February 1995. Outside our branch, geophysicist John Behrendt (in Gordie's former Branch of Geophysics) and Art Ford in Menlo Park—both productive, world-class experts in Antarctic geology (GS-15s)— were among those who were whacked. John found a home at the Institute of Arctic and Alpine Research at the University of Colorado, Boulder, and continued his Antarctic career. The Branch of Isotope Geology took many hits, even though most of its work was world class. Although Dave Nealey was not fired, he within the year left the USGS, back to Florida, where he and Myrna had grown up; I believe that he took a job with an Historically Black College or University (HBCU), but I never saw them again and I have been unable to locate him on the internet.

Legal actions instituted by persons who were fired began next, with some court cases continuing for several years. The legal cases had mixed results but were largely futile. The lawsuits certainly continued the agony of those involved. And the disruption to productivity was profound within all parts of the Geologic Division, where all scientists closely followed the legal procedures of their friends at multiple centers. Upper-level administrators, even many branch chiefs, now became part of the enemy to working geologists, a radical departure from pre-RIF days!

Curiously, and just before this trauma for all of us, I received a phone call on March 30, 1995, from Arthur O. Spaulding, President of the Thomas W. Dibblee, Jr., Geological Foundation, Santa Barbara, California. Art told me that the Foundation had selected me at its annual meeting to receive the Dibblee Medal, a national award for geologic mapping and field geology. The foundation was established to honor geologic mapping and to select a person each year to be honored for their quality and quantity of geologic maps. The foundation also existed to honor Tom Dibblee, an erstwhile USGS geologist and considered by many to be one of the most productive geologic mappers in the US, a man who at the time was eighty-three years old and lived in Santa Barbara. The foundation also had as a purpose to publish Tom's maps, primarily 7.5-minute (1:24,000 scale) geologic quadrangles in California. My application and résumé had been submitted, unbeknownst to me, by John Anderson. I was the second person to be so honored, after Lehi Hintze of BYU. A news release was issued on April 26, and John and I attended a ceremony at the American Association of Petroleum Geologists (AAPG) annual Pacific Section meeting in San Francisco on May 4. John, Art, and I gave speeches and attended a field trip led by Tom Dibblee. Dr. Dorothy (Dottie) L. Stout of Cypress College, in Cypress, California, a member of the Dibblee Foundation, wrote an article about the festivities that was later published. Doug Morton, Office Chief of Central Regional Geology from 1980 to 1984, also served as a member of the Foundation after he returned to southern California to map quadrangles; he doubtless was partly responsible for my getting the medal. This award was, of course, incredibly timely and doubtless part of the reason I was not RIFed; in fact, in August 1995 Gordie wrote me a nice personal memo of congratulations and to voice his support for geologic mapping. The next year the Foundation asked me to sit for the next two years on a three-person committee, with Lehi, to choose medalists. That first year (1999), my choice was Bart Ekren as Dibblee Medalist, and Lehi agreed; in the summer of 1999, I was there with speeches as Bart was given his medal. In 2000, we pushed Tom Steven for the medal, and in mid-June, I was back in California to help present the medal to Tom at another AAPG meetings. My term was up after that. The Foundation presented medals for about another decade, before Tom Dibblee and one of the principal workers in the Foundation died and the Foundation fell apart and ceased to exist.

Of course, if survival versus firing were the only story of the RIF, the USGS as an agency might have retained some form of integrity. But other changes were soon to come, and Gordie and Patrick were the presumed primary co-authors of these too. Among these changes was abolition of term limits for Branch and Office Chiefs, which in the past had been five years. That rule was for the benefit of excellent scientists, who were generally chosen for management jobs because it allowed them to return to the science they loved in a fixed time and at the same GS grade, once they'd paid their "dues" of administrative work. Of course, excellent scientists are not necessarily excellent leaders or administrators; some are and some are not, but at least those who are not will last no more than four or five years, too short to do lasting damage. Currently, however, excellent scientists are less likely to accept these jobs, yet failed scientists are likely to want them. So now we have many poor managers, who are a fixture that nobody can dislodge. And, of course, persons who could not cut it as scientists now aspired to these positions. Eventually the USGS became top-heavy with unproductive paper pushers. These people justify their existence by sending out streams of emails that say nothing, by pushing valueless meetings and committees, and by promoting workshops that accomplish little. After Dallas, each Director felt a great need to reorganize the USGS, probably because they were incapable of anything else.

Operation of projects was redrawn to remove much of the flexibility and incentive to publish. The funding for a project was now baked into the project design at the start of a project. In the past, it was the

Branch Chief's job to secure, maintain, or increase finances as conditions changed during the project's life and its success or failure was determined. In addition, since that time the financial support for the project includes a percentage that goes to the Office of Publications to pay for editing and publication, costs that a project chief previously did not have to worry about. This percentage currently is inadequate, so now editing a manuscript can take years, a major disincentive to publish. Publication costs mean that the final publication media, whether in-house reports/maps or as outside journal articles, had to be estimated at the outset of a project. No longer could a project chief determine publication type as the data results came in, nor could she encourage others peripherally related to the project (but making new noteworthy discoveries) to submit reports. Because the project chief now had to worry about costs of publications, she would have to make choices regarding whether field work and lab analyses were affordable, or whether publication was necessary at all. Furthermore, projects now had an ending date as well as a beginning date, and the sunset of a project could no longer be based on scientific findings. Furthermore, once her new project began, the scientist could not complete any final reports on her previous project. In science, much discovery comes as all the incoming lab and field data are assembled toward the end of the life of a project; some of those data, such as isotopic dates and geochemistry, take years to be determined, so they reach the project chief late. Then final reports are prepared. But now, with fixed sunsets and finances, projects had to be terminated, at times without allowing final conclusions to be released. These new rules are the kinds of things one would expect to be formulated by persons who were unproductive marginal scientists, which is exactly the types who now populate upper management! And all these changes came on top of more than two decades of Congress's inadequate funding of the USGS.

Another after-effect of the RIF was that scientists from other agencies were placed within the USGS, which was now to be the federal agency of science. Biologists from the National Park Service, the US Fish and Wildlife Service, the US Forest Service, and other agencies were moved administratively into the USGS. This had two results, presumably the intent of the Republicans: (1) places like the Park Service, which had a long record of research that created experts on bears, wolves, bison, etc., now lost much of their scientific credibility as an agency; and (2) housing all scientists in a single agency made it easier for politicians to starve scientific research to death simply by cutting funds.

Perhaps some small payback to Gingrich came on November 6, 1998, when he resigned his speakership in disgrace, after an election defeat amid an extramarital affair and eighty-four ethics violations, although he served no jail time for the latter. By then he had lost his popularity due in part to two government shutdowns (in 1994 and 1995) caused by budget impasses that the public blamed on him, and in part to a rise in Bill Clinton's popularity. Gingrich's legacy was the start of the culture wars, increased polarization of the electorate, Republican politicians' belief that their lies were justified, and a general undermining of democratic norms in government. For his part, Gordon Eaton preceded Gingrich out the door when Gordie resigned in 1997 over an attempt by officials in the Department of the Interior and by politicians to close the USGS office in Menlo Park, shortly after a new multi-million-dollar building there had been dedicated. The attempt was purported to save money for rent on the facility (Menlo Park is a high rent part of the country). At least Gordie went out fighting for a principle, showing he still had ethics! The Menlo Park center survived until 2020, when Trump closed it. Tom Casadevall, a former colleague from Mount St. Helens, became Acting Director when Gordie left.

Gordie's departure left the USGS in critical condition, but that illness was now exacerbated by a string of new directors with short terms of office. Because none of them was a career USGS employee, none of them had any idea of what the USGS used to be. Nor did they have leadership ability or moral authority. The directorship was just another stepping stone to something else in their futures. Each time a new director came in, the troops piled hope upon hope that this person would have some leadership ability, even credibility, that would right the ship. But it never happened. Most people who were picked as Directors were what I would call "collectors of administrative titles," who had spent their careers moving up by accumulating titles that made their resumés sound impressive to the uninformed and busy persons who sat on committees to appoint such people. Many of the new directors were unqualified, the only exception to this gross lack of credentials—as one might expect—was the initial person chosen by President Barack Obama, namely Marcia K. McNutt. But even she made no difference in the downward trajectory of the USGS as a credible science organization.

The first permanent Director after Gordie was Charles (Chip) G. Groat, born in 1940 and with a Ph.D. from the University of Texas at Austin. He served as USGS Director from 1998 to 2005. His previous position had been one of a series of two-year administrative positions at the University of Texas, El Paso. Chip, as noted above, had started his Ph.D. under Hoover Mackin as I was finishing up, but had had to switch to another advisor when Hoover died in 1968. When Chip came on as Director, I, like many others, was hopeful, and I wrote him a two-page memo congratulating him on his new assignment and describing the recent history, and even some pitfalls, in USGS management. I suggested he give me a call. He never did. He sequestered himself in his office on the seventh floor of the Reston office building, surrounded by his minions. All members of the USGS hoped he would show leadership and give them future hope on the direction their careers were heading, but he showed no such ability. John Sutter, who witnessed Chip's directorship, considered him a milquetoast who could not make decisions. He rarely got out to visit outlying offices, let alone to reassure employees that things would be all right after the disaster that had been inflicted. This lack of leadership, particularly in the face of new policies and uncertainties that stressed out all of us, was a major mistake. Nothing changed under his direction. When Chip left the USGS, Patrick Leahy took over in 2006 as Acting Director.

It is revealing to follow Chip's career after the USGS. He next went to the University of Texas at Austin as a professor and to direct the Energy and Earth Resource Graduate Program (another title that is less impressive than it sounds!). According to Wikipedia, the online encyclopedia (accessed in June 2019), Groat later led a study by the University of Texas Energy Institute on hydraulic fracturing ("fracking") for oil production, a practice then in the news because it commonly polluted groundwater, notably culinary water for small towns. Groat published a conclusion in early 2012 that "found no evidence that hydraulic fracturing...had contaminated shallow groundwater." Inasmuch as most of the University of Texas' and its alumni's wealth derived from oil and gas production, such a conclusion might be expected from a state as corrupt as Texas. Expected, perhaps, but hardly believable, based on virtually all objective studies of fracking. In July 2012, a public watchdog group discovered that while he was Director of the Energy and Earth Resource Graduate Program, Groat had received compensation from Plains Exploration and Production Company, an oil and gas producing company, and that he held 40,000 shares of stock in it. University of Texas university officials, who had not known this, commissioned an outside review and this forced Groat to retire in late November. In early December, the University of Texas issued statements on its website that announced (1) acceptance of the reviewers' conclusions that there were weaknesses in university ethics regulations, (2) the retirement of Groat, and (3) the immediate resignation of R.L. Orbach from his post as Director of the Energy Institute.

Things only became worse for the USGS when President George W. Bush, whose history of political appointments was indeed woeful, had his first chance to pick a USGS Director. He selected Mark D. Myers, who had a Ph.D. in geology from the University of Alaska, Fairbanks, then pursued a career in the Alaska oil and gas industry before he started accumulating titles. Apparently, Bush hoped that this person would favor drilling in the Arctic National Wildlife Refuge, but during Myers' term from 2006 to 2009, the issue never came up. Myers retreated into obscurity by retiring at the end of Bush's term as President. Furthermore, now USGS directorships had become regular political appointments that began and ended with presidential terms, something that had been scrupulously avoided in the past.

The aforementioned Obama appointee, Marcia K. McNutt, received a Ph.D. in geology and geophysics from the Scripps Institution of Oceanography. She was an expert in oceanography, including leading many ship expeditions when she was a professor at MIT, in collaboration with the Woods Hole Oceanography Institution, and she was subsequently elected to the National Academy of Sciences. Her term as USGS Director was from 2009 to 2013, during the first term of the Obama presidency. During that term, she was preoccupied with the environmental disaster known as the Deepwater Horizon oil spill in the Gulf of Mexico. Little else got done. During the remainder of Obama's presidency, from 2013 to 2017, Suzette M. Kimball was Director. She was a specialist in coastal processes, with a Ph.D. from the University of Virginia, but her career had been mostly in administration.

During the Trump presidency, from 2018 to 2021, James F. Reilly became Director. Reilly has a Ph.D. in geosciences from the University of Texas at Dallas. He was an oil geologist, then became a NASA astronaut, and participated in three space-shuttle missions; next he was an administrator at an online, for-profit university. When asked about climate change in his hearings for Director, he attempted to explain the view that uncertainties in climate science made it impossible to say anything definitive about climate change; this position is of course an utter lie. Under him, the USGS kept its collective head down, avoiding research on issues such as climate change—which it is supremely qualified to study—or other topics that might anger Trump and his oil-and-gas friends. Before the election of Joe Biden, Reilly thankfully left and David Applegate became Acting Director. Although he has a B.A. in geology from Yale University and a Ph.D. in geology from M.I.T., he has been a career manager, joining the USGS in 2004 to manage the geologic hazards response program dealing with earthquake and other hazards. The Senate confirmed Applegate as the eighteenth director on August 4, 2022.

In the past, new publications of the USGS contained many summary reports in which all geologists were interested, but little is published now except short reports that describe only a limited study, most of them in response to outside agencies that contracted the work and that required brief answers to their specific questions, rather than wanting regional scientific implications. Broad summary reports from the USGS are now rare, and there are far fewer journal articles by USGS authors than before 1995. Getting a manuscript through the USGS publication unit now takes years. The agency is underfunded and top-heavy with managers. It has become a silent nonentity, a bureaucracy.

1

The Pre-RIF USGS

What has the USGS lost because of Gingrich and Eaton? The agency, which in 2019 had a staff of about 8,700 people, was founded in 1879 to consolidate four geological surveys (the King, Hayden, Powell, and Wheeler surveys) that the US Congress authorized to explore the West and Great Plains, notably its geologic resources. One of these previous surveys was formally called the Geologic Exploration of the Fortieth Parallel Survey, led by Clarence King from 1867 to 1872 to study a huge area centered by the fortieth parallel between northern California and Wyoming (the fortieth parallel, or line of latitude, passes just north of Denver and Reno). King agreed to be the first Director of the USGS, but only for twenty months to get it started, then his hand-picked successor was John Wesley Powell, the one-armed explorer of the Colorado River/Grand Canyon and a brilliant renaissance man and writer who served as Director from 1881 to 1894. The agency, from its origin as a vehicle for exploration, has had a proud tradition and was led by a series of extraordinarily accomplished public servants. It has had its ups and downs with Congress, which has always been made up of politicians who did not understand science, and funding was rarely sufficient. When Ed Ruppel was Branch Chief, he had a framed print on his office wall of the Frederick Remington painting “A dash for the timber” that showed a dozen US cavalry men galloping on horseback to reach protection (presumably a grove of trees) while firing guns back at a large mass of horse-bound Indians in pursuit. Ed explained that the Indians were the US Congress and we (the Branch or the Geologic Division) were the cavalry hoping to survive another day!

The USGS that I joined in 1970 had three divisions: the Geologic Division, the Water Resources Division, and the Topographic Division. The Geologic Division (GD) was the largest and provided the scientific and administrative leadership. The Water Resources Division (WRD) was like a younger brother that got little attention, for its work was more nuts-and-bolts (that is, practical) studies of surface and groundwater resources and much less on the academic study of these topics. The Topographic Division consisted of engineers who put together and updated the topographic maps of the US, until about the 1970s when funding for updating was largely cut. The hydrologists of the Water Resources Division needed geologists of equivalent talent to help them assess water resources, but the Geologic Division had become more of an ivy-league college whose geologists generally turned up their nose at working for the WRD in doing mundane water-resource reports. Therefore, the GD and the WRD rarely worked together on projects, and the WRD necessarily had to go out and hire geologists, generally picking cheap young persons with little more training than a bachelor's degree. This resulted in the geology in WRD reports being given short shrift, meaning the overall hydrologic analysis was somewhat lacking. Later work on the Nevada Test Site (next chapter) reversed this trend, but that close collaboration largely ended with our retirement.

In the GD and WRD, promotions were based on quality and quantity of publications. The WRD reports typically concerned monitoring the volume and quality of surface water and groundwater, of great practical use to farmers, ranchers, and developers. Water resources are not particularly amenable to academic studies, although knowledge of where groundwater is coming from and where it is going is critical to its assessment. In the GD, academic studies ("pure science," as contrasted in the next chapter) dealt with many things such as plate tectonics, paleontology, metamorphic and igneous rocks, and many other topics. Of course, GD also did, like WRD, practical scientific studies. These included the geology of mining districts and generally how ore deposits formed. For mining geologists in private industry did not have the mandate to do much more than find mines and do local exploration, mostly through drilling, in the producing part of a mining district, so the regional aspects and reason for the mineralization was a task taken up by USGS geologists. Similarly practical work was done by branches that focused on oil and gas resources, coal, uranium, and geothermal energy resources. And little has more practical value than geologic mapping, which our regional geology branches concentrated on; quadrangle mapping, for example, not only found and explained resources of all types but it also identified geologic hazards, assisted engineering solutions for dams, roads, etc., and it provided solutions to a wide variety of environmental problems. Whether about academic or practical topics, the USGS routinely issued large in-house summary reports as USGS Professional Papers, Bulletins, and geologic maps. Academic goodies were issued as short outside papers in various journals. Scientists did not have to worry about funding details, which was largely the job of Branch Chiefs. Peer review and editing took only a few months because they were recognized as a high priority in the publishing process, and the USGS then understood the public's need to know its conclusions rapidly. Productivity was high in the old USGS.

The USGS also had a program of international geology, in which countries paid the USGS to send geologists, geochemists, and geophysicists to study large areas, primarily looking for metallic mineral resources. The Saudi Arabian government provided significant support for one such long-term program, and many mineral-rich South American countries had cooperative studies with us. Branches of Marine Geology,

Engineering Geology (it included studies of active volcanoes and active landslides, aswell as roads, bridges, dams, and hazards), Special Projects (mostly the Nevada Test Site, supported by the Department of Energy), and planetary geology (Astrogeology), had moderate staffs. And a significant effort by several branches focused on seismic faults and tectonic studies that explained earthquakes, with an eye to pinpointing hazard areas and with the hope (not yet attained) of predicting earthquakes.

The atmosphere in the GD and WRD was collegial, with any hierarchy downplayed and all people, from newly hired to nearly ready to retire, on a first name basis with each other. The only hierarchy was in the GS grade attained, and branch chiefs or above were looked upon as friends, not bosses. The main jobs of branch chiefs, in fact, was to find funds for, and be supportive of, their geologists, even to hire some if a rare "slot" came along. Project chiefs ran their own projects and wrote their reports for publication with no guidance from branch chiefs or other higher managers, whose job was to make sure the geologists got what they needed to be productive. Besides, any current branch chief knew that, under the five-year term limit on his position, he might soon be working for some geologist currently junior to him. This rotational method of management had the added advantage that a manager did not ossify in a job that could have lasted far more than five years, and new people with fresh ideas and energies were brought in. Many persons in the branch had already attained the highest grade they could get, the GS-15, based on science, and most scientists felt there was no reason to aspire to a branch chief job when they already had a dream job of just doing great science! (That was my case after getting the GS-15 in 1981.) Therefore, we all considered ourselves equals to each other and to branch administrators, in branches that had important and common goals.

Interestingly, unless you were a close colleague or friend, you typically had no knowledge of the alma mater, degrees attained, or pre-USGS experience of other USGS employees. Nobody addressed each other as "Dr.," either because one did not know or did not care. In all correspondence or if a person were meeting another for the first time, it was only "Mr." or "Ms." for all. In 1972, however, the USGS Pick and Hammer Club, an informal club that put on yearly skits that poked fun at others, especially administrators, published a ninety-six-page directory of scientific personnel of the Geologic Division. It included a small head shot and brief résumé, including degrees, of each person, at sixteen persons to a page. Allen V. Heyl, a geologist with Central Mineral Resources Branch, out of curiosity compared the GS grade attained to those whose highest degree was a B.A./B.S., M.A./M.S., or Ph.D. and found, perhaps surprisingly, that our highest GS grades were roughly equally distributed among those three degrees. In other words, the cream rose to the top of the milk regardless of whether a geologist had a terminal degree or not!

All of us loved that thought. In 1981, the American Geological Institute (AGI) celebrated the hundredth anniversary of the USGS (1879–1979) with a similar directory.

Before 1995, USGS authors dominated in most journals and books that published geologic studies, and in-house reports such as USGS Professional Papers (the report looked upon as the most authoritative) and Bulletins were prolific. Most of the in-house reports were final summaries—a task that included comparing one's area, rocks, and structures with those in other nearby areas or even in other parts of the world. Therefore, the USGS spearheaded many advancements in the geologic sciences, probably comparable in numbers and quality to all the geology departments in the country. In fact, new hires to GD were among the best geologists coming out of universities, for working with the USGS was almost as prestigious as joining the staff of a major university. Virtually all of us loved working for the USGS and had great pride in what we were doing. Each month, a large list came out giving new (in-house) publications of the USGS, and we looked through it not only with pride but, because those publications were free to employees, we always found reports that related to work we were doing and we ordered hard copies. Finally, it is worth noting that salaries of USGS geologists were comparable to professors in a geology department. Such camaraderie and productivity led to credentials that few other federal government agencies possessed—the "improbable USGS" as some termed us.

Such successes by the Geologic Division had the adverse effect that geologists considered themselves to be analogous to professors in leading geology departments. Thus, some of our attention strayed a little too far into academic issues that had less immediate value to the taxpayer. Members of the GD used outside (meaning non-USGS) journal articles to get wider distribution to the academic community, whereas nuts and bolts issues and local problems, published by in-house reports, became less popular among USGS geologists. That contrasts with the good state geological surveys, which devote themselves to the practical. As discussed below, I worked for the Utah Geological Survey (UGS) and consider it the best state geological survey in the country. Every scientific report that the UGS publishes has value to the taxpayer; these include identification of mineral and energy resources, geologic mapping, locations of flood hazards, groundwater resources and contamination, and discussions of geologic hazards such as seismicity, landslides, and others. Moreover, UGS in-house reports make a more concerted effort to reach Utah citizens than did those of the USGS by including many more photos and using less technical language. An especially effective method to reach taxpayers is their free quarterly, glossy color brochure (*Survey Notes*) that is aimed at Utah citizens and includes excellent photos. Each issue begins with an introduction by the UGS Director, with her/his photo and summary of what issues the UGS is tackling, and why. It also includes invitations to field trips to collect rocks and minerals, answers to questions about rocks or structures the public might ask, details on new dinosaur finds, and conclusions about new UGS releases.

When I joined the USGS in 1970, William T. Pecora, born in 1913 and having attained a Ph.D. from Harvard University, was nearing the end of his directorship (1965 to 1971). Like most directors before him, Bill was a career USGS employee; he had specialized in rare-earth minerals before moving into management. During his term as director, he created the National Earthquake Research Center in Menlo Park and the Landsat series of Earth-observing satellites, both practical measures that became widely successful. Bill was well liked and personable. He was succeeded by Vincent E. McKelvey (1971 to 1977), who had a Ph.D. from the University of Wisconsin and became a well-known USGS geologist specializing in deep-sea mineral deposits. Henry W. Menard, with a Ph.D. from Harvard, was a famous marine geologist who had advocated plate tectonics before it became popular. He had not previously worked for of the USGS, so was less effective than those who had. Instead, he came from Scripps Institution of Oceanography. His term (1978 to 1981) largely coincided with that of the Jimmy Carter presidency, one of mixed success, and Menard elected to step down and return to Scripps when Carter was defeated for re-election. Menard presided over the 1980 eruptions of Mount St. Helens, arguably the high point of public perception of the USGS. Yet immediately following the eruptions, scuttlebutt indicated that Menard could have used that success to push new policies and bring in new funding for the USGS, but during his visits to the Congress, it was said that he came unprepared in his testimony and that consequently only modest increases were granted.

Dallas L. Peck was another director with a Harvard Ph.D. who was a career USGS employee. He had specialized in volcanic rocks in Hawaii and plutonic rocks in the Sierra Nevada. His term (1981 to 1993) was a lengthy one of consistent, continued success, with expanded USGS projects in mineral resources, geologic mapping, hydrology, and global change. During his term, the USGS worked more closely with state geological surveys than before. Not only was Dallas an effective manager, he also was well liked and supportive and loyal to his scientists. When Dallas retired and Gordie Eaton, another (mostly) career employee who was likeable and accomplished, was chosen, all of us had high hopes for the new Director and a bright and prosperous future for the USGS. Those hopes were spectacularly dashed by Newt Gingrich, as I explained earlier.

So what was the high point in public perception and value of the USGS? Certainly, it could have been early in the history of the USGS, when geologic and topographic exploration resulted in new mines, newly identified mountain ranges, and other exciting finds on the western frontier. At the time of the four surveys that became the USGS, the Great Basin was largely unexplored, the last unknown hole in The West. Nonetheless, my preference for the high point of the USGS is more recent: during and after the eruption of Mount St. Helens. Mount St. Helens was the biggest news story of 1980, and the USGS did almost everything right during the eruptions, including saving many lives. Furthermore, Rocky Crandell and Don Mullineaux provided scientific credibility with their 1975 and 1978 predictions of this eruption.

Next came the eruption of Mount Pinatubo, a stratovolcano of nearly six thousand feet elevation on the island of Luzon, the largest and most populous island in the northern Philippines. About one million people lived near the volcano, with Manila about fifty miles southeast of the volcano, the then-leased US Clark Air Base about fifteen miles east of the volcano, and the then-leased US Subic Bay Naval Station about twenty-five miles to the south-southwest. The eruption started with small earthquakes in March 1991, followed by a powerful phreatic (steam) explosion on April 2, which opened three craters on the northeastern flank of the volcano. Although the volcano had not erupted for about five hundred years, until then the hypothesized recurrence interval had been closer to a thousand years. Ray Punongbayan, then head of the Philippine Institute of Volcanology and Seismology, contacted volcanologist Chris Newhall of the USGS on April 7 to ask for help, through the USGS Volcano Disaster Assistance Program. John Ewert, the USGS volcanologist on this team, arrived in the Philippines on May 23, and others soon followed to work with the Institute to set up monitoring stations, with aid from Air Force helicopters. The work was frenetic, for it included not only setting up a network of seismometers and surveys by helicopter-mounted spectrometers to gather emissions data but also the field identification of pyroclastic flows and then their recurrence intervals. The USGS scientists included many of the same people, including Rick Hoblitt, who had worked at Mount St. Helens. Gas studies run by Tom Casadevall, some using the concepts applied by the late David Johnston, were particularly effective. The Institute forecast an eruption and warned 25,000 people living or working near the volcano to evacuate. These included ships, planes, and other military equipment at the two US bases. The first magma was erupted on June 7 in the form of small flows, forming a volcanic dome, which continued until June 12, when a large eruption took place, with an ash cloud rising to 65,000 feet. As seismicity increased, Philippine officials convinced more than 65,000 people to evacuate; Clark Air Base, for example, evacuated most of its fourteen thousand employees on June 14. On June 15, a cataclysmic eruption of pyroclastic flows (ash-flow tuff) took place, with an eruption cloud rising to a whopping 115,000 feet, much higher than at Mount St. Helens.

Smaller eruptions continued until September. Mudflows triggered by spring rains and a typhoon contributed to the devastation. Between three hundred and four hundred people were killed, mostly from roofs collapsing under the weight of ash. However, conservative estimates were that between five and twenty thousand lives were saved and $250 million in property from the US bases were saved by that evacuation. The volume of juvenile material was estimated to be a cubic mile, about ten times the amount of magma erupted in various forms at Mount St. Helens. The lessons learned at Mount St. Helens turned out to be directly applicable to Mount Pinatubo. Not enough credit was ever given to the USGS for their immediate attention to the problem and for their aid and advice to the Philippine Institute. Yet geologists throughout the world were very much aware of this feat! I therefore conclude that the high point of the USGS was on the watch of Directors Menard and Peck, but that the short term of the former and the long term of the latter argue that the best days were largely due to Dallas.

Regarding excellence in geology, it is worth noting that there is no Nobel prize in geology. The Nobel prize, considered the most important award in science and announced with fanfare each year since the turn of the twentieth century, instead is for only physics, chemistry, medicine, peace, and literature; economics was added in the 1960s. Of course, peace and literature are not science, and many legitimate sciences are not represented. If geology were represented, I consider that the life work of Walt Alvarez and of Warren Hamilton, two geologists I have known, merited it; the combined careers of Pete Lipman and Tom Steven also would have been considered. And others before them who discovered or understood what became known as plate tectonics also would be worthy. All of us in geology have contemplated high awards and have our own favorite heroes. Geology has some superstars, just like in the fields that were selected by Mr. Nobel! And many are women.

CHAPTER 9

As a result of the RIF, I was moved into management. The first entries in my field notebook for 1995 were the details of a field trip that Gary L. Dixon and I led on May 9 and 10 for a half dozen officials of the Lake Mead National Recreation Area, then the fourth largest in terms of visitors (then eleven million per year, engaged largely in swimming, boating, and fishing) in the National Park Service (NPS). Lake Mead, 112 miles long and stretching across Arizona and Nevada, is the largest man-made reservoir in the US in terms of water capacity. It was

RELEVANCE IN SCIENCE:

GARY DIXON AND THE NEVADA TEST SITE

created by backing up the Colorado River by the 1935 construction of the Hoover Dam, the greatest engineering marvel of the day (during the Great Depression) just east of Boulder City and about thirty miles southeast of Las Vegas. However, its water level was declining (as of late 2023, its water level was at only about 34 percent of capacity, a fact of great significance to Las Vegas, and later to Gary and me. In 1995, the NPS was interested in the geologic controls of springs in the recreation area and in putting together signs and texts for the public about their geology and scenery. We described for them the geology and hydrology of the eastern part of the Las Vegas basin; then we took a NPS speedboat east on Lake Mead to describe the geology of the shoreline. My career had just been turned ninety degrees in part by a USGS geologist, Gary Dixon, whom I barely knew before 1995. And, of course, he did most of the instruction on this field trip, for though I was a field geologist/mapper like him, he knew the area. Gary was an expert on the geology and hydrogeology of Nevada, especially southern Nevada. I buckled down under his tutelage over the next decade and learned everything from him on these topics. The main lesson learned was the importance of science that is directly relevant to the needs of the taxpayer—which is how this project had always been run. For me perhaps that meant (as in the Rolling Stones song) that "You Can't Always Get What You Want" (mapping calderas) but perhaps "you'll get what you need" (selling useful science).

Geologists and administrators of the USGS, as with other scientists and administrators, have had a constant debate on the relative merits of a "mission-oriented" (also termed applied, relevant, directed, or societal) approach to doing science versus a "basic or pure science" approach. "Mission-oriented" science is widely thought to be narrowly focused ("directed") research aimed at solving specific problems, whereas "basic science" is an investigation to unravel general principles. The discussion in the Geologic Division (GD) has always come out in favor of basic-science studies, which are thought to develop a plateau of knowledge that then can be applied to address resources, environmental aspects, and other practical problems. Thus, most GD studies have attempted basic science, and those geologists who were tasked with doing mission-oriented science were considered the dullards of the organization. Most of the work of Water Resources Division (WRD) would be considered mission oriented, one reason why geologists in GD turned their noses up at it. The conclusion had always been that mission-oriented studies do not provide systematic analysis and synthesis, but instead depend upon fundamental geologic frameworks developed under "pure" research. Most USGS mission-oriented studies, in fact, are supported by outside funding agencies that pay for the USGS to solve problems that concerned them, and such funds may come with tight deadlines (generally a year), therefore negating any time to do general summary reports and significant analysis. In contrast, internal USGS funds historically paid for pure research. The Nevada Test Site was considered the classic mission-oriented project in the USGS. And because of that, scientists tended to denigrate it. Gary, who spent most of his career working on the NTS, was personally put down for his long-time involvement with mission-oriented work. Those who did not know him well cited his lack of a Ph.D. as a reason for disparaging him and his work. Of course, his work, by bringing in outside money, paid his salary and expenses many times over, making his career free to the USGS. Now, suddenly, I was put into this second-class-citizen status in the eyes of others. That did not bother me at all, largely because I enjoyed doing practical work that more directly aided the taxpayer. Of course, working in mining districts and doing geological mapping are two of the most practical things a geologist can do, and both of us did those, plus the "pure" science that those activities revealed.

Gary was to become my closest colleague and best friend for the next twenty-two years, until his death from cancer in 2017. He seemed like a big brother to me, although he was one-year younger. Gary was the sole geologist (Geologic Division) in the Las Vegas office of Water Resources Division, composed of several dozen hydrologists with whom he worked closely during most of his career. His role became that of a first-rate geologist who provided much of the geologic expertise and wisdom for his hydrologist colleagues, one of the rare instances of close cooperation between GD and WRD. He lived with his wife Wendy, an attractive, high-energy person, on several acres that they shared with two horses in (then) quasi-rural south Las Vegas. Their home and neighborhood, about a mile south of the end of the north-south Las Vegas Strip, soon (in the short time I lived there) became suburbanized by the growth of Las Vegas. Wendy had a magnificent career of her own, as a leader and manager, with a GS-16 grade, for the US Department of Energy (DOE), where she ended up supervising the Environmental Impact Statement for the future(?) Yucca Mountain Nuclear Waste Repository northwest of Las Vegas, a stressful job that took many years and filled volume after volume. She was a world-class dressage rider, and she alleviated the pressures of her Department of Energy job by riding in her arena, competing in many horse shows outside of Las Vegas, and teaching dressage. Wendy also was a symbol and role model for women's leadership and managerial competence at the male-dominated DOE. She brought other up-and-coming women managers in to work with her and inspired them, including one named Dawna Ferris, who later became my wife! Gary and Wendy were examples of cream rising to the top by dint of hard work and brains.

As had happened to me, so too the RIF brought Gary's previous project to an abrupt dead end. He had been enjoying a brief respite from his NTS work by mapping in southern California with Doug Morton (more on him later). Now, under the RIF reorganization, a return to his more practical ("societal") work would seem to be the ticket for the new USGS. For the last decade Las Vegas had been by far the fastest growing city in the country, and large Clark County in which it resided had become the fastest growing county. Las Vegas in 1992 was the second most popular tourist destination in North America (6th most visited city in 2019, with 2.9 million tourists). Las Vegas (Spanish for "the meadows") itself began as a stop on the Spanish Trail in the early 1800s because large springs in the middle of Las Vegas Valley had resulted in desert grasses for horses, mules, and driven cattle. As of 2020, greater Las Vegas has a population of 2.2 million people. The sprawl of homes now largely fills Las Vegas Valley, a broad alluvial basin that ends northward against mountains raised by the east-west Las Vegas Valley shear zone. The beautiful Spring Mountains on the west side of the valley has peaks (Mount Charleston) that rise to almost twelve thousand feet and supply springs on all sides.

But like most cities, it has environmental problems such as toxic waste and organic effluent in Las Vegas Wash, draining from Las Vegas into Lake Mead. And the water level of its main culinary source of Lake Mead had been dropping for some time, leaving obvious broad, white bath-tub rings above the lake level. Where culinary water had formerly been pumped, from the springs at "the meadows," the land had subsided due to groundwater withdrawal. Landslides, seismicity, flood hazards, and collapsing soils made up the remaining problems. Gary therefore proposed a new project called the Las Vegas Urban Corridor, managed out of Western Regional Geology Branch in Menlo Park (the RIF reorganization proposed new names for branches and branch chiefs, but for simplicity's sake I will retain the old names). Larry Snee, John Sutter, Gary, and Doug Morton had hijacked me administratively from Central Regional to Western Regional, doubtless to save me. I retained my Denver office, however. I became the unofficial co-project chief with Gary. Project products were to be the geologic maps of the 1:100,000-scale Las Vegas and Lake Mead quadrangles, as well as scattered 7.5-minute quadrangles and reports. Most of the 7.5-minute quadrangles were to be paid by outside funds, which Gary was an expert at getting, by water districts (Las Vegas and Mesquite) which required new well fields for culinary groundwater.

The project began in March 1995 and lasted through September 1998. Our job was to be managers only, but we allowed ourselves significant time out of the office to do mapping. For the first year, the job consisted of meetings and field trips with officials to inform agencies of what we were doing, to propose cooperative work, and to seek funds. We visited with members of the Nevada Bureau of Mines & Geology, the Southern Nevada Water Authority, Virgin Valley (Mesquite) Water District, NPS, University of Nevada, Las Vegas, and others. We employed between four and twelve geologists to do the work. We moved Ric Page into the project to head the Las Vegas quadrangle. We brought in Mel Kuntz on soft money, after he was reinstated to the USGS, to do the Lake Mead quadrangle. Some mapping and a report on groundwater flow in Coyote Spring Valley, northeast of Las Vegas, and funded by Southern Nevada Water Authority, were delinquent, so we took over several tasks, wrote the reports and maps, and shepherded them to publication (as required in contracts for outside funds). In the Mesquite area halfway between Las Vegas and St. George, the Virgin Valley Water District funded several 7.5-minute quadrangles, which we used to site big water wells by subsequently drilling some of the faults that were mapped.

> *Among the trees, so to speak, that would bear future fruit for many people was assisting with the Ph.D. field work of* ***David B. Hacker****, a Kent State grad student of John Anderson's, whom John had had to leave behind when he retired. Starting in about 1994, Dave had picked a study area that required mapping the Pine Valley Mountains, south of the Iron Springs mining district, north of St. George, and immediately west of New Harmony. It was, for me, a continuation southward of the long-term studies that Hoover Mackin and I had done at Iron Springs. As Hoover and Earl Cook discovered, the Pine Valley Mountains is underlain by the world's largest laccolith, of about the same age (20–22 Ma) as the other laccoliths at Iron Springs. Emplacement of the laccoliths was rapid and resulted in mountains. The rocks of these mountains*

> *included older (Cretaceous) shales that were weak and unstable, resulting in large gravity slides off the high flanks of the laccoliths. Hoover and Earl had recognized the slides, but most aspects were unknown because the range had not been mapped in detail. Some slides moved nearly 10 miles to the east into the Colorado Plateau, where Robert F. Biek, a particularly gifted geologist with the Utah Geological Survey, was mapping quadrangles, and the two met. I mentored Dave in his Pine Valley dissertation and accompanied him on some of his traverses, witnessing a field geologist of great talent. He hiked deliberately, almost slowly on his traverses, but he took everything in and seemed never to have to go back to look at things again, a trait that I unfortunately did not possess. Dave was easy going, immediately likeable, a good husband (his wife Jackie was a live wire and a businesswoman) and father, and funny. His dissertation and Ph.D. were completed in 1998. His study of gravity slides was ground breaking, and this expertise would be applied in later work with Bob Biek and me, as we took up the investigation, also after John Anderson, of the Markagunt Megabreccia (see next chapter). In 2000, Kent State University (Warren campus) hired Dave as Assistant Professor. Hiring somebody to your staff who earned their Ph.D. from the same place is not common in academia, but the faculty in the Geology Department recognized what they had in Dave and refused to let him get away. Dave was another in the Hoover Mackin and John Anderson line of terrific teachers. He runs the Kent State field camp in the Black Hills, as John Anderson had done. He rose through the ranks quickly and became full professor in 2018. The story of Dave Hacker and Bob Biek will continue below.*

In our second year in Las Vegas, in March 1996, things took a drastic turn and our work load more than doubled. Larry Snee in early 1996 became Branch Chief of my former Central Regional Geology, where my office was. As Branch Chief, he was a source of great strength for the whole branch during a chaotic and stressful period after the RIF. He displayed a ready laugh in a managerial era when any lightheartedness was sorely needed. By then, all projects formerly done by Special Projects Branch had been transferred to Engineering Geology Branch, then to Central Regional Geology. And that month Larry received an angry phone call from the US Department of Energy (DOE) manager in charge of the Nevada Test Site. In no uncertain terms, this guy (a martinet, I was to learn later) told him that he had finally reached the end of his patience over delinquent reports, and that management of the Nevada Test Site had to change! Or funding would go away! The buck stopped with Larry!

The Nevada Test Site (now called the Nevada National Security Site) is a restricted area of 1,360 square miles, slightly larger than the state of Rhode Island, located about sixty-five miles northwest of Las Vegas, in the southern Great Basin. It was carved out of a larger area—the Nellis

Air Force Bombing and Gunnery Range (now called the Nevada Test and Training Range)— run by the Department of Defense and extending north, east, and west of the NTS. Managed by DOE, the NTS was established in 1951 to test nuclear weapons. Much of the work there is run out of Mercury, a small government town in the southern NTS that then was a bustling place. From 1951 to 1962, about one hundred atmospheric tests were exploded, and the mushroom clouds that arose could be seen for a hundred miles. One of the last tests was the Sedan shot in 1962, intended to show that nuclear weapons could be used for peaceful purposes, such as creating bays and canals. Its crater is 1,300 feet wide and 300 feet deep, and one can drive up to its edge to a rickety viewing stand. However, it was soon recognized that nuclear fallout caused cancer, and westerly winds carried this fallout into the small city of St. George and many small towns, most of them in southern Utah; most fallout missed Las Vegas. For those areas where the fallout did not miss, cancer rates eventually became high, although it took years for this to become obvious; some families were decimated in the 1950s. Above-ground tests were curtailed as such ill effects became known, but as the Cold War continued, below-ground tests now became all the rage, numbering more than eight hundred before they ceased in 1992, by treaty. Yucca Flat, in the northeastern NTS, saw most of the above- and below-ground tests, and an aerial view of this broad basin shows that it is littered with dozens of small collapse craters from below-ground tests that were too close to the surface. Yucca Flat and many other places are sprinkled with debris from many tests, including the remains of buildings spaced at intervals from some above-ground shots to test and photograph shock waves and the scope of destruction. The craters and debris make the NTS a surreal place. Because of his long history of working at and managing the NTS, Gary would regale us with stories about DOE and military employees who thought up what to test, and where. These people were like kids in a free fireworks store who loaded up with the most powerful firecrackers (i.e., nuclear bombs) they had, then conjured up places to put them, including in vertical or horizontal shafts in bedrock.

The Yucca Mountain Nuclear Waste Repository, another DOE facility, is in the southwestern corner of the NTS and about eighty miles northwest of Las Vegas. Here, under a ridge named Yucca Mountain, a huge boring machine cut long, large-diameter, horizontal underground tunnels into Miocene ash-flow tuff. These tunnels were to store spent nuclear fuel from nuclear power plants nationwide, as well as other nuclear waste. But whether to put the facility into operation became a political football in Nevada. It always seemed to me that there were more pros than cons in moving waste here, and the DOE spent billions of dollars on the facility. Democrats such as Senator Harry Reid, and some Republicans, advanced their careers by opposing Yucca Mountain because, they claimed, spills of radionuclides would be inevitable, endangering Nevada residents. In 2011, perhaps as a gesture to Harry, President Barack Obama ceased all funding to the facility. Nuclear power plants around the country now resort to on-site wet or dry storage of nuclear waste in steel and concrete casks, a solution that would seem to invite disasters in the future, although likely years beyond the careers of the politicians who made such decisions.

The NTS and Yucca Mountain were selected because of their inaccessibility and their dry Mojave Desert environment. Nevada is the seventh largest state in the US and thirty-second in population, although the NTS was established when the state was the least populous in the country (Las Vegas was then only a small town). A particularly favorable feature of the NTS and Yucca Mountain is a water table that is commonly one to two thousand feet below the surface. Most below-ground nuclear tests were intentionally sited above the water table to avoid contaminating groundwater, and similarly waste storage at Yucca Mountain was to have been above any water table. The thick outflow and intracaldera ash-flow tuffs on both sites are good places to put tests and wastes. But, some north-south basin-range faults were potentially active. Cross sections, geophysics, and drilling were needed, the job of Special Projects Branch and its successors. In addition, nuclear particles from the tests will work their way downward to the water table. This water surface marks the top of a groundwater system that is moving southward to southwestward down the hydraulic gradient (generally parallel to the topographic slope), from the mountains of central Nevada where groundwater is recharged, to the springs in low basins in southern Nevada such as Death Valley and Oasis Valley where groundwater is discharged at springs. The USGS was asked whether this groundwater carried plutonium or other pollutants off the NTS.

The scientific studies of the NTS were by two large USGS projects, one by WRD and the other by GD, both funded by DOE. The larger of the two projects was a well-run WRD project, based out of the Las Vegas field office and employing most of its hydrologists—a project that cost more than $2 million per year. At the end of each summer, the manager of that project would submit a thick funding proposal to DOE. In return for acceptance of the proposal and payment, he would produce at least a dozen/year USGS published reports of the conclusions, known as "deliverables," that were due by the end (October 1) of the next fiscal year. The smaller GD project, totaling somewhere around $800,000 in 1996, was poorly run and its deliverables were overdue, as noted by the DOE manager when he called Larry Snee. The DOE manager who called Larry Snee told him that eight reports were long delinquent, and if something were not now done about that, DOE funding would be cut immediately to no more than $600,000 a year. The next people to whom Larry talked were Gary and me. And now I was to take over as the next NTS project chief for GD.

The division of labor between Gary and me now became for Gary to be the manager of record for the Las Vegas urban corridor project, run out of Western Regional Geology, and for me to be the manager of record for the NTS project for GD, run out of Central Regional Geology. As it worked out, however, we ran both projects together as co-project chiefs. In addition to becoming fast friends, we each brought different skills to the table as managers, and those skills meshed nicely. Gary was a natural leader and a master of oral communication, whereas I was good at written communication and producing deliverables. Larry was particularly incentivized in proposing this arrangement because the NTS project was a money-maker, and the Branch that ran it got a cut of that soft money; Larry still had an Argon-dating lab that needed funds!

The first order of business was to change the culture of indifference to one that respected deadlines. One excuse routinely given by those who could not produce in a one-year time frame was that "our science is too important to have a one-year deadline." After we countered that argument (it included some profanity), we concentrated on the backlog of deliverables. At the time, the Department of the Interior had a policy that allowed managers to insist that employees either do their job or be fired. I notified three employees that they had until the end of the fiscal year or I would terminate them. One of these three retired, and the others buckled down. Other deliverables were picked up by other staff whom we hired, or we did them ourselves. All were done by September 30, and consequently in the next fiscal year our proposed budget of $1.3 million was accepted. For the fiscal year after that, DOE accepted our proposals for $2.1 million, plus another $300,000 from the Yucca Mountain DOE program for the first of two summary geologic maps (and still another $300K the following year for the other map) on the Death Valley groundwater flow system. We supervised between nine and twenty-seven full-time or part-time USGS geologists for the NTS and Yucca Mountain work.

For another year, I therefore continued to fly in to Las Vegas for a week or so of meetings, field trips, and field work, and Gary would pick me up, we would do the work, and Gary would return me to the airport for Denver. I was loathe to give up my wonderful home in the mountains, and in addition Las Vegas was hardly appealing as a quiet, cool, thoughtful environment. But eventually, in May of 1997, I pulled the plug, sold that home, and moved to Las Vegas.

The NTS project allowed Larry Snee to hire eight geologists full time that he would have had to fire otherwise. One of the persons Larry hired back was Gary P. Landis, a brilliant geologist who had been fired in a typical act of lunacy characteristic of the RIF. Larry was able to bring in a mass spectrometer from Reston, no longer being used there because people in the Isotope Branch had been fired, and Gary was able to use it on some fundamental climate science using rare gases, mostly neon, helium, and xenon, that were collected from gas inclusions in amber and in ice cores that penetrated glacial ice that dated back (the deeper the core) to Pliocene time (amber dated back to Cretaceous). These gases allowed predictions of ancient climates up to pre-historic climates. Gary Landis ended up doing some of the best climate-change science in the USGS. Unfortunately, poor leadership has hamstrung the work, and the USGS has never been a leader in that topic.

> ***Gary Dixon*** *was born in Salt Lake City, graduated from East High School in Denver in 1961 and attended Western State College in Gunnison, Colorado from 1962 to June 1965, majoring in geology and history. The geology department at Western State was small but the professors were excellent, and Gary received the extra attention he likely would not have gotten at a large department. But as he worked his way through college, in 1965 he found a summer job as a temporary Physical Science Technician with the Special Projects Branch. As part of the Cold War, the Branch engaged in intense research to site underground nuclear bomb tests and to identify storage sites for radioactive wastes. Gary was soon overseeing the drilling of pre-test wells in which to place nuclear bombs and post-test wells to appraise the effects of the tests; his specialties became geologic mapping, hydrogeology, and volcanic geology. His budding leadership skills and ability to get along with everybody attracted the attention of his geologist supervisors, notably Bart Ekren, who soon became a mentor. Bart persuaded Gary to return to Western State to finish his degree, which Gary did from June 1967 to March 1969, and he received his B.A. in 1969.*
>
> *Gary returned to work, was promoted to geologist in 1970, and then began mapping 7.5-minute quadrangles, mostly in calderas and their outflow tuffs on the NTS. The initial mapping of the NTS attracted others who were among the best in the USGS, including Bart, Ernie Anderson, Bob Bucknam, Ken Sargent, Pete Lipman, and others. At the time, Yucca Mountain had not yet been chosen as the highest priority for a waste repository, so Bart and Gary also mapped calderas in northern Nye County, central Nevada, where one nuclear test was done. At about the same time, the USGS also decided to map all Nevada counties at reconnaissance scale (1:250,000). Bart and Gary led their study of the Cenozoic rocks of Lincoln County, in eastern Nevada, an important area for mineral deposits. Most maps and reports were on tight deadlines, for Bart and Gary were also siting nuclear tests. Gary soon had a wife, daughter, and son in Denver, so he never returned to school to chase advanced degrees. Perhaps it was the press of all these activities, but sadly his marriage failed. A lack of higher*

> *degrees did not matter then (it does now, unfortunately), for he soon moved up the promotions ladder. By 1973, he was supervising much of the work on nuclear tests and post-test analysis, and by 1977 he was the coordinator for many waste-management programs on the NTS. He was particularly good at applying hydrology to his work, and he worked closely with hydrologists of WRD. By 1981, he was the Acting Branch Chief of the Special Projects Branch, which then consisted of about forty-five geologists. But the Cold War was waning and the Special Projects Branch was disbanded as funding decreased, and besides most quadrangles had been mapped by then. He became program manager for the Radioactive Waste Disposal Program, moved from Denver to Las Vegas in 1983, and supervised seventy-five people and all activities by the GD on the NTS. He also was the Deputy Chief for program development of the Weapons Program. In July 1983 he married Wendy.*

In his various jobs, Gary represented the USGS in all collaborative work with DOE and other agencies through endless advisory panels and committees; he also led field trips for many geologists and non-geologists to the NTS, and held many interviews with print, radio, and TV news media. In 1989, he got a break from this to do mapping with Doug Morton in southern California. Their map area included several Air Force bases, from which they secured outside funding. This enjoyable project continued until the RIF of 1995. It included a study of the Landers earthquake (magnitude 7.3) of 1992, fortunately in a relatively unpopulated part of their field area north of Palm Springs, where it resulted in more than fifty miles of new surface faults and breaks.

Early entries in my 1996 field notebook reported on field-trip stops and conversations during a two-day field trip in March with WRD hydrologists Randell (Randy) J. Laczniak, Douglas A. Trudeau, and others, plus hydrologist administrator Gary M. Russell, in charge of the WRD Las Vegas field office. This, a long counter-clockwise trip around the outer boundaries of the NTS and Yucca Mountain sites, included a drive through Rachel, a tiny town just north of the NTS and the closest one can get to Area 51, just east of the northeastern corner of the NTS and the secret military base that is the subject of many conspiracy theories, the most persistent of which is that it houses captured aliens and their flying-saucer spaceships. (Rachel's only business to speak of is a gas station and bar called the Little A-Le-Inn [get it?], at which one can buy t-shirts, books, and highly-priced trinkets celebrating out-of-world space travelers.)

The two days enabled us to visit two great regional groundwater flow systems, the Death Valley on the west and the White River on the east. These flow systems were recharged in the Ely area of mountains in high central Nevada and flowed to low discharge areas (springs), respectively, either southwest to Oasis Valley and Death Valley or south to the Colorado River (at Overton, on Lake Mead). Movement of groundwater, like surface water, is downhill (southward) following the mostly north-trending basin-range faults through a process called fracture flow. Ike Winograd had discussed these things in several classic reports, and Gary had been taking the concept from there and expanding on it.

Geologic mapping of faults therefore gives us a first approximation of the directions of groundwater flow. Several early editions of the 1:100,000-scale geologic map of the Nevada Test Site had been compiled following completion of the detailed (1:24,000) 7.5-minute quadrangles that made up the NTS. Together with hydrologic studies, this map was instructive in describing groundwater movement. I put the final touches on that map and wrote the text in 1996, then handed it off to a new geologist to add Quaternary deposits, a person whom I thought needed senior authorship of a significant report. That was a mistake: she tinkered with it for two more years—not the way to handle DOE deliverables.

Field trips are the ideal way to focus all of us on geologic and hydrologic field problems. Such trips to NTS were perticularly entertaining. For there was one spot at a pass in northern Yucca Flat where we could look down, far in the distance, on runways and buildings of Area 51. We never did spot any flying saucers, and after a while I concluded that the alien/flying saucer theories were BS.

We had many meetings, visits to Menlo Park and Denver, and phone calls as we moved unproductive geologists off the project, hired new ones, and started new work. Outside funding enabled us to resurrect some of the maps and studies abandoned when the BARCO project collapsed, such as the Littlefield, Arizona 1:100,000 quadrangle with George Billingsley of the Flagstaff office. Southern Nevada Water Authority (SNWA), the research arm of Las Vegas Valley Water District, funded two quadrangles northeast of Pioche, in the White River groundwater flow system; SNWA was becoming interested in searching for upstate groundwater supplies for Las Vegas, because of the increasingly precarious culinary resources of Lake Mead; Bart Ekren and Ric Page mapped these two quadrangles together.

I started dating again during this time, though I often traveled back and forth to Las Vegas. Dawna E. Brown, Dawna E. Ferris, or Dawna E. Ferris-Rowley as she later called herself (mentioned above because she was mentored by Wendy Dixon) is almost five years younger than me, and was born and raised in Augusta, Maine.

> ***Dawna E. Ferris-Rowley*** *earned a B.A. cum laude in romance languages in June 1970 from the University of Maine, Orono. An extremely attractive woman, she soon married a man who gave her a last name (Ferris) but little else, and because of his philandering the marriage did not last long. Considering a career in*

teaching Spanish and French at the high school level, she left the state and earned a master's in French (cum laude) in 1972 from the University of New Mexico. With degrees in hand, she was off to Las Vegas, a move that unfortunately coincided with the local school system downgrading Romance languages and therefore teaching opportunities. Soon she was working at a succession of mafia-owned casinos, managing groups of gaming tables, mostly at The Dunes Hotel and Casino and The Stardust Resort and Casino, and at one time at The Frontier Hotel and Casino, all on the Las Vegas Strip (Las Vegas Boulevard). All have since imploded (literally been dynamited) to much fanfare, with Steve Wynn's Bellagio Hotel and Casino now on the site of The Dunes. Next, she decided on a career as an archeologist, and she found a job with the Bureau of Land Management (BLM) in Las Vegas from 1986 to 1989, at which point she returned to school at the University of Nevada, Las Vegas (UNLV). There, in June 1988, she earned another B.A., cum laude, this time in Anthropology/Archeology. In 1989, the Ely district of BLM hired her as an archaeologist, with her office in Caliente while she worked on a master's degree at UNLV in the same major, graduating (now magna cum laude!) in June 1991.

Dawna's master's thesis was, as noted above, a study of the historical archeology of the turn-of-the-century gold district of Delamar. It is an enthralling and beautifully written history that I came upon as I began my study of the Caliente caldera complex. All of us working on the caldera project lived in the trailer park north across the fence from the BLM compound, so we became friends with Dawna. She was the brightest person in town, has a great sense of humor, and each season all of us would get together for dinners and/or drinks with her on various occasions. She had bought a fixer-upper bungalow at the top of a cul-de-sac that climbed a wash in the canyon on the east side of town. And fix it up she did, turning it into a lovely home. Another parallel cul-de-sac, just to the north, had a legal house of prostitution at the head of it, and occasionally drunk and horny men would go up the wrong street and knock on her door! She had a mare paint horse, which is a variety of American quarter horse that has irregular patterns in two colors. This one was Silky Eyed Sadie, and she was mostly white with gray patterns and spots and she had beautiful blue eyes. Dawna rode her often, and kept her at various fairgrounds starting when she lived in Las Vegas. Dawna's love of animals is such that she could not bear to kill a rattlesnake that lived in a rock wall adjacent to a gazebo she built behind her home, against the canyon wall of BLM land. Besides (as she insisted), the snake never hurt anybody, and her cats got to tease it.

As she had found when she worked at the casinos, she was as good a manager as she was an archeologist, and she aimed in that direction starting in 1994 by enrolling in training in the fed-

eral government's Women's Executive Leadership Program. Three assignments resulted, first for three months in the summer of 1994 as Associate District Manager of the Eugene, Oregon office of BLM; then four months in the late fall and winter of 1994–1995 as a Special Assistant to the Solicitor in Phoenix; and finally for eight months in the spring and summer of 1995 as BLM Liason for the Yucca Mountain Waste Program, working with Wendy Dixon! Wendy and Dawna hit it off immediately. During some of these assignments, I house-sat her home, cared for the cats, and watched out for the rattler.

Dawna owned an old three-bedroom condominium in Henderson, a southeastern suburb of Las Vegas, in which she had lived while working for the casinos and going to school. It was now being rented, but the tenants were moving out. Both of us made the move to the condominium in March 1997. Dawna looked for a job nearby, eventually found one as an archeologist, then environmental protection specialist, at the Bureau of Reclamation in adjacent Boulder City, from September 1997 to May 1999. My office, as before, was in the WRD building in southwestern Las Vegas, but now I occupied it full time. As her "price" for all this, Dawna bought another paint-horse, this one pumpkin and white, known as Rumor Has It, and it was pregnant. We kept the horses at the county fairgrounds in Boulder City until Rumor was ready to give birth, when we transferred the horse to Wendy and Gary's barn. That was fortunate because we were away during the delivery but Gary was not, and when the colt's legs got tangled up during birth, Gary did what was necessary to save both mare and colt.

In 1997, Southern Nevada Water Authority (SNWA) paid soft money to the USGS to map quadrangles in the Pahrump area, an hour drive west of Las Vegas, where uncontrolled growth was lowering the water table. We also started a DOE project at Devils Hole, a large spring in Nevada and a forty-acre detached part of Death Valley National Park that contained the Devils Hole Pup Fish, an endangered species found only there. We started geologic maps of the Mesquite area and worked with hydrologists of the Virgin Valley Water District to site wells on the faults that had been mapped. We helped Ric Page in his geologic map of the Las Vegas quadrangle. In late August I met with Skip Cunningham in the Marysvale area to put finishing touches on my 1:100,000-scale summary geologic map of the volcanic field, published in 2002. In September 1998, Gary decided to retire. I was similarly inclined but wanted to finish some reports without this management albatross hanging around my neck. Ric and Mel took over the project.

In January 1999, when Gary retired, he started his own consulting company, Southwest Geology, Inc. As Wendy and Gary pondered their future retirement years, getting out of the heat of Las Vegas was prominent in their minds, so they bought a house on ten acres in Blackfoot, Idaho, constructed a big horse barn on it, and planted some of the rest of it in timothy hay for the horses. Wendy had a few more years before retirement, so she transferred to the Idaho National Engineering Laboratory (INEL, now called INL), a nine-hundred-square-mile DOE facility north of Blackfoot in eastern Idaho. The DOE here did research for many years on nuclear energy, and constructed fifty nuclear reactors. The site sits in high desert of the Snake River Plain, made up of many basalt lava flows, so some radioactive waste accumulated on site! Just southwest of INL is Craters of the Moon National Monument of the NPS, which contains the youngest (15,000 to 2,000 years old) basalt flows of the Snake River Plain. Before the RIF, Mel Kuntz mapped and studied the basalts of the Snake River Plain, including at Craters of the Moon National Monument. After he retired, Mel completed, for free, the final maps, including several 1:100,000-scale sheets.

I returned to science within the NTS project in October 1998. Larry gave me permission to move my office to the Dept. of Physical Science, Southern Utah University (SUU), Cedar City. That department had a staff of three geologists, graduating students with a B.S. in geology, and they were all in favor of me joining them. The Utah Geological Survey (UGS) also had an office on campus. My new office allowed close cooperation between UGS and NPS to work in their national parks in Utah, and with potential cooperators such as Washington County Water Conservancy District. My first project that fall was to work with Stan Hatfield, an Assistant Professor at SUU, on a study of Cedar Breaks National Monument west of Cedar City that required us and several of Stan's students to map two quadrangles, much of it paid for by NPS and both maps to be published by UGS.

As 1999 got under way, Dawna and I figured that marriage might work this time, so we made an appointment on January 15, 1999, with Judge Sarah K. Getker and her clerk for a witness, for our wedding at the million-dollar courthouse. Was Judge Sarah surprised to see me again! In early 1999, Dawna and I bought two adjacent lots of three acres each in a new development about five miles west of I-15 and a mile and a half southeast of New Harmony, Utah. We designed a house for that land and started construction while still living in Henderson. Dawna once again launched a search for a new job and found it as an archeologist with Zion National Park, in Springdale, Utah. She started in May 1999. We moved into our new house at Christmastime of 1999. We also built a barn, riding arena, and horse-turnout area, then we trailered Sadie, Rumor, and Baby Horse into their new digs!

In the summer of 2000, Park Service rangers found a female puppy in the back country of Zion National Park, where she had been running with her mother. They could not catch the mother but grabbed the puppy and brought her back to the building near the western park entrance where Dawna had her office. Dawna brought it home that night, unannounced. As an adult, Dawna commonly had dogs, notably Red Setters, and now we began the rewarding practice of dog ownership together. The puppy, mostly an Australian Shepherd, had black, brown, and white angular markings that looked like rocks made up of angular fragments, such as fault breccia, caldera breccia, gravity-slide breccia, and volcanic mudflow breccia. So her name became Breccia, the best dog we ever had. This was a high-energy "working dog," but our little ranch had plenty of room and horses to bark at, and Breccia forced me to take a mile-long walk each morning into the Pine Valley National Forest, which is west of our home, virtually across the street. Breccia soon joined me in field work, adding additional joy to all-day mapping traverses. In 2002, one of Dawna's friends in St. George had a large puppy (who was to become 95 lbs) that was dominantly Blue Heeler (also known as an Australian Cattle Dog) that was too exuberant for their small child and small backyard, so Dawna took him too! He was named after another geologic term Graben (an area downthrown along faults, but he also answered to Ben). The worst thing about dogs is that they do not live long, but in response to the terrible sadness of their deaths, Dawna would simply get another! In April 2011, it was another female Australian Shepherd that she named Oreo. And in January 2015 it was Charlie, a young Blue Heeler (so named because they herd cattle by nipping heels, and their dominant color is blue gray) that we got from Best Friends Animal Shelter in Kanab. In January 2023, we were told by Best Friends that a white and black small female puppy was a Border Collie that was named Leia, but when it grew up 6 months later it looked and acted more like a Jack Russell Terrier, unfortunately a high-strung yapper dog (but adorable). And in September 2023, from Best Friends we rescued a three-legged small female Blue Heeler that Dawna named Lego. Leia and Lego still are in residence!

The year 1999 gave me many opportunities, through field trips, to meet and collaborate with geologists at UGS and in business. The UGS office on the SUU campus was staffed by two excellent geologists: Robert Blackett, their geothermal-energy expert, and William R. Lund, their expert on seismic faults in Utah, including the Hurricane and Sevier fault zones. I got to know both men well, Bill a little more so. With a mere B.S. in geology from the University of Idaho in 1972, Bill went on to spend seven years as an engineering geologist in private industry, then thirty-six years with UGS, retiring in 2015. He published more than a hundred reports, some of which won awards for the best UGS reports that year. They concerned seismic faults and other geologic hazards such as landslides, rockfalls, and ground subsidence due to withdrawal of groundwater. During that time, he accumulated just about every award that one can gather from state agencies and local and national geological and engineering societies. (The list is a page long, including an award in 1995 for "Outstanding Public Employee," granted by the Utah House of Representatives.) He received the Lehi Hintze award in 2012 for "Outstanding Contributions to the Geology of Utah." At one time he was Deputy Director of the UGS, and his departure southward to run the two-man office was probably his effort to flee more administrative work. He is an articulate public speaker and great leader, one of the best geologists in Utah.

In late 1999, Larry Snee returned to research. My new boss was Robert S. Thompson, the worst Branch Chief I ever had. He had been hired after the RIF to be a manager. His specialty was Pliocene climates, so rather than manage, he spent all his time in his lab doing microscope and related lab work. Thompson later was put in charge of the entire USGS Climate Change program, perhaps a reason why it never amounted to much. Thompson's Assistant Branch Chief did all his chores, and as a result the branch had no leadership and just sailed along, rudderless. He did not know me or the USGS, nor did he care. He was, however, influenced by branch members whom I had moved off the NTS project and now were available down the hall, while I was in Las Vegas or New Harmony. Therefore, Thompson began to campaign to get me to retire by requiring me to move back to Denver, knowing full well that we had just built a home in New Harmony. I intended to retire, but I needed to get in thirty years to earn a full pension. And that anniversary was coming up soon, in August 2000. I finally told him that I would retire at this anniversary if he would just leave me alone.

Dawna and I were only a ninety-minute drive away from friends Manetta and Farrel Lytle, who lived in a small Nevada town, Rose Valley, east of Pioche and near the Utah state line. On weekends we would get together. This circle of friends also included Yolanda and Mark Henderson, who lived in Ely, where Mark was an archeologist with BLM. The excuse for getting together was field trips to one of Dawna's and Mark's specialties: rock art. These are petroglyphs, which are pecked or chiseled into rock, and pictographs, which are painted on rock. Most of these appear as ancient circles, animals, and strange stick-figure humans, dating to Fremont or pre-Fremont time. Some figures are sited so that they are illuminated at the summer or winter solstice. Rock art was particularly abundant in the Caliente-Delamar part of Nevada because this area has a lot of springs, some of which produced small lakes, and this was an east-west transportation corridor for ancient peoples across the Great Basin. Mark and Yolanda were Dawna's oldest friends, for they had met when she was at the University of New Mexico, where Mark earned his degrees in archeology. Yolanda, a law clerk in Ely, came from Spanish land-grant settlers who moved, long before the Mayflower, into the Taos area.

Farrel ran a consulting business dealing with the uses and interpretations of x-rays, working out of his home. He had retired from Boeing Aviation in Seattle, an X-ray specialist with as many publications as I had. He and Manetta came from Mormon stock in the Pioche area and adjacent Utah, where they were related to seemingly half the population. Farrel became interested in dating petroglyphs by using a non-destructive (that is, no samples were required) X-ray method to determine the chemistry of the iron- and manganese-oxide patina (coming from windblown dust) that formed over the pecked-in patterns and became thicker and changed in chemistry with time. He perfected this technique and eventually got reproducible results using a hand-held X-ray machine that he purchased himself, costing about the same as what they paid for their new Honda sport utility vehicle. Dawna, Mark, and I were too stupid to understand the theory behind the method. Together we published abstracts and some short papers, but mainstream science did not seem interested, even though other methods for dating rock art did not exist. Therefore, Farrel's ages could not be corroborated, so he could not prove that they were valid. Nor did he have any younger colleagues to whom he could teach the method.

Stan Hatfield was the best geologist in the three-man geology department at SUU, and he had an office not far from mine and a similar love of field work. He was doing field research with Bill Lund on the age and history of the Hurricane fault. In 2000, as I dealt with Robert Thompson, Stan and I together mapped the 7.5-minute quadrangles that contained Cedar Breaks National Monument, accompanied by our dogs, then published a summary report on the Monument and a road log of a field trip to it. I also participated in field trips on weekends with Stan, the other profs, and students to show them areas familiar to me. Stan, an inspiring teacher in the department, was the reason why most students became geology majors. But Stan had not attained tenure (he was an Assistant Professor), even though he had published quite a bit and the other two in the department did not publish anything. Worse, Stan was what is known in the teaching profession as ABD or All But Dissertation. He had never finished his dissertation, so did not have a Ph.D. The president of SUU, which is a state university, was at the time trying to upgrade the university (a decade or two before, it was a teachers' college), so he was on the warpath against profs without a Ph.D. That year, 2000, Stan came up for review in the spring and the administrative brain trust decided to terminate the best teacher in the department unless he could finish. He could not. He eventually found another job at Southeastern Illinois College, where we stayed in touch. As I contemplated retirement from the USGS, his loss presented an opportunity. I applied for his job as a nine-month appointment. I was accepted at a low salary, as Instructor. But it was sufficient for me to retire from the USGS at my thirtieth anniversary!

I started teaching about a week later, in late August 2000. That year

I taught Mineralogy, Optical Mineralogy, Igneous Petrology, Natural Hazards, and Environmental Geology. Once again, I was preparing lectures the night before for several courses simultaneously. I commuted to Cedar City from New Harmony, about a half-hour drive. I also took students and profs on two three-day-long camping field trips, for which Dawna and the dogs joined all of us, to the Marysvale volcanic field and to the Caliente caldera complex. Those trips were probably the highlight of the year for the students and us. I enjoyed the kids at SUU, but the intensity of preparing lectures on weekends and until late at night was a grind that became tiresome. Using the department as the platform, I proposed a research grant to the National Science Foundation. I chose to work with Farrel, in collaboration with several departments at SUU, on dating and interpreting rock art, and wrote an application for a grant of about $400,000. It was apparently the first NSF science grant proposal to come out of SUU. Unfortunately, NSF turned it down. I also applied for a permanent position with the department, which clearly resided in a geological paradise that I knew well but the other two profs did not. My pitch was to upgrade the department. I was passed by in favor of a young person with a new Ph.D., and the department continued in mediocrity.

A positive feature of my year at SUU was getting to know David J. Maxwell, in charge of the GIS/GPS (Geographic Information System/ Global Positioning System) Lab in my department, called the Dept. of Physical Science. GIS enables persons to use a computer to gather, store, and analyze all types of geographic data, using maps. Dave is a go-getter, but he only had a Bachelor's degree (he later got a Master's), so he was considered "staff," without opportunities to advance up the academic ladder. He also was an excellent and supportive teacher, so students loved him. Those students who took several of his courses in GIS, when combined with a major in one of the sciences, suddenly had a practical skill that enabled them to get a job upon graduation with only a Bachelor's degree. In fact, GIS was all the rage then, and local, state, and federal governments were hiring. Dawna hired one of Dave's students at BLM, and that person became one of her best employees. The two geology profs did not approve of GIS, preferring academic geology courses such as paleontology or others they taught. In fact, a Bachelor's degree in geology will, at best, get a person a job as a geological technician or field assistant; advanced degrees are needed in geology to become a professional. But most geology majors graduating from SUU did not go on to grad school. Once my consulting business got started, I paid Dave and his students to prepare my geologic maps for publishing. The partnership persists.

My parents continued to enjoy their retirement in Tucson, although Dad had had a relatively minor stroke that left one arm paralyzed. They discontinued their jeep trips into the desert and Mom took over more of the house chores, and whenever we visited all seemed well. Chris Rowley, Clark's son, also visited regularly for many years and was particularly close to Dad. Then disaster, for Dad died suddenly on December 24, 2000, at age eighty-three. Chris was there, and he and Mom talked with Dad in his last hours. Mom carried on at the house afterwards for several years until one day she nearly had an accident in her car. Shaken by the near miss, she thought hard about this and I think correctly determined that it might be better if she stopped driving. After that, in 2004, she sold the house and moved to St. George, about forty miles south of us in New Harmony.

After a while at Zion National Park, Dawna hoped for a job with a shorter commute and in an agency she preferred. In February 2001 she was offered the job of Assistant Field Officer Manager at the St. George Field Office of BLM. She held this job until July 2010, when she applied and was accepted to be the BLM Manager of the Beaver Dam Wash and Red Cliffs National Conservation Areas, working out of her same office. As the school year ended, in the spring of 2001, I started a consulting business, Geologic Mapping Inc., much of it collaborative work with Gary Dixon. We had designed a large office in our home. Surrounded by dogs that now sprawl in dog beds or directly on my office floor, and cats that curl up on unoccupied chairs or desks, and with views of the Pine Valley Mountains to the west, it is a peaceful way to do business! Classical music streams from my computer. On the down side, however, I am the house husband, who must deal with the horses and chickens, fix stuff that breaks, shop, cook the evening meal, and otherwise manage the farm. After a few years, Dawna decided she liked Andalusian horses, which are the royal horse of Europe, as seen in all the statues there. She bought four of them, bringing the total number of horses we had to seven (until Silky Eyed Sadie sadly died in 2012). Then Dawna chose to show the Andalusians, requiring much traveling to horse shows, not to mention entry fees, horse trailers, etc. Fortunately, eventually the horse shows stopped; we are now down to two horses.

CHAPTER 10

After working with Gary Dixon on USGS research at the Nevada Test Site (NTS), research that was practical and of near-term benefit to the public, becoming a geologic consultant came naturally. Besides, we had talked about working together long before we retired. When we led or went on field trips together, we were commonly joined by geologic or hydrologic consultants. Driving back to our office, Gary would look over at me and say disparagingly about them, "Well, that's the competition." Most of these consultants had weird ideas about how Las Ve-

GEOLOGIC CONSULTING AND FRACTURE FLOW

gas Valley had formed geologically or about many other basic geologic or hydrologic facts. Many also lacked ethics: when hired to represent a client in a dispute or lawsuit, some would ask the client what he or she wanted them to say or find. Clearly, if they were on the wrong side of a scientific argument, many were perfectly willing to distort the facts on or in the ground.

Gary was the best I knew at selling things that people needed, so I rode on his coattails and largely avoided that nasty little requirement known as sales and marketing. When we were with the USGS, all soft-money funds went to the USGS, of course. Now the funds went to us. When we were employed by the USGS, the main requirement was deliverables. As a consultant, however, reports were to be shorter so that the client did not have to pay for an expansive report, for we generally charged by the hour. Producing deliverables tended to fall to me, but that part, as opposed to selling, I enjoyed. Both of us generally had multiple projects going at the same time, many of them involving each other. We picked projects that sounded interesting. I paid my son, Scott, to create a website, www.geologicmappinginc.com, but our main selling strategy was word of mouth. Fortunately, we had one long-term "sugar daddy" on whom we relied for more than fifteen years, the Southern Nevada Water Authority (SNWA), who no longer needed the USGS. The old, happy, rock and roll song by the Dave Clark Five, "Glad All Over", expressed our contentment quite well.

Consulting, quite simply, is problem solving, then quickly issuing a readable, short report. Gary and I enjoyed that challenge, for we had confidence that we knew enough geology and hydrogeology that little would befuddle us. And if it did, we could simply go to the field and map our way out of it. We were open to any geologic problem a client could throw at us, although much of it was hydrogeology. We made and applied geologic maps to delineate water resources, most of it for water districts in Nevada, Idaho, and Utah. At least half my time, however, was spent working for the Utah Geological Survey and some for the Colorado Geological Survey and Nevada Bureau of Mines & Geology to make geologic quadrangle maps—the most enjoyable jobs although they paid much less. Small projects done that added a little fun and variety, but not much money, were siting culinary water wells for private parties and water districts, in which we used the principles of putting them on faults that we identified on published geologic maps or areas we had mapped ourselves. I sited thirty to forty wells, and Gary did at least that many. To the best of my knowledge, neither of us ever had a duster (dry hole), yet ours were desert states where finding drinking water is hardly a sure bet. For those projects that contained something of scientific interest, we published our findings on our own time and expense. The stories given below are just a small number, from more than twenty years of such work, of the most interesting projects. The stories provide examples of how, thousands of times each day, science solves local problems for businesses and governments.

The SNWA—the research arm of Las Vegas Valley Water District—is responsible for procuring water resources for Las Vegas and many outlying communities in Clark County, all growing rapidly at the time we were doing our work. Most drinking water came from Lake Mead, though it was decreasing in size due to climate change, which lessened the flow of the Colorado River drainage upstream from the Hoover Dam. Therefore, the SNWA decided to seek additional water from groundwater upstate (central Nevada, near Ely). This decision came from Pat Mulroy, the General Manager at the time. With a B.A. and M.A. from UNLV, on paper she might not have seemed like the best person for the job, but she had risen in the ranks precisely because she was smart, ethical, decisive, and politically savvy. We did not deal with people at her level but instead with a series of hydrologists and geologists under the direction of Kenneth Albright, another smart person with a sense of humor. Our main contacts ended up being two excellent hydrologists, Andrew G. Burns, and his deputy, James M. Watrus. For geophysics, we brought in Edward A. Mankinen to do gravity surveys and Darcy K. McPhee to do audiomagnetotelluric (AMT) lines, both contracted from the Menlo Park office of the USGS.

Our primary project was the geology of the White River and Great Salt Lake Desert regional groundwater flow systems. This mapping was to support a digital groundwater flow model and a well field, to be created largely by the hydrologists at SNWA. Eventually we focused on a large area of about 25,000 miles^2, mostly in Nevada but spanning the Nevada-Utah state line, which required a geologic map and cross sections at 1:250,000-scale. This major project had started when we were still with the USGS, most notably with a geologic map of the southern half of the White River system, senior authored by Ric Page. That 1:250,000 map of about 6000 mi^2 had largely been compiled from existing maps, though sizeable parts came from new detailed (1:24,000) published and unpublished mapping that we and others had done as part of the BARCO project before it went belly up. But the main thrust of the project came after we retired, as we pushed our mapping and compilation northward along the regional flow system. Andrew, Jim, Gary, and I spent many days doing field work to understand the hydrogeology, to site monitoring wells, and to refine maps and cross sections. We worked with SNWA biologists and surface-water hydrologists, and we had meetings and led field trips with officials of regulatory agencies and interested parties, such as the Bureau of Land Management (BLM, which had to do an Environmental Impact Statement of the entire area), the National Park Service (NPS), the US Forest Service (USFS), the USGS, Utah Geological Survey, the US Fish & Wildlife Service, and others. Meetings in Las Vegas were common. We also represented SNWA at water-rights hearings with the Nevada State Engineer, the organization that adjudicated water rights. We wrote many reports over about fifteen years, most released by SNWA and available to the public, but without widespread distribution. As when we were with the USGS, we wanted no company-confidential reports and we published as much as we could with the Utah Geological Association or other outlets. The SNWA reports included about a dozen cross sections across the area, some of which Bart Ekren created.

In 2016, management at SNWA changed when the Great Recession of 2008–2009 finally resulted in the collapse of the housing market and cessation of the explosive growth of Las Vegas. Funding at SNWA became tight, about one hundred persons lost their jobs, and the project was terminated. Nonetheless, Gary, Andrew, Jim, and I knew that the results of the entire study needed to be disseminated to a broad audience. Considering the effort and money that the study had already cost, Gary and I proposed to write it for free, although by then Gary was dying of cancer. The SNWA agreed to pay the Nevada Bureau of Mines & Geology (NBMG) for publication of the report, maps, and sections. The final report, released in 2017, consisted of a 146-page hard copy and digital report on the geology and geophysics of this huge area in Nevada and Utah, as well as the geologic map and cross sections.

In 2000, a year after Gary retired from the USGS but before I did, Gary senior authored a 125-page digital report and a 1:100,000-scale geologic map with cross sections of the Mesquite basin, created for the Virgin Valley Water District and published by them in 2002. That report, co-authored by our hydrologist colleague, Terry Katzer, along with the two 7.5-minute geologic quadrangles we had mapped of the Mesquite area while with the USGS, were the bases for a series of wells that Mike Johnson, the chief hydrologist of Virgin Valley Water District, Gary, and Terry sited for the rapidly growing little city of Mesquite. I summarized the findings in a report (listing Mike as first author), and in May 2002 during a section meeting in Cedar City we led a Geological Society of America field trip to see the geology and several new water wells drilled by the district, which each now produced at an average of 1,500 gallons per minute (gpm).

My first job as a consultant, however, was not hydrogeology. It started about a week after the Southern Utah University (SUU) school year ended. It was a study of the geology on which the earthen dam of Piute Reservoir rested, following a Utah state study of dam sites that had found that an active fault cut across the alluvial terraces on which the dam rested. A consulting engineering geologist called me because the dam and reservoir were in my dissertation area, north of Junction. That geologist was David B. Simon of Simon Associates LLC, Salt Lake City. I said "Sure, sounds like fun," and off I went to do the detailed geology and help Dave examine the 10+-foot-deep trenches across the fault, dug by locally hired trenching machines. Trenching faults is the way in which engineering geologists determine the age and potential dangers of a young fault that might be active (most faults in the Great Basin are long dead). The trick is to locate young unconsolidated alluvial deposits that contain evidence of faulting, then trench them at 90° to the fault; the fault, if it exists, can be seen in the trench and various methods can be used to date movement, such as the age of the alluvium itself, applying C^{14} dating of wood in the fault zone, etc. Our project lasted only a week or so, was a good education, and introduced me to an outstanding geologist who kept me laughing. Our work showed that the dam had to be moved. That project marked the beginning of a long friendship that included other collaborations.

During the project we stayed in Panguitch which, as a tourist center (being the "gateway" to Bryce Canyon National Park and other areas), has since grown somewhat, from having only one bar and, as I remember, only one café. This café was the Flying M Restaurant. That it had a liquor license and a salad bar made it almost a restaurant. One night over dinner there, I asked Dave if he knew about a recent controversy in Salt Lake City, where he had his office, that the press had covered extensively. It was about the recently completed construction for the 2000 Winter Olympics of a southern extension for the Salt Palace Convention Center in downtown (just north of 200 South Street, an east-west street, and five blocks south-southwest of the State Capital). As reported in the *Salt Lake Tribune*, the site was on a major fault. But Salt Lake County had built the extension anyway. As I rambled on, Dave looked at me as if I were a complete idiot, then interrupted, grinned, and exclaimed, "That was me!" Back in 1998, Dave's consulting firm, Simon & Bymaster, LLC, had been hired to investigate the site before any building went up, as required by local construction codes. A geologic map published in 1985 by the USGS (in fact by two friends in my former Branch—Willie Scott and Ralph Shroba) showed a large, north-trending, range-front basin-range fault, the western splay of the Warm Springs fault, as going directly through the site of the Salt Palace extension. So, to confirm or disprove a fault, Dave had dug eleven east-west trenches across the site, and three of them, in the center of the site, showed the fault. Over dinner, he told me this classic story.

> ***Dave Simon*** *received a B.A. in geology in 1976 from the University of California, Santa Barbara, then went to work in Reno, Nevada, for Harding Lawson Associates, a large environmental engineering and construction services company. In his spare time between 1977 and 1979, he did graduate work at the University of Nevada, Reno. Several other firms doing civil and geotechnical engineering geology—in other words applying geology to planning, design, and construction of engineered works such as buildings, roads, etc.— subsequently employed him. One of these firms was in Salt Lake City, and from there Dave started his own company in 1997. The Salt Palace job was one of his first projects and his most famous. Because of this job and subsequent ones, he became well known as one of the best and most ethical engineering geologists in Utah. Inasmuch as Salt Lake Valley, in the Great Basin, is bounded by active (seismic) faults along the west side of the Wasatch front (the Colorado Plateau east of Salt Lake City), there are many problematic building sites in the area. Of those problems, putting a building on an active fault—meaning a fault that has moved during the Holocene time period (the current Epoch, from 11,700 years ago to present) without proper engineering—is considered reprehensible. Although he is Jewish, Dave became a favorite of the Mormon Church whenever they built what they call "meeting houses" and what others call "churches," which are huge look-alike ugly things, each with a small pointy steeple rising from it. Over time, the Church contracted him to study construction sites of about fifty potential meeting houses—until he did the engineering geology of one in the town of Stockton, Utah, where he found collapsible soils. The architect on the project decided that he knew better than Dave, so that church was built by standard construction despite the identified risks, and Dave's reaction when he found out was, "You guys built that?" From 2002 to 2004, Dave was National Treasurer, then National President, of the Association of Engineering Geologists, in an era in which*

> *previous presidents had ignored the finances of this, the primary American society for engineering geologists. Dave put it back in the black. Dave received the Hintze Award in 2020; Bill Lund and I nominated him.*

On the Salt Palace extension project, three trenches that Dave dug, some as deep as thirty feet, crossed the western splay of the Warm Springs fault zone. Those trenches showed that, as with most big faults, it was a complex fault zone at least fifty feet wide. Radiocarbon dating showed that the faults displaced paleosols (old soils) 8.6- to 6.4-thousand-years old, although some faulted soils were considerably younger. From these data, Dave interpreted the recurrence intervals (the time between faulting events) as being between two thousand and three thousand years. In most trenches that engineering geologists dig, faults (if present) are not necessarily obvious because the unconsolidated alluvium may consist mostly of sand and gravel that flake away from the fault fracture. But Dave's trenches were in fine-grained lake beds of Lake Bonneville, the Pleistocene version of the present Great Salt Lake that filled Salt Lake Valley during the various Pleistocene glacial pulses; during its high stand, Lake Bonneville extended as far south as the latitude of Cedar City. The fault zone was so magnificently displayed that any elementary-school child could have seen what it was. Dave, being ethical but not naïve, brought other geologists in to look at his trenches, just to confirm the obvious. And, of course, he knew his employers were not going to like what everybody saw and verified, so he told all those he brought to see the site to keep the findings to themselves until his employers decided what to do. Among his visitors were geologists of the Utah Geological Survey, which had been built up to what I consider the best state geological survey in the country partly due to the best Director and State Geologist the Survey ever had: M. Lee Allison (1948–2016). Lee, an ethical people person with a sense of humor, was a classic example of one who "did not suffer fools gladly." But he was clearly more naïve than Dave, for when he saw the fault zone for himself in the trenches, he notified the *Salt Lake Tribune*, which did a big story, with photos, in a 1999 Sunday edition, no less! Within a few months, Lee had been fired from his state job as Director. The 2000 winter Olympics had already been scheduled by now, and the new convention center had already been planned, for the site was a priority as a huge money-maker for the county. If the county had accepted the trench evidence and had followed its own ordinances, setbacks for a large fault zone running down the middle of the site would not have allowed any large building there. Therefore, Salt Lake County brought in a second engineering firm to serve as arbiter, hoping that Dave was wrong; but this firm confirmed Dave's interpretation.

To obfuscate, Salt Lake County brought in a third engineering firm, a big one. This one, having read the tea leaves, came up with a bogus interpretation of what was clearly visible in the trenches, then proceeded to do an exhaustive study and report, in which it completely ignored the pertinent geologic data and refused even to cite Dave's report. A fourth party, which was used as a technical reviewer of the false report, likewise knew the political realities and therefore confirmed the totally misleading report. The existing data, including those facts published before Dave's work, included a general geologic-hazards map published in 1989 by Salt Lake County that copied the fault of Scott and Shroba (1985) that went through the Salt Palace site. But now (1997), Salt Lake County re-created their map to show only the main splay of the fault. This fault was now drawn only north of the Capitol. Later, in 2001, an east-west survey financed by a federal earthquake program went along 400 South Street (two blocks south of the Salt Palace) with a truck mounted probe (called a CPT) that penetrates the ground and sends an electronic current, resulting in a cross section of the ground. This cross section showed that the 27,000-year-old Bonneville Lake beds have been offset about ten feet (and the 13,000-year-old beds above it by about three feet) by the active splay of the Warm Springs fault. Of course, these offsets are only tiny amounts of the likely displacement on the fault when you go back a few million years. (As a footnote: One of the principal geologists in the third—bogus—study by the big engineering firm was a senior engineering geologist whom Dave had gotten to know and had respected greatly at Harding Lawson. One day, as they were working on their independent projects at the Salt Palace site, Dave asked him matter-of-factly "[w]hen did you become a whore?")

Little has been published about the project except for Dave's abstracts at geology technical meetings. Award-winning novelist and geologist Sarah Andrews (1951–2019), a long-time friend of Lee Allison (after she received bachelor's and master's degrees in geology, and after her first job with the USGS), in 2002 published a fiction mystery-crime book entitled *Fault Line* that borrowed some plot lines from the Salt Palace story. The protagonist of the story, as in most of Sarah's dozen other novels, was feisty "straight-talking" forensic geologist Emily Hansen. Sarah, who taught geology courses at Sonoma State University, was well liked in the geologic profession, but tragically she, her husband, and son (all pilots) died on July 24, 2019, in Nebraska in a crash of their twin-engine Beechcraft. Before this, in a nice "author's note" at the end of her book, Sarah wrote that there was an "ethical hero" in the real story (she could have been talking of Dave or Lee or both), then urged public support for state and federal geological surveys, where public servants "exchange much handsomer incomes for the government's modest ones just so they could spend their time advancing the science." Other parts of the Salt Palace story are lacking in the literature, and nothing of the controversy is currently available on the internet. As might be expected when geology is disregarded, a year or two ago, another piece of undeveloped land owned by Salt Lake County on the southeastern corner of the Salt Palace block has been targeted for a new hotel. The word soon out among the geologic community was "Don't go near the place." And the 7.5-minute quadrangle that includes the Salt Palace was only recently published by the UGS, erasing an obvious blank spot in a UGS mapping program that has concentrated first on urban areas, especially along the seismic Wasatch front.

In the early summer of 2001, while working with Gary on the study of the White River regional groundwater flow system, I got a call from Vincent Matthews, III, the Director of the Colorado Geological Survey, asking me to map the geology of a 7.5-minute quadrangle in Colorado, using Statemap federal funding from the National Cooperative Geologic Mapping Program. I was to be hired for a six-month appointment as a Physical Science Researcher III. Vince had hired several others during the last several years to map other quadrangles in the Colorado Springs area. These included Chet Wallace (RIFed USGS geologist), Jon P. Thorson (an excellent mapper and consultant on metallic mineral resources and general geology, living in Littleton), and Richard F. Madole (an outstanding Quaternary geologist, formerly of the USGS, hired in our branch by Ed Ruppel, but who later left to start a consulting business). There was a lot of camaraderie among these and others, most of them already friends, and we visited each other in the field and solved problems together. Vince, a man of about my age, had earned his 1971 Ph.D. from the University of California, Santa Cruz, had been an Associate Professor in Geology at the University of Northern Colorado from 1971 to 1976, then an oil geologist, and had then assumed his present position from 2000 to 2013. While Director, his research topic was earthquakes in Colorado, presumably in hopes of receiving federal funding as others had done in Utah, but there were not many active faults in the state (though there are zillions of dead faults). Attractive, friendly, and with the gift of gab, Vince's overall legacy, unfortunately, did not do much for the Colorado Geological Survey, which was subsequently downsized. This is a common affliction of state geological surveys when Republicans take over statehouses and legislatures, for that political party seems to care less about helping private citizens, small local businesses, and governments avoid landslides, swelling soils, collapsing soils, seismicity, and other geologic hazards. Instead, Republicans care more—indeed fear—that such studies might lead to regulations, thus limiting profits for developers. State Directors must spend much of their time reaching out to the public and local politicians, and the good ones did. Vince, however, did not make these activities a priority of his job.

But back to the story. I was to map the Cheyenne Mountain quadrangle, just south of Colorado Springs. It contained the eastern scarp of the Front Range, with the High Plains on the east. A road, Colorado Highway 115, trended north-northeast in the central to eastern part of the quadrangle, on the High Plains; north-south Interstate Highway I-25 was a half dozen miles to the east, mostly outside the quadrangle. Most of the quadrangle, between these two highways, was part of the Fort Carson Military Reservation, the two-hundred-square-mile home of the Fourth Infantry Division and several other units of the US Army. Fort Carson, which is the town on the base, houses nearly 14,000 military people, with all the amenities of any town. Cheyenne Mountain itself, west of the military base, is an impressive mountain that reaches an altitude of about 9,500 feet, with a relief of almost 3,000 feet down to the plains. The mountain is made up of Precambrian granite that is about 1.7 billion years old. The Cheyenne Mountain Complex, which is a US Air Force station attached to Peterson Air Force Base in Colorado Springs, was excavated underground into the mountain during the Cold War to be impenetrable (by missiles) to any adversaries. The Cheyenne Mountain Complex housed elements of the North American Aerospace Defense Command (NORAD) that monitored the airspace of the US and Canada as a control and warning system against air and missile attacks. The Complex is a self-contained system of five acres of tunnels, and the top of the mountain contained a large array of radio towers. By the time I arrived, well after the Cold War had ended, the Complex had been reduced in status to an alternative center manned only by a skeleton crew.

I started field work in early June 2001. The mapping was delightful, for the rocks had a lot of variety and were well exposed, and a huge fault zone formed the mountain scarp. The substantial relief, heavily timbered and beautiful, added to the fun. Large canyons that cut into the Front Range were where I started. One, carved by Little Fountain Creek, contained a ranch owned by Louisiana State University that was a geological field camp to teach students how to map, so I enlisted the help of one of its former professors. Access to the top of Cheyenne Mountain was by a gravel road with many switchbacks, and one day I asked someone to drop me off on top and I then hiked and slid down the mountain during one traverse back to my car at the bottom, near the tunnel entrance to the NORAD facility. This person dropped me off for other traverses too. The work on Fort Carson was much easier, for hills were small and most rocks belonged to the Pierre Shale, a thick Cretaceous shale that contained many fossil clams and ammonites and was easily accessible thanks to the many roads, some of them leading to shooting and artillery ranges. After a month, I went home and compiled what I had done, then returned to Colorado for two more weeks of field work before returning to New Harmony. By late August, I was back to finish field checks and to use USGS stereoscopic plotters at Building 25 in Denver to transfer linework from aerial photos to stable map bases. That time included a field review with Vince and others.

On September 11, 2001, I got my normal early start, intending to do the last field work on Fort Carson. While washing breakfast dishes, with the television on at 7 am (local time) in my rented condominium, I watched the horrifying picture of the first terrorist (al-Qaeda) plane hitting the north tower of the World Trade Center in New York City. It was not clear at that moment whether this was "merely" a terrible accident, so I finished the dishes and got in my Toyota 4-Runner and headed for field work. By 7:15 a.m., I was waved through the Fort Carson gate, as usual. The second plane hit the south tower either on my trip to the gate or shortly after that, at which time it became clear that what was happening was a terrorist attack. Both towers collapsed within about two hours, killing more than 2,750 people on the ground and in the planes. I spent a long day doing the geology, mostly short visits to outcrops to clean up loose ends and thus listening to the car radio between stops. It was in the evening right before dark when I headed back to the gate. They waved me out quickly, but the line of cars coming in stretched for a half mile or so as guards checked entering cars, including their undersides, and the people driving them. Military facilities seemed to authorities to be vulnerable to additional attacks, although those never materialized. The next day I drove back to New Harmony, having completed all field work.

In December, I submitted the manuscript map at their deadline. Though it hadn't been requested, I included an expanded text to summarize the geology of the Front Range, for this map area had many clues to the history of this controversial range. The report and map went through review and editing over the next few months, but for some reason Vince kept it on his desk until July 2004, when it was released (the publication date was 2003). I did one more map, the Cascade quadrangle in the Front Range northwest of Colorado Springs, the next year (June 2002 to June 2003), but a young staff geologist with the Colorado Survey did more work (mostly digital) than I did and I had him be senior author. For some of the field work, I brought our beloved Australian Shepherd, Breccia, with me as my companion. The Cascade map was also released in July 2004 (with a publication date likewise of 2003). I did not hear from Vince after that, so apparently I had been dismissed for an unknown reason. At about that time, Vince also let Chet go. About five years later, out of the blue, Vince left a voicemail that the Cheyenne Mountain map was not only an excellent job, but that he had found the text to be well written. It was nice to know that he had finally read it!

Mid-2001 to mid-2002 consisted of much field work and many field trips. Some were field studies in Coyote Spring Valley (fifty miles northeast of Las Vegas), paid by SNWA, in which Gary Dixon and I mapped faults that were the flow paths of the White River flow system. For the May GSA meeting in Cedar City, I prepared manuscripts and led or co-led field trips to the Marysvale field, to the gravity slides west of the Pine Valley Mountains (led by Dave Hacker), and to the Mesquite basin and well field (noted above) with Mike Johnson and Gary. The Bureau of Land Management (BLM) hired me to do a quick geomorphological study/report on a BLM archeological site north of Mesquite, then another short study of a quarry that Farrel and Manetta Lytle had recently discovered, one from which Indians had for thousands of years mined obsidian for arrow points in the base of a rhyolite lava flow on BLM land east of Pioche. Then followed a year-long study and large report on the hydrogeologic controls of two endangered plants on BLM land east of Death Valley. I also started a three-year study for BLM, on the hydrogeologic analysis of a large Environmental Impact Statement (EIS) being done by a large consulting company for a power plant to be built on BLM land northwest of Mesquite (though this was never built because the Nevada State Engineer allowed them too little groundwater for cooling the turbines).

Washington County Water Conservancy District (WCWCD), in St. George, supplies (sells) the water resources to Washington County, a large county forming the southwestern corner of Utah. Most county residents live in the St. George metropolitan area, which then was the second fastest growing city in the country. (After 2008 it became the fastest growing.) According to the 2018 census, St. George city, the county seat, has a population of 87,000 people, with about 172,000 in the overall metropolitan area. The General Manager of WCWCD then was Ronald W. Thompson, a lawyer who earned his law degree from BYU, and one of the most influential persons in Utah politics and governance. His deputy was Barbara Hjelle, also a lawyer, with bachelor's and master's degrees in biology before the law degree, all from the University of New Mexico. Barbara was formerly the city lawyer for the polygamist community of Hildale, Utah, adjacent to the other polygamist community of Colorado City, Arizona, though she is not Mormon. The WCWCD's chief hydrologist is Corey Cram, a professional geologist with a bachelor's degree in geology from SUU and a master's degree in hydrogeology from Penn State University, although he was pressed into service as a manager, earned another master's in business, and now is Associate General Manager for Project Development. This trio, all bright and hardworking, seemingly did most of the work at WCWCD, and because of them the district remained on top of the high growth, with none of the water shortages that have plagued other districts in the state. I would commonly tease each of them, as they were always overworked and underpaid, that SNWA employed another hundred people to do the things they did! With the retirement of Ron and Barbara in 2018, Zach Renstrom became General Manager, maintaining the same standards. My work with them started with a phone call from Barbara in February 2002. She wanted me to study Pah Tempe.

Pah Tempe Hot Springs is a high-discharge (about 5,000 gallons per minute) spring complex along the Virgin River upstream (east) from the town of Hurricane, which is at the western base of the Hurricane Cliffs (Colorado Plateau). The Cliffs are uplifted along the north-trending Hurricane fault zone, which passes beneath the town of Hurricane. The river has cut a deep canyon in the Hurricane Cliffs, and the many springs issue in the canyon from fractures created by the fault zone. The springs are artesian, which means that the water rises out of its orifices, the biggest springs being in the river and progressively smaller flows coming out to a height of twelve feet above the river. The spring complex contains the highest total dissolved solids (TDS, i.e., salt) of any large spring in Utah, about 9,500 milligrams/liter (a level greater than 1,000 mg/liter exceeds national drinking water standards). Fractures of the Hurricane fault allow rapid rise of groundwater from great depth, with the heat from the geothermal gradient, and most of the salts were leached from hot gypsum-bearing rocks at depth. Thus, Pah Tempe Springs is a major source of natural pollution of the Virgin River, so the salty river water downstream from the Hot Springs has no value for drinking or for crops. Furthermore, this natural pollution was a major contributor to deteriorating TDS values still farther downstream, along the Colorado River. This is especially critical because, by treaty, the United States is required to maintain drinking water standards for the Colorado River where it passes into Mexico. The US Bureau of Reclamation published an exhaustive study of the springs in 1973, with an eye to constructing a desalination plant, but the plant was never built. Therefore, rather than allow all good water in the Virgin River upstream of the springs to become polluted, the WCWCD intercepted this water above Pah Tempe Hot Springs and pumped it directly into a five-foot-diameter pipe to one of their reservoirs five miles west of Hurricane. The pipe was placed in and beneath the stream channel in the canyon, which therefore required dewatering by pumping and then blasting a trench.

Construction of this pipeline, from 1983 to 1985, led to a controversy about which WCWCD sought my advice. The source of the problem was that people had been bathing in these presumably beneficial sulfurous, 105°F waters (now in a concrete pool, fed by the two highest springs twelve feet above the river) ever since its discovery in 1776 by the Escalante-Dominguez expedition. At one time, for people who suffered from arthritis and other diseases for which other relief was minimal, a small business in the form of a sanitorium was built near the upper two springs, and this business later became the Pah Tempe Hot Springs Resort, which charged the bathers entrance fees. A new owner, who had a short fuse, bought the business in 1985, and claimed that his two springs had dried up due to the blasting (which had happened before he became the owner). He began collecting data and preparing a lawsuit against WCWCD, but either he lied to magnify his perceived damages or his record keeping or memory was faulty. He hired first one geologic consulting firm, then another, and they parroted not only his faulty flow data but also his goofy non-scientific interpretation as to the hydrogeology and origin of the springs. It is likely that flows in the two springs returned within a few days or weeks after the blasting, but now good objective records needed to be obtained. The WCWCD used its long-time civil engineering firm, Alpha Engineering Company, to do that. The WCWCD also paid the Utah Water Resource Division office of USGS to do a hydrologic study. Barbara Hjelle of WCWCD wanted me to interpret the hydrogeology. In 1992, the largest (5.8 magnitude) earthquake since 1975 in Utah—the St. George quake with an epicenter at Washington, Utah, an eastern suburb of St. George—occurred on the Hurricane fault. That dried up the two springs for a short time too. And in mid-2000 and twice in 2001, the pipe ruptured, apparently due to corrosion by the acidic hot spring water. This required more dewatering, lifting the pipe out of the trench to patch weld it, then suspending the pipeline above the river in 2002. All these things infuriated the new owner, as well they might. In 2005, I produced several reports as expert witness, when the lawsuit was scheduled for trial, but at the last minute the new owner settled with WCWCD out of court. By then we had compiled sufficient data and interpretations to contradict his "data." The owner went bankrupt in 2011 and WCWCD bought the property and water rights. Unfortunately, on the question of what to do with the salty spring water, it appears that the spring system is too big and diffuse to dewater by pumping, and nothing short of a desalination plant has been suggested as a viable solution.

In 2002, WCWCD had two other projects for me. One was to analyze the hydrogeology of the Gunlock well field, northwest of St. George, which needed a new pipeline to Santa Clara, a western suburb of St. George. To construct a pipeline required an Environmental Analysis (EA), one step down from an Environmental Impact Statement (EIS), which is generally required to get through environmental regulations for larger, more elaborate projects. Here, water from the south-flowing Santa Clara River, coming off the western side of the Pine Valley Mountains, was stored in Gunlock Reservoir, part of which was sitting on Navajo Sandstone. Just below the dam at the southern end of the reservoir, wells into the Navajo were pumped and the culinary water put into the pipeline. Gunlock Reservoir is an artificial-recharge reservoir that recharges the Gunlock well field, at the time St. George's largest. The Gunlock fault, a large north-trending basin-range fault almost as big as the Hurricane fault, had raised the western side of the Pine Valley Mountains. The fault is about a mile east of the well field. I observed that the yields in the wells were progressively larger as they came progressively closer to the

Gunlock fault. This is because most groundwater was moving southward (downhill) along fractures due to the fault, and these fractures were more abundant and larger closer to the fault. This movement of groundwater along faults and their related fractures (called fracture flow) is the same mechanism we studied, looking for nuclides, below the Nevada Test Site. My Gunlock analysis became a chapter in the EA.

The other project was to analyze a new culinary well, Sullivan's Well, drilled by WCWCD northwest of Hurricane, which showed a nice combination of artesian flow and a greater yield than they had anticipated. My report showed that the well penetrated rocks that had been shattered like glass when they had been folded long ago, so again southward fracture flow made for greater yields and a source area of the water in higher topography to the north made for a higher head at the well site, and thus artesian flow.

I had gotten to know geologists and directors of the Utah Geological Survey from the first time I set foot in Utah. In the early days, it was below average in quality among state geological surveys. It is now the best. The change began in 1981 when ***Genevieve Atwood*** *was appointed Director, the first female to be a State Geologist and Director. She had earned her B.A. from Bryn Mawr College and a master's in geology from Wesleyan University. As the story goes, in 1974, upset over Watergate, she went into politics as a Republican. She won terms to the state legislature in 1974, 1976, and 1978—and these political terms were her secrets to success (other than brains and hard work) when she became Director. Because now that she had the ear of the people with the money, she built programs. And she hired well. One hire was Grant Willis, whom I got to know at Marysvale and, in an act of genius, Genevieve hired Lehi Hintze part time after he retired from BYU. She left in 1989, married the great geophysicist Don R. Mabey of the USGS, and they started their own consulting business. Simultaneously, she began teaching geomorphology and geology at the University of Utah, earned a Ph.D. from there in 2006, and after Don died in 2016, she continued in earth science education as a volunteer and emeritus professor. Genevieve in 2018 received the Lehi Hintze Award for Outstanding Contributions to the Geology of Utah.*

The next step in the evolution of the UGS was the appointment of ***M. (Merle) Lee Allison*** *(1948–2016), who was Director from 1989 to 1999. With a Ph.D. from the University of Massachusetts, Amherst, he was well liked, as were Genevieve before him and Rick Allis after him. More important, he recognized the economic importance of geology to state citizens. All UGS publications now—and indeed the entire emphasis of UGS—have practical importance to Utahns, and he received good support from the state legislature as an advocate and as a charismatic leader with*

> *a great sense of humor and boundless energy. He also could recognize talent, and many of the senior geologists now at UGS are ones he hired. However, as noted above, he ran afoul of politics during the Salt Palace fiasco, so his next stop was Director of the Kansas Geological Survey (1999–2004) where, in addition to reviving that agency, he was instrumental in preventing "creation science" from being taught in public schools. Such enlightened views do not fare well in Kansas, however, so his next stop was Director of the Arizona Geological Survey from 2005 till his death in 2016 from a head injury from a fall in his home. He received many awards and commendations, for he was much loved and missed. He also was a leader in global efforts to develop and share digital data in geology at all three State Geologist jobs, on many national committees, and in close work with the Association of State Geologists. In many ways, he modernized all forty-eight state geological surveys. One story about him in his obituary published by Eos (visited online on 7/28/2019) is from his stay at UGS. He wanted the national Geological Society of America (GSA) meeting to be held in Salt Lake City, but this idea met with resistance from "the beer-loving segment of the geoscience community based on the perception that the city was dry." So, Lee got some Salt Lake businesses to supply kegs of beer that he had flown to a GSA meeting in another city, then advised the group that free beer was available at the UGS booth. In a few years, sure enough, the national meeting was in Salt Lake City!*

Richard G. Allis was Director from 2000 to 2018, bringing UGS almost to its current position, thanks to his steady leadership. Rick is Australian, with a Ph.D. in geophysics from the University of Toronto. Again, with emphasis on work that serves citizens, he has kept up funding until about 2016, when it began slipping because of Trump. Rick maintained a lively research program of his own, bringing in grants on innovative concepts on development of shallow geothermal energy resources in Utah using new technology. He was followed in 2019 by Bill Keach, with a B.S. in geology from BYU and an M.S. in geophysics from Cornell University. Bill had been in the oil and gas industry, involved in seismic interpretation and reservoir modeling (including for seventeen years at Halliburton) and, before his appointment as Director, his specialty was teaching applications of new technology to geologic studies at BYU and the University of Utah. It is early in his term, but he seems to be cut from the same cloth as the last three great directors there!

Such was the UGS organization that I joined through contract in July 2002 (the start of its fiscal year), when Grant Willis, by then in charge of the Mapping Program, approved money for me to begin work on compiling the Beaver 1:100,000-scale quadrangle, then the Cedar City 1:100,000-scale geologic map, to be published in preliminary form (2005 and 2006 respectively) by UGS. As with the Colorado Geological Survey, the money for my work came from Statemap part of the National Cooperative Geologic Mapping Program funds (my 1986 report on geologic mapping needs in the US came back to benefit me!). I had done much of the mapping in the Beaver quadrangle with the USGS, including having recently compiled the map at the same scale of the overall Marysvale field in the last year of my USGS career; it was published in 2002. Much of the digital file of that USGS map could be lifted from the USGS map, the southern part of which became much of the Beaver quadrangle, and much of the northern part became the Richfield 1:100,000 quadrangle, which Lehi Hintze published with UGS in 2003. Grant and I estimated that it would take us a mere two years to create and publish the preliminary Beaver quadrangle...but I am still working on the final version as I write in 2024! After that came the Cedar City 1:100,000 quadrangle, much of which I had already mapped when with the USGS, so I had a head start on it, too; Grant and I estimated that I could prepare it in three years. New field work on the Beaver and Cedar City maps began in late 2002 and occupied most of my time for the next several years. I paid Dave Maxwell and his students at SUU to create the digital (GIS) linework, and I included them as coauthors. The student with whom I collaborated most on the two maps was Garrett Vice, as I will discuss below.

I also began mapping two 7.5-minute quadrangles in the Mineral Mountains, on the northwestern side of the Marysvale field west of Beaver, Utah, thanks to funding from Robert (Bob) E. McDonald, a private mining and oil and gas consultant in Ojai, California. This mapping began with Bob in late 2001 and continued with him during the summer of 2002; in the late summer and fall of 2003, I continued it on my own. During the fall, I brought a USGS trailer onto BLM pinyon/juniper land at the base of the Mineral Mountains and hired Garrett Vice (see below), Bart Ekren, and Ernie Anderson to join Breccia and me for a week to help work out some of the thornier issues in the complex geology of this area. I stocked ice chests with T-bone steaks, salad, potatoes, and beer; we had a blast climbing up and down this high, rugged range! These two quadrangles were within the Beaver 1:100,000 quadrangle. The Beaver quadrangle, northeast of the Cedar City quadrangle, started in July 2003 and took me two years. The Beaver map area contains most of the eruptive centers of the Marysvale volcanic field.

> *I met geology student **Garrett S. Vice** in 2002, through the recommendation of Dave Maxwell and Bill Lund. Garrett, at the time working on his B.S. in geology at SUU, had taken GIS (Geographic Information Systems) classes from Dave and I paid him to do much of the GIS digital work on my Cedar City and Beaver maps. It seemed to Dave and me that he ought to be able to see the rocks he was going to put on my geologic maps, so I brought him to the field with Breccia and me, slowly at first because I had almost never used a field assistant before. Breccia, now relegat-*

> *ed to the back seat instead of the shotgun seat, got her revenge by leaning over his seat back and licking Garrett on the face; in a year or so, after we added Graben (Ben) to our family, Ben was doing the same. One of the most outgoing persons I have ever met, and perhaps recognizing that the SUU geology department had severe limitations to him ever becoming a good geologist, Garrett inserted himself into the lives of Bob Blackett, Bill Lund, and me. Soon all of us were paying him to do minimum-wage jobs while showing him how professional geologists operate. He soaked up all this knowledge like a sponge, plus he had a great sense of humor and was smarter than any of us! He was about five years older than the rest of the students in his class because he had taken time off to earn money for college and to travel, so he was mature and motivated. And his wife Candice was just like him! Soon he was accompanying me on weekends and, thanks to some nifty scheduling, on some week days too, plus much of the summer. Less than six months after beginning as a field assistant, he was helping me to map and I had to pay him more! During spare time on class days, he slowly put the map together. And he kept getting better and better as a field geologist, a job he clearly loved. When I was with the USGS, I had brought in Carleton students to be field assistants or lab assistants, and those kids were universally smart and motivated. Garrett was as good or better, and he became second author of the two 1:100,000-scale preliminary maps.*

From about 2002, the phone started ringing with calls from private landowners or small water districts who wanted me to site water wells. When we were with the USGS, Gary and I had done them for SNWA, Virgin Valley Water District, and others. And Gary continued them as a consultant. Such problem-solving is fun, at least if the client has a significant piece of land that might be crossed by a fault, but I charged by the hours spent and did not make much. I would compare a map of their property with a published UGS or USGS geologic map of the area from my library. If no geologic map existed, I would use my large collection of aerial photos, or buy or borrow them, and then do field work to make my own map. I would access the Utah Department of Natural Resources Water Rights website to see whether other water wells nearby had been drilled and what the likely depth to the water table was. Over the years, I tackled about thirty such projects, some requiring multiple wells. I also did a half dozen projects that involved writing reports dealing with water rights, intended for litigation or trials. Some included interference by a new well on a client's existing well, where the new well decreased the flow in the client's existing well. One project concerned litigation against a person living near a spring owned by a client in St. George, in which this person piped the water without permission, then "dug out" the spring with a backhoe, attempting to increase the flow but instead destroying the spring.

One of my first wells was for a local engineering firm contracted by the owners of a three-square-mile subdivision made up of twenty-acre lots ("horse property") west of the town of Kanarraville and northeast of New Harmony. Of course, they had already done the layout of roads and lots and had even sold a few lots before they worried about their common water supply! They had sought out a water witcher, then drilled two of his sites (one dry hole and another only 5 gpm) before they called me. I spent a couple of days looking at my aerial photos and doing field work, and found a fault that looked good. I planted a flag on the spot and they brought in a Cedar City-based drilling company, brothers Roger and Jay Grimshaw. I expected the water table to be at about two hundred feet but the drillers had gone to about four hundred feet without finding water, a significant concern. Then they lost the drill bit in the hole! They tried to retrieve it but could not. Over the years, I enjoyed working with drillers, most of them good, most of them hard working and salty, and all of them ethical. The Grimshaws had all those traits, and they owned that the loss of the bit was their fault and, noting that we had not hit water yet, they said that if I picked another site elsewhere on the property, that was fine and they would not charge the engineering firm for a second hole. I conferred at the site with the engineers and drillers, finally said "let's not second guess ourselves," and proposed they move to the western side of the drill pad, closer to the west-dipping fault and about twenty-five feet away from the old hole (so as not to hit it) and try again. They got about 170 gpm at about 450 feet. Years later, in August 2020, Jay hired me to write a report that explained the geology of his possible geothermal property at Newcastle, which Mary Siders and I had mapped in 1987.

In early 2004, the WCWCD asked me to provide information on a proposed artificial-recharge reservoir to be perched on the Navajo Sandstone. The reservoir, to be called the Sand Hollow Reservoir, was to be 1.5 miles long north-south and 0.6 miles wide, on BLM land southwest of Hurricane, that the WCWCD had acquired to be a well field. Some USGS hydrologists out of the Utah Water Resources Division (WRD) office in Salt Lake City had done a good deal of the testing regarding what yields would likely be, as the reservoir was dug but before it was filled (by a pipeline from the Virgin River but later to be by a pipeline from Lake Powell). The WCWCD had previously sited a half dozen wells, and they had been tested at yields that ranged from 200 gpm to nearly 2,000 gpm. This order of magnitude in yields was surprising, inasmuch as all wells were in the Navajo Sandstone and within a mile of each other. Ron Thompson always scheduled an exit meeting when projects were over, meetings in which he and other members of the staff quizzed the consultant about many things about the completed project. I always welcomed those meetings, which were good humored and which enabled us to speculate beyond our written report on questions that were surprisingly perceptive (for a BYU lawyer!). Ron told me about an exit meeting with the USGS project chief in which Ron asked him why a well on the western side of the projected reservoir yielded 200 gpm, whereas a well a few hundred yards off the northeastern side of the reservoir, less than a mile away, produced at 2,000 gpm. The hydrologist simply and somewhat evasively responded that such things can happen. Ron knew better. Then Ron asked him where he would put more wells. The hydrologist answered that it made no difference, for they would all be in Navajo so they would come in at somewhere between 200 and 2000 gm.

When Ron proposed this project to me, he asked me the same question. I told him that I would bet him twenty bucks that the low yield well was far from a fault, and that the high-yield well was on a fault. We proposed to map a thirty-five-square mile area, with the reservoir in the middle, at a scale of 1:12,000, then to propose additional well sites based on what the maps showed. I brought in Gary Dixon to help. To do maps that showed flow directions (potentiometric-surface maps), I asked Dr. Frank D'Agnese, a hydrologist who had worked for the USGS WRD out of the Nevada state office in Carson City before he became a consulting hydrologist out of Tucson. In December 2004, we turned in our maps and a fourteen-page digital report. Our geologic map showed a half dozen basin-range faults and, sure enough, I had been correct in my answer to Ron's question. We proposed about a dozen additional production wells around the perimeter of the reservoir. The WCWCD started drilling some of our wells a couple of years later, and our highest-priority well (because it intersected one of our mapped faults) was tested at 2,500 gpm. The reservoir, but not the well field, was handed over to Utah, where it became Sand Hollow State Park, one of the most popular parks in Utah because of swimming and fishing, as well as its beautiful setting, nestled among the red rocks of Navajo Sandstone. Few visitors know it is St. George's biggest well field.

In May 2004, Gary Dixon brought Bart Ekren and me into a large project for which we were employed to map (at 1:100,000 scale) the geology of an area in the Smoke Creek Desert and parts of adjacent ranges in Nevada and California. Our digital geologic map, cross sections, and report were to aid hydrologists, led by Terry Katzer, with a ground-water flow model of the basin and with a production water well-field for a water-cooled, coal-fired power plant to be constructed by Sempra Energy, one of the largest energy companies in the US. The Smoke Creek Desert, nearly a hundred miles north-northeast of Reno, was well named: it contained a dry lakebed left over from the Pleistocene, and at times when the wind blew—which it often did—clouds of white lake clay dropped visibility to a few hundred feet. The ugly town of Gerlach, near a gypsum mine and wallboard plant, and with a population of about five hundred then (considerably less now) and an altitude of 4,000 feet, was at the northern end of the basin. The only claim to fame in the area is that Gerlach is near the site of the Burning Man festival, which has taken place for a week in the heat of late August for more than thirty years, in the Black Rock Desert, the next basin north of the Smoke Creek Desert. This basin also was the site of a Pleistocene lake, so participants share similar dust experiences to Gerlach residents. This counter-cultural art event, with an apparent spiritual awakening guaranteed for all, now is attracting about 70,000 people (and a leave-no-trace trailer home/tent city) each year and culminates in the burning of a fifty-foot wooden effigy, "The Man." But when it rains, as it did for several days in 2023, the lake bed turns to deep mud that traps most of the revelers!

Field work on the project continued during 2004, generally at one- or two-week intervals. I brought in Garrett Vice as junior geologist, whose work during this summer built up his bank account sufficiently for him to afford grad school, which was to start in the fall semester of 2005. Our geologic reports were submitted to the client in 2006, but the power plant project was abandoned soon after that, because of the Great Recession and because coal-fired plants were being phased out due to economics. The groundwater evaluation and plans for water wells were sold to the large private LSC Housing Development northeast of rapidly growing Reno, and our large final CD report was issued in 2010.

> *To start the Smoke Creek project, Gary and I flew into Reno, where we collaborated with* ***James E. Faulds****, an outstanding field geologist at the Nevada Bureau of Mines & Geology (NBMG), affiliated with and on the campus of the University of Nevada, Reno (UNR). Jim had gotten a Masters at the University of Arizona (advisor George Davis), then earned his Ph.D. from the University of New Mexico in 1989. He was Assistant Professor at the University of Iowa from 1992 to 1997, then moved west to NBMG. Jim, with whom Gary and I became friends once he got to Nevada, also was a professor in the Department of Geological Sciences and Engineering at UNR. From 1988 to June 2012, the Director of NBMG and the State Geologist was Jonathan G. Price, so of course we talked to him about the project. Jon, with a Ph.D. in geology (1977) from the University of California, Berkeley, was a bright and likeable economic geologist. However, he spent much of his time sitting on committees, and for a year each was even the President of the Society of Economic Geology, the most prestigious national society for economic geology, and of GSA, when he should instead have been building up the NBMG. On the good side, however, he emphasized work by NBMG in gold districts, for Nevada is the largest gold-producing state in the US. But his outreach to citizens and the state legislature was limited. Furthermore, his mapping program consisted of only two bedrock geologists, Christopher D. Henry and Jim, and they were only mapping 7.5-minute quadrangles rather than the 1:100,000-scale sheets that Utah was pro-*

ducing. Considering the size of Nevada, mapping the state at this scale would take nearly forever. Jon hired Chris (Ph.D., University of Texas, Austin, 1975) in 1993 and Jim in 1997. Fortunately, Chris and Jim are two of the most able and productive field geologists in the country. When Jon retired, Jim took over as State Geologist and Director of NBMG in July 2012, and things changed radically for the better, although by then the NBMG had been caught in the general malaise of science in Nevada and elsewhere following the Great Recession and then Trump. But now regular newsy e-mails to the public became abundant, garnering support. Among the specialties that Jim emphasized, in fact his main line of research (with grants to NBMG), was in geothermal energy, for Nevada has many hot springs in areas of recent volcanism. Jim supervised Garrett Vice's master's thesis on a geothermal-energy topic, after which Garrett was hired by Exxon/Mobil in Houston.

As for family matters during this period, my mom put her home in Tucson on the market in December 2004. When it sold a few months later, I moved her into a two-bedroom apartment at The Meadows, a friendly retirement complex in St. George, about an hour's drive from our home in New Harmony. She made friends easily there but soon found that most of them had dementia or Alzheimer's and were not there mentally. Those that were played cards and did the other things that retired people did. Fortunately, Mom remained mentally fine, and even kept up her finances and correspondence with no help from me. She visited us on many weekends and most holidays, a good and happy respite for all of us. Mom died in her apartment at 1:30 a.m. on December 20, 2012, at the age of ninety-four. We talked memories and she asked me for, and received, a sip of red wine a few hours before that. Then she fell asleep.

My wife Dawna's mom, Evangeline Brown, continued to live in the family home in which Dawna had grown up, located in Augusta, Maine. She had a "boyfriend," Robert Murphy, whom she could not marry because her primary means of support, besides their Social Security checks, was a pension to Dawna's dad that continued even though he had died many years before, within a year after his retirement, access to which pension depended on her not remarrying. We visited Van and Bob a few times, and they visited us once. Dawna had a sister, Linda, who survived on welfare nearby in Augusta. Van and Bob, however, remained mentally alert and active during their long retirement, near others of an extended family (the Paquins) of French-Canadian descent (Bob was Irish). Bob died on December 25, 2009, Van on February 24, 2012, and Linda on March 17, 2014.

In July 2005, Grant Willis hired me for a two-year project to geologically map and publish the Goldstrike quadrangle (1:24,000 scale) and the Utah part (two to three miles west of 114° longitude) of the adjacent Docs Pass quadrangle, Utah. This map area was an unmapped hole within the St. George 30' x 60' quadrangle (1:100,000 scale, south of the Cedar City quadrangle), Utah, and I was also to assist (as second author) with that quadrangle, which was being compiled and mapped by ***Bob Biek****, Senior Scientist with UGS. This was my first collaboration with Bob, arguably the best and most productive geologist in the Mapping Division of UGS, about seventeen years younger than I am and one year older than Dave Hacker. Bob had been hired in 1996 by Grant and by Director Lee Allison, and was probably the best hire they ever made. From 1992 to 1996, he was a geologist with the North Dakota Geological Survey, where he mapped seven 7.5-minute quadrangles and Theodore Roosevelt National Park, among many other duties as assigned. He had earned his B.S. in geology from University of California, Berkeley, then an M.S. in 1988 from Northern Illinois University, where his thesis was mapping the Nephi 7.5-minute quadrangle in Utah under the direction of Professor Malcolm P. Weiss (1922–2002), a legendary field geologist whose research was mostly in central Utah. The Nephi quadrangle was funded by UGS, and that finished product (published in 1991 by UGS) and the field-trip review that was required for the funding identified a quality geologist, and when Grant got the opportunity, he hired him. As of 2021, when Bob retired after twenty-five years with UGS, he has published thirty-eight 7.5-minute geologic quadrangles, of most of which he is sole or senior author, and twenty-four reports. But he is best known for four 30' x 60' quadrangles, of three of which he is senior author. Inasmuch as these, as well as some of the 7.5-minute quadrangles, deal with the world's largest landslide complex on the land surface, I will continue his story about that slide in the next chapter. I quickly became friends with Bob, as soft-spoken as he is a terrific field geologist, and Bob and I collaborated extensively thereafter. His wife, Lisa, is an artist, primarily a painter, and Bob had similar interests, expressed by throwing and glazing pottery and making mosaics for hot plates and trivets for fun, then selling them. In 2021 they moved to Fort Bragg, California, where he is finishing his maps under contract with UGS. Bob received the Hintze award for "outstanding contributions to the geology of Utah" in November 2024; Bill Lund and I nominated him.*

The southern half of the Goldstrike quadrangle consists of the Goldstrike gold district, which had a recently active open-pit mine. Even though access to the mine was good, the rest of the quadrangle was accessible only by scary jeep trails through country with a lot of relief that had mostly been burned over by forest fires. I hired Jonathan Boswell, another SUU student, as a field assistant, and we camped out most of the time because it took hours to get home. Ben and Breccia were in the back with the camping gear, occasionally poking their heads into the front seat to give us kisses. Again, I supplied T-bone steaks and potatoes, which we cooked on a campfire, plus we had salad and chocolate-chip cookies; the dogs got the bones as well as their dog food. The mine was along large west-northwest strike-slip faults, and others of this type occupied the rest of the quadrangle, so I sampled extensively to look for other gold deposits. One day, near the end of a long traverse, we had to cross a deep canyon to get back to the car, and we mostly slid down the slope through oak brush to get to the bottom. Just before we reached the bottom, we came to an area of rock hidden by oak that was chock full of quartz veins, a giveaway (so I thought) for a possible gold deposit that nobody had discovered before. We gathered many samples, then all of us scrambled to the bottom and up the other side, thinking we might have hit paydirt! Unfortunately, when the analyses came back six months later, though gold values were present, they were minimal. The preliminary Goldstrike map area was open filed in 2007. The preliminary St. George map, containing my map area and other western parts I had compiled, was open filed in 2007 and published in final form in 2009. Bob had written an expanded text, and the map received an award. I always considered it one of the best 1° x 2° sheet ever done, comparable to Ric Page's Las Vegas sheet.

The years 2006 through much of 2008 were busy ones, as Utah and the US boomed in the days leading up to the Great Recession in the fall of 2008. Land developments were being proposed all the time, and I was tasked with siting water wells for at least a dozen of them. From July 2007 to June 2008, Grant Willis hired me to map the two-to-three-mile-wide strip of Utah west of the Cedar City quadrangle to the Nevada border, to be published with the Cedar City quadrangle. It included the Stateline gold district and I found two calderas that formed the eastern end of the Caliente caldera complex. That map, at 1:24,000 scale, was open filed in July 2008. The SNWA cooperated extensively with the UGS, which has an excellent program in hydrology, in the White River and the Great Salt Lake Desert flow systems that straddled the state line and, in the fall of 2009, we all led field trips and published reports in the Utah Geological Association annual guidebook.

The year 2008 was the beginning of the Great Recession, the worst worldwide economic downturn since the Great Depression of the 1930s. Few escaped it, and for some it was a disaster. The US went into recession in July, Fannie Mae and Freddie Mac went bankrupt in August, and the stock market crashed in late September. As with the warning signs leading up to the terrorist attacks of September 11, 2001, George W. Bush was largely asleep at the wheel. Looser banking regulations pushed by Bush had exacerbated the recession. On September 29, the Dow Jones Industrial Average fell almost 800 points, then the largest daily point drop in US history. The immediate cause of the crash was because the Congress rejected the TARP (Troubled Asset Relief Program) proposal, designed to provide funds to businesses (for jobs!) impacted by the recession. When Congress woke up in early October, they passed TARP to the tune of $700 billion. When President Barack Obama assumed office in January, the country was finally able to appoint qualified economists to deal with the catastrophe. They passed a hard-fought stimulus package ($787 billion, later revised to $831 billion), known as the American Recovery and Reinvestment Act, in February 2009, and the Dow bottomed out in March. Economic recovery was slow, essentially invisible until 2010, because the Republicans had argued vociferously against a higher stimulus, and in the end no Republicans from the House of Representatives voted for it and only three Republicans from the Senate did. The Trump administration lied about this history and accused Obama of causing the Great Recession. In fact, the Obama administration saved the country from a depression!

With the recession, construction of homes and businesses stopped in southern Utah; contractors and engineering firms drastically cut employees or went under. Some of those projects for whom I sited wells were never built and the wells never drilled, but the only non-payment I suffered was when a golf course in Apple Valley, east of Hurricane, went belly up. In the fall of 2008, the Utah Geological Association (UGA) and UGS announced that I was to receive the Lehi Hintze Award for "outstanding contributions to the geology of Utah." Grant Willis and Bob Biek proposed me for it. Lehi F. Hintze (1921–2014), for whom the award had been named in 2005, and Grant were there to present it to me, a real highlight at a time when it looked as if the world were falling apart! Lehi, whom I considered the best geologist in Utah, had retired from BYU in 1986 after a thirty-year, spectacular career as professor in which he was much loved as he taught students mapping and stratigraphy. His mapping and report writing for UGS continued until 1997.

Washington, an eastern suburb of St. George, has a north-trending, basin-range fault, named the Washington fault, running through it. The fault has been geologically mapped for about forty-four miles, from northern Arizona to just north of Washington. At least parts of the fault had been trenched and shown to be Pleistocene, although not yet proved to be active (that is, showing evidence of movement in the last 11,700 years). A ten-to-fifteen-foot-high, west-facing fault scarp is spectacularly visible about fifty feet south of the four-lane I-15, which in Washington goes east-west across the fault. Houses, not to mention I-15, other roads, and businesses, have been built on, below, and across the fault scarp for years, with no setbacks, let alone engineered structures, because developers did not recognize that the low mesa on the east is the upthrown side of a fault. In June 2008, I joined David R. Black of Rosenberg Associates engineering firm, St. George, and Dave Simon in a proposal to study a thousand-acre parcel owned by SITLA (State of Utah School and Institutional Trust Land Administration) land that straddled the fault just north of the interstate. Rosenberg is one of the best and most ethical engineering firm in St. George, and Dave Black was one of the best engineering geologists in the state. The SITLA manages land ("school land" sections) given to Utah at statehood, about 6 percent of Utah, to generate revenue for its citizens. At this point, SITLA was considering the sale of the property so that it could be developed for homes, at the time during a housing boom. But because the fault had not only been mapped but was well known by 2008, it could no longer be ignored. The first step was to map the half-a-square-mile area in detail (1-inch equals 300 feet), which I did after approval of the project in the fall, and Rosenberg put together and submitted our map and report in February 2009. It showed the fault and several splays crossing the SITLA property. We recommended that we trench the fault to provide clear evidence that the fault was active, but SITLA chose to put no more funds into the study to discover what seemed likely, and the land remains undeveloped to this day.

Following hard on this study was another one on the Washington fault, but this one south of Washington city. By now we were clearly in a recession, and pretty much all construction in the US had ground to a halt. To save jobs and businesses, the huge stimulus package ($787b) that President Obama passed, over the objections of Republican politicians (yet the Utah delegation showed up in force for all ribbon cutting!), provided money for construction projects across the country. Some part of that huge pile of money came to southern Utah, where it funded construction by the Utah Department of Transportation of a four-lane beltway, the Southern Parkway that by-passed metro St. George by going around its southern side, and of a new airport for the city. Obama's stimulus money kept many construction and engineering firms and their employees solvent. The Southern Parkway appeared to cross the Washington fault, and that was where we geologists came in, with Rosenberg again taking the lead with a proposal. When it was accepted, the first task was to do a geologic map of the apparent area where the projected roadway crosses (we found that it did) or passes near (it also did) the fault. In 2009, Rosenberg published an initial report, with my geologic map (1-inch equals 400 feet) of a ten-square-mile area. As before, we recommended digging a dozen trenches that Dave Simon would do, and this time our recommendation was accepted, along with additional geologic mapping that I did. This work kept us busy through the winter of 2009 and spring of 2010, with a final report in the fall of 2010. Dave's trenches, and the isotopic dating of the deposits cut by the fault, showed for the first time that the Washington fault was active (Holocene), with one surface-faulting earthquake about 1,500 years ago and an earlier one about 7,000 years ago. A projected interstate interchange had to be moved off the fault. Dave Simon senior authored and published these findings in 2015.

Anderson Junction is about eighteen miles northeast of St. George and just northwest of the small town of Toquerville, where Utah Highway 17, on the east, joins I-15. The Junction is about midway, from north to south, along the thirty-mile-long and fifteen-mile-wide Ash Creek basin, which is drained by perennial to intermittent Ash Creek from its headwaters and tributaries in the northern end of the basin, southward to where it joins the Virgin River. The Junction, west of the Hurricane fault scarp, is underlain by the great 2,000-foot-thick Navajo Sandstone. Therefore, as at Sand Hollow, the WCWCD decided to put an artificial-recharge reservoir east of I-15 on top of the Sandstone to recharge this aquifer, with wells around the edges of the reservoir to pump the aquifer. The reservoir was to be 0.75-mile-long by 0.5-mile-wide, and to be recharged by Ash Creek and several tributaries that drained the Pine Valley Mountains. In September 2009, the WCWCD asked me to map the area geologically, discuss the hydrogeology, and propose wells for the wellfield. The small Cottam wellfield, also pumping in Navajo Sandstone west of I-15, was included. I brought in colleague Gary Dixon. I mapped a five-square-mile area at 1 inch = 500 feet in the fall and winter of 2009 to 2010, and I suggested locations for four production wells and four monitoring wells.

However, this ended up being a project that never went away, for an Environmental Analysis (EA) by BLM was needed. Being perpetually understaffed, BLM does not do these themselves but contracts them out to big consulting firms, virtually all employees of which possess the twin characteristics of being incompetent and unable to write. In mid-2012, I was back at it to assist the engineering firm and WCWCD to provide additional hydrogeologic information and alternative scenarios for the EA. The overall project, however, remained dead until June 2016. Eventually Barbara Hjelle tasked me with a report on the hydrogeology of the Ash Creek basin. Completed in October, it dealt with the source and destination of groundwater and spring water. I suggested that some of the southward flow of groundwater passed beneath the Virgin River and likely on to the Grand Canyon. Also, I concluded that the water resources in the basin were far greater than previously estimated, largely because of the role of fracture flow, whereby groundwater was moving southward along fault zones that defined the eastern and western sides of the basin. The conclusions had to be published, which I did with the Utah Geological Association in 2018 at no charge to the WCWCD.

In the spring of 2010, the UGS had some extra energy-lease money (the price of oil was high) and they funded about ten energy-related projects in Utah, for bid by geologists outside UGS, with the reports to be published by UGS. I proposed to create a geological map (at 1:12,000 scale) and to analyze the geothermal-energy prospects for the Sulphurdale geothermal-resource area south of Cove Fort, Utah, on the northern side of the Marysvale volcanic field. At the time Sulphurdale, long ago a sulphur mine, then a ghost town, was the site of the smallest of three geothermal power plants in Utah. But Enel North America, Inc., had recently bought the plant with the intention of tearing it down and replacing it with a larger one using binary technology. Before the old plant had been erected and before power had been produced, the area had been studied by the University of Utah Research Institute using soft money, but that was in the 1970s and the mapping was at 1:24,000 scale. I brought in Edward F. Rutledge, a consulting geologist and geothermal expert from Payson, Utah, and I collaborated with Gary Dixon and Chet Wallace. In mid-2011, Ed and I, with my dogs, finished mapping about fourteen square miles, using five cross sections. The map suggested that the area was underlain and supplied its heat by an underlying rising, probably partly molten pluton. The UGS published it in 2013.

I was expert witness in litigation of several water-rights issues involving interference in the New Harmony area in 2008 to 2010. Interference takes place when pumping a new water well causes a drop in yields or water levels in nearby springs or older wells, and this affliction is widespread in Utah where water rights are generally granted with little thought to groundwater availability, and where over-appropriation or "mining" of groundwater is common. We won in court. Another interference job began in early 2011, when a well partly owned by Elliott F. Christensen, a former consultant for the Church of Jesus Christ of Latter-Day Saints (LDS or Mormon Church) was accused of causing interference. Elliott's land containing a large well lay a half mile from Sawyer Spring, about three miles south of New Harmony. His well supplied the drinking water for the community of Mountain Springs about a mile east of Sawyer Spring, which owned the rest of the well. A person who owned a spring just east of Sawyer Spring claimed that Elliott's well interfered with his spring, and he paid an unscrupulous consultant to back him up. I mapped several square miles and built a case for Elliott with a report, using pump data not only for Elliott's well, Sawyer Spring, and the man's nearby spring, but also for several big production wells that the Mormon Church owned northeast of Sawyer Spring that supplied center-pivot irrigation systems to grow alfalfa. The interference case was thrown out of court. As I finished Elliott's case, I realized that this was the best example I'd ever seen of conduit flow along a fault zone because the man's spring and several wells (but not the one on Elliott's land or Sawyer Spring) interfered with each other because they were tapping the same fault. Yet this fault and faults west of it also prevented flow across them (barrier flow) from Elliott's well (west of the fault) to the man's spring. The remarkable thing about this study was that we had pump records, unavailable for most cases of interference because wells are rarely aligned in correct places. I brought in Gary Dixon, Andrew Burns, and Jim Watrus to keep me honest by reviewing what I wrote, and we published the report free of charge with the Utah Geological Association (UGA) in 2013. Elliott was second author.

The Black Mountains, north of Cedar City and south of Minersville, is almost entirely made up of Tertiary volcanic rocks, which I mapped in 1969 and 1970. The exception to the volcanic rocks is Black Mountain, in the northern Black Mountains. This mountain, which rises to about 7,000 feet elevation, is made up of Paleozoic rocks (older than about 260 million years old). It rises about 1,000 feet above the surrounding hills, like a giant piston about a square mile in area. In other words, it had been uplifted 3,500 feet with respect to the surrounding (overlying) much younger volcanic rocks, along north-south basin-range faults and cross faults. A small mine on Black Mountain had shipped a small volume of silver ore many years ago. These bits of evidence triggered an investigation by a self-employed economic geologist, Keith Norlin of Australus LLC, on behalf of some investor colleagues. They reasoned that the uplift was due to an underlying pluton, and this fit with a model they were developing for silver-lead-zinc mineral deposits. In late 2011, they ran geophysical (electric) profiles across the area and drilled two holes from which they extracted drill core. Keith had learned that I had previously mapped the area and asked that I do detailed mapping to test their model. I settled on an area of six square miles, which my dogs and I mapped in April 2013 at 1:6000 scale. My work, issued soon after as a report and map, confirmed their uplift model. The study added details I had missed in my old, less detailed mapping: I found that the sharp uplift, which probably took place about twenty million years ago, had generated a Miocene landslide that covered an area of at least ten square miles. Unfortunately, my study was not enough to get them to put any more money into site development. But, free of charge to Keith, I published the map and a summary report with UGA in 2014, with Keith, Dave Maxwell, and Dave's son/student (who did the map GIS) as coauthors. In August 2019, an economic geologist from Arizona called me to talk about this prospect, for his model indicated that it contained gold deposits. I pointed him to our report. Some mineral localities intrigue people for years.

The year 2016 was the first of several bad years in which I lost some of my closest friends, resulting in writing memorials to them. My wonderful friend Gary let colleagues know that year that he had been diagnosed with colorectal cancer, although he never let us know how serious it was. I continued to call him regularly (generally every other week), as I had done for years, and we collaborated on our big final report on the White River and Great Salt Lake Desert flow systems for the Nevada Bureau of Mines & Geology, as well as on the report on the Ash Creek basin for WCWCD. For most of the concepts were his, which he had patiently taught me for nearly twenty years. Garrett Vice had scheduled a talk and an Exxon/Mobil recruiting trip to Montana State University, not far from Gary's home (Blackfoot, Idaho) and that of Bart Ekren (White Sulphur Springs, Montana) in the fall of 2016, and I drove up to join him. We stayed in a motel in Blackfoot, spent the afternoon and evening with Gary and Wendy, then drove on to White Sulphur Springs the next day to see Bart. The Sawyer Spring report went into press while Gary was sick, and he died in January 2017. The Ash Creek report was published by WCWCD before Gary died, with the UGA version published after he died. Our memorial to Gary came out in 2017.

John Anderson died in October 2017 of old age (87) in Seattle, at the apartment he and Linda had shared for a decade near his daughter, Janet. Bob Biek and I kept him abreast of our progress on his/our giant landslides (which I describe in the following chapter). Our memorial to him came out in 2018. In late 2019, Bart Ekren's daughter Sally Brainerd called to tell me that Bart had died of a recurrence of prostate cancer in August, near her family in Livingston, Montana. Sally had found my phone number among Bart's stuff that she was sorting through, under his instructions to contact me. Our memorial to Bart came out in 2020.

With his retirement in 1992, Paul Williams, Jean, and Marjorie returned from Saudi Arabia, after an absence of ten years, to their home in the mountains west of Golden (west of Denver). But to unwind he needed to return to mapping, which he did until 1995 with the USGS covering the field work (expenses only) to produce geologic maps of a 30-minute area (four 15-minute quadrangles) in the Cassia Mountains of Idaho. Then Paul and Jean moved to Santa Fe and again, also as an Emeritus Geologist and free to the taxpayer, Paul senior authored the Albuquerque 1:100,000 quadrangle, which was published in 2007. Before it was published, they moved back to Colorado to be near their kids, settling in an apartment in downtown Golden. In October 2017, Jean was on her morning walk in Golden, near their apartment, when she collapsed, perhaps from a heart attack, and died right after she phoned Paul. After Jean died, Paul moved to a small retirement-apartment complex in Driggs, Idaho, near his daughter Marjorie, her husband Stanley Edwards, and their high-school-aged children. Art Ford, Ken and Linda Pierce, I, and others stayed in regular contact with him by phone or visits. Paul worked on a book about his family history, which can be traced back to the Mayflower. Stan is a multilingual consulting agronomist who travels the world, advising on organic farming, and Marge is a consultant and teacher of deaf children. Paul died of old age, aggravated by respiratory disease, on March 20, 2020. Our memorial to him came out in 2020.

Although I continue to do consulting work as of 2024, most projects are small. I only do the interesting ones. As the Great Recession settled in, during 2009 and 2010, President Obama's stimulus money dried up, and the price of oil continued to fall (Utah Geological Survey gets a percent of their revenue from oil- and gas-lease money). Then came Trump.

Although UGS did not have to lay off employees, retirements were encouraged, with staff not replaced; field funds were curtailed. Yet it was at precisely this point that one of the most exciting finds occurred in Utah: three of the world's largest terrestrial landslides! Though the credit for the initial discovery goes to John Anderson and his students, Bob Biek and Dave Hacker put the whole thing together. But I was in the middle of it. The UGS had little money, yet I was happy to work free. That story is next.

CHAPTER 11

SLC
Marysvale volcanic field
Map Location
MOUNT BELKNAP CALDERA
MOUNTAINS
TUSHAR
Milford
257
Minersville
20
130
VALLEY
VALLEY
RED
HILLS
PAROWAN
Paragonah
Parowan
CEDAR
Cedar City
113
15
Marysvale volcanic field
extensional deformation
60 miles
compressional deformation
breakaway zone
sidewall breakaway
bedding-plane slip surface
Mount Belknap pre-caldera uplift
ramp
Panguitch Lake
debris-avalanche deposits
former land surface
Sidney Peaks area
Isom Fm
Wah Wah Springs Fm
Brian Head Fm
Leach Canyon Fm
Isom Fm
Brian Head Fm
Claron Fm

Geology is the most practical of the sciences. Few people in the US are not affected each minute of each day by what geoscientists discover: metals; petroleum products for energy, for fibers in clothes, and for plastics; and gem and other rock products for many uses. Enough clean water is a requirement of life, and geologic knowledge helps protect the public from natural geologic and hydrologic hazards. Enabling this relationship between people and geology is best done by geologic mapping. Our studies of the Nevada Test Site and Las Vegas

GEOLOGIC MAPPING AND THE LARGEST GRAVITY SLIDE ON LAND

area, described in Chapter 9, were largely funded by outside organizations. We found through those studies that the products such organizations most desired were 1:24,000- to 1:100,000-scale geologic and geophysical maps. Such maps, by skilled field geoscientists, provide the geologic framework of an area and allow collaboration and analysis by those of many geologic specialties. Geologic maps are the basis for broad interdisciplinary studies of the earth's crust.

Discovery by mapping is not always a simple linear exercise in the sense that scientists do not always piece things together one after another in an orderly or timely way. This chapter will chart the process of a single discovery—that of the world's largest gravity slides on land—to show you what I mean. For in this example, the discovery of the slides, by new mapping, came long after initial mapping of the Marysvale volcanic field had been completed. Discoveries of most mineral deposits currently take place after—and at times long after—initial mapping missed them. In other words, mapping must be done multiple times, using multiple or new concepts, different persons, and more detailed map scale. In fact, USGS mapping is commonly done with that in mind: publishing our maps at a regional scale will generally prompt a mining company to ask for detailed mapping of particularly favorable areas. Or years after initial mapping has been done, concepts change and we know more about what features, such as landslides or low-angle faults or porphyry copper deposits, are recognizable.

In a 1998 book on his Antarctic trips, John Behrendt contrasted inductive versus deductive reasoning as an approach to research. Inductive reasoning, or a bottom-up approach, involves gathering data during research, then formulating the overarching hypotheses to explain them. John's geophysical work in Antarctica, where little was known about the mountains or ice sheets he visited, falls into this category, as does most geologic mapping. Deductive reasoning, or a top-down approach, takes a hypothesis and tests it by gathering observational data, resulting in a modification, or even the destruction, of the hypothesis. He pointed out that most research proposals for funding involve applying a deductive method, and the disadvantage of this is that the committee that must decide about which project to fund inevitably finds itself choosing which research problem is the more important. Our work on the world's largest gravity slide was inductive at first, as our field data indicated a gravity-slide origin, then became deductive as we filled in the details and tested the overall favored origin. We did not have to worry about proposals for money because funding for us was minimal during most of the effort.

What are the reasons why some discoveries are not made during initial mapping? One is that map scale may be not what is needed to solve a problem. Alternatively, researchers may come and go without the time needed for one to finish all mapping, or cessation or decrease of funding may not allow completion of systematic mapping and analysis. A scientist may not have the required specialty or sufficient talent to make the discovery. Or a perfect outcrop where evidence is obvious may not be found. Concepts may not exist in the literature that alert a scientist to the necessary evidence. Or, more simply, a scientist may be looking for something else and may miss evidence for something that is unexpected. In other words, he/she sees what she/he is looking for and may miss the other stuff.

This chapter takes you on this ride, and shows you the length of time it may take for a problem to be identified, then solved. Therefore, this chapter provides a classic example of how scientific inquiry, discovery, and publication of results may take place, and how inquiry and discovery will continue, as with all scientific studies, and advances will go on far beyond today. In short, there is no "end" to a scientific study! The old song "The Weight" by The Band is one of my favorites, yet it has nonsensical lyrics that have been over-interpreted on the internet, thus as surreal and as enigmatic for a long time as our *gravity slides*.

To summarize the history of concepts that led to finding the world's largest gravity slides, the first step was to map the Iron Springs mining district, southwestern Utah. This began in 1939 to support the war effort. The leader of this study, Professor J. Hoover Mackin, had been recognized for his exceptional talent and therefore was hired by the USGS for this important work, requiring him to be put on leave from the Department of Geology, University of Washington. Along with industry-backed geologists and geophysicists as well as colleagues from the USGS, the result was rapid recognition and exploitation of what soon became the largest iron-producing district in the West. Blast furnaces to turn ore into steel were built. The steel resulted in tanks, aircraft, ships, etc. that now could be constructed relatively near the ports on the west coast. More important, Hoover also discovered how the iron deposits formed, which made finding more ore bodies easier. But the immediate objective was to find and extract the ore, so writing about scientific discoveries (including how they formed) had to wait. After the war was won and Hoover returned to Seattle, he began to put his hypotheses into print. He had concluded that the iron deposits resulted from emplacement in Miocene time, later dated at between 22 and 20 Ma, of laccolith plutons that rose rapidly above huge batholiths at great depth and produced mountains at the surface. Next, his main job was to enlist and teach graduate students, first at the University of Washington and then at the University of Texas, to add to the work. He and his students found that, as one of the plutons, the Iron Mountain laccolith, rose, the sedimentary rocks that were forced upward by the pushing molten plutonic mass (piston-like, as at Black Mountain or the bulge at Mount St. Helens) became progressively over-steepened, such that they failed and produced large landslides off the steep flanks. These landslides, which traveled at least a half dozen miles off the mountain, were of course the same age as the plutons. Hoover gave the deposits of broken (brecciated) rock the name gravity slides. This name was to invoke a different thought picture, for these landslides were different from what one thinks of as landslides. The features called landslides, in fact, make life miserable for highway engineers and housing developers in mountainous and even hilly country, and they lie at the base of slopes due to gravitational failure of the slopes, resulting in unstable, unconsolidated masses of broken rock. They are young (Holocene or late Pleistocene) and are easily identified in the field and on aerial photos. The gravity slides off Iron Mountain were older (consolidated), much larger, and not so easy to recognize.

One of Hoover's first graduate students, Earl F. Cook (1920–1983), was sent south to map the Pine Valley Mountains, west of New Harmony, whose upper part Hoover had recognized as another laccolith, but one that was much bigger than usual. Earl's mapping of more than eight hundred square miles, including basins to the east and west, was of reconnaissance scale, two inches to the mile (1:125,000). Because of its large area, some of the work necessarily resulted from interpretation of aerial photos rather than footwork. Earl's 1954 Ph.D. dissertation was published in 1957 by the Utah Geological Survey (then called the Utah Geological and Mineralogical Survey). In 1960, Earl extended his reconnaissance work by publishing a reconnaissance map and report on all of Washington County, an area of about 2,400 square miles. The Pine

Valley laccolith has an areal extent of about 80 mi^2 and a pluton thickness of about three thousand feet, and Earl recognized it to be the largest laccolith in the world. Earl discovered that the northern flank of the laccolith contained lava flows that erupted from the laccolith. This extrusive phase, in addition to the intrusive phase, would come to be typical of other laccoliths in the general area, which fell along a northeast-trending line or belt of plutons known as the Iron Axis.

It is unfortunate that I never met Earl Cook, for he was a brilliant scientist and teacher. Later in life, Earl was Director of the Idaho Bureau of Mines and Geology (1957–1964), then Professor of geology and Dean (for ten years) of the College of Geosciences of Texas A&M University. In 1976, he published an authoritative and fascinating book, *Man, Energy, Society*, that was the text for a class he taught at Texas A&M. Although the title might suggest that it was a history of the oil industry, it was far more than that, for it touched on many issues, including problems of population growth, of availability of energy resources, and of pollution catastrophes that were then starting to become clear. Even though the book was written before climate change had become widely recognized, it urged humankind to rely on renewable energy. Among the interesting aspects of the book was Earl's idea to count food as an energy resource. This led to his review of the historic path of humankind, near the beginning of which was human's discovery of fire. This discovery allowed better digestion of plant and animal food, resulting in some leisure time (from fulltime hunting and gathering) for thinking, cave painting, etc. Further discoveries, such as use of dogs, growing crops, building permanent homes and towns, and the like, led to more leisure time, resulting in the start of civilization. Fast forward to the Greek civilization or the Renaissance and beyond, to today, where getting and processing sufficient food to survive is for most people not a full-time job, allowing time to pursue art and, of course, science!

Another early student of Hoover's was H. Richard Blank, Jr., who finished his dissertation in 1959 on the Bull Valley Mountains west of the Pine Valley Mountains. In this area, on the western part of the Iron Axis, he mapped another intrusive (laccolith) and eruptive complex, which had resulted in a small iron mining district that was similar in origin to Iron Springs. And Paul Williams, working with Hoover and Earl from the beginning, was instrumental with them in developing the ash-flow concept and its meaning in understanding the geologic history and resources of the Great Basin. Later (1967) Paul summarized his findings on ash-flow distributions and their caldera sources in another Hoover-supervised Ph.D. dissertation.

Two of Hoover's last grad students, John Anderson and I, were the only Mackin students who climbed out of the Great Basin to map in the Colorado Plateau to the east. When John finished his Ph.D. dissertation in 1965, after mapping 250 square miles (scale 1:62,500, one-inch equals one-mile) in the northern Markagunt Plateau, the dedication to the printed copy he gave me said: "For Peter Rowley—Now let's get down to work and find out what the geology really is!" I finished my Ph.D. thesis in 1968, after mapping six hundred square miles (also at 1:62,500) farther east, in the southern Sevier Plateau. We both mapped Miocene laccoliths, but neither of us found any gravity slides. With his dissertation done and a new job at Kent State University in Ohio, John began getting his students to map outward from his dissertation area. I also began mapping outward and, with colleagues Tom Steven and Skip Cunningham, we all tackled the huge, metal-rich, mountainous Marysvale volcanic field.

In early 1970, James Judy, one of John's students mapping south of John's dissertation area, found large areas of crushed, brecciated, and broken slabs of rock that were capping high parts of the Markagunt Plateau. Then other students, including Thomas A. Iivari, found similar deformed rocks. One of these high parts was immediately east of Brian Head, the highest point on the Markagunt Plateau (11,300 feet) and the site of popular Brian Head Ski Resort, adjacent to Cedar Breaks National Monument. Clearly these broken rock masses belonged to one or more gravity slides of Miocene age. But where did they come from? There were no higher areas from which they could have been derived, so clearly the Markagunt Plateaus had risen to its present height *after* the gravity slide had been emplaced. And, even though it was an unconsolidated mass of rock pieces when it had moved, time had now hardened it into a consolidated deposit (i.e., it was solid rock). We considered the possibility that the slides came from the laccoliths we had mapped, which resembled and were only a few million years older than the Iron Springs laccoliths, but field evidence was slim and such a hypothesis was speculative. In the 1980s, John mapped areas in collaboration with USGS geologists Ed Sable, Florian Maldonado, and David Moore, and they all came up with hypotheses about the gravity slide(s). In 1990, John's beloved wife, Linda, suffered a stroke, and John took early retirement from Kent State. He could only spend a few more years assembling what he knew, and in 1993 published a summary in which he described the best exposed parts of this broken rock body and called it the Markagunt Megabreccia. He estimated the size at about two hundred square miles. In 1995, the USGS RIF took place and the BARCO project, which supported John, was canceled. John's USGS colleagues, including me, were forced to do other things or, in Ed's case, retire. Many questions about the Megabreccia remained.

After the RIF, Dave Hacker began a Ph.D. study of the Pine Valley laccolith, plus three other laccoliths extending north into the edge of the Iron Springs district and west of the Pine Valley Mountains into the Bull Valley district. His effort demonstrates the importance of a new generation of mapping, for Dave was attacking the areas that had been previously studied by Earl Cook and Dick Blank. But Dave was doing it with new eyes and through detailed (1:24,000-scale) mapping. Dave was John Anderson's last graduate student, but soon Professor Daniel H. Holm of Kent State had to take over the official supervision of the thesis. I supervised the field work because I knew the area and was soon to live there. Dave mapped more than six 7.5-minute quadrangles, about three hundred square miles. In the interim between the work of Earl Cook and Dave, several theses (not supervised by Hoover) and published reports disputed all or parts of Hoover's and Earl's laccolith interpretation. Dave confirmed that virtually all of Hoover's and Earl's interpretations held up. Then Dave found that not only had Iron Mountain produced gravity slides, but so had the Pine Valley laccolith and several others. During the rapid rise of laccolith magma, as close to the surface as about 650 feet, the unstable mountainous domes had failed. The largest gravity slide, off the Pine Valley laccolith, had traveled west for twelve miles to cover an area of about sixty square miles to a thickness of as much as 1,800 feet. In addition, the slides removed part of the roof of the upward moving laccolith magma chamber, like pulling the cork out of a bottle of champagne (as at Mount St. Helens), resulting in eruption of ash-flow tuffs and/or lava flows. Dave Hacker, with his 1998 dissertation and subsequent publications, had now become the world's expert on laccoliths and gravity slides! Not surprisingly, not all his ideas were accepted. One older geologist, who had also published on laccoliths a decade earlier, found himself as a technical reviewer of some of Dave's manuscripts, and he rejected publication because Dave had moved into new mechanisms and models of emplacement. New ideas are hard for some people to accept, especially if they come at the expense of one's own models. But that's science.

In 2006, a decade after the BARCO project was shut down, the Utah Geological Survey began geologic mapping of the Panguitch 30' x 60' quadrangle (1:100,000 scale), funded in large part by the USGS Statemap part of the National Cooperative Geologic Mapping program. It was to be at least a three-year effort at that funding level. The Mapping Program of the UGS, under the leadership of Grant Willis, had long concentrated on geologic maps of this scale, although they continued mapping 7.5-minute quadrangles in areas of geologic complexity, resources, hazards, or urbanization. And by now most of the State had been mapped as 30' x 60' quadrangles, though they had a long way to go in 7.5-minute quadrangles. Bob Biek was assigned to be the UGS geologist. Happily, the Panguitch quadrangle contained all the exposures of gravity slides that John and his students had found. Bob, almost unique in his astounding productivity, had a different way of doing a 30' x 60' sheet: he mapped and published all or most 7.5-minute quadrangles within it. Even though some 7.5-minute quadrangles had been done, including by John and me, and other detailed maps also had been published within the Panguitch sheet, Bob went back through those areas too, and he gave them a detailed second look.

Among the areas where Bob started in the southeastern part of the sheet was an area that George Davis and his students (University of Arizona) had mapped just north of Bryce Canyon National Park. There they found thrust faults that were Miocene, synchronous with the volcanism of the Marysvale volcanic field. They interpreted the thrusts as having been formed by the huge weight (gravitational loading by eruption from stratovolcanoes of one lava flow and mudflow on top of another) of the eruptive center of the Marysvale field. This weight was directed downward, then outward toward the edges of the volcanic field (but only exposed in that one place near Bryce) by thrusts rooted in horizontal, soft evaporite (gypsum) beds in the Jurassic Carmel Formation, at great depth. Oil company seismic data made available to George provided some of the evidence and suggested that the amount of horizontal sliding (thrusting) of the volcanic rocks on these weak beds may have been about 1,500 feet to a mile.

Next Bob moved farther west, near the town of Panguitch and into the southern Markagunt Plateau, where John, Ed Sable, Florian Maldonado, Dave Moore, and I had worked. In fact, I spent most of September 2008 finishing the Brian Head quadrangle, a freebie to UGS because what little National Park Service funded for our field work was passed on, at my request, to co-author Stan Hatfield. The map area contained, in addition to part of Cedar Breaks National Monument that NPS was interested in, critical exposures of the Markagunt Megabreccia, including the places noted before that are east of Brian Head peak, where a horizontal shear plane placed megabreccia on top of intact rocks (that is, undeformed rocks that represented the Miocene erosion surface on which the gravity slide moved). Bob's 7.5-minute quadrangles included the Haycock Mountain and Panguitch Lake maps, just east and southeast of the Brian Head quadrangle. There he began finding more well-exposed megabreccia that the others had missed because they had not undertaken a systematic study of all outcrops. Bob and I shared evidence and field visits.

In October 2008, Bob discovered the southern leading edge of the gravity slide on the southern flank of Haycock Mountain. A mass of Isom Formation, its lower part pulverized, overlay a horizontal shear plane that placed the mass on white undeformed sedimentary rocks of the Oligocene Brian Head Formation. In missing these exposures, the others had interpreted that the leading edge was about ten miles farther north, near exposures described by John in his 1993 report. Bob finished the mapping of the Haycock Mountain and Panguitch Lake quadrangles that field season and the next. In July 2009, Bob invited me to spend a day with him looking at his evidence. At the end of that day, we hiked a couple miles to the exposures of the leading edge of what he called his "discovery site." The underside of the upper transported plate, just above the shear zone, contained north-south grooves, called slickensides. Directional indicators in the plate, called riedel shears, demonstrated southward movement of the plate. I marveled at what he had found and we talked about where he needed to go from here. This was a eureka moment in that the slide had to be significantly larger than what John and the others had concluded; Bob had made a breakthrough and this news had to be published now rather than waiting until the maps came out (UGS published the Brian Head quadrangle in 2013 and the Haycock Mountain and Panguitch Lake in 2014). Bob gave a talk at GSA that fall (2009) and presented his first preliminary UGS version of the Panguitch sheet, which showed the 7.5-minute quadrangles that he had completed. He gave another GSA talk, then a UGS progress report in 2011. By 2012, he put together a preliminary version of the entire Panguitch sheet, cross sections, text, and other parts. I was quoted in the text as saying that we had no evidence that it got into the Beaver 30' x 60' quadrangle, just north of the Panguitch sheet. (I had published that map in preliminary form with the UGS in 2005.) In July 2012 the Panguitch field review was held, and that version of the preliminary map was published soon after.

The day after the field review, Bob asked Dave Hacker, because of his expertise with the Iron Springs slides, to join us as we visited the discovery site and another outcrop of the shear plane about a half mile to the west. As we were talking, Bob or Dave looked down at our feet, just below the shear plane, to see rounded pebbles of the Isom Formation, 26–27 Ma, and the Needles Range Formation, 30.5 Ma. This undeformed bed, a Miocene stream channel perhaps twenty feet thick and wedging out to the east and west, was sitting on the undeformed Brian Head Formation. Of course, the gravity slide had to post-date (that is, be younger than) the Isom Formation. Another tuff, the Haycock Mountain Tuff defined by John in his 1993 report and undeformed, had been deposited on top of the slide soon after movement. Bob and the dating experts contracted by the UGS dated the zircons there at about 21.6 Ma.

In May 2013, Bob's summary of the gravity slide in the UGS newsletter Survey Notes gave the size of the gravity slide as 300 mi^2. By then, Bob had started on another Statemap-sponsored 30' x 60' sheet, the Loa quadrangle just east of the Beaver quadrangle. At that time, I was mapping my own small landslide at Black Mountain as part of my consulting study for Keith Norlin, which I interpreted to have slid off a laccolith at about 22 Ma. In August 2013, after Dave had taught the Kent State field camp, he and Bob spent a couple of weeks together on the slide. One important and well-known concept, whether for a landslide or gravity slide, is that the place where the slide rocks separated from those that remained in place and moved downhill was a steep (high-angle) surface known as a breakaway, which joins a low-angle shear surface in weak rocks below. Furthermore, the shear plane in the part of the slide below the breakaway is at considerable depth. We interpreted this shear plane to be within the Brian Head Formation, a white lake-deposited weak ("greasy") shale (bentonite) bed just below the thick volcanic pile; the slide failed (slid) catastrophically along this weak bed, well above the much deeper thrusts that George Davis and his student found. The high-angle breakaway feeds into the shear in the weak bed, which is gently dipping in the direction of movement. Farther south in the direction of the movement of the slide, this gently dipping failure plane, however, will climb up (called a ramp) to the surface (it "daylights") and then the slide mass moves along the land surface away from the point of failure. When it moves along the surface, it is more deformed than elsewhere, therefore more apparent, and that is the part we had been mapping to date. In August, Bob and Dave found the first evidence of the ramp part of the basal fault about fifteen miles north-northeast of Brian Head. Here, the slide plane (ramp fault), which appeared to trend east-west, was dipping north, heading to significant depth. That meant that the gravity slide extended northward. I joined them for a day to visit the ramp. Suddenly we realized that the size of the slide (Markagunt Megabreccia) was well beyond what we had understood in 2012. This realization began a period in which the three of us prioritized our investigation of the slide. Dave and Bob began giving talks, mostly at GSA meetings, that described new data as they came in. These announcements, in turn, stirred the profession's interest.

We began to investigate areas we and others had already mapped, including John's dissertation area, which is north of the ramp. Consequently, another discovery, made by Dave the same month while he was on his own about five miles north of Panguitch, was a thin (mostly less than an inch thick) black glass, known as pseudotachylyte, running along a low-angle shear surface, and equally thin dikes rising above it. This glass, a rare occurrence in landslides, gravity slides, and faults, represents a zone where dry rock melted due to the frictional heat of movement. I then found an outcrop that contained a vein of pseudotachylye about the width of a finger, from the gravity slide at Black Mountain. That laccolith and slide also was one of those incorporated in the much bigger Markagunt Megabreccia. The more we looked in areas to the north, the more we saw other small or subtle examples of deformation that we had missed before. For example, in these volcanic rocks (mostly mudflows of the Mount Dutton Formation, which are at least five thousand feet thick), occasional or locally abundant clasts (pebbles, boulders, etc.) of volcanic rock were fractured, with pieces of the clast having moved along these fractures (they were thus small shears/faults). These are known as jigsaw clasts because the pieces fit back together as in a jigsaw puzzle. In other places, we found rounded pebbles that resembled diagrams from rock mechanics called strain ellipsoids that looked as if they had been put in a vice and squeezed until they sheared. These sheared pebbles would commonly erode out and be left on the land surface, although the squeezing had involved hundreds of square feet of rock. Occasionally we also saw small shears in outcrops. Yet most of those huge volumes of rock north of the ramp contained no evidence of movement. And, of course, the entire gravity slide was consolidated, quite unlike the unconsolidated nature of a Quaternary landslide. Furthermore, the entire area had been cut by post-slide basin-range faults, which can be easily mistaken for the shears formed during Markagunt Megabreccia gravity sliding. Those facts, plus our realization of the huge scale of this landslide, was why we had missed it at first. The basin-range faults, of course, had later deformed the landslide mass into the basins and ranges of the present topography.

Before Dave had left for Kent State to begin teaching again, the three of us had agreed that we needed a short report to announce what we had found. Dave volunteered to put one together, intended for *Geology*, a prestigious journal published by GSA whose maximum published article length was four pages. The Utah Geological Association (UGA) had already announced that its 2014 guidebook would have as its theme the geology of southern Utah, and Bob figured he could also put another article in that; I also thought I could put the story of my Black Mountain slide there. Then, after Dave left, Bob and I got back to more field work. We now realized that the basal shear plane in areas north of the ramp became deeper as we moved northward toward the Tushar Mountains in the Beaver quadrangle, and that entire mountains, thousands of feet thick, as well as John's dissertation area, had moved southward as part of the gravity slide. Bob worked westward in the Red Hills, in the Great Basin west of the Markagunt Plateau, where Florian Maldonado had found and published on deformed rocks previously. Bob confirmed that these rocks had moved southward along the paleosurface south of the ramp; in fact, UGS geologist Tyler Knudson found slide breccia as far south as the outskirts of Cedar City. In October 2013, Bob showed the dogs and me what he had found in the Red Hills. In October and November, I looked for evidence of the slide in the southern Beaver quadrangle; it was subtle, as occasional shears and jigsaw clasts, because of the great depth there of the main basal slide plane.

Where could the breakaway be? It must be in the Tushar Mountains, and the decades of USGS mapping that Tom Steven, Skip Cunningham, John Anderson, and I had done, as shown in the Beaver quadrangle and other published maps, must have missed it! I had assembled the latest versions of our geologic maps in 2001 and 2005, so I knew the entire area, and as I looked at those maps in October 2013, I had an "aha" moment, as recorded in some notes and diagrams that I put dates on and saved. Tom and Skip had mapped a caldera, which they named the Big John caldera, on the southern top of the Tushar Mountains. They had suggested that it was the source of the Delano Peak Tuff Member of the Bullion Canyon Volcanics. I had never liked their caldera because they had found no thick intracaldera tuff phase of that member, as was normal in calderas, but it was in a poorly exposed area in which I had spent little time, and I had therefore trusted the judgment of these two excellent geologists. Yet now the map pattern suggested something else: the northern "margin" of their "caldera," which was well exposed, faced south and dipped steeply in that direction. The southern "margin," also well exposed, of course faced north and dipped steeply in that direction. The eastern and western "margins," which they showed as joining the others, were partly covered by younger rocks, and their locations were poorly exposed and uncertain. It seemed possible to me that the northern and southern margins were not connected. In other words, the northern "margin" could better be considered the breakaway, whereas the southern "margin" could be an antithetic (dip in an opposite direction) fault that, with depth, feeds into the breakaway, a common feature of landslides. Field work with my dogs in April and May 2014, following the melting snow drifts as we reached progressively higher into the Tushar Mountains, confirmed the hypothesis.

During his Christmas break, Dave put together a draft of a paper. Bob and I did author input. Dave submitted it to *Geology* in May, after which it received peer reviews. The reviews were favorable, for a lot of evidence had accumulated in forty years; the reviewers (one was George Davis) had good constructive advice that we followed closely. The GSA approved the report in August 2014 and it was published the next month. In the report, we named it the Markagunt gravity slide; John's name, Markagunt Megabreccia, was retained for the deposit itself. Our evidence indicated that it had formed catastrophically, that is, failure of the whole thing happened at once, then it traveled at speeds that were unknown but likely pushing 100 mph, somewhere in time between 22 and 21 million years ago. South of the ramp, it traveled 18.5 miles along the Miocene land surface to what later became Haycock Mountain. We estimated then that its north-south dimension is more than fifty-five miles, that it covered an area of more than 1,300 square miles (larger than Rhode Island), and that its volume was about 500 cubic miles. As such, it was almost identical in dimension to a 50-million-year-old feature in northwestern Wyoming called the Heart Mountain gravity slide, which had been famous for many years and was considered unique as the largest subaerial landslide on Earth (submarine gravity slides in the ocean basins have been found that are bigger, and much bigger gravity slides are on Mars, including some that came off the lower flanks of Olympus Mons—the largest known volcano in the solar system—apparently formed billions of years ago when Mars had an ocean). We had no sure idea of what had triggered our slide, but mapping by Skip Cunningham and Tom Steven had long ago confirmed that the large Mount Belknap caldera of 19 to 18 Ma cut (postdated) the breakaway (their "Big John caldera" margin) on the southern crest of the Tushar Mountains, so we speculated that emplacement (rise within the crust) of a huge granitic pluton that fed the Mount Belknap caldera may have lifted up the area into mountains, followed by eruptions of a Mount Belknap rhyolite ash-flow tuff (Joe Lott Tuff Member) and then subsidence of the caldera. The UGA guidebook was published in the early fall, with Bob's and my report. Bob's article contained many more pages than Dave's, and he filled them with photos and detailed descriptions of key outcrops.

In the spring through fall of 2014, my dogs and I (occasionally on weekends with Dawna, who wanted the exercise) remapped the Big John "caldera" and provided the field data supporting that it was not a caldera. In addition to the breakaway, we found subsidiary breakaways, appearing as parallel (synthetic) faults inside (south of) the main breakaway, as well as additional parallel antithetic faults. The breakaways are major structural features, as befitting our new interpretation that this was one of the world's largest gravity slides; at places the main breakaway was a quarter-mile wide zone of broken rock, now consolidated and resembling a fault zone. We were in beautiful country, near and in places above timberline; in one place, at timberline, we could see mountain goats as white specks in the distance. Some of the mountains are within the Mount Holly Ski Resort, with ski runs cut through aspen, ponderosa pine, and Douglas fir forests.

In August, 2014, after field camp, Dave made his annual trek here and Bob joined us for four days of joint field work. We visited critical outcrops in the Tushars that I had found, including taking a hairy road up to the crest of the Tushars (a popular goat viewing spot) where the Mount Belknap caldera cuts the breakaway. These annual get-togethers were events of discovery for all of us, as we hashed out the meaning of the many outcrops we visited together. As with my association with many other persons mentioned in this memoir, it is always a delight to interact with talented people you trust, all of us focused on defining and solving problems. We ended each annual visit by specifying what we had to do in the future, and what needed to be published.

In September 2014, Bob led the UGS field review attended by geologists and the public, preparatory to submittal of the final Panguitch map for publication. As the weeks went by into the fall, I made traverses starting at the southern crest of the Tushar Mountains and traced the breakaway fifteen miles down the eastern flank of the Tushar Mountains into Sevier Valley. Then I followed it ten miles west of the crest, then northward another twenty miles along the western flank of the Tushar Mountains almost to Sulphurdale, then ten miles west across hills in the northern Beaver Basin. There the breakaway passed south and west ten miles across the Mineral Mountains, although all field evidence for that had been eroded away, down to the granite core of the range. Then I interpreted that it went southward for thirty miles beneath Escalante Valley. After each trip, I sent emails to Bob and Dave giving details, almost gloating that I was doing field work in this incredible country and they were not! One morning in mid-November, high in the Tushars, the dogs and I were caught in a blizzard and we abruptly terminated field work in high country. But the task of finishing lower areas awaited, including the Black Mountains, where I soon hypothesized that the eastern half of this range consisted entirely of gravity slide. In late December, on a field trip to show Bill Lund, Garrett Vice, and Tyler Knudsen the gravity slide in the central Black Mountains, we barely escaped getting stuck in mud on our way out of the area, marking the end of the 2014 field season.

In early 2015, the UGS published the final version of the Panguitch sheet as a CD. Bob had done a masterful job of authoring a beautiful digital map. It had an expanded text of 162 pages, 100 figures (line drawings and photos), and five tables. Dave and I joined the long list of coauthors. We now estimated the size of the Markagunt gravity slide at 1,600 mi^2, making it the largest on the land surface of earth. The American Association of State Geologists awarded the map the 2016 Charles Mankin Award for the year's best report by all state geological surveys. We were proud, but we knew that science does not stop, and that there is never a "final" version. There were discoveries yet to be made, with this report now the existing stepping-off place for the next revelation. To add to his amazing accomplishments for 2015, Bob also senior authored the preliminary map, after two years, of the western part of the Loa quadrangle. Bob told me that he had found a large gravity slide on it, but we did not think that it was particularly noteworthy. This slide mass covered an area of about a square mile and was located several miles northeast of the town of Antimony. I was a co-author with him on the Loa map because he compiled my two 7.5-minute quadrangles of the eastern Monroe Peak caldera. Dave was not a co-author, for he had yet to work his magic on this map as he had with the Panguitch sheet. But that was to come, and that story is next.

In the spring and summer of 2015, I continued revising the Beaver quadrangle to show the Markagunt gravity slide in the Tushar Mountains, Mineral Mountains, Black Mountains, and places in between. Bob continued the Loa mapping, as time allowed, because he always had commitments to map a 7.5-minute quadrangle or two in a suburb close to Salt Lake City. During Dave's August visit, we looked at some of the outcrops I had found in the Circleville area of Sevier Valley, concluding that the Markagunt gravity slide did not get as far east as the Sevier Plateau. On another day, we visited eastern parts of the Tushar Mountains north of Junction, but found no evidence of gravity slides there either. As the three of us summarized where we were in the study, Bob suggested that we propose to GSA a week-long field trip that they would sponsor, known as a Thompson Field Forum, to be held for volcanic-landslide experts and students to visit the Markagunt slide to provide help on its interpretation.

After that August visit, I returned home to horse chores. My colleagues then worked together for a few days, and for some reason zeroed in on the rocks in the southeastern Sevier Plateau, even though that area had just been published as part of the Panguitch sheet. There, in Cottonwood Canyon, on the eastern side of the Plateau, they stumbled upon smeared ash-flow tuff and Mount Dutton Formation mudflows above a shear plane on the weak, greasy underlying Brian Head Formation. Suddenly, in one outcrop, they had found evidence for the southern leading edge (on the land surface) of another gravity slide, resembling the southern edge of the Markagunt gravity slide at Haycock Mountain! And with that one outcrop, they made sense of a head-scratcher that had plagued me when I did my dissertation. That problem area was in east-draining Pole Canyon, just southwest of the town of Antimony. There, long ago, I had mapped Brian Head Formation that was nearly vertical, overlain by gently east-dipping undeformed volcanic rocks. I recognized it as an angular unconformity (a surface of erosion that represents not only a large passage of time but an older deformation event overlain by typical deposition of horizontal beds) but attributed it to an old normal fault that had folded and faulted the Brian Head Formation, after which volcanism resumed and covered the old broken rocks. In the late 1970s, while working on the Marysvale project with Tom Steven and Skip Cunningham, I had returned to my thesis area and carried the unconformity between the deformed rocks and the gently dipping rocks westward to the top of the Sevier Plateau, just east of Mount Dutton and Mount Pierson. Coupled with the outcrops in Cottonwood Canyon and the gravity slide northeast of Antimony, Bob and Dave now recognized this as another gravity slide east of the Markagunt gravity slide! We called it the Sevier gravity slide. I consequently spent the fall of 2015 revisiting my old friend the Sevier Plateau, confirming Bob and Dave's diagnosis of the Sevier slide, and mapping what I had not seen in the previous fifty years. As with the Markagunt slide, nearly half of which was in the Beaver quadrangle, at least the northern half of the Sevier slide was also in the Beaver quadrangle. The southern half of the Sevier slide was in the Panguitch sheet, but that map was too far along in press for us to correct it. By the late fall, Bob had a draft proposal to GSA for a Thompson Field Forum. (In 2022, Michael J. Braunagel completed an Ohio State University Ph.D. dissertation, working with his advisor W. Ashley Griffith [see below], on the gravity-slide outcrops on the eastern side of the Sevier Plateau.)

The GSA approved our proposal for a Thompson Field Forum, to be held in September 2017; naturally, we now added the Sevier slide to the list of field stops! We had room for about thirty persons, including students. Of course we asked the lead investigators of the Heart Mountain gravity slide, Professors John P. Craddock (the son of Cam Craddock, John Anderson's former advisor) of Macalester College in St. Paul and David H. Malone of Illinois State University, with whom we had been in touch. The announcement of the time and place for the Forum was published by GSA in November 2016, and included a request for people who were interested to call us. My 2016 spring and summer field season continued with more field work in the Sevier Plateau, then in the Mineral Mountains, mapping the margins of the two gravity slides. In May, a reporter for the local NBC television affiliate, KSL in Salt Lake City, contacted Bob for his cooperation in filming a short (three-minute) segment for one of their local evening news shows. Bob and I spent the day with TV reporter John Hollenhorst, who filmed and recorded us with his TV video camera in the Panguitch area. His film included a view of the pseudotachylyte site that Dave had found. As we were driving back from this site along the dirt road to the main highway, we came upon a rattlesnake crossing the road. We stopped and John videoed the snake. The short, tightly edited version of the day's filming aired a month or two later (John sent us a copy), and the snake was worked into the show as John narrated how the Miocene gravity slide rushing across the country at 100 mph would have killed all life in the area. A lengthy newspaper article followed. But that was the extent of the general news' interest in the slides, although Dave's 2014 report in *Geology* elicited quite a bit of interest in science magazines.

In June, Bob and I spent two days tracing the Sevier slide in the Loa quadrangle: we found exposures where the evidence was subtle (but clear) all over the place in Grass Valley east of the Sevier Plateau and in the foothills of the Awapa Plateau farther east, and the gravity slide that Bob had found previously, where the evidence was obvious, now was recognized to be part of the Sevier slide. At our annual get-together, Bob, Dave, and I worked together along the crest and western flank of the Sevier Plateau and in the Piute Reservoir area, finding the Sevier slide in all those areas as far north as the edge of the Monroe Peak caldera (23 Ma), which postdated the slide. The slide is several million years older (25 Ma, as we found later) than the Markagunt gravity slide. As with the Markagunt slide, the northern breakaway of the Sevier slide is cut by the southern side of a younger caldera, this one being the Monroe Peak caldera, suggesting that there is a connection between triggering the slide and the caldera. But with a two-million-year age difference, even this hypothesis was a stretch.

Slow but methodical field work (although only a handful of days at a time because of our other commitments) followed on the Markagunt and

Sevier gravity slides during the spring and summer of 2017. Our brief August get-together with Dave was spent in the Loa quadrangle, including on a possible eastern lateral fault boundary of the slide, which Dave had discovered previously. Bob then began, as lead author, an elaborate field guide for the trip, which became a handout for Forum participants, then was modified for publication by GSA. As is typical, as field and lab work continued past the tenth anniversary of our work on the slide, we were discovering new data and rejecting our past interpretations on a monthly basis. My part in the publication process was to revise the Beaver 30' x 60' quadrangle, which Dave Maxwell blew up into quarters we intended to publish one after the other at 1:50,000 scale. The first one was the southwestern quarter, and it was published in 2019. Dave Hacker wrote a lengthy report for a 2017 book on shallow intrusive bodies in which he compared the Markagunt slide with those at Iron Springs and the Pine Valley Mountains.

The Thompson Field Forum was an enjoyable six-day opportunity to show experts our slides and to get feedback from talented men and women whom we now came to know, and we used the meetings to bring others into the research. One of these was Professor W. Ashley Griffith, a rock mechanics expert at Ohio State University who could help create models of the slides. For one field-forum day, to get to Bob's discovery site on Haycock Mountain, Bob had gathered about a dozen members of an ATV (All-Terrain Vehicle) club in Panguitch to bring their vehicles and transport the participants to the site. By then, many Panguitch citizens had heard of our huge landslide, for Dave, Bob, and grad students were local heroes who had become well known in Panguitch (one shop sold a "Gravity Slide" flavor of ice-cream cones), where they camped each field season. At the time, our latest estimate for sizes was that the Markagunt gravity slide covered an area of about 2,000 mi^2, roughly the size of the state of Delaware and an order of magnitude greater than what John Anderson had estimated. The volume was about 700 mi^3. We estimated that the Sevier slide covered about 600 mi^2.

One may wonder where the funding for our gravity-slide research came from, for some may think that science is always well supported. Bob had funds from the Statemap program, first for the Panguitch map, then the Loa map, for the first half dozen years, but that money was gone. He continued to be funded largely by the same program, but it was for preparing other maps, so he had to tackle those deliverables first. That allowed Bob about three weeks each year to work on the gravity slides, an arrangement approved by Grant Willis and UGS Director Rick Allis because they recognized that he was arguably the most productive geologist they had and they knew that he would meet all his deadlines. Dave, who had been granted full professorship in 2017, brought his own field vehicle, a twenty-five-year-old Toyota 4Runner, to Utah and largely paid his own way, although as Director of the Kent State geology field camp he received some salary and per diem from the Geology Department of Kent State because his visits to see us were along the way home. Both Dave and Bob camped out in tents in an RV (Recreational Vehicle) campground in Panguitch during all their field work on the gravity slides. Starting in 2016, Dave had master's students doing their theses on the slide, funded by the Geology Department. Dave Malone and students have joined us by contributing zircon ages for the slides, free of charge. Ohio State University is considered the main research university in Ohio, and Kent State was not, so Ashley Griffith proposed a NSF research grant with Dave Malone and Tiffany Rivera of Westminster College in Salt Lake City (our Ar-dating expert, later moving to the University of Missouri), along with Dave Hacker, as junior authors of the proposal, and it was eventually accepted in 2020 after the election of Joe Biden as president. As for me, I wrote off field expenses, my contribution for the Thompson Field Forum, and my share of GSA publishing charges as business expenses.

Since the Thompson Field Forum, field work and our annual field visits with Bob, Dave, and at times Ashley, Dave Malone, others, and their students continue. In the fall of 2017, I returned to the western Black Mountains, shown as intact in my published (1978) Thermo 15-minute quadrangle, and confirmed evidence that the entire Black Mountains had moved as part of the Markagunt gravity slide. That provided the evidence that the Markagunt slide covered 2,000 mi.^{2} Field and office work to repair the Beaver sheet continued: the southwestern quarter of the Beaver quadrangle was published in 2019, the northwestern quarter was published in 2020, the southeast quarter was in press, and the northeast quarter was submitted for review in 2022, although given a delay in GIS (digital work) and finishing of the entire draft map in 2023, both of those quarter maps do not need to be released separately because at the end of the fiscal year on May 31, 2024, we finished our deliverable, the entire Beaver quadrangle; it is now in review. Thanks to the Biden administration, the UGS provided me some modest funding. In 2020, I also re-examined the mineralogical data on hundreds of thin sections I had accumulated in the Sevier Plateau ever since my dissertation work there, focusing on some ash-flow tuffs that predated the Sevier gravity slide and one near-identical tuff that postdated the slide. Argon ages done by Tiffany Rivera found that the one tuff (deformed) that predated the Sevier slide had an age of 25.2 Ma, whereas the tuff that postdated (undeformed) the slide had an age of 25.1 Ma, about as good as you can get in dating the slide! (The age of the Markagunt slide, as noted, is about 22 to 21 Ma.)

During the months leading up to our 2019 field meeting, we came to believe that we had three gravity slides and that they are younger from east to west. Our evidence for the western slide was that our isotopic ages on it were far too young (younger than 19.5 Ma) to be the Markagunt gravity slide. At our August 2019 meeting, we visited it and agreed to call the western one the Black Mountains gravity slide. The GSA field guidebook was, of course, by then in press so it was too late to change our interpretation; at 121 pages, it was published in December 2019. And, of course our published southwestern and northwestern quarters of the Beaver quadrangle just became obsolete as well! (Such obsolescences are acceptable outcomes in publishing cutting-edge science.) The likely location of the breakaway on the Black Mountains slide is in the central Mineral Mountains, where the granite batholith that underlies the central to northern Mineral Mountains is about 19 to 18 Ma; this is our estimated age of sliding. This batholith must have been the culprit for uplift and triggering, although we are uncertain of any exact mechanism. In May 2020, Bob senior-authored a three-page article in the UGS *Survey Notes* that was the first announcement of the three-slide concept. Under this new interpretation, we estimated the entire slide area at 3,000 mi^2; we call it the Marysvale gravity-slide complex. The revised estimate of the size of the Markagunt slide is 1,350 mi^2 of that, whereas the size of the Black Mountains slide is at least 1,000 mi^2. A reporter for the *Salt Lake Tribune* interviewed Bob and me and published a summary of Bob's short article on June 29, 2020. Bob retired from the UGS in 2022 but continues producing reports with support from this agency. In late 2022, he senior-authored a 66-page summary, intended for the layman, on the entire slide complex, that was published by the UGS. During the spring of 2024, Bob, Dave, and I finished stitching together the entire Beaver quadrangle at 1:62,500 scale and have drafted preliminary cross sections through it. Bob is revising the Panguitch sheet to show the Sevier gravity slide. Science never stops!

CHAPTER 12

Most people agree that the modern "environmental movement" began with publication in 1962 of the book *Silent Spring* by Rachel L. Carson (1907–1964). The book dealt with the impact of synthetic pesticides, primarily DDT, and how they accumulated in the food chain. Her inspirational life story, in which she became a national hero, would have seemed unlikely from her beginnings as a shy bookworm raised on a small family farm in western Pennsylvania. But she was precocious and brilliant, and before her teen years she started writing stories (some of

CLIMATE CHANGE

which were even published) about animals she saw. She earned her master's degree in zoology from Johns Hopkins University in 1932, but had to forego a planned doctorate to take a teaching job to support her family during the Great Depression. An aquatic biologist, she was hired as a nature writer and scientist by the US Bureau of Fisheries (later renamed US Fish & Wildlife Service). After her first bestselling book on marine life, she went out on her own as a prolific author of many more well-received books and articles in the 1950s, including one that won a US National Book Award. She turned to conservation writing in the late 1950s, but before her well-researched *Silent Spring* came out, she had contracted breast cancer. When she began pre-publication book excerpts and talks, a fierce, well-funded, public-relations campaign against her began, paid for by the chemical companies, including DuPont, and continued unabated for years. (Not much has changed, it seems. Bruce Hornsby's wonderful song, "Look Out Any Window", tells of the well-financed campaigns by deniers of the ill effects of many well-documented hazards, from cigarettes to climate change. Recently deceased rich-man David Koch spent much of his life funding deniers.) Many of the attacks against Carson were not against her science, which was undeniable, but were personal and highly public. She simply continued teaching and speaking despite her advancing illness, and eventually their vicious campaign backfired. One of her last public appearances was in 1963 when she testified before John F. Kennedy's Science Advisory Committee, which issued a report that year that backed her claims and issued new policies against pesticides. She and her book inspired the widespread grassroots environmental movement of the 1960s and 1970s. It resulted in the creation in 1970 of the US Environmental Protection Agency (EPA) under President Nixon; in 1972 the EPA banned DDT in the US. In 1980, President Jimmy Carter posthumously awarded Carson the Presidential Medal of Freedom.

Depletion of the earth's resources has a popular analogy— "the tragedy of the commons," an old concept that was revised and then famously discussed in an article published in 1968 by Garrett Hardin. This relates a story by which a shared and unregulated resource, in this case a plot of town land— "the commons" in many English villages—was set aside for the common use of townspeople for grazing their cows. The tragedy is that, eventually and nearly always, individual townspeople act in their own self-interest rather than for the common good, and in time the commons are overgrazed and the use collapses. This analogy is widely used in environmental science for finite resources on earth. The "commons" is used as a synonym for the oceans, surface water, atmosphere, etc., or in analyzing human psychological behavior with respect to taxes, politics, etc. where people soon figure out how to game the system. When that happens, the resource is depleted, overgrazed, overfished, overpopulated, or otherwise destroyed. At the present time we see depletion of salmon populations in the northwest, overfishing of the Grand Banks, mining of aquifers, drying of the Aral Sea, depleting Owens Lake to supply water for Los Angeles, dropping shorelines of the Great Salt Lake, global warming, extinction of plants and animals, pollution by fracking, overlogging/burning of forests, resistance from overuse of antibiotics, air/water pollution, ozone depletion, and more. Or, as warned by the aforementioned Bruce Hornsby in "Look Out Any Window, look out for the backroom boys that say the smoke is gonna blow away."

Archaeology and history provide many examples of collapse of civilizations by overuse of resources or from other reasons. The fascinating 2005 best-seller *Collapse* by Jared Diamond documents many terrifying failures (and a few long-term successes) of societies that pass into ruins, some of which could have been saved by decisions that they did not make ("failures of group decision-making," namely groupthink). The inhabitants of Easter Island, for example, descended into cannibalism. The end of the Maya, although too far back in time to know most details, saw accelerating warfare and apparently increased human sacrifices. Close to the present time, the "answer" to an exploding population in Rwanda was genocide, and the misery in Haiti—as opposed to better results in the Dominican Republic on the same island—is well known. Diamond noted (p. 341) that "a society's fate lies in its own hands and depends substantially on its own choices." He provides an analysis of why and what can be done, much of it clearly intended for our looming crisis of climate change, which like many other collapses, "takes the form of a slow trend concealed by wide up-and-down fluctuations" (p. 425). Among the significant "whys" is a political "lust for power," particularly displayed by Presidents Trump and Putin, and the book provides other examples "of leaders too self-absorbed in their own pursuit of power to attend to their society's underlying problems" (p. 439). One of Diamond's conclusions (p. 498) is that "the world's environmental problems *will* get resolved, in one way or another, within the lifetimes of the children and young adults alive today. The only question is whether they will become resolved in pleasant ways of our own choice, or in unpleasant ways not of our choice, such as warfare, genocide, starvation, disease epidemics, and collapses of societies."

The environmental movement of the 1960s and 1970s took many forms. At first it focused on pollution, and it resulted in legislation to promote cleaner air and water. President Lyndon Johnson between 1963 and 1968 signed almost three hundred conservation measures, plus $12b in funding to support them. The Wilderness Act of 1964 set aside federal lands from commercial interests. The first Earth Day was in 1970, amidst much public support. In 1970, President Nixon signed the National Environmental Policy Act (NEPA), which required documents (EIS or EA) for all "major federal actions that significantly affected the quality of the human environment." There were several spectacular environmental crises at this time that turned the public onto the side of earth awareness, such as the Cuyahoga River catching fire near Cleveland in 1969. From 1970 to 1976, an additional spate of environmental legislation took place, including setting the federally enforced quality standards for air, water, and drinking water. By the late 1970s, the movement addressed toxic wastes, resulting in the Superfund Act in 1980. As environmental issues became more popular, environmental organizations such as the Sierra Club grew rapidly or were founded to lobby for additional research and regulations. By 1981, the EPA had a staff of nearly thirteen thousand and a yearly budget of $1.35b.

An important aspect of the environmental movement of the 1960s and 1970s, but arguably little since then, was concern over overpopulation. This topic was hardly new to the environmental movement, for Thomas Robert Malthus (1766–1834) had proposed his "law" of overpopulation in 1798, by which he stated that the curve for population growth was exponential, whereas the curve for food-supply growth was linear. When the curves crossed, starvation soon resulted. In 1968, Stanford biology professor Paul R. Ehrlich (born in 1932) and his wife Anne H. Ehrlich (though Anne was not acknowledged as such by co-authorship, she did much of the work on the book) published their popular small book *The Population Bomb.* They pointed out that the Earth reached a population of one billion in 1850 (this figure is in dispute, for some have suggested that it happened closer to 1800) and it doubled to two billion by 1930. Projecting this exponential curve into the future (for example, another doubling was just around the corner when they published), they concluded, would result in worldwide famines and societal upheavals by the 1970s (later editions said the 1980s). They proposed immediate action against population growth, and in the 1960s and 1970s this good advice was acted upon by many people, including myself (a vasectomy), and by some organizations, although not by governments. The book's overconfident (even arrogant) alarmist style in its proclamations received criticism that increased as the years went by and as the disasters did not happen. The milestones in world population since their book came out are as follows: 3 billion people by 1960, 4 billion by 1974, 5 billion by 1987, 6 billion by 1999, 7 billion by 2011, and 8 billion in November 2022 (the population of the USA is about 330 million). The population growth curve is complicated but has not been exponential for some time because world growth slowed in the late 1960s. The Ehrlichs updated their book and its projections more than once since 1968.

> *A similarly influential book, but less alarmist and arrogant, is* The Limits of Growth, *published in 1972 by seventeen researchers, many associated with MIT, from what they called the Club of Rome. Four of these persons were given authorship, with* ***Donella H. Meadows*** *and* ***Dennis L. Meadows*** *the first two. The book was a computer simulation of assumed exponential economic and population growth that dealt also with food production, industrialization, pollution, and consumption of non-renewable resources. It projected several future (by the latter part of the twenty-first century) scenarios, only two of which were predicted to be disasters (namely, overshoot and collapse). Donella and Dennis are graduates of Carleton College, Dennis in my class ('64) and Donella in the previous class. A book of biographical information published for our fiftieth anniversary (June 2014, which I attended) gives some information on them. Donella Hager was the lab instructor for Dennis' first chemistry course, and they dated soon after, then got married upon Dennis' graduation. Donella earned her Ph.D. from Harvard and Dennis his from MIT, after which he joined the MIT faculty in 1969. Both moved to faculty positions at Dartmouth University in 1972. Sadly, they got a divorce in 1984 and Donella died from meningitis in 2001. Dennis received four honorary doctorates and retired in 2004. Their book, like that of the Ehrlichs, received significant criticism (the typical computer lament of "garbage in, garbage out," for example) but also significant praise over the years. It was updated in 1992 as* Beyond the Limits, *and several times since then. In these revisions, they proposed several scenarios predicted a hundred years into the future, but with optimism ("the world faces not a preordained future, but a choice").*

Renowned biologist E.O. Wilson (1929–2021), a Harvard professor for nearly fifty years, wrote thirty-five books and four hundred technical articles, twice winning the Pulitzer Price for nonfiction. One of these Pulitzer winners and a *New York Times* bestseller was a 2016 book *Half-Earth: Our Planet's Fight for Life*, in which he makes overpopulation the central cause of climate change. He noted that "Reproduction is obviously necessary, but it is a bad idea, as Pope Francis I has pointed out, to continue multiplying like rabbits." He warned against "[a]n Earth packed wall-to-wall with people." Yet, on a positive note, he observes that "[I]n every country where women have gained some degree of social and financial independence, their average fertility has dropped by a corresponding amount through individual personal choice." Educating and advancing the rights of women is Wilson's favored method of solving overpopulation, which in developing countries has happened in "one or two generations" (June 2019 *Population Connection* 51, no. 2, p. 10–20).

Aaron Timms, in a substantial essay in the June 2020 issue of *The New Republic* ("Making Life Cheap: population control, herd immunity, and other anti-humanist fables"), takes a significantly different tack on population in the time of climate change. He is diametrically opposed to the view advocated by many, and notably Donald Trump, that controlling immigration is the first step all countries should do to constrain population pressures. (Of course, Trump's position has nothing to do with a thoughtful analysis of overpopulation or climate change, but instead he is using racism to divide the electorate by claiming that immigrants are "taking our jobs" to appeal to anxious older white people.) Timms also is opposed to another step advocated by most—that of constraining reproduction. Both steps would result in an aging population (geriatric society, as now in China, due to their former one-child rule) that would not only put a strain on the economy, "stretching welfare states to the breaking point," but he also assumes that older people have fewer innovative ideas or less get-up-and-go—when so little time is left—to come up with workable solutions to many complex problems due to climate change. Instead, solutions require collective innovative thought and action as well as a vibrant economy (more taxpayers). Furthermore, Timms notes, "immigrants play a disproportionate role in American entrepreneurship" and "immigration makes countries far richer, more diverse, more dynamic, and more interesting than they would otherwise be." There is significant evidence that Timms' position on immigration is true. In an interesting side note, the Society for Science & The Public, in their February 1, 2020, issue of *Science News*, published a list of three hundred of their top scholars among American high-school seniors who received cash awards and scholarships based on their applications for science grants. For fun, I looked over the list and found only sixty-six persons (22 percent) who had common Anglo-Saxon names that I assumed were white or black students; the rest, although of course also American, had names that indicated, to me at least, an ancestry from China/Japan (most abundant), India (perhaps next most abundant), the middle East, and Africa. Other evidence is the growing percentage in our hospitals of M.D.s and in our prestigious universities of Ph.D.s, many with non-US accents, that must represent first or second-generation Indians, Africans, and middle Easterners. The reason, which seems obvious, especially for people in professions that require the M.D. or Ph.D., is that immigrants are highly motivated to endure the long slog to get their education and training. White Americans, it seems to me, are often less motivated.

Although the environmental movement of about 1960 to 1980 was positive for society, there were setbacks throughout this time, and they became increasingly common starting in the 1980s. Professors Naomi Oreskes of Harvard University and Erik M. Conway of the California Institute of Technology in 2010 published their readable, well-researched, but overall depressing classic *Merchants of Doubt* about how a corrupt, self-promoting, greedy series of once-prominent scientists had capitalized on their formerly good reputations by taking money from big businesses to obscure truth about a series of environmental disasters. Their "studies" and reports were done in collaboration with, and supported by, captains of the polluting businesses and industries that caused the disasters, and with the full support of Republican presidents and politicians. Although Oreskes and Conway are experts in the fields about which they were writing (Oreskes is also a geology professor), the prominent scientists who led campaigns against doing anything about the several disasters were not experts in these fields. The polluting businesses injected hundreds of millions of dollars into "think tanks" whose job was to obfuscate scientific facts and delay action, and such methods fit right into the right-wing Republican ideology of anti-regulation at any cost. The businesses also approved grants to university researchers to do non-objective research (meaning it was not peer reviewed) and reports slanted to the business that provided the grants. The news media (whose publications have never been required to be peer reviewed) were a major part of the problem because few journalists understand the principles of science and because they think in terms of fairness, in which both sides of an argument be given equal time. As Oreskes and Conway pointed out (p. 32), equal time may work in a two-party political system but "it does not work for science, because science is not about opinion." False claims—in other words research not subject to critical (peer) review—that are made contrary to the actual scientific results, are not themselves science, so they "do not deserve equal time in a scientific debate." Furthermore, as these authors also noted (p. 34), there "are always uncertainties in any live science, because science is a process of discovery." Such uncertainties are misunderstood by the public and the media, and they are capitalized upon by the professional doubters as signs that the science is uncertain about important issues such as evolution, climate change, DDT, etc. In other words (p. 268), "science does not provide certainty" or proof, for it "only provides the consensus of experts, based on the organized accumulation and scrutiny of evidence."

The controversy about tobacco smoking had many parallels with Rachel Carson's fight against DDT. In this case, a massive public relations campaign by the tobacco industry began soon after it became well known in the early 1950s that tobacco caused cancers that were commonly fatal. The tobacco industry knew these things, as well as other evidences about cancer and addiction as they became available during legitimate research, but it obscured and obstructed the evidence in courts of law and in courts of public opinion. Not until 2004, half a century later, did a federal case find the industry guilty under the RICO Act (Racketeer Influenced and Corrupt Organizations Act). In 2006, a district judge concluded that the tobacco industry had executed a scheme to defraud, and successful lawsuits of more than $200b followed. Yet today, the tobacco industry continues to make spectacular amounts of money.

Man-made acid rain was another topic covered by Oreskes and Conway. Discovered in the mid-1950s in the forests of New England, by the 1970s acid rain was known to be caused by air pollution, notably from coal-fired power plants. One acid-rain committee after another under Ronald Reagan's presidency obfuscated facts, despite the objections of the Canadian government, which funded some of the legitimate studies. No legislation addressed acid rain during the Reagan administration. It was not until 1990, under the administration of George H. W. Bush, that amendments to the Clean Air Act used "cap and trade" to control acid rain. Cap and trade, as opposed to a carbon tax, seems to be a practical compromise method to curb atmospheric pollution, and at one time even Republicans supported it. (Now, Republicans simply deny that pollution is a problem.) Under such legislation, a cap on greenhouse-gas emissions would be determined by science and assigned to each company; the cap would become more stringent each year. Companies pay penalties if they exceed the cap. But if they cut their pollution quickly, they can sell their allowances to companies that pollute more, or bank them for future use. The object is to encourage companies to cut emissions and to create a market price for pollutants. California, which adopted such legislation in 2013, as well as the European Union and South Korea, have approved cap and trade. In 2009, the US House of Representatives approved cap and trade, but the US Senate has since blocked it from reaching their floor for discussion or vote.

In their chapter on Earth's protective ozone layer in the stratosphere, Oreskes and Conway noted that the first alarm was given for depletion of ozone in 1970, when an American company announced that it would develop a supersonic transport (SST) that would fly inside the ozone layer.

In fact, as noted by Oreskes and Conway, William W. Kellogg (a member of the National Academy of Sciences and the father of my colleague Karl Kellogg) chaired a panel that investigated this topic. At about this time it became recognized that the decomposition of chlorofluorocarbons (CFCs), which are used in spray cans, air conditioners, and refrigerators, destroy ozone. More research followed. Then the British Antarctic Survey discovered severe ozone depletion in the atmosphere over Antarctica in 1981, and after confirmation announced it in 1985. Additional study showed that an "ozone hole" covered all of Antarctica. Even we, during our last two field seasons on The Ice, noticed that we were more prone to sunburn than during our previous field seasons, and later field work in Antarctica has been done with greater use of sunburn lotions and head coverings. Well-funded disinformation campaigns, some affiliated with the Reagan administration, went into high gear in the 1980s. The 1995 Nobel Prize in Chemistry to the scientists who originated the stratospheric ozone depletion hypothesis, however, largely ended the professional doubters' arguments.

The environmental movement came to a halt with the election in 1980 of Ronald Reagan. His brand of conservatism included the politically expedient platforms of pro-business and anti-regulation views at any cost (i.e., the idea that true prosperity cannot be achieved with even a modicum of regulations), negating the Nixon and Carter ideas that the country can prosper even when preventing environmental degradation. James G. Watt, an ideologue and right-wing Christian, became Secretary of the Interior in 1981 and immediately affected the environmental movement with such advocacy positions as urging oil and mineral extraction on all government lands and by suggesting "de-authorization" of congressionally authorized national parks. In 1983 he even barred the Beach Boys band from playing on the National Mall in Washington D.C. on Independence Day because they were "the wrong element" by their encouragement, he claimed, of drug use and alcoholism. In a speech in 1983, he mocked affirmative action and minorities; this uproar had a greater effect than the previous objections to his views, leading to his resignation in about three weeks! Donald P. Hodel, an evangelical Christian who served as Secretary of the Interior from 1985 to 1989, was not much better: he made fun of ozone depletion in 1987. The EPA in 1981 was run by Anne Gorsuch, who had the same views as Watt and Hodel; she cut EPA's budget by a third, leaving it underfunded and understaffed. Another of Reagan's appointees was Edwin Meese, III, who was approved as Reagan's Attorney General despite a long list of ethics lapses, then served from 1985 to 1988, when he had to resign following a report by a government independent counsel that he had provided government help to friends in the Wedtech Company, not long after he had done the same for the Bechtel Corporation. (Trump awarded Meese the Medal of Freedom in October 2019.)

The Ronald Reagan presidency had a major negative impact on new policies to stabilize population growth, as Reagan and his advisors recognized that he could use an anti-birth-control and anti-abortion message to capture voters who held right-wing religious views. Advances to counter overpopulation before Reagan included the following: both a United Nations (UN) fund for reproduction assistance and a World Bank fund for health and population assistance became operational in 1969, birth control became legal for anyone because of a Supreme Court decision in 1972, a USAID Office of population began supporting reproductive health training in 1972, abortion became legal in the US because of *Roe v. Wade* in 1973, the first UN-organized International Conference on Population in Bucharest on family planning took place in 1974, and the first home pregnancy testing kit was marketed in 1978. In 1984, the second International Conference on Population, dealing with family planning, was held in Mexico City, but under Reagan the US position was to state that it would not fund any programs facilitating abortion. This position was formalized in the same year in what became known as the global gag rule or the Mexico City policy, prohibiting foreign nongovernmental organizations (NGOs) who receive global health assistance from providing legal abortion services or referrals, even if done with a NGO's own non-US funds. This forced NGOs to choose between comprehensive sexual and reproductive health care/education without US funding or to comply with the policy and receive US funds. This policy was enforced by all subsequent Republican administrations and rescinded by all subsequent Democratic administrations. In 2017, Trump expanded it to recipients of any US global health funding, including HIV and AIDS funding, and even programs that support clean water and sanitation. Then Trump expanded it again in 2019 to organizations that do not even receive US aid.

The Keeling curve provides a simple way to get the hair to stand up on the back of your neck. It is a graph kept at the observatory near the top of Mauna Loa (eleven thousand feet in elevation) in Hawaii that since 1959 has measured the accumulation of CO_2 in the atmosphere. Initial funding was from the International Geophysical Year (IGY), after which funding was low until the late 1980s, when it received institutional support. On a fine scale it has a sawtooth pattern, each tooth marking a year in which the tooth (high) part shows the amount of CO_2 each winter after leaves fall off deciduous trees in the northern hemisphere, and the gum line (low) marks the summer in which leaves take up CO_2 for photosynthesis. But the overall curve has a steady rise, showing an increase in the mean CO_2 each year, such that by 1965 it was clear that the curve was accelerating upward (exponential), as explained in an appendix to a report prepared by the Presidential Science Advisory Committee. The mean level was 316 parts per million in 1959, 354 ppm in 1990, and 409 ppm in August 2019; by this rate it will be at least 500 ppm by 2050, about double pre-industrial (the 1800s) levels. The curve was begun and maintained by a then-young Charles David Keeling (1928–2005), for which he later received the National Medal of Science. Yet the relationship between CO_2 and trapping heat and keeping it from escaping to outer space had already been established in the mid-nineteenth century, and in the early twentieth century CO_2 was recognized as being released to the atmosphere from burning fossil fuels (starting in the early 1800s with coal-powered steam engines of the industrial revolution) and for its effects in altering the Earth's climate. Furthermore, by the 1960s American scientists were beginning to warn political leaders of the problem. President Lyndon Johnson, among others, heard the message (the 1965 report was on his watch) but he failed to act, perhaps due to seemingly more pressing issues such as the Vietnam War. President Nixon also heard the message and continued Johnson's start at environmental reforms, but he also was distracted by other things (like Watergate).

The biggest and best funded of the anti-truth and anti-science campaigns, as documented by Oreskes and Conway and others, was/is of course against climate change. And on this topic, another book amplifies the sobering realities of the debate, a classic first published in 2006: *Field Notes from a Catastrophe* (my edition is from 2015) by Elizabeth Kolbert, staff writer for the New Yorker and a bestselling author. Her book is a series of chapters on some of the most compelling evidence for human-made climate change, featuring many interviews of some of the major scientific experts in the field. She writes, for example, that the connection between CO_2 levels and warming was being noted by 1979, much of it by climate scientist James Hansen of NASA, with a prediction that doubling the pre-industrial level of CO_2 (then about 270 ppm) would increase the average surface temperature by about 4°C. The National Academy of Sciences researched and expanded upon this during the last days of the Carter administration. But, as the Reagan administration began, professional doubters watered down the report. Nothing would happen on climate-change research until the Reagan presidency ended. Yet James Hansen of NASA put climate change on the national agenda with his 1988 testimony in the US Senate. Later work showed that temperatures in the Arctic are rising twice as fast as the global average. The jet stream, so important in causing our weather in North America, is apparently weakening due to the rising temperatures, and its wavy pattern has become more extreme, allowing warm air to move north and cold air to move south to a greater extent. Therefore, some winters now are becoming colder—which Trump has claimed to be evidence *against* global warming.

George H. W. Bush was inaugurated in January 1989 and gave lip service to climate change. In fact, his 1988 campaign promise was to "do something" about global change and to fight the "greenhouse effect" with the "White House effect" (Emily Atkin, "The blood-dimmed tide," *The New Republic*, October 2019). In 1990, the results of the first Intergovernmental Panel on Climate Change (IPCC), convened to assess the state of climate science, and gently repeated the conclusions made in 1979. But the climate doubters, in reports of their own, drowned out the IPCC conclusions. By 1995, however, the second IPCC put far more effort into the research and the report: they forcefully concluded that human activities were affecting global climate. The climate deniers attacked, but the change in politics, to the years of President Bill Clinton (1993–2000), allowed this important report to be released and to be influential. The Clinton years also allowed adoption in December 1997 of the Kyoto Protocol, from an international treaty in Kyoto, Japan. The Protocol extended a 1992 United Nations (UN) agreement that commits countries to reducing greenhouse gas emissions, based on scientific consensus that global warming is happening and that it is caused by humans. It was signed by President Clinton in 1998 but, to become binding, it had to be ratified by the US Senate; it was never sent to the Senate. It had an unhappy history, especially after the election in 2000 of George W. Bush, who stated then that he opposed it.

As research on climate change accelerated, new reports as well as subsequent IPCC reports made the case for climate change and future implications ever stronger. The fifth IPCC report came out in 2013. Yet, despite the overwhelming evidence, politics were not favorable for doing anything, let alone giving the reports the attention they deserved, until Joe Biden became president in 2020. The sixth report in August 2021 was the most encyclopedic (with 234 authors) and forceful (better science now, resulting in assessments with "high confidence"), with additional reports on impacts and mitigation done in 2022. Alyssa Battistoni wrote a review (*The Nation*, June 3 to 10, 2019) of two books (one was *Losing Earth* by Nathaniel Rich) on climate change. In *Losing Earth*, Rich suggested that in 1989 John Sununu, White House Chief of Staff under George H.W. Bush, with the influence of the oil and gas industry, ended the last chance for the US to do something about climate change. This was when Sununu killed an international agreement to reduce carbon emissions, as reported by Meehan Crist in the May 2019 issue of *The New Republic*. Yet Battistoni noted that there are plenty of other villains to go around, especially oil and gas types. Nathaniel Rich had been quoted earlier in a special issue on climate change (September 24 to October 1, 2018, p. 14) in *The Nation* that "human nature" is to blame for our inability to deal with climate change and that "we had an excellent opportunity to solve the climate crisis" in the 1980s, for "almost nothing stood in our way—nothing except ourselves."

The election of George W. Bush in 2000 brought eight more years of climate denial: although he admitted that climate change was real, Bush asserted that the cause—whether by humankind or by nature—was in doubt, which is not true. Bush ran his first presidential race on cutting taxes, and his priority was just that, with legislation in 2001 cutting $1.3 trillion. He added to that later in his term. Next, despite many warnings, the country was caught unprepared in the al-Qaeda attack of September 11, 2001 (nearly 3,000 deaths, not including the many who have since died or are still to die from cancer and respiratory diseases caused by the collapse of the World Trade Center), and a month later Bush declared his global "war on terror" which required the invasion of Afghanistan (the longest war in American history). Not satisfied with just one war, in March 2003 he asserted that Iraq under Saddam Hussein possessed stockpiles of "weapons of mass destruction" (nuclear weapons) and ordered a massive air and land war. By the end of 2003, no weapons had been found but the US had suffered the deaths of 580 US soldiers, two thirds of whom died after his "mission accomplished" speech in May, on the deck of an aircraft carrier while wearing a flight jacket, pretending that he had piloted a jet onto that deck. The deaths of Iraqis during and after this war were horrendous and vast in number. Along with two wars, the lifting of banking regulations, and other monetary mismanagement, the balanced budgets of the previous Clinton administration now turned into huge federal deficits. This led to the global meltdown in world financial markets and the Great Recession of 2008, in which eight million people in the US lost their homes and the unemployment rate climbed to 10 percent.

With respect to the environment, I have often wondered what would have happened if Albert A. Gore, Jr., the Democrat, had won the 2000 election. When Bush withdrew from the Kyoto agreement on climate change soon after his inauguration, he asserted that CO_2 limits were not justified owing to the "state of scientific knowledge of the causes of, and solutions to, global climate change," which he termed "incomplete" (Kolbert, p. 160). The rest of the Bush presidency was more of the same, as typified in July 2003, when Senator James Inhofe called global warming "the greatest hoax ever perpetrated on the American people." Bush's campaigns, as other Republican campaigns before and since, were heavily funded by the oil and gas industries, notably by the largest of these companies, ExxonMobil. Rex Tillerson (born 1952), with a bachelor's degree in civil engineering, became CEO of ExxonMobil from 2006 to 2016, where he continued the campaign of funding disinformation on climate change. It is not surprising that Trump picked him to be the first (2017 to 2018) Secretary of State in the Trump administration. Tillerson's actions led to budget cuts, low morale, and the resignation of many career diplomats. Tillerson's most noteworthy positive accomplishment was when he referred to Trump as a "fucking moron" to colleagues, an incident that the media reported.

Hurricane Katrina, perhaps the worst major environmental disaster in the US in the last fifty years, should have generated an environmental wake-up call for the country. Once again, Bush and his political appointees were asleep. The hurricane made landfall as a category-5 storm (the highest rating) in the last days of August 2005. It came ashore east of New Orleans, but most of the destruction was the result of the failure of poorly maintained levees in and around that city. About 1,800 people died, many of them poor who had few means to escape beforehand and were then overwhelmed by the speed of the storm. The devastation was assessed at $125b, making it the costliest weather disaster in US history, a record that would be equaled but not surpassed by the destruction in Houston from Hurricane Harvey in late August 2017. When Katrina came ashore, Bush was vacationing at his ranch in Texas, and his appointed director of FEMA (Federal Emergency Management Agency) was Michael D. Brown ("Brownie"), who got his job by being a long-time lawyer friend of a person who partly managed Bush's presidential campaign. Bush, Brown, colleagues, and some local politicians did little in the first two days, while people died clinging to rooftops. Yet Bush stated on September 2, "Brownie, you're doing a great job." Once the videos showed otherwise, it took only a week or so for Brown to resign in disgrace.

Journalists and news-media owners at one time understood that they were to be the watchdogs to political power and corporate greed. Unfortunately, newspapers in recent years have undergone major decreases in staff size in the face of competition from the internet and in being bought by corporations interested mostly in the bottom line and their stock price. Other news-media outlets, notably the all-important TV media, have also changed to become more business oriented, financed by ads paid by big corporations. (Fox TV, of course, is hardly an objective news organization.) Perhaps for these reasons, the general media, although not the scientifically-literate media, have long muddied and minimized the issue of climate change by their "fairness" doctrine (for example, that climate change is an opinion in which both sides must be given equal time), their general ignorance about science, and their greed in seeking corporate advertising. As part of their fairness doctrine, they will frame the debate by classifying people who "believe" in climate change and those who do not. Yet climate change is not a belief but a fact, a difference that the news media seems not to understand.

Bill Moyers, a TV commentator, critic, and journalist, dealt with news-media ignorance and passiveness with respect to climate change in a speech in April 2019 (excerpts in the July 15–22 issue of *The Nation*) in which he observed that "the Goliaths of the US news media, those with the biggest amplifiers—the corporate broadcast networks—have been shamelessly AWOL [in amounts of accurate coverage], despite their extraordinary profits." Furthermore, he noted that "many outlets have ignored or misreported the climate story and failed to counter the tsunami of deceptive propaganda unleashed by fossil-fuel companies and the mercenaries, ideologues, and politicians who do their bidding." Therefore, from reporters in the field or office covering Houston flooding or Paradise burning, "we rarely hear the words 'global warming' or 'climate disruption" in their reports. The big backstory of rising CO_2 levels, escalating drought, collateral damage, cause and effect, and politicians on the take from fossil-fuel companies? Forget all that. Not good for ratings, say network executives." Speaking from my own experience, local TV newscasts in Republican states such as Utah are much worse than national telecasts: the words "climate change" are apparently off limits. Nonetheless, in 2004, the *National Geographic* declared that global warming "got respect," in other words that ordinary Americans finally came to believe that the issue was important. Opinion polls since then have shown that an increasing percentage of citizens agree that human-made climate change is taking place; 2019 polls show that Americans agree by a clear majority. Of course, virtually all scientists of any discipline had concluded this by the early 1990s. Scientists are curious, and the easy availability of climate-change reports or their summaries on the internet or science magazines allows all scientists to keep track of the debate. I personally know of no scientist of any type who does not accept that climate change is happening exponentially, is due to human actions, and is not only unbelievably scary, but the most pressing issue of our time!

President Barack Obama was like a nutritious filling in a sandwich between two slices of cheap white bread that had no nutritional value. Because both Bush and Trump led careers of incompetence, they could not recognize in their political appointees what excellence or even competence looks like. So they picked hacks who were friends or whom others told them to pick (in the case of Bush), and sycophants (in the case of Trump). President Obama understood the concept of groupthink and picked persons for his cabinet and other positions who had been successful and accomplished and who could be relied upon to keep him honest by arguing alternative positions when they were needed. For an eight-year period, from January 2009 to January 2017, we had people with brains dealing with domestic and international problems. The first job needed was to fight the global financial crisis inherited from Bush, accomplished with a major stimulus package and with new financial regulations to prevent this from happening again. Obama's primary achievement was the Affordable Care Act ("Obamacare"). But the Tea Party, an extreme right-wing citizens' movement within the Republican party, was founded early in 2009 to advocate lowering taxes and the national debt and to oppose government-sponsored health care, along with advocacy of lots of nutty ideas and conspiracy theories. It rapidly became a political force and in the 2010 midterm elections, Tea-Party politicians took control of the House of Representatives. In 2012, Obama was elected to his second term, but the House remained in Republican hands. Obama ended the US military presence in Iraq, found and killed Osama bin Laden in 2011, and negotiated agreements with Iran and Cuba. On the environmental side, in 2009 he allocated funds to encourage domestic renewable energy production. Obama described global warming as the greatest long-term threat facing the world, but in his second term, when he attempted legislation, the Republican congress blocked him. And in the 2014 midterms, the Republicans won control of the Senate too, ending any significant Obama legislation. He therefore used executive actions to do what he could, including establishing improved vehicle fuel-efficiency standards; of course, Trump easily reversed all of them. During Obama's second term, representatives of 196 countries (the only significant emitters that refused were Iran and Turkey) negotiated and adopted by consensus the UN Paris Agreement of 2015, with the goal of keeping global average temperatures below 2°C (the actual goal) above pre-industrial levels and of pursuing efforts to limit increases to 1.5°C (the aspirational, hoped-for goal). At the time of the Agreement, it was calculated that global average temperatures had already risen 1°C (1.8°F) above pre-industrial levels. Virtually all scientific investigations indicated that, beyond 2°C, a threshold would be reached (a "tipping point") beyond which it would be nearly impossible to reverse environmental catastrophes, including collapse of the ice sheets.

Senator Pat Moynihan once said, "Everyone is entitled to his own opinions, but not to his own facts." This was long considered a humorous way to express things, for most people know that an adversary will not make up facts. But then came President Donald Trump. The Democrats were powerless to stop Trump until the first mid-term election in November 2018, when they took back the House of Representatives and Nancy Pelosi took over as Speaker. Yet any legislation of significance by the House has been blocked by the Speaker of the Senate, Mitch McConnell. Trump then repeatedly called any suggestion of climate change a "hoax." On environmental policy, Trump announced in June 2017 that the US would withdraw from the Paris Agreement. His appointees to the EPA and virtually all other agencies were politicians and fossil-fuel executives who had derided EPA. In fact, almost all these appointees had conflicts of interest in the positions to which they were appointed. Yet all were approved in robotic votes by the Republican Senate majority. Subsidies to fossil fuel industries, including the coal industry, were increased. All references to "climate change" were forbidden on agency websites and reports, and research on or related to climate change was suspended or canceled. Protections were weakened for endangered species, stream/river and ocean pollution, drinking water, and other standards. Trump had no Science Advisor to the President, and most science that might address climate issues or regulations on industry was decreased or eliminated. By June 2019, Trump had rolled back 83 environmental regulations. In short, almost everything in the realm of environmental protections was suspended. A 2018 poll of political scientists, administered by the American Political Science Association, ranked Trump in last place among all presidents. On Presidents' Day, February 2024, the third poll of these political science academics, now called the Presidential Greatness Project, again ranked Trump the worst of all 45 presidents.

Under the Trump administration, weather disasters took place with increasing severity and expense. Yet the climate implications of these disasters seem to have escaped the notice of the mainstream news media. Weather, which is brief in time and local in size, is different from climate, which is long term and regional. Rising sea levels from melting Arctic and Antarctic ice are resulting in increased flooding and storm surge related to storms. With rising temperatures, more moisture evaporates from the oceans and warmer air can hold more water vapor: tornadoes in the mid-American continent in the last several springs and summers are greater in size and frequency, and hurricanes are stronger. Each year produces new average, hottest-ever-temperature records, reported on TV local news shows and weather bureaus, whether for cities, countries, or the world. On July 26, 2018, for example, an internet summary from an article in *The Washington Post* noted that temperatures in Ouargla, Algeria reached 124°F on July 5, a likely record for the continent of Africa, and in Quriyat, Oman, 109° was the low for June 28, the hottest low temperature ever recorded on Earth. Perhaps the most visible and frightening weather or weather-related disasters are hurricanes in the late summer and early fall and forest fires in all seasons. Below, I mention a few climate-related events that caught the attention of the media.

Hurricane Harvey, tied with Hurricane Katrina as the costliest weather disaster ($125b) in US history, made landfall as a category-3 storm in late August 2017, on Trump's watch. The source of the problem was that the speed of the hurricane slowed and stalled just inland: rainfall, with subsequent flooding, was the main source of the destruction. Many areas in Houston had 30 inches of rain, with 60 inches in some places. It was the third so-called five-hundred-year storm to strike Houston since 1979. The death toll was 107 people, and about 300,000 structures and 500,000 vehicles were destroyed or damaged. The fourth costliest ($92b) weather disaster in US history was Hurricane Maria, a category-5 storm that in early September 2017 passed over the Dominican Republic, the US Virgin Islands, and then slowed over Puerto Rico. The death toll, mostly in Puerto Rico, was originally given as a relatively small number, but the lack of electric power for extended periods, coupled with a terrible and incompetent relief effort by FEMA, resulted in about 3,000 deaths, many when ventilators and oxygen machines lost power or medicines ran out. Trump disparaged this claim and stated that Democrats inflated the number to "make me look as bad as possible." The awful response by FEMA, including ignoring distribution of relief products housed in several island warehouses, can be considered racist, egged on by Trump in his hatred of Hispanics, and Trump responded to universal criticism of his response to the Hurricane by proclaiming that the federal aid authorities did a "fantastic job."

In addition to a heavy hurricane season and many forest fires, two large, back-to-back fires in northern California stand out in 2018. The Carr Fire, named for Carr Powerhouse Road, where a flat tire on a travel trailer being towed by a car caused sparks on the pavement, which then started the fire, took place from July to August. The fire burned nearly 230,000 acres (359 mi^2), mostly in a National Recreation Area, but it also reached the large town of Redding, where 38,000 people had to be evacuated. It destroyed 1,600 structures (1100 were homes) and damaged 300 others. It was California's sixth most destructive, with eight fatalities and about $1.5b in insured losses. It saw tornadoes of hot air and flames that rose 17,000 feet into the air and tossed a forty-foot-long shipping container up and around. The Camp Fire, named for Camp Creek Road where fire was first seen, destroyed 95 percent of the densely-populated town of Paradise and the communities of Concow and others during most of November. It was the deadliest (85 fatalities) and most destructive fire in California history, destroying 18,800 structures, mostly homes as well as five public schools, and burning over 153,000 acres (about 240 mi^2). Total damage was estimated at $16.5b, and many of the homes were not insured. Paradise is on a flat-topped, tree- and brush-covered ridge, the classic wildland-urban interface, west of the Sierra Nevada. Warnings had been issued in a 2005 report by the California Department of Forestry and Fire Protection (Cal Fire) that the area was vulnerable to ember-driven conflagration during east winds. With a heavy grass cover due to a wet spring, an unusually dry fall, and katabatic ("Santa Ana") east winds of 25 to 35 mph, the sparking of high-power electric lines operated and poorly maintained by Pacific Gas & Electric (PG&E) apparently started the fire. This electric company has since proposed bankruptcy to avoid paying lawsuits in full. And most recently, instead of investing in their infrastructure or clearing brush near power lines, this private company on October 10 shut off power for two days to more than 700,000 residents of northern California during dry high winds; one wonders whether this is how the company will deal with Santa Ana winds in the future. Also, in the case of Paradise, too few roads west out of town were available for all residents to escape; the speed of the fire was such that most of the town was destroyed in the first four hours. Years of fire suppression, especially in the canyons north and south of town, were partly to blame, as were policies that encourage house construction in fire-prone areas. Both fires were in districts that voted overwhelmingly for Trump. Of course, these fires are just two noteworthy examples among many, as fires have continued throughout the West in subsequent years. Wikipedia noted that, as of October 6, in 2021 California had a total of 7,883 fires that burned about 2,500,000 acres and destroyed 3,600 buildings.

North-moving Hurricane Dorian, a category-5 storm, made landfall in the southern Bahama Islands, a British Commonwealth, on September 1, 2019. The hurricane slowed over the central Bahamas and stalled for thirty-six hours over the northern Bahamas, obliterating nearly all structures. The damage estimate in the Bahamas, where at least 70,000 people were left homeless, was at least $3.4b. The direct death toll was seventy-seven persons, but 245 were missing. It was the most intense storm on record ever to hit the Bahamas, with winds clocked at as much as 185 mph. As the storm passed beyond the Bahamas, moving north along the Atlantic coast, Trump warned that it would impact Florida, Alabama, and the Carolinas. Apparently, he thought that Alabama is on the Atlantic Coast. When the Birmingham office of the National Weather Service pointed out that it would not disturb Alabama, Trump (by then informed that Alabama is on the Gulf coast) used a felt-tipped pen to modify the weather bureau's hurricane track to show that it might veer west to impact Alabama. He kept this argument going for about a week. The National Oceanic and Atmospheric Administration (NOAA) admonished its weather service for contradicting the president; we later learned that Wilbur Ross, the rich Secretary of the Commerce Department, where NOAA resides, had ordered NOAA to rebuke the weather service. The hurricane continued up the coastline, clipping the outer banks of North Carolina, then moved northward and made landfall again, still as a hurricane, in Nova Scotia and Newfoundland, with more damage. It fizzled out near Greenland on September 10. Agencies in the US initially offered to take many homeless Bahamian people to the States until building could begin again. Most of these homeless persons had only the clothes on their backs left to them. Trump, fully knowing that passports had been washed away and apparently noticing from news videos that many people from the Bahamas are black, told them that they could not enter the States unless their passports were in good order.

Australia in recent years has witnessed horrendous drought and bush fires, yet its conservative politicians ("Liberals") in power, and even some who are liberal ("Labor"), had until 2022 denied climate change. In other words, the Liberals won elections by being pro-coal and anti-climate change. In 2019, as the fire season got underway in the spring (mid-November), with their past winter the hottest on record, with temperatures in the mid-90s, and with high winds and drought, already 129 fires had taken place and a total of 2.5 million acres had been blackened, notably near and north of Sydney. Then Prime Minister, Scott Morrison, his government poorly prepared for the tragedy, had long championed (because it was beholden to their support) the fossil fuel (notably coal) and mining industries, and thus has made a career out of denial and lying about facts. His Deputy Prime Minister, Michael McCormack, on ABC radio on Monday November 11, 2019, blamed "raving inner-city lunatics" for linking the fires to climate change. As the fire season raged, the internet recorded that six-million acres burned by mid-December, ten-million by late December, twelve-million by December 30, and eighteen-million by January 12, with a final tally on March 2, according to *Science News* of March 28, of forty-eight-million acres. Along with this, the Science News article listed thirty-four people killed, 6,000 buildings destroyed, and an estimated 1.5 billion animals killed by the fires. The 2020 World Climate Change Performance Index ranked Australia last for their climate policies among fifty-seven countries it monitors. During the fires, Morrison notably insisted that this was no time to talk about climate change!

Climate disasters, of course, have continued after Trump was defeated for the presidency. The National Oceanic & Atmospheric Administration (NOAA) on January 12, 2024, announced that the planet's warmest year on record, "by far," was 2023 (it is possible, if not likely, that 2024 will exceed this, when the numbers come in). The Associated Press, in an internet story accessed on September 12, 2023, noted that the US just set a record (in 2022) for billion-dollar weather disasters in a year—with four months yet to go. Of course, heat is probably the biggest killer across the globe, although it is difficult to quantify because deaths may not be directly related to heat or that heat deaths get misdiagnosed. For example, in Maricopa County, Arizona (includes Phoenix), deaths directly related to heat have increased ten-fold in the last decade, with 645 listed for year 2023; about half of these are homeless persons (source is the internet, accessed October 20, 2024). George Packer in the July/August 2024 issue of *The Atlantic* wrote a long essay about Phoenix that dealt with heat deaths (with mention of a May 2023 study that predicted that a temporary electrical blackout of the energy grid would likely kill one percent of Phoenix's population), dwindling water supplies, politics, and stupidity. The third costliest ($113b) weather disaster in US history was Hurricane Ian, which made landfall as a category-4 storm in southwestern Florida on September 28, 2022, and killed 161 people. Beginning in March 2023 and continuing through the fall, Canada had more than 6,500 fires—in all its provinces—which consumed 45,700,000 acres (71,000 mi^2, about the size of North Dakota, and about 5 percent of Canada's forests) following drought, and the air pollution was at times extreme in Canada and the northern US. We all remember the fire on the island of Maui of August 8 to 11, 2023, in which most of the town of Lahaina on the northwest coast was burned, with a loss of about a hundred lives, the worst American death toll from wildfires since 1918. The Smokehouse Creek fire, from February 26 to about March 26, 2024, scorched about 1,100,000 acres in the panhandle of Texas and adjacent Oklahoma. Of course, in the 2024 hurricane season, Hurricanes Helene and Milton made landfall two weeks apart in late September to mid-October. Hurricane Helene landed in the big bend of Florida as a category-4 storm and tracked north, causing much damage and most deaths (225, with many missing) in western North Carolina; damage has been estimated at $45b. Hurricane Milton landed on the western coast of Florida south of Tampa as a category-3 storm and tracked east across the peninsula, with damage estimated at $30b and at least 17 deaths. Displacement of US citizens from high- to low-hazard states, following the loss of their home insurance or finding it difficult to sell or rebuild their home, has not been addressed because it has only started, but such movement of humanity is much worse already in some parts of the world. Disputes over access to culinary water, as in Arizona and some parts of the West, also are only beginning.

International wars provide a good excuse to do little about climate change, and the world is experiencing two of them. On February 24, 2022, after a buildup that began in late 2021, Russian dictator Vladimir Putin invaded Ukraine. Putin expected his war to last only a few days and he told his troops that they would be greeted as liberators, but the Ukrainians under the leadership of their president, Volodymyr Zelensky, have fought bravely for well more than two years, resulting in a stalemate, with Russia occupying almost 20 percent of Ukraine. The lion's share of the support for Ukraine has come from Europe and from the US under President Joe Biden, but the Republican-controlled House of Representatives refused more support for many months of 2024. The second war started on October 7, 2023, with an attack on Israel by Hamas, the Palestinian Islamist terrorist group, on the Israel border east of Gaza that killed about 1,100 Israelis, most of them civilians, and took 250 hostages. Benjamin Netanyahu, Prime Minister of Israel, a Trump wannabe, apparently was distracted by his attempt to remain in power and did not see the warning signs. But in response, he ordered the Israeli military to bombard and invade Gaza, with significant Palestinian casualties (at least 40,000 deaths estimated as of the fall of 2024), most of them civilians, and most of those women and children. Hezbollah, the Lebanese Islamist terrorist group, began bombarding Israel on October 8, 2023, resulting in retaliation by Israel that escalated (airstrikes, a limited invasion of southern Lebanon) in September 2024 and continues.

Action to combat climate change has been minimal except in Europe, California, and a few other places, which are rapidly developing solar and wind energy. And of course, some countries, notably Brazil and India (and the US until the election of Biden), are reversing progress made in previous administrations. Yet much of what needs to be done is well known, including sea defenses, drought- and heat-resistant crops, cap-and-trade or carbon taxes, more rapid change to electricity grids that do not require fossil fuels, etc. Planting millions or billions of trees is a low-technology first step. These things will initially be expensive, and private companies are holding back, hoping that federal funds will be available. No laws have been passed providing negative consequences or economic incentives, so nothing happens. Clearly needed is government investment on research on carbon capture, energy storage, CO_2 absorption, and other ways to deal with greenhouse-gas emissions, but there is little sign of that happening. Direct carbon-capture technology seems to be most needed, but so far we have only ideas. In fact, the entire world economic system must change, as noted by Ben Ehrenreich and Mariana Mazzucato in back-to-back articles in the April 2021 issue of *The New Republic*. The economic system must veer away from "perpetual economic growth" (GDP growth) (p. 18), in which "capitalism … is narrowly aimed at maximizing the returns to shareholders" (p. 25). In 2023, Oreskes and Conway published the book *The Big Myth* that deals with the same arguments. And how will countries deal with the mass migration of peoples, especially from poor countries? By more religious and ethnic wars, including nuclear wars? International cooperation, far beyond what the United Nations is attempting, will be necessary. The book *Collapse* provides historical evidence that there is little collective will to stop societal destruction even when rulers and citizens knew it was coming.

Books, both fiction and non-fiction, on the consequences of doing nothing are proliferating. One short book—*The Future We Choose: Surviving the Climate Crisis*, published in 2020 by climate-change experts Christiana Figueres and Tom Rivett-Carnac (who led UN negotiations for the aforementioned 2015 Paris Climate Agreement), attempted to offer an optimistic choice to countries that are not interested in experiencing the horrible consequences of when the Earth passes the tipping point. (According to the COP 28 UN Climate Change Conference in 2023, a tipping point is when a tiny change can cause events that are "large, abrupt, and hard to reverse," an analogy being when you lean back in a chair, then pass its tipping point, the chair goes over backwards!) Such a tipping point is likely later in this century, at which juncture little more can be done to save Earth. They note that "[i]f the first duty of government is to protect its people, then across much of the world the form of democracy we have become used to is failing." A main reason is that "[i]n many countries today, corporate interests have captured our democracies." The authors urge that global greenhouse gas emissions be halved by 2030, then cut to zero ("net zero") by 2050! In April 2022, the IPCC issued a more urgent warning, that unless greenhouse gas emissions cease rising by 2025, humanity will not be able to limit the warming of the planet to 1.5°C (*The Nation*, September 18 to 25, 2023), a cry that seems to have been met with a yawn by the world.

In 2023, David Lipsky's book *The Parrot and the Igloo* updated the climate-change denialism given in *Merchants of Doubt*, in fact showing that denialism has increased, even before the Trump presidency, then through those four years. A particularly alarming scientific conclusion that he gives comes from Greenland ice cores, some a mile deep that give a continuous annual record, as reliable as tree rings, that go back 110,000 years (starting before the last ice age and continuing through it, until to-day) and show that "a crazily jumping climate has been the rule, not the exception" (p. 174). They show that "[i]n the Greenland [ice core] record, there are eras where temperature changes of fifteen degrees in a half century, [even] periods when heat leaps eighteen degrees in a decade" (p. 174) according to the writings of Richard B. Alley, Pennsylvania State University Professor of Geosciences. Alley's book, *The Two-Mile Time Machine*, noted on page 4 that "the ice age came and went in a drunken swagger, punctuated by dozens of abrupt warmings and coolings" that were "larger, faster, and more widespread climate changes than experienced by industrial or agricultural humans." It was only in the Holocene (beginning 11,700 years ago) that stable temperatures allowed civilization to develop, especially starting about six thousand years ago.

In *Merchants of Doubt*, Oreskes and Conway make two analogies with respect to climate change, in their book's conclusions. In one, they ask the reader to imagine a giant banquet at which hundreds of millions of people have eaten for many years. Finally, one day a person shows up with a bill. "[T]he diners are in shock." They debate the issue but finally they decide that, if they simply ignore the person with the bill, "he will go away." That is where the issue of climate change now stands. In the other analogy, the authors point out that when we buy a house, we do most of the work except a title search, which is beyond our capabilities. Therefore, we hire somebody to do that. In the same way, the public can do most things in life by themselves, but they cannot do their own science. They hire others to do that, and commonly they also will ask the National Academy of Sciences or some other respected scientific organization to investigate pressing issues and provide a concluding report. "So it comes to this: we must trust our scientific experts on matters of science, because there isn't a workable alternative." "For even if modern science does not give us certainty, it does have a robust track record. We have sent men to the moon, cured diseases, figured out the internal composition of the Earth, … all on the basis of modern scientific knowledge."

In an editorial on the legacy of the aforementioned David Koch (an ardent foe of any talk of climate change) on the day he died (August 23,

2019), Ryan Cooper of *The Week* stated: "In the decades to come, climate change will be the issue that determines the fate of not only American politics but human society. And it will be the issue on which the history books judge everyone, not just David Koch." Emily Atkin, in a report suggesting future outcomes to the impending disaster (*The New Republic*, October 2019), hypothesized that "future historians will look back at the bleak landscape of resource rivalry and mass climate migration in the twenty-second century, and marvel at how long, and how extensively, world leaders were briefed on the specter of climate chaos, only to shirk their basic responsibilities to provide and govern for posterity over an entire political generation." In the October 2, 2017 issue of *The Nation*, right after Hurricane Harvey, Mark Hertsgaard noted that a simultaneous storm (but one unreported in the US) produced much greater devastation in India, Nepal, and Bangladesh. The author repeated a quotation he had previously made, namely that, "To refuse to act against global warming is to condemn thousands of people to death and suffering today and millions more tomorrow. This is murder, even if Trump's willful ignorance of climate science prevents him from seeing it." Hertsgaard advocated for "no more treating climate denial like an honest difference of opinion."

> ***Noam Chomsky*** *(born 1928) is a philosopher, author, researcher in linguistics, historian, social critic, and political activist who has joint positions as Professor Emeritus at MIT and Professor at the University of Arizona. He has used common sense and a vast knowledge of many fields to put his finger on societal problems and their likely solutions, in books, lectures, and interviews on a huge variety of topics, including Trump and climate change. His ability to define and zero in on the crux of a serious problem is like what I saw with Hoover Mackin, indicating to me that Chomsky also is a genius. He has a wide and enthusiastic following, especially among college students, the liberal press, and liberal TV shows.* The Nation *(January 7, 2019, p. 3) quoted Chomsky in a 2017 interview with Amy Goodman of* Democracy Now! *regarding the policies of the Republican party under Trump and Paul Ryan with the comment: "Has there ever been an organization in human history that is dedicated, with such commitment, to the destruction of organized human life on Earth?" His answer: "Not that I'm aware of."*

Experts who have written books and articles on climate change, especially when they predict the future, have tended to use their writings to try to scare the hell out of the reader, hoping that this will motivate him/her or their government to do something. This approach may be certainly justified, for each year that goes by these authors see little happening, and they know better than most what awaits. However, it seems to me that hyperbole is self- defeating, in the same way that the scare tactics of *The Population Bomb* strained the scientific credibility of those authors. As with population increases, climate change will proceed in ways that are difficult to predict, other than being terrible! And the effect on many persons, facing this horrifying world catastrophe, is to freeze into inaction. The mental effects of climate change are becoming more severe, as Katherine Kornei related in 2023 (see Notes), with a deep rage among many young people on why their elders are not doing something, and suicides on the increase. An article in *The Economist* (October 14, 2023, p. 50-52) documents overreaching by "green" parties or liberal governments, some of which enact or propose inappropriate and unneeded extreme solutions, such as bans on gas-powered engines and cars, mandating that heat pumps be put in homes, fines for people with gas-guzzlers, taxes on eating meat, stop flying, compulsory car-sharing, etc. Such advice is not only elitist (most people do not have the resources to abruptly retire their cars, install heat pumps, etc.) but it has provided great ammunition and opportunities for populists such as Trump, who want only to get elected, and to do so are turning the issue into another culture war ("the lefties now want to take away my car and my hamburgers, in addition to my guns"), thus stifling debate and stirring up uninformed voters. Such individual effort is pointless and frustrating. Instead, the effort should be collective, toward increasing efforts toward governments and corporations on big things, such as increasing clean energy, a carbon tax/cap and trade, decreasing carbon dioxide and other warming pollutants, education. Voting for Democrats, running for office, peaceful protests/boycotts against fossil-fuel providers and polluters, organize/unionize, helping increase awareness to the problem, talking to your neighbors, etc. need to be done first. Al Gore, in his book *Our Choice* urges that people change the way we think, as we did with smoking cigarettes.

Speaking of Trump, he showed his climate-change chops on December 22, 2019, just in time for Christmas, when he gave a speech to Turning Point USA. This organization is a far-right nonprofit organization that among other things maintains a website that documents and exposes "college professors who discriminate against conservative students, promote anti-American values, and advance leftist propaganda in the classroom." Trump's topic appears quite commonly in his MAGA rallies, for he considers himself an expert on it (he has stated that the noise from "windmills" causes cancer). Trump left us with the following:

"I never understood wind. I know windmills very much. I have studied it better than anybody. I know it is very expensive. They are made in China and Germany mostly, very few made here, almost none, but they are manufactured, tremendous—if you are into that—tremendous fumes and gases are spewing into the atmosphere. You know we have a world, right? So the world is tiny compared to the universe. So tremendous, tremendous amount of fumes and everything. You talk about the carbon footprint, fumes are spewing into the air, right spewing, whether it is

China or Germany, is going into the air."

The speech continues in this vein, a small example of a typical Trump "word salad" speech about, allegedly, wind turbines and renewable energy. I do not know whether the audience applauded the speech or perhaps gave Trump a standing ovation, but either certainly would not be unexpected.

On November 23, 2019, sixteen-year-old Greta Thunberg from Sweden gave a very short address to the United Nations Climate Action Summit in New York City. Most of the talk is as follows:

"This is all wrong. I shouldn't be up here. I should be back in school on the other side of the ocean. Yet you all come to us young people for hope. How dare you? And yet I'm one of the lucky ones. People are suffering. People are dying. Entire ecosystems are collapsing. You have stolen my dreams and my childhood with your empty words. We are in the beginning of a mass extinction, and all you can talk about is money and fairy tales of eternal economic growth. How dare you? For more than 30 years, the science has been crystal clear. How dare you continue to look away and come here and say you're doing enough when the politics and solutions needed are still nowhere in sight? "

Trump, who could only find the time to attend the UN gathering for ten minutes that day, was still able to find the time to mock this child for her views. Yet there was nothing controversial in what she said. She did not pretend to be an expert on climate change. She just stated the obvious better than the experts could. For that, *Time* Magazine named her 2019 Person of the Year.

CHAPTER 13

LIES AND INSANITY: TRUMP AND THE CONSEQUENCES

The years spent writing this memoir coincided with the second half of the presidency of Donald Trump, plus four years after his re-election defeat. The insurrection of January 6, 2021, Trump's last attempt to prevent the transfer of presidential power, and the increasing intransigence of Republicans since then indicate that American democracy is up for grabs, and the election of November 2024 confirms that. The first Trump presidency devastated science, but a second presidency under Republican control may demolish science and any hope that we can counter climate change. Under authoritarian regimes, business and most of the arts can survive but science is at great risk, for science depends upon truth. As much as is my disgust for Trump, it is much stronger for the people who voted for him and the elected or appointed Republican politicians, all of whom stymie any consequences for what Trump has done. But I am baffled by the motivations of voters. To this end, I have immersed myself in books and articles that deal with aberrant behavior. This chapter records simple history that I watched unfold. Future consequences would seem to be predicted by the classic song by Bob Dylan, "A Hard Rain's A-Gonna Fall."

1 Trump

Let me first itemize some observations about Trump (I deal with Trump voters and the Republican party separately below) and his effect on science. As background, we need a snapshot of his upbringing, character/relationships, business dealings, presidential campaigns, and actions when President.

Trump, born in 1946, came from great wealth and grew up in New York. His father, Fred Trump, Sr., made his money in real estate and left most of it—at least $400m—to Trump (though Donald claims he is a self-made businessman) over the father's lifetime, generally through a series of schemes that allowed them to avoid gift or inheritance taxes. In college, Donald spent most of his time chasing women, and therefore learned nothing; one professor called Trump his worst student ever.

Mary L. Trump, the niece of Trump and born in 1965, published the book *Too Much and Never Enough* in mid-July 2020. It is a family history tell-all book that portrays Trump as the world's most dangerous man. Mary has a Ph.D. in psychology, so her observations about several generations of a dysfunctional family seem trustworthy. Her father, Fred Trump, Jr. (1938–1981), was the older son (by eight years) of Fred Trump, Sr., whereas the younger son is Donald. Fred, Jr. became an early, path-breaking commercial pilot for TWA; he was a gentle and kindly soul with many friends who came forward at his death. Fred died at forty-two years old of diseases related to alcoholism, due to his upbringing by his cruel, tyrannical father, who died in 1999. Mary refers to her grandfather as a "high-functioning sociopath" who wanted Fred, Jr. to take over the hugely successful real-estate business that Fred, Sr. had built up, significant parts through shady if not illegal business practices. Fred, Sr. practiced emotional abuse on both sons, in effect destroying Fred, Jr. Donald, however, realized what he had to do to please his father, including giving up his humanity. The result was a con artist and what Mary called "a terrified little boy" whom his father repeatedly bailed out. The lessons he learned were to sacrifice emotions and to succeed at all costs. His older sister did his homework and he paid somebody to take his SATs for college. Mary claimed that Trump was never loved by his parents and became a sociopath too, with no empathy for others, whom he considered only to be pawns in his own success. Mary noted that Donald is indifferent to people in need or distress, considering them to be losers. She concluded that he is humorless and, in fact, he sees little joy in life; his emotions are negative and are dominated by self-pity, and he lacks leadership ability, shirking instead of acting on tough decisions by lapsing into denial.

Have you ever witnessed a small spoiled brat throw a tantrum in a grocery store? Well, we have seen similar behavior by Trump throughout his four years as president. He wears a perpetual pout interrupted by smirks and grimaces, he lies, he craves attention, he makes excuses for his failures, he has no self-control, he brags and gloats, he whines and whimpers, he has regular tantrums, he throws his food against White House walls, etc. He is convinced that strength is demonstrated by being a bully and, like most bullies, he is a coward and picks on the most defenseless, such as immigrants from central America fleeing criminal gangs, or even vocal children like Greta Thunberg mentioned in the previous chapter. He appears sadistic, for one of his pleasures is humiliating others, especially politicians who are as weak and amoral as he is.

More than twenty women have accused Trump of sexual misconduct, including rape. Only one of the women he molested, columnist E. Jean Carroll, kept at it sufficiently long enough to get to trial, and on May 9, 2023, Trump was convicted of sexual assault and defamation, and he was fined $5m. Then, within a few days, he defamed her again (and continued at it), and in a jury trial for that Trump on January 26, 2024, was assessed a penalty of $83m, mostly in punitive damages.

Just before the 2016 presidential election, he bought the silence of two women with whom he had affairs, so this news did not come out until well after the election. This secret likely allowed his victory. In August 2018, Michael Cohen, Trump's attorney, who paid off one of the women, Stormy Daniels, pleaded guilty to eight criminal charges and was sentenced to three years in prison. In March 30, 2023, at the behest of Alvin Bragg, Manhattan District Attorney, a Manhattan grand jury voted to indict Trump on thirty charges concerning hush money, election interference, and business fraud related to this case; he was arraigned in April, with a trial on March 25, 2024. On May 30, he was convicted on all thirty-four counts. He awaits sentencing, but this has been postponed until after the Presidential election in November 2024 because he is the Republican candidate.

Despite what he claims, Trump has shown little competence at economics or business. He has used his connections—and fudged data—to borrow large amounts of capital, much of which he has refused to pay back. Being constantly bailed out by his father meant that he never had to face reality or make tough decisions. His "yes men" did the work and took any blame. Using the money his father left him, adulthood saw him cutting one deal after another, many of them ending in losses and bankruptcy: his limited tax returns from 1985 to 1994 showed net losses of $1.7b, among the greatest of any person in US history (*The New York Times*). One of his "great businessman" specialties was employing small-business people to do construction and maintenance, then not paying them; he dared them to sue him or he had them agree to settle at pennies-to-the-dollar owed.

Trump has worked closely with Russian oligarchs and mobsters, who cultivated and flattered him by inviting him to Moscow, then bailing him out after the bankruptcy of his Atlantic City casinos when Vladimir Putin replaced Russian President Boris Yeltsin in the early 1990s. Those relationships included money laundering because rich Russians needed safe places to hide their cash; they bought luxury apartments in Trump's high-rise apartment complexes or other types of real estate. Trump has avoided paying taxes by various real estate schemes. He was prosecuted in federal court for the fraud that was Trump University, but the fine was only $25m on April 10, 2018. His Donald J. Trump Foundation, claiming to be a charity, used the funds for his personal use; the Foundation was found guilty by the New York Attorney General, ordered dissolved with all money going to charity, and fined $2m in November 2019. On August 18, 2022, Allen Weisselberg, the Chief Financial Officer of the Trump Organization, pleaded guilty to fifteen felonies against the Organization and, in a deal in which he was sentenced to five months in Rikers Prison and had to pay $2m in back taxes and penalties, he agreed to testify against Trump; at the trial, the organization was convicted on multiple charges of tax fraud and falsifying business records, but Weisselberg did not implicate Trump (i.e., he committed perjury). He, Trump, and his sons became defendants in a $250m (later increased) civil lawsuit brought on September 21, 2022, by New York State Attorney General Letitia James, who alleged a decade of fraud (thus liable for interest penalties from the date on which it started). On February 16, 2024, Trump was assessed penalties of $355m, plus nearly $100m in interest, for a total of $454m. Weisselberg has yet to be sentenced for his perjury.

In 2015, Trump began a presidential campaign as a Republican but more accurately as what is dubbed a populist, which is an ideology that supposedly represents the common man against the elite, but the dictionary definition instead points out that it is a political demagogue who pushes hate and simplistic solutions to complex problems. For example, Trump promised to bring back jobs in coal and manufacturing, to lower taxes, increase wages, decrease costs of prescription drugs, bring troops home, build infrastructure, solve the opioid problem, etc. Virtually none of those happened during his presidency except for lowering taxes *for the rich.*

The only other promise kept by Trump during his 2016 campaign was stacking the US Supreme Court with three deeply right-wing-religious, pliant, and partisan Republicans whose primary professional legal record before confirmation was to oppose abortion, the only reason they were chosen. They joined three others of the same philosophy and incompetence. *Roe v. Wade*, a basic right for women for 50 years, was struck down by these six.

Most of Trump's speeches invoke overt racism, including finding minorities of all types to blame, tactics used to great effect in Nazi Germany. Another tactic is what is called "strategic misnaming," also known as "Swift-boating" (from the 2004 campaign, when George W. Bush, a Vietnam-war draft dodger, used a bunch of Republican Vietnam War veterans to lie about the courage and patriotism of John Kerry, Bush's presidential opponent and a decorated Vietnam war hero when he was a Swift Boat commander). The point is to accuse a political rival of engaging in some corrupt practice that he himself is constantly engaging in.

Once he became president, Trump continued his obsession with being the top daily headline. Thus, he gave endless comments and "tweets" about any prominent national or international news story. These daily utterances have proved that he does not know much about even basic facts in geography, history, economics, science, etc., so he makes things up. Most comments are unintelligible word sandwiches, which jump from topic to topic with no overall point except to bad-mouth opponents or express self-pity. *The Washington Post* fact checker column reported that Trump obtained the level of 10,000 "false or misleading claims" on April 26, 2019, only 827 days into his presidency (updated to 16,241 during the first three years of his presidency)! He makes up lies even when they are silly, pointless, or easily fact-checked.

Trump promised to "drain the swamp" but instead brought in the most corrupt administration in the recent history of the country. Trump packed his cabinet and virtually all political-appointment positions with sycophants. He required only servitude, adulation, loyalty, and praise, although occasionally some provided expertise. Yet post-2020 tell-all books that came out from many of these persons reveal many disturbing truths about Trump's intelligence and character, as well as the news that they had to dumb down and severely abbreviate oral or written morning briefings or other transfers of information that Trump as president needed to know.

Trump's ignorance of the basic concepts of business and economics resulted in the US having the greatest national debt in its history by the end of Trump's term. This was caused by Trump's only major piece of legislation—his December 2017 tax law that decreased taxes on the rich by $1.5t (the tax rate for the four hundred wealthiest Americans became lower than for many other income groups). Such a decrease (Reagan called it "trickle-down economics") is only a Republican gift to the rich and has never produced significant jobs, as real economists know.

Trump impulsively embarked on trade wars, largely by imposing tariffs, never favored by legitimate economists, with China, Mexico, Europe, and other countries on a seemingly-monthly basis, apparently because he was angered at their presidents. He thinks that tariffs are paid by the country he targets, unaware of the fact (known by most people and all economists, though they will not correct him) that they are instead paid by the consumer who buys that foreign product. His 2024 presidential campaign has promised that tariffs will go up on most imported goods, a recipe for inflation. Trade deficits, which Trump obsessed over and worked hard (and unsuccessfully) to overcome, are hardly indicators of the economy, as Trump thinks. Trump also thinks that the Dow Jones Industrial Average is an indicator of the economy, so he pressured the Federal Reserve to keep interest rates low to counter stock-market declines. Of course, his economy ended up going into recession with his mismanagement of the COVID-19 pandemic. Jobs that white people will not do, such as non-union meat processing and picking produce, became difficult to fill because of his immigration policies, bankrupting many small agricultural businesses and hurting the economy. Jobs in coal and manufacturing are decreasing because of economics, despite his claims to the contrary and his false promise to restore them. His impetuous, almost daily nonsensical comments or tweets on economic things prompted wild gyrations on the world's stock markets, resulting in unstable business climates.

A significant part of Trump's speeches, interviews, tweets, and other methods to rile up the base or to get attention revolve around hate. Hate of immigrants and support of white nationalists continue to dominate Trump's speeches, for those traits appeal to his racist voter base. His executive orders denying or hindering immigrants from central America and Mexico and from Muslim countries are an assault on our historic national character, as expressed by the Statue of Liberty. A related domestic policy of Trump derived from his most repeated (about 200 times) campaign promise in 2016 to build a "beautiful" wall along the entire Mexican border to prevent immigration; furthermore, he promised that Mexico would pay for it. But the border is more than two thousand miles long, and as of May 2020, the amount of new wall built on "fresh ground" is apparently about three miles. Mexico never paid a dime.

Another way to get the base to a fever pitch is the gun issue. On October 1, 2017, a lone gunman with an AR-15 and a bump stock (making it a machine gun) killed 58 people and wounded 489 in Las Vegas. Massacres by guns, of course, happen on average once a day in the US, before and since Trump, but Republicans singularly refuse to do anything about it, even after each mass killing. The gun-murder rate in the US is twenty-five times higher than that of any other developed country.

Trump's "foreign policy" was characterized by actions that are overwhelmingly whimsical and irrational, traits that reflect his willful ignorance of foreign countries and their histories. He knows only stereotypes of persons of other nationalities, and he applies simplistic solutions to longstanding international problems. North Korea is such an example. To ingratiate himself with dictator Putin of Russia, he trashed US relationships, trade agreements, and NATO and other alliances with all our allies, even picking fights and making personal slurs, one by one, against nearly all their presidents. Trump also expressed admiration for all presidents who have throttled democracy in their countries, such as those in Brazil, Egypt, Saudi Arabia, Hungary, India, Israel, the Philippines, and Poland. He praised dictators in Russia, North Korea, Turkey, and Saudi Arabia. He clearly wants to join this club.

The year 2023 was a banner year for Trump as far as indictments go (three of them). On June 13, 2023, Trump and others were indicted by a federal grand jury in Miami for taking (and hiding) classified national defense documents from the White House after he left office; the case, however, was dismissed by a partisan judge, but it may resurface eventually. On August 1, 2023, Trump was indicted by a federal grand jury in the District of Columbia related to the January 6, 2021 attempts to overturn the 2020 presidential election, as well as calls to assassinate Mike Pence and members of Congress; many of the charges were dismissed by the partisan U.S. Supreme Court based on their decision that a President has "immunity" (i.e., above the law) from prosecution if he/she has violated the law in an official capacity. Some charges have been refiled for violations not in an official capacity. On August 24, 2023, Trump and others were indicted in Georgia for attempts to overturn Biden's victory in Georgia; the trial is in doubt because partisan politicians in Georgia have postponed decisions on it until after the 2024 election.

Trump's war on science included large funding cuts in his budgets of many science and health agencies, including NASA, EPA, NIH, and the USGS; furthermore, budget documents of federal agencies were prohibited from using words such as "science-based" and "evidence-based." Trump's Executive Order 13875 required one third of all federal advisory committees, which are staffed by scientists and academics, to be dissolved. He has always dealt with the world as a zero-sum game (i.e., one person's gain must be balanced by another's loss, so the total is zero), a childlike view that eliminates win-win trade or other deals that require negotiations and diplomacy. Thus reaching agreements on a wealth of science-related problems, from climate change to epidemics, were/are unlikely under a Trump administration. To use the words of Noam Chomsky in an interview on Memorial Day 2020, the presidency is "in the hands of a sociopathic megalomaniac who's interested in nothing but his own power, electoral prospects—doesn't care what happens to the country, the world."

2

Trump Voters

It has been said that Americans will go to the ends of the earth to fight wars "for democracy" but cannot be bothered to walk across the street to vote. And many, it seems, mindlessly pull the lever for the party for which their parents or their tribe vote, or they obey authoritarian family or religious leaders. Trump voters, however, represent a new phenomenon, for they seem to be motivated by things that are different and more hateful than in the past. Polls show that most are white, middle-aged or old males who have no more than a high-school education. In her campaign, Hillary Clinton called these voters "deplorables." If she had not said this, she might have won her election with Trump. But she is correct, and many of these people—as opposed to the opportunistic politicians they vote for—are driving the bus. In other words, the voters are the main problem. The persons who took part in the insurrection at the Capital on January 6, 2021, are typical of one end-member of Trump voters. These had brushes with the law and their main emotions are racism, resentment, and hate. The Proud Boys and others of that type took the lead on January 6, resembling petulant children who like to put on costumes and hats and march with others, where their numbers allow them to be bullies.

As Clinton implied, not all Trump voters are deplorables (see subchapter on Pat and Fred below), for a significant number are indeed lonely and marginalized outcasts from society who want to belong to a group, the price of which is conformity of thought. Such people are especially vulnerable to street gangs and cults, as well as to con artists. Beyond those excuses, the reasons many others voted for Trump are mysteries to me, and pundits and political scientists have a wide array of theories. Below I will tick through some of these. For, when asked outright to give reasons for their vote, Trump voters' answers make no sense, nor will they admit to any racist or religious hate.

Jim Hightower, from a liberal group in Texas that included the late Ann Richards and Molly Ivins, proposed in the August 2018 issue of his *Hightower Lowdown*, that the Trump voter saw Trump "merely as a handy, blunt-force club to whack a two-party system that no longer speaks to—much less for—them." In a November 2018 article in *The New Republic*, Garret Keizer suggested that one rationale for why certain people vote for Trump is that they are nihilists, who believe in nothing. In other words, many Trump voters are using Trump as a raised middle finger to America. These, as well as the insurrectionists, are Clinton's *deplorables*.

Noam Chomsky summarized these persons in an interview with the *Huffington Post* in February 2016, in which he stated that Trump is "evidently appealing to deep feelings of anger, fear, frustration, hopelessness," perhaps also resulting in the current epidemic in suicides, alcohol abuse, drug and opioid overdoses, and the spike in mortality in working-class whites. This is the grievance model, and an apparent excuse not to vote responsibly.

But there are other reasons. Some jobs that people without a college education are doing can now be done more efficiently by robots/computers/AI, or are otherwise decreasing in number. Wages and prestige for these jobs have been flat or falling for about forty years (ironically, largely due to efforts and laws against unions and minimum-wage increases by Republican regimes). And worse, a minimum-wage job is no longer sufficient to support a family and home. Perhaps some of these people feel stuck in a rat race and vote for Trump because they have nothing to lose. Ironically and in contrast, immigrants and blacks are far worse off, yet they seem to have a positive outlook on the future because they believe that through hard work and education their lot or their descendants' lots will improve. Yet physical labor and working with one's hands, whether as a family farmer/rancher, a person on an assembly line, a cook, truck driver, artist, musician, day-care worker, construction worker, or a person engaged in a myriad of other important and valuable jobs, are critical for society and deserving of respect. And most of these persons love their jobs, may not want a "desk job," and take well-earned pride in what they do.

Another analysis about Trump voters (and the reasons why polls missed predicting the election of Trump) appeared in an article ("A minority president: Why the polls failed, and what the majority can do") published online on November 22, 2016, by George Lakoff, Emeritus Professor of Cognitive Science and Linguistics at the University of California, Berkeley. He contrasted nurturing versus authoritarian family models of upbringing. A nurturing upbringing includes an emphasis on empathy and learning to think for oneself, critical thinking, and a normal worldview where facts matter. In the seventeenth and eighteenth centuries this type of thought led to the Age of Enlightenment, which touted reason and logic in contrast to, say, the religious fears and superstitions of the Dark Ages. In contrast, a strict-father family model is one in which a father or an authority figure thinks he knows best and thus has the final word; children or underlings obey or are punished. All persons who are part of this model conclude that the father has the moral authority as well; in other words, he is doing what they (and often by extension God) consider to be appropriate. This model claims that you will prosper under the model but, if you do not prosper, that is evidence that you are weak, undisciplined, lazy, and undeserving. What you become is up to you.

Morally, those who dominate should dominate, and the rich deserve their wealth. The military has a similar message as officers prepare enlisted men for war by teaching them to obey blindly the person directly above them in the chain of command. Many of those who leave or retire from the military carry into civilian life this way of doing things. This form of patriotism means that you should never question authority. It works especially well in Russia and China, where public addresses that announce bad news for the average citizen are accompanied by the national anthem or other patriotic music to reinforce the message.

The strict-family model has a hierarchy and commonly a religious flavor: God is at the top, men below, and nature below that. Furthermore, men are above women, whites above non-whites, rich above poor, Christians above non-Christians, and so on. This model is the one advocated by right-wing white evangelicals, in which you should fear the final authority, which is a wrathful God. For if you do not follow, God can send you to Hell, where you will burn for all eternity. Yet sinners get a second chance to become "born again," or saved, if they confess their sins, "accept Jesus as their Lord and Savior," and follow rules that they believe God has established. Fear dominates their life. For some, right-wing religion is the dominant reason in voting, such as Fox talk-show host Laura Ingraham. Journalist Anne Applebaum provided an explanation in 2020 in *The Atlantic*, whereby such persons consider America doomed, perhaps because it is Godless, and therefore "[a]ny price should be paid, any crime should be forgiven, any outrage should be ignored if that is what it takes to get the real America, the old America, back." Furthermore, some of these persons consider Trump to be a messiah figure.

Lakoff points out additional implications of the strict model. For some people, corporations and CEOs are essentially strict fathers. Most owners of corporations embrace laissez-faire free markets, in other words those that are nearly free of government regulations, taxes, fair employee wages and benefits, etc., because the goal is to maximize profits for the owner. These are Republican values, of course. Whereas nurturing-family offspring tend to become scientists, MDs, teachers, social workers, academics, nurses, and the like, those from strict families commonly go into business, marketing, and sales. One secret of marketing/advertising is reaching the subconscious thoughts of potential customers. Ads on TV or in print use repetition, commonly the same few words. Trump's repeated use of "crooked Hillary" is an example. Use of fear or hate, which also generates unconscious thought, is also effective in marketing. Therefore, videos of nonwhite people shooting whites are effective in making the consumer/voter want a strong strict father. Trump constantly claims that blacks, Hispanics, Muslims, and other non-whites are invading the country to rape and kill you, and to take your job. Trump's boasting and bullying are intended to present a father figure. His claim that "I alone can fix it" is in the same vein. Manipulation of the press, whether to intimidate them or to make them into whipping boys, is another part of the message. These tactics are totally different from the way Democrats think and sell themselves. They feel that just laying out facts will persuade people of their arguments. And polls use questions based on facts. The Trump method, which has been used by demagogues throughout time, has produced a rock-hard base constituency for Trump that is not recognized in polls.

Professors Marc Hetherington and Jonathan Weiler, in their 2018 book *Prius or Pickup*?, provide another analysis of Trump voters, with many of the same arguments as Lakoff. They also present a case for two types of people. One type is what they call "fixed," who are persons wary of social and cultural change and therefore more set in their ways, more suspicious of outsiders, and more comfortable with the predictable. The other type is "fluid," and refers to persons who tend to support the current rapid change in social and cultural norms, are excited by things that are new, and welcome people who are different. The people in between, who are lesser in number, are "mixed." The two types came to be that way by how they were raised, but they began to take extreme positions as either Republicans/conservatives or Democrats/liberals during the tumult of the 1960s and 1970s, with the Vietnam War, protests against the war, the Nixon presidency, and so on. And with time, the two views became increasingly intractable, as they are now. To detect whether a person is fixed or fluid, one need ask only four questions about their childhood upbringing, namely whether it is more important to: (1) have respect for your elders or be independent, (2) be obedient or self-reliant, (3) have good manners or curiosity, and (4) be well behaved or considerate. The first answer to these four questions, of course, is that advocated by a fixed parent. The authors also use the terms strict father (an example being Trump), and a nurturing one (an example being Barack Obama). Fixed people tend to think that, except for family and friends, others—especially those who differ in race, religion, and ethnicity—are not fair and cannot be trusted. This specter of nationalism is dangerous, for it led to the rise of Nazi Germany in the 1930s and 1940s. After World War II, the European Union (EU) was created largely to prevent future nationalism. But then, in Great Britain, Boris Johnson's "Brexit" (leaving the EU and returning to the British imperial model of old) was sold on racism, nationalism, and "making us great again," even though leaving the EU would do little to prevent immigrants from entering Britain. Fixed people also tend to think that the world is a dangerous place, that the past was better than the present (e.g., "Make America Great Again"), that the Bible is the word of God, traditional gender roles are best, blacks are lazy, etc. And as the title of the book suggests, a fixed person is more likely than a fluid person to buy a pickup truck than a Toyota Prius. A mixed person, sadly, is closer in philosophy to a fixed person when it comes to issues of race, religion, and immigration, one reason why Trump won the election of 2016 even though polls suggested otherwise. Furthermore, in stressful times such as Trump has made the present out to be (beware the terrorists and unsavory immigrants who are invading!) in his campaigns, high-minded ideals in a "mixed" person give way to self-preservation, in which an authoritarian figure is more desirable.

Tragically, as these books and others point out, the country does not win when about half of the electorate views the other half as mortal enemies. Polls show that parents do not want their daughters to date or marry a person from the other side, for example. Furthermore, deeply held beliefs do not change even if the person who believes them must do mental gymnastics to rationalize and hold onto them. This is particularly true for the fixed person, who must explain to himself/herself a president who has enormous character issues. Therefore, they fail to acknowledge facts because it would be psychologically painful to do so. Those who voted for Trump in 2016 were psychologically likely to do the same in 2020 and 2024. The book noted that, in an argument where one cites overwhelming facts to counter beliefs (such as their opposition to climate change or vaccines), rebutting a fixed person with more facts tends to make that person dig in to his or her beliefs even more. The staunchest defenders of misperceptions are those on the political right, explaining why right-wing media attracts much larger audiences than left-wing media. In an online 2020 article in *The Atlantic*, taken from their new book, social psychologists Elliot Aronson and Carol Tavris discussed the clinical name for the problem—cognitive dissonance—which underlies the reluctance of many people to admit mistakes or accept scientific findings, even those that could save their lives. In other words, humans are deeply unwilling to change their minds, to admit they were wrong, or to accept new information. Or, as Earl Cook long ago (1976) pointed out in his book (*Man, Energy, Society*, p. 447), "[m]en and women usually believe what they want to believe, even when the evidence is overwhelmingly against them." This contrasts strongly with scientists, who are trained to change their minds on a dime as they come upon new data—as we had to do with our landslide discoveries.

Anne Applebaum theorized in an August 9, 2020, piece in *The Atlantic* online that, for Trump voters, "being Republican was part of their identity. Images relating to God, patriotism, and the Republican party were all around them. Cumulatively, those messages were much stronger than any dislike for Trump." Commonly, their version of patriotism is centered on white supremacy and blind loyalty. There also is the pull of the group ("I and all my friends have always been Republican"). She concludes that "[i]n this post-mass-media era, sowing division is far easier than creating unity, giving an advantage to politicians who seek to win by creating scapegoats and enemies."

Thomas Geoghegan in *The New Republic* pointed out the dangers of class snobbery by college-educated people, noting that 70 percent of US adults (those twenty-five and older) do not have a four-year degree. The author reminded us that Trump in the 2016 election famously stated "I love the poorly educated," and the author suggested that if Hillary Clinton had said that, she would have become president. Geoghegan also noted that Trump, by bragging about his own moral deficiencies, let working people know that "they could vote for a president who was incapable of looking down on them." Furthermore, many US working-class people "are stuck in authoritarian workplaces where they learn to take orders, and therefore they pass on that culture to their children." In the world, knowledge is power and data govern everything, and Democrats seem to rub it in. European wage workers are in a more democratic workplace, and individuals who work with their hands can advance within it. Of course, Geoghegan sadly points out the irony that most wage workers vote straight Republican, generally against their own interests and in favor of the party that does them the most harm. A solution where corporations invest in their own workers, as in Europe, seems a long way off.

Finally, it is worth repeating the words of H.L. Mencken (1880–1956), at one time a widely read journalist, satirist, and critic who stated, in the *Baltimore Evening Sun* on July 26, 1920, "[a]s democracy is perfected, the office of president represents, more and more closely, the inner soul of the people. On some great and glorious day, the plain folks of the land will reach their heart's desire at last and the White House will be adorned by a downright moron."

My former grad-school roommate, Bob Kier, discovered a lesser-known quotation from Ulysses S. Grant, (https://www.victorianvoices.net/ARTICLES/CENTURY/Century1898A/C1898A-Grantsspeech.pdf): "[I]f we are to have another contest in the near future of our national existence, I predict that the dividing line will not be Mason and Dixon's, but between patriotism and intelligence on the one side, and superstition, ambition, and ignorance on the other." These words were said in a speech at the Annual Reunion of the Army of the Tennessee in Des Moines, Iowa on September 29, 1875, as recorded by James S. Clarkson (1842–1918), a Republican politician who wrote a first-hand account during the dedication of Grant's Tomb on Riverside Drive, New York City on April 27, 1897. At the time, Grant's speech was considered to be one of the great speeches of the era.

3 Pat and Fred

Handy-man Pat Michel Aragon was both a Trump voter and a good friend from Cedar City whom I knew since November 2009. One of the mistakes that Dawna and I made when we built our ranchette on six acres in mile-high New Harmony was to select a Santa Fe style house with a flat roof. Flat roofs work fine in dry and hot parts of the Desert Southwest, where snow is rare. Furthermore, our local builder had little experience with flat roofs, so it leaked during heavy rains and moderate to heavy snowfalls. The house walls are thick, using foam blocks (1' x 2' forms of Styrofoam and steel, with an interior into which concrete is poured). Their disadvantage is that any water entering the blocks from a bad roof can go in directions that are impossible to predict. I brought in handymen who suggested fixes, none of which worked. We replaced the roof at considerable cost, which decreased leaking significantly but did not stop it. Eventually we found Pat, and slowly but surely, he found solutions, one leak at a time.

Highly intelligent, Pat was adept at analyzing problems, then using his experience to solve them. Pat (born November 15, 1954) was powerfully built and sported a beard. In his younger years, he was a biker and got into some trouble now and then in bars in Las Vegas before he and his wife, Dawn, settled down, raised kids, and had grandchildren on whom they dote. Pat was born in Austin, Colorado and grew up in Delta, not far from Grand Junction, and is of Spanish and Apache ancestry. His ancestors originally had an early Spanish land-grant property and home in New Mexico, before ownership paperwork was needed; white settlers later came and filed the correct paperwork and took the home away from them. Pat was taught many life lessons, including a love of fishing, from his grandfather, Toribio Aragon, who was an Apache and had married an Irish woman. Pat's dad, George, had a career, eventually in management, working at a Colorado home for mentally disabled persons in Grand Junction; he also leased fields (a truck farm) where he grew tomatoes near Delta. Pat learned hard work and compassion from his dad, who died in 2020. Pat and I soon discovered that we shared a love of the state of Colorado and the Denver Broncos, the professional football team. As we tried to determine where the roof leaks were, I insisted on working alongside him, as we talked Broncos and the human condition. Some leaks were caused by a bad layer of stucco or by windows that had not been properly framed, so we had a lot of time working together. He or I might spray water on the stucco while the other tried to determine where the water went. We repainted the house, applied sealant in all parts imaginable, and eventually, over the course of several years, managed to rid the house of all its leaks. Recognizing what a gem I had in him, I paid him more than others did, and I used him for the many other ranch repairs that my wife Dawna requested. Some of his solutions showed a high artistic bent, and his work was done with quality and pride. I continued working alongside him, and after each job at the end of the day, as I paid him, we had a beer together and talked many things, including morals, children, and dogs and cats. He was a highly loyal, moral, and ethical person, more religious than I am. Once Trump was elected, however, we learned that talking politics was not a good idea. But about once a week, even if we did not work together, we talked on the phone. Sadly, Pat died on April 27, 2024, from a cancerous brain tumor that was only discovered in February. I thought he would beat it because of his inner strength, and then we talked every day or two, but this good person finally passed!

One afternoon after 3 p.m. in early December 2019, a medium rain had started and I went out a little early to bring the horses up to the barn for their dinner. Galga, our youngest horse at about fifteen, a white Andalusian mare, was not standing by the gate to our large turnout area, waiting to go. I walked down into the turnout area and found her down on the ground. Her blanket was all dirty, for she had been rolling around in pain, a sure sign of colic, an affliction that horses, dogs, and rarely people (including Dawna several months previously) get. The small intestine becomes twisted, preventing food from passing through the digestive system. The cure for horses is to get them up and walk them, and most of the time the intestine untwists. If that does not work, surgery is the only option, as Dawna had required, but surgery means putting the horse in a trailer and taking it to a vet who has facilities for horse surgery. In a driving rain, I walked the horse up and down the driveway for two hours, but she kept going down and rolling. I changed into dry clothes and called Dawna at work, but she was unavailable, apparently starting the hour-long drive home. The horse became weaker and finally could not stand for very long. At nearly 6 pm, Dawna drove up in the dark and the rain, joined me, and gave the horse a hypodermic shot for pain (Banamine) that did little good. I had called, but no vets were available after business hours, but even if one had been, the horse could not have gotten into a trailer. Dawna had to make the call to end the horse's life, for it was her beloved "Princess." So I shot it with a .357 magnum revolver, a traumatic and sad moment. I had purchased the gun from a great geologist in our Branch, Jack Harrison, when he retired. Jack mapped in Montana, working closely with Chet Wallace, and carried it for bears. After my mountain lion incident in Nevada, I had decided I needed it.

Now I had a dead horse alongside our driveway. I spent the entire next morning calling people to see if they could come winch the horse onto a trailer and haul it to the Cedar City landfill. That was how we had dealt with another horse, Sadie, who had died. Nobody was available, so I tried Pat, and he agreed to take up the hunt. In a half hour he called back and said he had a friend, Fred Bronsema, but Fred said that he could not make it until the next day. I called Fred and scheduled it, asked him how much he would charge for that. Not knowing my financial status, he quoted $100 to $150. I got the clear impression from talking to this clearly caring person that, because I was a friend of his friend, he would have done it for free. I told him that quote was too little and that I would happily pay at least $400. Fred owns Hero Services, LLC, a successful small Cedar City business that sells septic systems, pumps septic and recreational-vehicle tanks, rents portable toilets, and does whatever other odd jobs he can to keep him and several loyal employees afloat. This is a type of business that does things that few others want to do, and does it successfully because of a high work ethic. Pat noted that Fred must weigh 500 lbs and was about Pat's age, and when Fred showed up the next morning, I estimated his weight at closer to 400 lbs, but his stomach hung over his belt and he was clearly restricted in any physical activity. He had a long gray beard and a big smile, and he wore a blue MAGA hat! His loyal friend and employee, Frank Young, was in the passenger seat of Fred's GM Excursion SUV. We talked about Pat, other friends, life, and morality, all while Frank did most of the work. I found Fred to be highly intelligent and unbelievably loyal to friends, clearly a person with a heart of gold, much like Pat. Frank and Fred positioned the trailer and hooked up the winch. I asked Frank how he would get the horse out of the trailer when he got to the landfill. He said he did not know but they would figure that out at the time, "besides, the horse is our problem now!" Within fifteen minutes, Galga was on the trailer, and I got my checkbook and asked Fred how much. He said that the $400 I mentioned would work, but I was still aghast about this wonderful favor they were doing me, and I upped it to $450. I shook his hand and said "I really appreciate this." He said, "I know you do" and drove off with that big smile on his face!

That night, I woke up in the middle of the night and could not get back to sleep. I was reliving the short time with the two angels of mercy whom I had met only that morning. I came downstairs, read a book, then penciled about two pages of notes and thoughts that were bothering me. My problem was that I was reassessing Trump voters. Perhaps not all of them, but I knew from published interviews that many other Trump voters are reasonable, smart, and wonderful human beings, just like Pat, Fred, and Frank. So why the hell did such caring, empathetic, intelligent persons vote for Trump and were likely to do it again in 2020 (and 2024)?! I started making excuses for their political positions. Perhaps they were too busy scratching out a living to spend time reading about current events. I wrote down all the excuses: perhaps, perhaps, perhaps! Or, perhaps I just do not know, and I wonder whether they themselves do!

On a Friday evening in late June 2020, I made a call to have our equine veterinarian, Dr. Cayla Moore, put to sleep our twenty-year-old white Andalusian mare Ebanista, while I cradled the horse's head in my lap. The mare, on which Dawna had won ribbons, had something called founders (laminitis), where her feet are painful due to too rich a diet. Dr. Moore had treated her with pain killers, which Dawna and I administered four times a day, for nearly a month but there was no improvement, and on that day the horse would not get up, outside her stall at the barn. Again, I called Fred. He had been delivering port-a-potties to a large, rapidly moving brush fire north of Milford for two days, but he got to our home with his two sons, Marcus (about twenty or twenty-two), and Josh (about eighteen), just before dark. Both were impressive young men, as hard working as their dad, which is unusual for young men in Utah. Getting the horse out of her stall was a little trickier this time, for I had to take down some metal fences, then we had to winch the horse out of her corral, then down to the driveway. The two sons and I hauled on the metal cable while Fred positioned the trailer expertly, then manned the winch. I did not have much time to talk with Fred this time. But I learned that his wife, Jennifer, was a nurse at a clinic in St. George that deals with about two hundred persons per day who are on drugs, and Fred told me to make the $500 check out to her. Fred was wearing a different baseball hat this time, rather than his MAGA hat, out of respect for me, I think. I again told him how grateful I was, and he gave me his smile and headed off. A year later, our white Andalusian former-stallion Ebanista (Ernie), contracted laminitis, and in January 2024 our thirty-year-old mare Rumor had fallen in her stall and could not get up, and I had to shoot them both, then call Fred and Frank. The only thing good about these terrible events was catching up with Fred.

Clearly there is a great need for Trump voters and non-Trump voters to talk more to each other. Although Pat and I avoided politics in our many conversations, I can see that he has not gathered any evidence to support his view of Trump. Articles by others who have asked Trump voters their reasons report similar results to mine. They even note that commonly the Trump voter, having no reasons for his/her position, soon resorts to a question, presumably taught by Fox News to avoid unpleasant admissions, "Why do you hate America so much?" Therefore, the explanations seem to be brand loyalty (some always buy Chevrolet, others Ford), also known as identity politics, or loyalty to others ("Don't rock the boat") in your family or group. Some are one- or two-issue persons (against abortion, for guns) with which Republicans side.

As his 2023 indictments added up, Trump's popularity steadily increased, suggesting on the other hand that Trumpism is a cult. In fact, there are similarities in the message from Jim Jones to his cult "Peoples Temple," and one hopes that the outcome will not be similar, such as the mass murder-suicide of 900 cult members on November 18, 1978, at Jonestown, Guyana. Whatever it is, the result is what seems to be a combination of mob rule and of a tailgate party at a National Football League game, where you can depend upon everybody being of the same mind, known as confirmation bias. Nonetheless, in the words of Martin Luther King, Jr., "to ignore evil is to become an accomplice to it." And they are accomplices—whether they know it or not!

The Republican “Party”

Early in my life, I voted Republican, following my parents and not paying much attention to what Republican politicians were doing, and how it differed from what they were saying. Therefore, I am ashamed to say that I voted for Nixon (first term only) and Reagan. After that I better researched the policies and values of political candidates. And when George H.W. Bush tried to foist Danny Quail on the public for Vice President, I voted Democrat and have generally continued that since. All of us considered ourselves to be moderate to liberal Republicans, which then was the mainstay of the party and included morally-decent leaders such as Nelson Rockefeller, Gerald Ford, and Robert Dole (before he went far right during his run for the presidency).

4

Noam Chomsky’s analysis of the Republican party, while not differing from what most of us already know, is inciteful in its brevity and truth. He noted in one of his 2019 interviews with Amy Goodman of the TV show *Democracy Now*! that the party has, for more than fifty years, served only extreme wealth and corporate power. But such a constituency does not make up much of the voting population, so Republicans had to pretend to advocate for other things that were initially not appealing to them. After all, pushing for policies that benefit the rich and powerful at the expense of the other 95 percent of the population had to be obfuscated by focusing Republican talk and campaigning on other things. Cultural issues became their favorite. The most obvious of these was the “southern strategy,” as used by Richard Nixon, especially in his second campaign, to turn the south to the Republican party by using dog whistles (e.g., “States rights,” “law and order’) of subtle racism. Lee Atwater applied the southern strategy even more openly in the campaigns of Reagan (rhetoric about “welfare queens”) and George H.W. Bush (ads about Willie Horton, a black murderer in Massachusetts who was given weekend passes from prison until he committed more crimes). Then the Republican party pretended to be opposed to abortion and even contraception, and in favor of gun rights. Racism, abortion, and extreme gun rights are now the mainstay of the party, and their constant talk drowns out and conceals the real agenda, which is making the rich richer. The party has few policy positions, and certainly no legislative victories, on topics that might benefit the broader public.

Timothy Shenk, in comments in a book review in *The New Republic*, noted the long-strained marriage between conservative “thinkers” and “the right” grassroots activists. The former included such people as George Will, William Buckley, Bill Kristol, and others who kept silent and went along with the comments and actions of the latter, namely the Birchers, Joe McCarthy, Phyllis Schlafly, Jerry Falwell, Trump, QAnon, Proud Boys, Tucker Carlson, Steve Bannon, and others. Yet even the thinkers "typically looked at democracy with contempt." The partnership, of course, was meant only to achieve power, thus "to convert cultural grievance into tax cuts for the rich." Although some of the thinkers can no longer stomach the result, the alliance remains intact. Senator Mike Lee of Utah, one of the most obsequious to Trump, was foolish enough to state this position toward democracy out loud, when he tweeted on October 9, 2020, that "democracy isn't the objective; liberty, peace, and prosperity are. We want the human condition to flourish. Rank democracy can thwart that." His position is in effect fascism, which is a belief that a certain class of people is superior to other classes, and that therefore an authoritarian, nationalistic, and militaristic right-wing system of government that wallows in racism, grievance, and nostalgia is needed; furthermore, fascism declares that the state is more important (thus all powerful) than individual citizens, and it should be placed in the hands of a dictator holding power by a cult of personality.

Beginning with the 1980 start of the Ronald Reagan presidency, economic conditions have increased the percentage of the poor, destroyed the post-World-War-II American Dream, and stoked the fears of non-rich Trump voters. An important 2.5-page summary of Social Security, written by Steven Pressman, was published in the December 2019 issue of *The Washington Spectator*. It noted that this important safety net, established by President Franklin D. Roosevelt in 1935, has been tinkered with by Republicans since Reagan took office. Little by little, small changes here and there have eroded its value as a resource for non-wealthy retirees, and the wealthy have benefitted from these changes. Pensions ("defined benefit plans" with a fixed monthly amount after retirement), on which wage earners relied for retirement before the 1980s, also have been drastically decreasing in number or at best have been replaced by "defined contribution plans," in which employees now are responsible for contributions, which employers match. An almost-universal health-care system, the Affordable Care Act ("Obamacare") that was passed in March 2010, was the most important triumph of the Obama presidency. Yet little by little, Trump and many Republican governors of states have chipped away at the Act, so currently for-profit health-care plans that many persons cannot afford returned to prominence—just in time for COVID-19! Student debt, due to cutting aid to college students, and a lack of inexpensive child care are also hurting the middle class. Currently, as widely reported by a Federal Reserve study, about 40 percent of US families lack $400 in ready cash to deal with an unexpected expense. A Pew Charitable Trust survey noted that one-third of American families have no savings, including 10 percent of those making more than $100,000 per year, and that 70 percent could not cover a $2,000 expense. Most wage earners who contemplate retirement have little besides Social Security in store.

Republican Newt Gingrich, as noted, took control of the House of Representatives during the first mid-term election (1994) of the Clinton presidency. But he was much different from other majority leaders, for he instigated a single-minded war for power at the expense of both the country and the two-party electoral system. As bad as Gingrich and his colleagues were, the Obama presidency brought Republican extreme partisanship to a new level. The January 2010 ruling (five to four) by the Supreme Court of *Citizens United* allowed corporations (most of which support Republicans) and other outside groups to spend unlimited amounts of secret campaign money, so that now political campaigns continue every day, all year. One of the best examples of partisanship was the declaration by Senate leader Mitch McConnell, Jr. in October 2010, right before the mid-term elections of President Barack Obama, that "the single most important thing we want to achieve is for President Obama to be a one-term president." A large tactic of their partisan behavior was obstructionism, beginning when McConnell advocated that the Republican party had nothing to gain from compromising with Democrats, even on policies they had advocated previously, because that might be looked upon as a victory for Obama. Thereafter, Republicans thwarted anything proposed by the Democratic party, although it took a long time for Obama to realize that. The most extreme example of obstructing Democrats was McConnell's refusal to allow Obama to select a Supreme Court justice a year before Obama's term ended.

George Packer, in the June 2020 on-line and print issues of *The Atlantic*, noted a deterioration over time in response to the three major crises experienced by the US in the twenty-first century. The first was the terrorist attack of 9/11/2001, when the US came together as a country. The second was the 2008 financial crisis, brought about by policies of the George W. Bush presidency, and it was eventually resolved with significantly more partisanship than during 9/11. Though this forestalled a deep depression, the bankers came out fine. The pain was felt and persisted in the middle class and the poor, and inequality worsened, leading to the growth of populism (both parties lost credibility) and the ascendancy of Trump. Trump's message was in many ways a repudiation of the Republican party, but "the conservative political class and the new leader soon reached an understanding" because "they shared a common goal: to strip-mine public assets for the benefit of private interests." "Republican politicians and donors also wanted government to do as little as possible for the common good and could live happily with a regime that barely knew how to govern at all, and they made themselves Trump's footmen."

"If lying was his means for using power, corruption was his end." And now we had "an empty government led by a con man and his intellectually bankrupt party." The third crisis was, of course, the pandemic. And the leaders of this government "were too corrupt or stupid to head off mass suffering." "From the president [Trump] came willful blindness, scapegoating, boasts, and lies ...when a government doctor tried to warn the public of the danger, the White House took the mic and politicized the message."

Ed Burmila, in the November 2019 issue of *The New Republic*, provided another analysis of Trump's Republican "party." He noted, of course, that the Republican base is a strange coalition of extremely rich people and uneducated white people (the author calls the latter the "economically anxious"). The older citizens of the Republican party are dying off, and that drives it to hold onto power by any means possible. For there are lots of examples of other societies where minorities have retained power, one recent example being apartheid-era South Africa, a so-called democracy in which a white minority of about 15 percent of the population held power for decades. During the Trump era, as previously when conservatism ruled, the issue that drives the voter base to the polls is fear that is stoked by demagogues. This includes the made-up fear of Communism in the McCarthy era and the made-up fear of terrorism in the Trump era. But the most basic fear today among Trump voters is that "real Americans"— here meaning older white, Christian, conservative males—might lose power. The Republican objective therefore is a broad attempt to attain minority white supremacy. It is a fight between what the Trump base considers the "other" undeserving group that outnumbers their far-more-deserving group (i.e., fascism). Therefore, the Trump base deems any way to combat it to be fair, especially neutralizing checks and balances in local, state, and federal politics. Deep down, no matter what happens, this minority believes that they alone deserve to govern and they are entitled to manipulate with impunity all democratic institutions to secure power. As Burmila further notes, Trump essentially proclaims at his rallies that "elections that don't produce the result you want are fake and illegitimate."

Part of the reason for the rise of Trump is the news media. The print media is in rapid decline as the internet and social media syphons off advertising dollars. One reaction to the loss of news analysts and opinion editorials is to focus on the sensational rather than doing "truth to power" and substantive analyses. The TV media, especially local news, are particularly superficial. With the rise of video from cell phones, there is an emphasis on action involving auto deaths, house fires, tornadoes, mass shootings, etc. Then there is Fox TV, whose corporate goal is to provide Trump voters with "news" and "analysis" that the viewer wants to hear (e.g., there is no climate change, Trump has never lied, masks do no good in preventing virus transmission, etc.). Perhaps a more important failure of the news media are the false equivalences they apply to different statements and policies of Democrats and Republicans. This is their "fairness doctrine" of giving equal time to each side, elaborated earlier. And Trump is normalized as just another politician, with no mention of his criminal career, to say the least!

At the present, there is no Republican "party." For the party rejected character as a qualification for a president when they nominated Trump. Biden was elected by a sizeable margin in 2020, but Trump and most Republican members of the House and Senates refused to accept this (Trump called it the "Big Lie," following the same use in Hitler's autobiography *Mein Kampf* in reference to Jews). The rule by and for all people (i.e., democracy) is being countered in red states where Republicans control both state houses by use of extreme gerrymandering and deleting tens of thousands of people from voter rolls who are more likely to vote Democrat. *Project 2025*, published in April 2022 as an 887-p. book written by the conservative think tank Heritage Foundation and debated during the 2024 campaign (Trump claimed he "knew nothing about it"), is a frightening wish-list for the next Republican administration on how to downsize and change federal government, as well as to address many of the culture wars, authored by disciples of Trump and his aides from his last administration. As Noam Chomsky pointed out, Republicans represent a "radical insurgency that has abandoned normal parliamentary politics. Republicans represent a 'party' of power by the white minority and of hatred of Blacks, Mexicans and other Hispanics, American Indians, Muslims and most non-Christians, gays, women, young people, and indeed the future. They are anti-intellectual, anti-science, and anti-government." The bottom line, here, is that most Republican voters and politicians are opposed to democracy, and that they game the system to cheat their way to minority-white rule.

5 Impeachments and Investigations

Trump was impeached for the first time by the US House of Representatives on December 18, 2019. By then, due to the November 2018 mid-term elections, Democrats had taken control of the House of Representatives, with Nancy Pelosi as Speaker, whereas the US Senate remained in Republican hands, with Mitch McConnell as Leader. Trump was impeached for the second time on January 13, 2021, one week before his term expired. Also, the Special Counsel Investigation of Russian Interference in the 2016 US election, headed by Robert Mueller, preceded the first impeachment.

Links between Russians and the Trump election campaign to aid Trump and help defeat Hillary Clinton were obvious during the campaign leading up to the 2016 election. The FBI started an investigation, and in May 2017 the Department of Justice appointed Robert Mueller as Special Counsel to lead the investigation and issue a report. The investigation dragged on, reviewing as much evidence and interviews as could be gathered despite refusals by the White House to honor requests for evidence. On March 22, 2019, a 448-page report ("Report on the investigation into Russian interference in the 2016 presidential election") was submitted to William Barr, Attorney General. Barr refused to release the report to Congress, but summarized it in a four-page letter that misrepresented it. In April, after much legal maneuvering, however, a heavily redacted version of the report was released to Congress and the public. The report documented clear Russian interference in the 2016 election (but little go-to-hell proof of collaboration). More importantly, it found ten examples of obstruction of justice by Trump, but said it could not accuse Trump of a crime because of a Justice Department rule that a sitting president could not be charged. Clearly Mueller was telling the Congress that it was their job, not his, to take the evidence forward to impeachment if they so choose. It referred many of Trump's collaborators to various courts of law, and all of them were convicted and sent to prison (most were later pardoned by Trump). Although many Democrats advocated impeachment, Nancy Pelosi knew that she did not have the votes in the House to approve impeachment.

Nancy Pelosi's position on impeachment changed overnight when a new misdeed by Trump surfaced from Ukraine, and this time the interference was intended to corrupt the 2020 election.

The new revelation was a telephone call on July 25, 2019, between Trump and the president of Ukraine, Volodymyr Zelensky. A week before this phone call, Trump had withheld $391m in US foreign (military) aid, long before approved by Congress, to Ukraine so that it could defend itself in the ongoing incursion by Russia into Ukraine territory. On August 12, an unidentified whistleblower contacted the Intelligence Community Inspector General with "urgent" concerns about what Trump had said in that phone conversation. These concerns were passed to the Office of the Directory of National Intelligence on August 26. Soon, all congressmen and the public had heard what had been said. When House members returned from their 2019 summer break, after listening to voters, the *Washington Post* on September 23 carried an op-ed column written by seven of Pelosi's freshmen members in swing districts (as opposed to safe districts) who had changed their positions and now advocated impeachment. The next day, Nancy Pelosi announced that the House would begin a formal impeachment inquiry.

On September 25, Trump finally gave in and released a memorandum that gave a summary of the Trump-Zelensky phone conversation. (Trump hid the entire transcript by making it classified.) In the conversation Trump explicitly said, "I would like you to do us a favor." The favor for which Trump asked was to dig up dirt on Hunter Biden, the son of Joe Biden, who was the frontrunner and most feared (by Trump) of the many Democratic candidates for President. Hunter had been on the board of a Ukraine gas company, and Trump apparently wanted Zelensky to announce that a corruption investigation was planned. Trump—despite the Mueller report—had not learned his lesson and clearly was at it again, thus impeachment was now needed. And even a child could understand the Ukraine incident, so there was hope that Americans could too.

Over the next few weeks, in formal hearings by the House, witnesses came forward to testify, but the White House denied subpoenas for documents and testimony by persons in the White House who had intimate knowledge of clear obstruction of justice. Nearly all Republican politicians, with Senator Lindsey Graham and Representative Kevin Nunes among those most obsequious toward Trump, remained silent or rabidly defended Trump's acts. But by mid-October, public-opinion polls showed that voters favored impeachment. On December 18, the House of Representatives passed two articles of impeachment, Article I for abuse of power and Article II for contempt of Congress.

The impeachment articles, as a report, were submitted to the Senate on January 16, 2020. The Senate, run by Republicans, refused witnesses or evidence from being subpoenaed, so the "trial" had neither! The Republican majority denied any evidence about Trump's crimes, so the spectacle was that of senators and lawyers presenting opposing political opinions. In fact, Senate Republicans worked totally in concert with the

White House defense lawyers. Conviction required a two-thirds (67) vote, but during the vote on February 5, 2020, Article 1 received 58 votes for and 42 against (the only Republican to vote in favor was Mitt Romney) and Article 2 received 57 votes for and 43 against to acquit.

On November 3, 2020, Joe Biden was elected president (Kamala Harris, the Vice President, was the first woman to be so elected), although confirmation required that all ballots, including mail-in ballots, be counted. By December 7, all major media declared Biden the winner. On December 14, electoral ballots were turned in, to be counted by a joint session of Congress (presided over by Mike Pence, the Vice President) on January 6, 2021. Biden/Harris received 306 electoral votes, vs. 232 for Trump/Pence. Biden/Harris received 51.3 percent of the popular vote, winning more than 81,000,000 votes, the largest win count in history and a popular vote margin of more than 7 million votes over Trump. Trump claimed voter fraud and never conceded the election, and from November 7 did everything in his power (and beyond his power) to overturn it. This included 63 lawsuits, all unsuccessful. With options dwindling, Trump demanded that Pence declare the vote void, for he had fake electors to bring in. When that failed, he browbeat Pence anyway to cancel or thwart the ceremonial count on the 6th somehow.

To threaten Pence, Trump's final act came when he and his most loyal supporters ("the Crazies") concocted a plot to bring insurrectionists (Proud Boys, Oath Keepers, etc.), many of them heavily armed, into Washington, D.C. Over several weeks, Trump asked Trump voters to come to his speech, at noon on January 6, 2021. This inflammatory send-off speech was at the Ellipse, a large park just south of the White House, and it riled up the crowd of thousands. In the speech, he made many false claims and concluded with "[I]if you don't fight like hell, you're not going to have a country anymore." Yet armed militant factions of Trump's mob had skipped Trump's speech and were already at the Capitol, scouting out the places to breach the defenses. Once sent off, the crowd marched on the Capitol, many chanting "Hang Mike Pence" and even erecting a gallows outside where Pence and members of Congress had gathered. A pitched hand-to-hand battle took place, in which this lynch mob overwhelmed the Capitol police and the Washington police with greater armaments and numbers and broke into the Capitol, which had been locked for the recount. At least two thousand rioters entered the Capitol, some vandalizing it and looting offices. The Capitol was locked down and congresspeople shepherded to safety. Five persons died, including one of the rioters who was shot, and many others who were hurt, including 138 police officers. Trump watched the insurrection on TV (as did the rest of the nation), but he did nothing to stop it until about 4 pm, when threats by his non-Crazy lawyers about taking power from him by the 25th Amendment finally forced him to tell the mob to "go home, in peace." He nonetheless continued to assert that the election had been fraudulent, a position from which he has never budged. Clearly the riot was a last-ditch attempted coup by Trump to overturn the 2020 election, although the "smoking-gun" evidence did not come until more than a year later. More than one thousand rioters were charged with crimes, and the trials started in June 2022. Almost all those charged were convicted. It was not until the early hours of January 7 that the electoral votes were counted and Biden/Harris declared the winners.

Reacting to the insurrection, Nancy Pelosi and the House on January 13, 2021 (one week before Trump's term expired), passed one article of impeachment, namely incitement of insurrection. Ten House Republicans voted with all Democrats for this, for a total of 232, vs. 197 Republicans against. After another sham "trial," on February 13, 2021, Trump was not convicted, for the Senate voted 57 to convict and 43 to acquit.

In the November 3, 2020 election, the Democrats had taken the House and there was a tied vote (50–50, which can be broken by the Vice President in power) in the Senate. But a majority vote does not pass things in the Senate because of the filibuster. And, worse, two of the Democrats were Joe Manchin of West Virginia and Kristin Sinema of Arizona, who admire the filibuster and some of the things on Republican agendas. Therefore, it was a frustrating first year to the Biden/Harris presidency. Democrats, however, created a US House Select Committee to Investigate the January 6 Attack on the US Capitol to attempt to find out who had planned and financed it. The US House established a committee, chaired by seven Democrats and only two Republicans, Liz Cheney and Adam Kinzinger. The chairman was Bennie Thompson, with Liz Cheney the Vice Chair. More than a thousand witnesses were interviewed and many thousands of documents studied. Public hearings went from June 9 to October 13, and the report was released on December 22, 2022, convincing all reasonable citizens that Trump had planned the coup. Three of the four 2023 indictments of Trump noted in the Trump subchapter are the result so far.

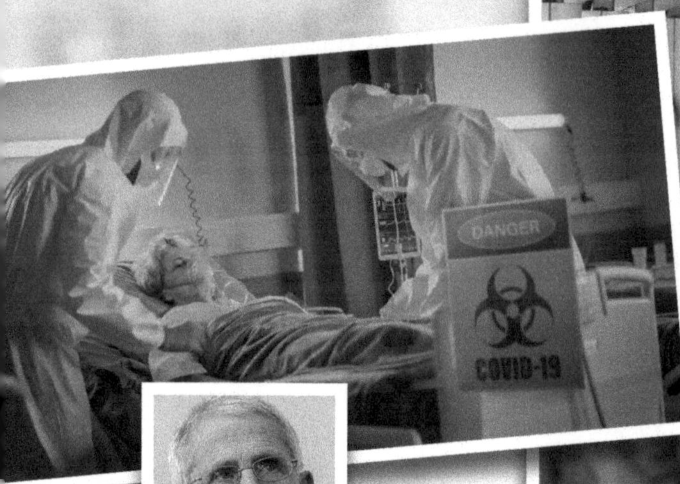
DANGER
COVID-19

CHAPTER 14

12
Minneapolis

KENDRA JAMES
JASON WASHINGTON
DEONTAE J. KELLER
JUSTICE FOR JUSTIN HOWELL
I CAN'T BREATHE

BLACK LIVES MATTER

Many Americans, whether Republican or Democrat, rationalized to themselves at the start of the unexpected Trump election victory in 2016 that perhaps the country could survive four years of his presidency, "but God help us if a real national or world catastrophe happens on his watch!" The COVID-19 pandemic virus that engulfed the world, starting in late 2019 and continuing through the rest of the Trump term and beyond, was such a crisis, affecting both lives and economies. And partway through that, Black Lives Matter protests be-

COVID-19 AND BLACK LIVES MATTER

gan in response to police killings of black persons. Trump could not deal with either. The US officially went into recession on June 9; the drop, from a robust economy, had taken only about three months. It was a classic "perfect storm" of disasters. As George Packer stated in his June 2020 article in *The Atlantic* mentioned above, "[t]he crisis demanded a response that was swift, rational, and collective. The United States reacted instead like Pakistan or Belarus," other failed states. Through indifference to suffering, incompetent messaging, and letting the virus get out of control, at least 200,000 unnecessary deaths (documented below) of Americans and millions of jobs and homes were lost. These citizens, namely Trump and his political colleagues, according to the words of John Dickerson in the July/August 2020 issue of *The Atlantic*, displayed "a vision of organizational design befitting fruit picking or hotels that staff up for vacation season, not the kind required for the sort of catastrophes presidents inevitably face." In the words of Waylon Jennings in one of his many wonderful songs, "Are You Sure Hank Done It This Way," "[w]e need a change." And thankfully, Joe Biden defeated Trump during the height of the pandemic and recession.

Telling the Trump story (Chapter 13) illustrates how democracy and science can cease to exist. Telling the COVID-19 story illustrates how science must be obeyed. The pandemic descended upon a nation whose citizens, many already in poor health, were in a third-world health-care system made worse because for three years the Trump administration had been taking a wrecking ball to it. The health care system before was called "Obamacare," which was passed early (March 2010) in the Obama administration. It attempted to provide country-wide universal health care, yet the best they could do politically was set it up within a system that maximizes profits for the pharmaceutical and related industries. As soon as results were in on the midterm elections in Obama's first term, when Republicans took over the House, Republicans voted 60 times to repeal Obamacare, although not successfully. Once Trump came to office, parts of it were dismantled incrementally. The result, according to George Packer ("The legacy of Donald Trump," *The Atlantic*, January/February 2021, p. 9–12) was that 2.3 million Americans lost their health insurance before the pandemic, not to mention those who lost coverage during the pandemic when they lost their jobs. Republican-controlled states deleted even more benefits. Health costs therefore returned to pre-Obama levels, just in time for the pandemic, with costs far higher than in other countries (and health outcomes far lower). Employer-provided health benefits are expensive, making businesses who provide them less likely to provide good wages and more likely to outsource jobs. For example, to keep drug costs high, Big Pharma has controlled the US Congress such that only the US and New Zealand allow pharmaceutical companies to market directly to consumers. According to the *Washington Post*, nine out of ten of the biggest pharma companies spend more on advertising than R&D (Research and Development), and an estimate in 2018 stated that 70 percent of TV news ads are by Big Pharma. With US wages essentially flat over the last forty years, and with many wage earners provided with no health care, there is a general mental and physical malaise in the country among those without college degrees, resulting in a fall in life expectancy, relative to Europe and other rich companies, since the 1990s, and most notably between 2014 and 2017. This fall in life expectancy is due especially to increases in suicides, alcoholism, drug use (including the opioid epidemic), and other health choices. Regarding the health of Americans, poor people commonly cannot afford a healthy diet or do not have physical access to shops with healthy food, and many of those who can afford it prefer an unhealthy diet, so 11 percent of Americans are diabetic and more than 40 percent are obese, both far in excess of the same health issues in Europe; cardiovascular diseases (which account for one in four deaths; almost half of Americans have high blood pressure) and respiratory diseases also are far more common here than in Europe.

Government ineptitude in preventing pandemics peaked early in the Trump presidency, even though Obama had in 2015 established a pandemic-preparedness office that had printed plans for cheaper ventilators and 20 million reusable face masks; in fact, incoming Trump officials even met with outgoing Obama officials on how to handle epidemics. Yet in May 2018, John Bolton, the national security advisor at the time, "streamlined" the National Security Council, including eliminating its pandemic preparedness office, forcing the resignation of its head, Rear Admiral Timothy Ziemer. This meant that protective clothing was in short supply and widely scattered in availability as 2020 dawned. Even later in 2020, with no improvement in getting these supplies, the Federal Emergency Management Agency (FEMA), which could have helped if so ordered, did little. In fact, at times Trump doled out scarce equipment only to those Republican governors who were friendly to him. In 2019, in its general purge of pesky scientists who tend not to be yes men, the administration decided no longer to embed an epidemiologist from the Centers for Disease Control and Prevention (CDC) in China's CDC. This prevented early recognition and resulting decisions on the seriousness of the disease (which originated in China) from the administration. Furthermore, at the time of the outbreak, the CDC had been undergoing serious budget cuts ($12.7b in 2010 to $8b at the time of the outbreak). In addition, faulty test kits were manufactured by the CDC early in the outbreak, whereas other countries successfully employed huge numbers of tests for, and extensive tracking of, the virus. Red tape and wrangling between the CDC and the Food and Drug Administration (FDA) further prevented deployment of test kits, and the problem still was a bottleneck in many places by late summer. Nonetheless, the dominant factor in the perfect storm was Trump's absolute lack of awareness of the magnitude of the problem for the first two months of 2020 combined with his complete absence of leadership then and continuing through 2020. In contrast, governors of many states, including some Republican states, exerted excellent leadership as they shut down their states, but Trump could not even learn from their example! Unfortunately, a half dozen governors of Republican states took their cues from Trump and did little, with disastrous results. At various times, in fact, Trump argued that the federal government should play little role in equipment and policy; in March 2020, he told governors to "try getting it yourselves."

The specific dates of the timeline of the disease and of its effect on the economy, as well as the response by the White House, are enlightening. The dates come from the internet, *Wikipedia*, an article by Laurie Garrett on how Trump and Xi deserve blame in the May 2020 issue of *The New Republic*, the July/August 2020 issue of *Mother Jones*, and a January 31, 2021, *Mother Jones* Special Issue. Dr. Anthony Fauci, a physician and the director of the National Institute of Allergy and Infectious Diseases from 1984 to 2022, published a short retrospective, notably some of the political aspects, in a 2024 article in *The Atlantic* that was taken from his book. I also add some of Trump's comments from 18 videotaped interviews by Bob Woodward in his book *Rage* (published September 2020) that showed that Trump knew the seriousness of the disease yet continued to downplay it—at the cost of many thousands of lives. To save space, I have not included the many well documented days when Trump played golf rather than deal with the pandemic.

We start on December 29, 2019, when Li Wenliang, a doctor (ophthalmologist) at Wuhan Central hospital, in the city of 11 million people of Wuhan, China, sent out a cell-phone text message to warn other doctors to protect themselves from a virus (he thought it was SARS) he had just witnessed. These comments were passed on (the term is "went viral"), therefore violating the start of a cover-up, through censorship and obfuscation, by China of what Li Wenliang and a few others had correctly recognized as a highly infectious disease. In fact, later leaked documents indicate that the disease was recognized by December 8 and perhaps by November 17, if not earlier. Four days after his message, the Chinese police summoned Li Wenliang and forced him to sign an admission of making false comments. Over the next few weeks, his warnings were verified and he became a world hero, but it was a terrible victory: he had caught the disease and he died on February 7, aged thirty-three. On December 31, the city of Wuhan confirmed that health authorities were treating dozens of cases of pneumonia, apparently emanating from the wholesale seafood market in the city. On January 3, after Dr. Robert Redfield, Director of the Center for Disease Control (CDC) received a phone call from his counterpart in China (who reportedly sobbed on the call), Trump was formally notified of the virus by Alex Azar, Director of Homeland Security. From this point it took Trump more than seventy days before he began taking the virus seriously. By January 4, the World Health Organization (WHO) began tracking the cases in China, issuing a report the following day. On January 11, the first death was reported in China but two days later, and continuing until January 20, the Wuhan Commission on Health happily promulgated the myth that no new cases had occurred. By January 13, similar cases were reported in Thailand, then within a week in Japan and South Korea. On January 17, screening procedures began at three US airports. On January 18, Wuhan authorities announced that 40,000 people would be allowed to gather for the traditional Lunar New Year celebration; Trump on this day, during a golf weekend at Mar-a-Lago, received his first major briefing on the pandemic from Azar, although Trump was instead preoccupied with lifting a ban on flavored vaping products. On January 20, the first case was found in the US, in Washington state after an American in his thirties had returned from Wuhan five days earlier. On January 22, in response to a question by a reporter, Trump said "[w]e have it totally under control. It's one person coming in from China. We have it under control. It's going to be just fine."

On January 22, a travel ban from the city of Wuhan was ordered, although by then five million people had left the city, some carrying the virus. On January 23, 2020, Chinese authorities locked down Wuhan, as by then 570 people had become infected and seventeen had died. Trump tweeted on January 24 that "[i]t will all work out well." However, Anthony Fauci recalled in his 2024 article that a "code red went off in my mind" when he saw photos showing that China was quickly erecting a 1,000-bed prefabricated hospital and "my anxiety took a sharp turn upward." Something was going on, and on January 26, China placed more than fifty million people under some form of quarantine, with thirty provinces reporting a total of 2,744 cases and eighty deaths. Chinese health authorities announced that same day that transmission was spread by person-to-person contact, and that some spreaders were asymptomatic. That day, the US had five known cases. Joe Biden, in an op-ed piece in *USA Today*, stated "[t]he possibility of a pandemic is a challenge [that] Donald Trump is unqualified to handle as president." Trump held a rally in New Jersey on January 28, without mentioning the coronavirus, and knowing full well that such rallies can be super spreader events. White House trade advisor Peter Navarro warned in a memo on January 29 that immediate action was needed to prevent deaths of between one and two million Americans in a pandemic. The WHO declared a public health emergency on January 30, and Trump stated "we think we have it very well under control." On January 31, Trump suspended entry to the US from anyone who had been to China in the past fourteen days, if that person were not an American citizen. Trump by then was fixated on raising the drawbridge on the moat surrounding the castle, a response that would prove largely ineffective. For the Huns, so to speak, were already in the castle. On February 2, the first death from the virus outside China was in the Philippines, but Trump said "we pretty much have shut it down." By then it was clear that any national stockpile of protective gear was inadequate due to underfunding by the administration. On February 5, 2020, a cruise ship with 218 infected persons was quarantined off the coast of Japan. Anthony Fauci, who with colleagues had started the work toward a vaccine in the first week of January, now became the public face of the disease.

In an audiotaped interview with Bob Woodward on February 7, 2020, for Woodward's book *Rage*, Trump told Woodward that "you just breathe the air and that's how it's passed... . It's also more deadly than even your strenuous flu." Clearly it had registered with Trump by then that the coronavirus was serious. But his reactions to securing protective gear, ventilators, etc., were half-hearted, and his public statements did not reflect his knowledge about the seriousness of the disease. For example, at a MAGA rally in New Hampshire on February 10, he said that "it's going to be fine." And at that time, he said that the virus was no more virulent than seasonal flu. On February 19, under questioning by Woodward, Trump admitted that "I wanted to always play it down. I still like playing it down because I don't want to create a panic." Although he did not clarify "panic," it seems clear that the "panic" he wanted to avoid was in the stock market, which he needed to keep at a high level to help his re-election. Therefore, those people who believed him exposed themselves unnecessarily to infection, if not death. In not leveling with the American public, it is likely that many tens of thousands of people died.

To continue with our timeline: on February 8, 2020, one day after CDC testing kits reached state and local labs, most of them reported that the kits were not working properly. On February 11, the virus got its name, COVID-19 (the "19" comes from 2019, the year in which it was discovered). On February 10 and 14, Trump told a lie, as part of his happy talk, that the virus would die in hotter weather, as early as April. Also on February 14, the first death was recorded in Europe. On February 19, the virus arrived in Iran, where it soon devastated the country because its leaders allowed group prayers and would not accept some preventive measures. They subsequently denied and covered up the virus' presence and its increasing toll. On February 19, 20, and 21, Trump held rallies in Phoenix, Colorado Springs, and Las Vegas, respectively, but made no mention of the virus. On February 23, cases exploded in Italy, and Navarro in a memo to Trump asked for $3b to fight the coronavirus, including $600m for protective equipment. At a news conference that day, Trump inexplicably played doctor when he proposed injecting bleach into oneself and disinfectants into the lungs of coronavirus patients. Trump tweeted on Monday January 24 that "[t]he Coronavirus virus is very much under control in the USA" and that the "[s]tock market [is] starting to look very good to me." In fact, on that day, a Monday, the Dow Jones Industrial Average lost 1,032 points, the start of a bad week in which it declined every day, for a total by the close on Friday of 3,583 points; during that same week, the S&P 500 lost 11.5 percent. On February 25, Trump said "people are getting better, they all are getting better." On February 26, the first case was seen in Brazil, even as cases increased rapidly in Europe. That same day, Trump announced that Vice President Mike Pence was to oversee coordinating the White House virus response and said the number of cases was fifteen (a lie) "and the 15 within a couple days will be close to zero." On this day, Redfield testified before Congress also that the risk to the public was low. On February 27, Trump stated "it's going to disappear, one day it's like a miracle it will disappear." On February 28, as cases spiked in Europe, we had what we thought was the first death in the US, in Seattle (later we learned that there had been a death earlier than this in California); Trump stated "and you will be fine." He held a MAGA rally in Charleston that evening, in which he called the virus a "hoax" and said "we are doing great." He told Sean Hannity on March 4 that "[i]t's not that severe." At an Oval Office meeting, Anthony Fauci told Trump that we were in the early stages of developing a COVID vaccine, so shortly, on March 6, Trump toured the CDC sporting a red MAGA hat and stated that "anybody who wants a test will get a test" (a lie), and while surrounded by masked medical doctors declared himself to have a keen medical mind. By March 7, the world (greater than 90 countries) had a count of 102,000 infected persons and 3,500 deaths. On March 10 Trump said "it's really working out, and a lot of good things are going to happen." Trump's coverup of the seriousness of the virus was remarkably like that of Chinese President Xi Jinping in the early days of the virus, as emphasized by Laurie Garrett.

March 11, 2020, was a pivotal point in the Trump presidency. It was not only when the WHO declared COVID-19 to be a global pandemic and the NBA (National Basketball Association) suspended its season, but also when Trump gave his first speech to the nation on the pandemic, from the Oval Office. In it, he declared that he would block for thirty days all travelers from Europe except those from Great Britain, then said this order would not apply to US citizens; this demonstrated a lack of understanding that the virus did not respect national boundaries and besides, the horse was already out the barn door and gone. He also stated that his administration was "responding with great speed and professionalism." In fact, hospitals and especially retirement homes were short of personal protective equipment (PPE, namely masks, gowns, etc.) and supplies (ventilators), and testing was limited; these shortages were not being remedied and Trump had tasked few people or organizations with improving the supply. The speech made it clear that Trump was out of his depth, and that he could no longer lie his way out of the situation or blame others. The stock market, which Trump had hoped to mollify with his speech, was not fooled and the Dow lost 1,400 points. The next day, March 12, the Dow plunged another 2,300 points as Trump continued to insist of the virus, "it's going to go away." He also claimed that travelers to the US were being systematically tested, which was not true. On Friday March 13, in response to a question from reporter Yamiche Alcindor, he stated "no, I don't take responsibility at all" for his bungling. Fox TV had until then made every effort to downplay the seriousness of the epidemic during the previous two months, and hosts of many of their shows called it a "hoax" (a word that Trump used commonly at MAGA rallies). On this same Friday, however, Trump declared a national emergency and said that he would make $50b available to states; the Dow rose 2,000 points. On March 15, Europe in essence shut down. On March 16, a Monday, we got the first indication that the lightbulb had finally gone on for Trump, as the Dow fell nearly 3,000 points (13 percent, its biggest drop in history) and his demeanor suddenly changed: he offered excuses such as "this came up so suddenly." Yet when asked how he would rate his response to the crisis, he said "I'd rate it a 10. I think we've done a great job." The next day he had recovered his composure and began to rewrite two months of history when he stated "[t]his is a pandemic, I felt it was a pandemic long before it was called a pandemic, all you had to do was look at other countries."

On March 19, 2020, as California Governor Gavin Newsom issued the first statewide stay-at-home order, China claimed that it had recorded no new infections the previous day, whereas Italy (a tiny country in comparison) the same day exceeded China in deaths. As many health experts noted shortages in PPE, Trump said governors should be buying the equipment because "[y]ou know, we're not a shipping clerk." On March 19 and 21, Trump hyped the anti-malaria drug hydroxychloroquine as a COVID-19 treatment, after it was being promoted by Laura Ingraham and others at Fox News, as noted by Anthony Fauci in his 2024 article. This advice was made against the objections of health professionals, and on March 23 an Arizona man died from taking the drug. In fact, this hype continued, as summarized in November 2020 by the *Journal of Medical Internet Research*, whereby between March 1 and April 30, Trump urged quack remedies, especially hydroxychloroquine and chloroquine, which he called "miracle drugs," in eleven tweets and sixty-five times in White House briefings. Another quack "miracle" drug, pushed by Republican politicians and Fox news (especially in late 2020 and in 2021) was ivermectin, which we give our horses to de-worm them. On March 20, worldwide deaths from the virus had exceeded 10,000 and Trump announced a sixty-day ban on foreign immigration. The UK locked down on March 23. The following day, India shut down and the 2020 Tokyo summer Olympic Games were postponed. Yet Trump said he would like to see the US open back up by Easter. On March 25, the US Senate passed a $2 trillion aid plan by a vote of 96 to 0. David Corn in his 2020 article noted that a huge amount of this sum was handed over to banks to manage (essentially ensuring that their preferred [i.e. already wealthy] customers would be rewarded) and that Treasury Secretary Steven Mnuchin got a pot of $500 billion to dole out with few strings attached. On March 26, a record 3.3 million people filed for unemployment funds and global cases exceeded 500,000. On March 27, the US surpassed Italy as the country with the most coronavirus cases, at 82,000, with New York state alone at 23,000 known cases and 365 deaths; nearly half of Americans were under lockdown, announced by governors, although many Republican governors held out. On April 2, worldwide cases were greater than 1 million, with 50,000 deaths; the US had 20 percent of the cases and 5,000 deaths. On April 7, for the first time, the US reported 2,000 deaths in a twenty-four-hour period. Yet Wuhan lifted its lockdown on April 8. The world recorded a death toll of 100,000 people on April 10. By April 13, world cases had swelled to 2 million, of which the US had 600,000 cases and 25,000 deaths. Amid evidence of testing delays, Trump blamed the disaster on the Obama administration. On April 14, as Trump continued to look for excuses, he stated that the WHO had been "slow to respond" and in response halted $400m in funding to this organization. On April

15, US citizens started receiving stimulus checks for $1,200. The US death toll passed 30,000 on April 16. On April 19, nursing-home deaths totaled more than 7,000 in the US, although this figure was a crude estimate because most states kept no records; China announced that it had no new cases.

Some states and large cities, nearly all of them run by Democratic governors and mayors, imposed stringent lockdowns that resulted in bell-shaped curves of both daily new cases and deaths. The reason for lockdowns (confining persons to their home), which were first applied (brutally) in China, is to force or ask citizens to endure short-term economic pain as the price for getting the pandemic under control. The most spectacular US example of lockdowns was in New York state under Governor Andrew Cuomo, who held long daily (including weekend) press conferences at which he discussed the daily state statistics and argued forcefully for people to behave responsibly. He and those others who displayed similar leadership made a remarkable difference! His bell-shaped curve of new cases per day peaked at above 10,000 cases per day by early April, and his bell-shaped curve of daily deaths peaked at almost 1,000 in mid-April. By late July, the long tail of the New York curve had an average of about 700 new cases and 15 to 20 new deaths per day. By late July, Cuomo and two fellow governors required a quarantine of fourteen days for any person from a less responsible state entering the collaborative tri-state area of New York, New Jersey, and Connecticut.

The worst aspect of the pandemic was its politization. How, one might ask, could Trump politicize an epidemic? Perhaps he took his cue from protests to state and city lockdowns that erupted on April 13 in Ohio and on April 15, 2020, in Michigan, but soon also sprang up elsewhere. For it was a great way for him to distract citizens from the shocking numbers of daily infection cases and deaths. Many Trump voters termed themselves "liberate" protestors, a percentage of whom wore MAGA hats and carried firearms, and staged many rallies, although their numbers were small. They argued that lockdowns should be broken, that masks infringed on their constitutional liberties, and the whole COVID thing was a conspiracy. Some resorted to violence by attacking mask wearers or people employed by businesses that required a mask to enter the store. Other "liberate" protestors threatened health workers through social media or confrontations at their homes. On April 17 and other times, Trump tweeted encouragement to these crazies and egged them on at news conferences, including one time when he stated that the "great people of Pennsylvania want their freedom" and other times tweeting "liberate Minnesota" and other states (April 17). Such people showed an exceptional amount of hate of other persons, notably those who were and are most vulnerable to the virus.

Tennessee, South Carolina, and Georgia announced plans on April

20 to reopen from their late and limited (voluntary) "lockdowns." Yet no recommendations from the CDC for reopening had been given for these or most other states, whose cases continued to accelerate in number. On April 21, Georgia provided more specifics, namely that most businesses should re-open by May 1. (The insanity of the Georgia governor reached a new, even comic, level on July 16, when he issued orders that people in counties, cities— including Atlanta, where the Democratic mayor was desperate to reduce virus cases—and towns were prohibited from requiring masks!) At a daily press briefing on April 23, Trump wondered aloud if disinfectant could be used internally to kill the coronavirus "by injection inside or almost a cleaning?" The next day the maker of the disinfectant Lysol issued a statement that "under no circumstance" should disinfectant be injected into the human body. On April 24, COVID-19 cases in Brazil began to spike, a country led by a president (Bolsonaro) who was, if anything, worse than Trump. Following a *New York Times* article that reported that Trump spent his days watching TV news and skipping coronavirus task force meetings, Trump tweeted "I am the hardest-working President in history." The US reached milestones on April 28 of 1 million cases and 57,000 deaths, yet many states reopened on May 1; to hasten reopening, the White House shelved a CDC written draft (leaked to the press) of guidelines on May 6. Unlike other advanced countries, the US had no consistent national guidelines. Yet those countries that had implemented real lockdowns had bell-shaped curves of cases and deaths; the result of those lockdowns for Spain, Italy, Denmark, Germany, and New Zealand was that on May 11 they could already ease lockdown restrictions. Israel and Iran also announced easing on that date, but neither had bell-shaped curves and clearly reopened too early, as was also the case in nearly all Republican states in the US. On May 13, following warnings by Anthony Fauci the previous day on states reopening too soon, Trump rejected this advice saying, "[t]o me it's not an acceptable answer, especially when it comes to schools" because "young people are very little affected by this." The world death toll was 300,000 on May 15. Cases surged in India on May 16. By May 24, worldwide cases exceeded 5 million.

May 25 was Memorial Day, and across the country, maskless revelers, mostly white people, packed together in celebration, even as confirmed cases shot upward in twenty states. Trump played golf all weekend but also broadcast angry tweets at Hillary Clinton and Nancy Pelosi and again voiced his latest conspiracy theory, this time calling for a murder investigation of MSNBC host Joe Scarborough, a frequent critic of Trump despite being a Republican. Trump made no mention of the milestone 100,000 US deaths reached on May 27. Newspapers noted that this number represents more deaths than America suffered in the Vietnam, Korean, and Iraq Wars combined.

During the Memorial Day revelry, an event occurred that would add a shocking and horribly depressing accent to this springtime of tragedy. This was the killing by police in Minneapolis of a black man, George P. Floyd, Jr. (born in 1973), a former truck driver and security guard but now largely employed by various charitable groups. It happened during his arrest, allegedly for passing a counterfeit $20 bill. After white police officer Derek Chauvin and his white colleagues had subdued and handcuffed Floyd, Chauvin knelt on Floyd's neck for 8 minutes and 46 seconds while Floyd repeatedly pleaded for his life saying he could not breathe, and then died. Chauvin's colleagues, standing there, did nothing. But these cops made a big mistake, for the whole thing had been recorded by at least two high-quality videos and several recordings. The videos were soon released by the news media and played repeatedly (and watched by persons confined to their homes by the COVID lockdowns). There was nothing equivocal in these videos, for the entire act was clearly visible. It sickened every person who saw it. Chauvin and his colleagues acted as if they were exterminating a bug by stepping on it. Yet this murder was only the latest in a recent string of killing of unarmed blacks, particularly males, by white police officers. Such acts had occurred frequently in the US throughout the twentieth century and before. In fact, the generally accepted figure for lynchings between 1865 (when Confederate soldiers founded the Ku Klux Klan) and 1950 is nearly 6,500 black men, women, and children (Kali Holloway, September 21/28, 2020, issue of *The Nation*). In the case of Chauvin, the police officer, he received a sentence of life in prison. Floyd's lynching came on the heels of the killing of Breonna Taylor, a twenty-six-year-old, African-American emergency medical technician who was sleeping in her own apartment in Louisville, Kentucky, on March 13 when seven white Metro Police Department officers broke in to exercise a no-knock search warrant (a mistake), then shot twenty times into the darkness, killing Breonna with eight shots. These officers got off free. And on February 23, Ahmaud Arbery, a young black man jogging in a Georgia suburb, was shot by father and son white racists, with the aid of a white neighbor who made the mistake of using his cell phone to make a video of the murder. Eventually, all three received a sentence of life in prison.

Protests for George Floyd and against police brutality began soon after the release of the tapes; by May 31 they were accelerating in number across the country. Soon they had reached four hundred cities in the US, in all fifty states. The primary chant was "Black Lives Matter," a movement and rallying cry that had begun in 2014, but had never resulted in the scale that now engulfed America. Other countries, who had minorities of their own, many of whom are black, also erupted. Most demonstrations were peaceful, in the spirit of Martin Luther King, Jr., although some businesses were burned, notably in Minneapolis and in

Portland, WA. Inaction by the Justice Department led by Attorney General William Barr, as well as racist statements by Trump, made things much worse. In the early evening of June 1, just after giving a speech in the Rose Garden declaring that he was the "law and order president," Trump ordered police, many on horseback and firing rubber bullets and tear gas, to clear peaceful protestors from Lafayette Square, a park adjacent to the White House. Then he and his people arrogantly strode through the park and across the street, where Trump posed in front of St. John's Church for a photo op, with a Bible (upside down) raised in one hand. In late July, Trump brought to Portland, WA, a federal police force of heavily armed officers from the Department of Homeland Security (hardly a police department but now used by Trump as his own federal police force), dressed in camouflage but whose name and agency badges were missing, purportedly exercising "law and order." The media interpreted this as an attempt to trigger riots and resurrect his campaign for re-election, which was foundering because of the fall-out from the bungled response to COVID-19.

The US officially entered recession on June 9 because of the lockdowns. The first clue that the stock market was in trouble came on February 20, a Thursday. Although the Dow Industrial Average fell only 128 points that day, virtually all other world stock markets also fell; banks and businesses clearly had entered a crisis. The next day the Dow fell 227 points. The crash took place the next week, as noted earlier. The International Monetary Fund (IMF) announced on April 14 that all G7 nations had entered or were entering a "deep recession" and the IMF projected it to be the worst since the Great Depression. One third of the world population had been placed on lockdown. Western nations expected employment to decline about 10 percent. Job losses in the US reached 16 million during the three weeks ending on April 4. On May 8, unemployment in the US was 14.7 percent, and 5.4 million Americans lost health insurance when they lost their jobs. On July 30, statistics showed that the US Gross Domestic Product (GDP) had dropped 32.9 percent during the second quarter, the worst contraction since records began to be kept in 1947.

On June 11, 2020, two million cases of COVID-19 and 113,000 dead were recorded in the US. On June 28, the world toll exceeded 10 million cases and 500,000 deaths. Cases and deaths continued at exponential rates to new daily highs in most places in the US and many parts of the world, although most timelines of such depressing news were suspended in June or July. On July 23, the US attained 4 million cases and 144,000 deaths. On July 25, the US reported 65,000 new cases and 914 new deaths; total global cases were 16 million (a jump of 1 million from four days earlier), with 310,000 new cases that day. On this day, Texas had 6,000 new reported cases and Florida had 12,000 new reported cases. That same day Brazil had 50,000 new cases, bringing cases there to a total of 2.4 million, and 1,200 new deaths. On June 29, South Dakota Governor Kristi Noem announced that she and others would attend Trump's July 3 speech at Mount Rushmore and promised that they "won't be social distancing." New US cases per day exceeded 50,000 for all except two days of the month, and for seven of the previous ten days they had exceeded 60,000 new cases per day That day, the world count of deaths was 650,000. In an Independence Day speech, Trump falsely claimed that "99 percent" of COVID cases are "totally harmless."

On July 7, the Associated Press reported that the Paycheck Protection Program, a relief effort meant for small businesses hit by the pandemic, had loaned about $273m to over a hundred companies owned or operated by major Trump donors. On July 13, Families USA, a non-partisan think tank, noted that 5.4 million people had lost health insurance between February and May. On July 16, the US broke a single-day record of more than 75,000 new coronavirus cases. On July 21, in an interview with Woodward, Trump said "the virus has nothing to do with me. It's not my fault." On July 31, with the national unemployment rate at more than 10 percent, the CARES Act's expanded benefits for unemployed people expired. On August 6, US unemployment filings topped one million persons for the twentieth consecutive week (a run that had begun in the last week of March), shattering records. Nonetheless, over the next week 365,000 people attended the Sturgis Motorcycle Rally in South Dakota. On August 11, Russian President Putin announced approval of a coronavirus vaccine, the first vaccine—but without clinical trials (Putin never took it, as far as we know). On September 16, Attorney General Bill Barr said that pandemic lockdowns and stay-at-home orders are "the greatest intrusion on civil liberties in American history." On September 22, the US death toll surpassed 200,000, which represented more Americans dying in any discrete mass casualty event besides the Civil War, World War II, and the 1918 flu pandemic. On September 28, the worldwide death toll surpassed one million. On September 29, airlines braced for mass layoffs unless a new stimulus bill passed; Disney announced 28,000 layoffs. In the presidential debates, Biden criticized Trump's COVID response, saying "[t]he president has no plan;" in response, Trump mocked Biden for wearing a mask in public. On October 1, while attending several fundraisers maskless even though White House advisor Hope Hicks had tested positive, Trump declared that "the end of the pandemic is in sight." On October 2, 2020, Trump tweeted that he and Melania had tested positive for COVID-19. He was taken to Walter Reed National Military Medical Center, given experimental medicines and expert care, and pulled through.

With the election of Joe Biden, leadership of the CDC was finally wrested from Trump. Help was on the way, but cases and deaths could hardly be stopped quickly! President-elect Biden declared a COVID-19

Advisory Committee on November 9, 2020, a day on which the US had 15 million cases. By December 14, the country surpassed 300,000 deaths, which translated to 961 deaths per day since the first death on February 6, 2020, and 50,000 deaths in the previous month, with an average of 2,403 daily deaths in the previous week. In December 2020, COVID-19 vaccinations became available for emergencies! But now we had antivaxxers—some of them red-state governors—vying for attention. By January 1, 2021, the US had 20 million cases. By January 19, 2021, we had 400,000 deaths, and by February 22, 500,000 deaths. On March 11, Biden gave his first national address on the one-year anniversary of Trump's speech, vowing that vaccines would be available by May 1 for all adults. On May 6, authorities estimated that deaths had been undercounted and that the actual number probably exceeded 900,000. By July 4, 67 percent of adults in the US had been vaccinated, but a sizeable number refused, and these were the ones that made up most of the future deaths. On September 8, the US reached 40 million cases, and on September 20, official deaths exceeded 675,000, the number of American deaths caused by the 1918 Spanish flu. On October 1, 2021, Reuters reported that worldwide deaths had exceeded 5 million. In February 1, 2022, the US fatality rate was 63 percent higher than that of other wealthy countries, due largely to lower vaccination rates. On May 4, 2022, American deaths officially reached 1,000,000. In June, the CDC approved a vaccine for children less than five years old. By September, 2022, a new vaccine that would protect against the latest virus strains was available for all adults.

Throughout the pandemic, it has been widely noted that Trump voters who have opposed masks and vaccines have many excuses when asked, nearly all based on misinformation or conspiracy theories, with the rest of the excuses given as "individual freedom." Everyone has the freedom to make choices, but anti-maskers/anti-vaxxers have corrupted this excuse by their lack of understanding of the role and rights of US citizens given in the US Declaration of Independence, Constitution, and Bill of Rights. They seem to think that Thomas Jefferson gave them the "right" to do anything they want, and any person who asks them to don a mask or get vaccinated to prevent infecting others is somehow out of line. Most are white US citizens who have never known oppression, so they perceive that they are being oppressed by being urged to be responsible and civic-minded and wear a mask or get vaccinated. They conclude that their rights are more important than those of others, a typical expression of majority white male privilege. As the founding fathers clearly stated, all rights come with responsibilities for the common good, nor is one free to do things that impinge on the rights of others. In fact, the attitude of anti-maskers/anti-vaxxers is that they are rejecting all forms of authority in the name of "freedom," much as brats will reject the advice of parents. These are the same "adults" who become infuriated if anybody tells them to wear a helmet when riding a motorcycle, not to smoke, not to drop out of school, to wear seatbelts, to try to lose weight, to lock up their guns, to get health insurance, etc. They believe they are simply exercising their freedom to mess up their lives through irresponsibility. In fact, some were so stubborn that they declared that they would rather die than wear a mask or get vaccinated! And many did. Perhaps they concluded that shots or masks would be an outcome worse than death because this would be an admission that they had been wrong. Even though more than a million Americans died from the pandemic, there is no evidence that more than a few died from COVID vaccinations.

A July 27, 2020, opinion piece in *The New York Times* by Nobel-prize winning economist and professor Paul Krugman provided an important indictment of anti-mask Trump voters and of the Republican party that encourages them. Regarding the surprising opposition to wearing a mask, he marveled at "the intensity of right-wing anger against relatively trivial regulations…It's the principle of the thing: many on the right are enraged at any suggestions that their actions should take other people's welfare into account." Also, that "what they call freedom is actually absence of responsibility." And in the larger context, "the right has made irresponsible behavior a key principle," for they are "sacrificing the future because they weren't willing to show a little patience." He concluded that "[t]he modern U.S. right is committed to the proposition that greed is good, that we're all better off when individuals engage in untrammeled pursuit of self-interest. In their vision, unrestricted profit maximization by businesses and unregulated consumer choice is the recipe for a good society."

An editorial by Win McCormack in the September 2020 issue of *The New Republic* pointed out that many Trump voters claim that wearing masks and doing social distancing "is a violation of their constitutional right to do whatever they want, wherever they please, even if it endangers other people in the community or themselves." Such a claim, derived from "Trump's fusion of irresponsibility and incompetence," and fomented by Trump's words, is a "misinterpretation" of Thomas Jefferson and other founding fathers who emphasized "the obligation of citizens to participate in the political life of their community and attend to its collective needs." McCormack quoted a former Republican congressional staffer Drew Holden, who stated that "[c]onservatives shouldn't conflate the ephemeral necessity of collective sacrifice in pursuit of the greater good with an assault on individual freedom, particularly in moments of crisis."

Margaret MacMillan, an historian at the University of Toronto, offered some lessons from the COVID-19 crisis in the May 9, 2020 issue of *The Economist*. She suggested that the pandemic turned "a spotlight on the dark sides of our world." It exposed the frailty of international supply lines and world trade, the withdrawal of the US as a world leader at the expense of a hostile China, and a reduction in trust of governments. She noted that "those societies that survive and adapt best to catastrophes are already strong. Britain rose to the challenge of the Nazis because it was united; France was not and did not." As the virus played out, some "populist demagogues played to baser fears and fantasies," as in Brazil, India, and the US. Returning to the words of George Packer, we at least learned that our "essential" workers and heroes are mostly people in low-paying jobs that require them to be on the front lines and put their lives at risk, at no extra pay. These are not only nurses and doctors, but also shelf-stockers, delivery drivers, long-haul truckers, warehouse workers, health aides and staffers, meat workers, fruit and vegetable pickers, etc. Packer hoped that "[w]e can learn from these dreadful days that stupidity and injustice are lethal; that, in a democracy, being a citizen is essential work; that the alternative to solidarity is death."

As a postscript to the COVID-19 saga, the February 2024 issue of the Elsevier journal *Biomedicine* and *Pharmacotherapy* (v. 171, by Alexiane Pradelle and others) estimated that taking the drug hydroxychloroquine killed at least 17,000 gullible people. Furthermore, the 2023 book (see Notes) by Peter Hotez (M.D, Ph.D., Baylor College of Medicine), who assisted with vaccine development and was a frequent spokesman for its efficacy, stated (p. 46) that "the trajectory of COVID-19 deaths more or less overlaps with European actions up until the time vaccines were widely available," after which American deaths peaked, in contrast to those in Europe. He estimated (p. xii) that "in 2021–2022, more than 200,000 unvaccinated Americans needlessly lost their lives to COVID-19 because they refused vaccines." This estimate is similar to what others have calculated (Hotez, 2023, p. 46–48). This appalling number must be considered a minimum because it reflects only those who refused vaccinations and does not include deaths after 2022. These deaths are directly attributable to Trump, to most Republican politicians, to most conservative news media, and to people who are unable to think for themselves. This is but the tip of the iceberg in the Republican war on science, as Hotez's book claims.

One of the greatest merits of science has been its benefit to human life and health. Germ theory played out during the 1800s. Each discovery, then and in more than the hundred years since then, has built upon previous discoveries or previous failures. As we saw from 2020 to now, as new data about the pandemic came in, ideas and hypotheses changed and were quickly announced and published. Some seemingly contrary data were not the result of indecision or incompetence, but were how science works, for it rarely proceeds in a straight line. This is just the path of basic science. Even now, seemingly every week we hear of science giving us a new drug or new surgical procedure that lessens or prevents suffering or saves lives. The instantaneous emergence in late 2019 and rapid progression of COVID-19 presented a huge challenge. To put it euphemistically, Trump failed in that challenge, especially in distribution of remedies and in the messaging.

CHAPTER 15

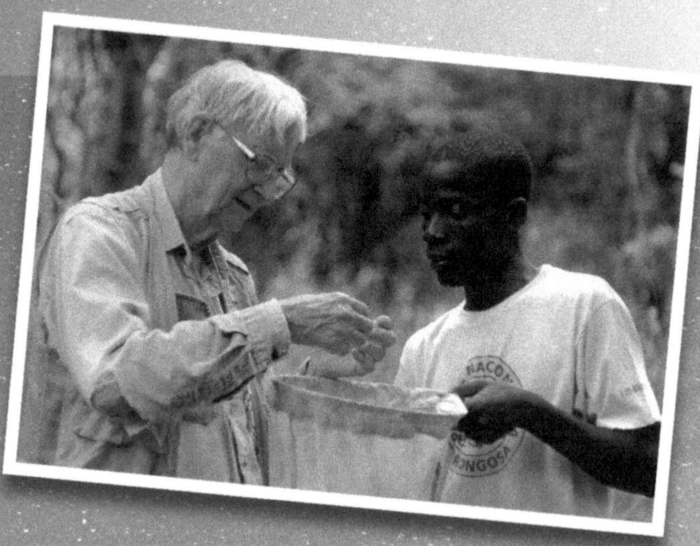

Modern science has been under assault since the presidency (1981–1989) of Ronald Reagan, and this assault has increased with each successive Republican president. Its culmination—and the nadir of science in my lifetime—has been the first four-year reign of Donald Trump. He and the Republican party have used lies and pseudoscience (what I call "biblical science") in their march toward fascism. They are not alone in doing this. Other authoritarian "populist" world governments and dictatorships have likewise targeted science in order to attack truth—both

THE FUTURE OF SCIENCE

because science has had widespread credibility and because it is the basis for many things, such as climate change, evolution, and other topics that they have politicized for various reasons including, in the words of former Vice-President Al Gore, because they are "an inconvenient truth." Climate change, in fact, is their biggest target. Although every one of us is partly to blame, climate change's chief culprits are the coal, oil, and gas industries—and they are major contributors to Republican election campaigns. Other matters, such as abortion, evolution, banning books, prayer in school, and of course racism, are cultural issues largely meant to distract unwitting and religious voters from matters of global and indeed planetary importance.

An article on the internet (accessed in April 2024) by Josh Baker of *Bolde.com* concerned "Sixteen reasons so many Americans don't trust science anymore." An internet search of similar key words will bring like opinions, most taken from polls of citizens. Some of the reasons are valid, but most are based on misperceptions of what scientists do—part of my purpose for this book. Among the valid reasons are that scientists are not good communicators of what they do and why it improves the lives of people. Furthermore, scientists may seek the truth but that does not mean that they always find it: some scientists have been wrong or their inventions (e.g., the atomic bomb) have turned out to be harmful. Other reasons Americans have lost trust are based on stereotypes, such as how can you trust a geek in a white lab coat when even glamorous and loved movie stars and sports heroes are selling hokum in TV ads.

Misperceptions that have taken hold include the desire by the American public that the scientist respond with immediate explanations and black and white answers (as opposed to equivocating on complex problems, or reversing their advice as new tests or data come in), which social media can then transmit or news media can publish as a sentence or two of "truths." Simplistic solutions are much more satisfying, especially to those who are not knowledgeable on the subject. The public craves certainty, but scientists cannot deliver that. So for some persons, it is easier to believe demagogues or conspiracy theories that promise to solve all problems. Also, when controversial issues such as pollution, the ozone hole, or climate change are debated, the corporations that are responsible will fund politicians, lobbyists, corrupt scientists, or news-media types to lie, and the public becomes confused and frustrated. Finally, some scientific conclusions are opposed by personal or religious beliefs that are based on ignorance.

More insidious than the attacks on science or the lack of trust, and more pervasive under Trump, is the denial of facts and truth. An expression from US Democratic congressman Patrick Mahoney of New York (attributed to Jonathan Swift, an Anglo-Irish essayist of several centuries ago) is that "a lie gets halfway around the world before the truth gets its shoes on." Republicans have used this concept to good effect. Even worse, Trump stated in some of his Make American Great Again (MAGA) rallies that "just remember, what you're seeing and what you're reading is not what's happening." This is an example of the 1970 George Jones song about an unfaithful wife "[t]ell me my lying eyes are wrong." Or the Chico Marx line in the comedy film *Duck Soup*, "well, who ya gonna believe, me or your own eyes?" And it is nearly identical to the words written by George Orwell (1903–1950) in his 1949 book on totalitarianism, *Nineteen Eighty-Four: A Novel*, "[t]he Party told you to reject the evidence of your eyes and ears. It was their final, most essential command." The most obvious recent example of this is many Republicans' demands that you not believe your own eyes about the January 6, 2021, insurrection, which we all watched for hours on TV, and which they now want us to believe was just a bunch of tourists visiting the rotunda (with weapons and wearing bullet-proof vests). The term currently used for their claim is "gaslighting," a form of long-term psychological manipulation to convince another of a lie by sowing doubt in a victim's mind so that he questions his own sanity, memory, or reasoning.

Modern attacks on science include real assaults on humans, murder, injury, destruction of careers, and personal threats. An article by Sujata Gupta in the December 7, 2019, issue of *Science News*, noted the prison time given scientists in Turkey, Russia, Brazil, and Iran, and the false accusations of treason in the US during Trump's term. By 2019, Syrian President Omar al-Bashir and his Russian allies had carried out 583 attacks on medical facilities in their civil war, resulting in the deaths of 912 medical professionals.

Astronomer Carl Sagan (1934–1996) devoted much of his 1996 book *The Demon-Haunted World* to the difference between science and pseudoscience. He pointed out that there are far more astrologers than astronomers, for pseudoscience is far more emotionally satisfying than facts, perhaps because "[i]t caters to fantasies about personal powers we lack and long for" or because "it offers satisfaction of spiritual hungers, cures for disease, promises that death is not the end" (p. 14). I am reminded of a political cartoon by Wiley Miller (for his series *Non Sequitur)* published in 2017, of an information booth at which one side offered "Facts," with no takers, and the other side (manned by a clown) offered "Shmacts," and had attracted a long line of customers. Sagan quoted Henri Poincare, a French scientist who died in 1912, as saying, "we know how cruel the truth often is, and we wonder whether delusion is not more consoling" (p. 368).

Denial of the truth is more obvious in the debate about climate change, for it has become polarized between Republican and Democrat voters. It has now reached the point where, for most Republicans, even recognizing that each citizen has a role in this, let alone that each of us in a small way is responsible, is a step too far. In fact, it has triggered in many people, especially Republicans, into the politics of the armed lifeboat, a paranoia where the only solution is to secure your own borders, food, energy, etc. from invaders. A disturbing trend is that some people figure we are doomed, so why do anything? All American citizens are susceptible to the ostrich approach (hiding your head in the sand) on this issue. In a 2024 editorial for *The Nation*, the writer and environmental activist Samuel Miller McDonald warned that for some persons who experience the anxiety and despair from repeated extreme weather events are turning to dictators who make false promises of security, to denying reality, to "succumbing to apocalyptic mysticism," or on "focusing their ire on scapegoats" such as trans people, Jews, the UN, or wokeness, all because the "real culprits are too powerful or inaccessible."

For those scientists who have chosen a career in climate science as a specialty, the assault is dangerous and troubling. In an article in the July/August 2019 issue of *Mother Jones*, David Corn detailed the psychological toll on American climate-change scientists during the Trump presidency, many of them in federal agencies but some in academia. Examples of the assaults include censoring their reports, "editing" conclusions during the technical review stage or just prior to publication, withdrawal of funds, denial to attend professional meetings, etc. And in climate-change research uniquely, the scientist doing the work knows that time to do something is running out, yet apathy among the public and Trump officials prevails. Depression, rage, loss of sleep, and grief are symptoms of such assaults that many climate specialists experience. One such scientist interviewed by Corn distinguished between denialism and "bystanderism," the latter exhibited by those who say they care about climate change but take no action, an affliction that currently is commonplace amongst almost all of us. This scientist is not sure which type of person is worse! Martin Luther King, Jr. had another saying for this phenomenon, although for him it applied to racial injustice: "[i]n the end, we will remember not the words of our enemies but the silence of our friends!"

The defense of science or scientists is not done in scientific journals, which do not have op-ed pages, and rarely in science-news magazines such as *Eos*, *Science News, New Scientist, Scientific American, or National Geographic*. Generally, support and arguing for science is done by journalists writing in objective news magazines with large readership, such as *The Economist* or *Time* magazine, in newsletters distributed by charitable societies, in documentaries, expanded TV news shows, or in newspapers. Not many scientists are activists about scientific issues that affect society. But whether from scientists or citizens, Paul D. Tonko (an engineer before he became a member of the US House of Representatives—a Democrat from New York) argued in an opinion piece in the December 2019 issue of *Eos*, "[l]et me be clear: if you care about scientific research, funding, and the continued prosperity of humanity, political apathy is no longer an option."

An opinion piece in the June 2019 issue of *Eos* debated how Trump and many populist countries "prioritized the reduction of scientific freedom, further fostering a culture in which reliance on fact is somehow considered unpatriotic." This editorial, of course, gets into a major problem now facing science, namely denial of facts. If this problem is not resolved, science faces a dim future. Any resolution will of course take time, and about all the article could offer was encouragement on how to educate future scientists, including that the "need to imbue students with a central value: adherence to the scientific method is, in itself, good citizenship." More to the point, grade-school education is paramount on not only climate science but on American history and civics for future voters to understand why political change takes so long or can be stifled in a democracy but can be improved by an educated electorate. High-school classes in economics are needed not only to improve students' future family finances but to learn simple national concepts such as inflation, tariffs, supply and demand, and even tax strategies for billionaires so that future voters can question authority. It is also critical that students receive training before they graduate from high school on the meaning of objectivity in interpretation of data, the importance of critical thinking, and the significance of a free press. That same article also suggested that higher education must include more training in scientific integrity and how to handle political meddling. Not specifically proposed there, but clearly needed, is for scientists to reach out to the public with clear information, not arrogance and obfuscation, and for them to advocate for their research concerns; clearly the public would likely find science less controversial if they were better informed.

Perhaps as much as science we also need a new ethos of truth, and we need to start that in the USA. The beauty of science is that it is true (as far as we know at the time), whether we like it or not. The beautiful song "Imagine" by John Lennon has such aspirations. A four-stanza poem by my colleague Tom Steven, also described the problem.

Truth, by Tom Steven

Truth is relative, we know,
But relative to whom?
Who decides what greater good
Is served by which falsehood

The politician lies and lies
As though it were his right,
The Generals know no other way
To keep us safe, secure

Lawyer, businessman, and clown
Preacher in his pulpit, too,
Tell us what will make them bucks
Not what is right or true

Are we really total dolts
Who swallow all that's said?
Or are the liars also fools
Whose victims are themselves?

Following publication of her new book, *Why Trust Science?*, Naomi Oreskes was interviewed in the Winter 2020 issue of *Reports of the National Center for Science Education*. She noted that "[m]any people resist scientific conclusions in specific areas that they interpret as threatening their values or beliefs," conclusions such as evolution. And "that when [such] people are disparaging science, their motivation is rarely the truth." Oreskes noted another trait of science and scientists, namely that "[i] f science is a learning process, then it is inevitable that sometimes our ideas will change." We saw this process in action during the coronavirus pandemic, as health officials made new and ever-changing recommendations (attacked by some politicians to indicate that health providers did not know what they were doing) on how to avoid illness as therapies (vaccines, etc.) were being developed on a near-daily basis in 2020 and since then by numerous groups of scientists and drug companies to combat the virus. In pseudoscience and in dogmatic religions, of course, there is little change in ideas.

My dissertation advisor, Hoover Mackin, stated in the classroom that "science is pure skepticism." Such skepticism includes self-criticism and testing of your ideas against other ideas (multiple ruling hypotheses), in contrast to pseudoscience, fundamentalist religions, or conspiracy theories, which are uncritical of themselves. Therefore, as Sagan put it (1996, p. 27, 230) science polices itself and "science has built-in, error-correcting machinery at its very heart."

Robert T. Pennock in his 2019 book, *An Instinct for Truth: Curiosity and the Moral Character of Science*, discussed the central tenet of science: discovering nature's truths. Pennock argued that "science has an inherent moral character" (p. 264). In fact, the phrase "instinct for truth" in his book title comes from a letter by Charles Darwin, in which he recognized that he, along with other scientists, have some virtues that enable them to enter such a profession. Pennock noted that these virtues, such as honesty and curiosity, are "so central that it hardly seems possible to be a real scientist without them" (p. 97). But science has been taught to students in many places as facts to be memorized, and "not enough as a method of thinking" (p. 290). Pennock emphasized that doing science is enjoyable, namely a "romantic adventure tale" and a "fascinating pursuit" in which "scientific discovery is...exhilarating" (p. 116) in the mold of the astronauts in the TV series *Star Trek*. Pennock contrasted science with religious dogma, noting that religious faith and hope do not work in science, and if you are looking for hope, you came to the wrong place in science. In fact, curiosity and wonder about the natural world may be frowned upon by some dogmatic religions, which "consider...that curiosity has been seen as a sinful rejection of God's power and authority" (p. xxvii). Pennock mentioned Richard Feynman (1918–1988), a member of the Manhattan Project that developed the atomic bomb, who in 1965 received the Nobel Prize for his work in quantum electrodynamics. Pennock found that Feynman had many of his same views, as shown in a 1974 commencement address in which Feynman urged students to employ the concept of integrity in science, calling for "utter honesty" in their experiments and scientific reporting (p. 36). Feynman found science to be fun (he sometimes played his bongo drums at parties), and in the words of Pennock (p. 281), Feynman showed that "scientific curiosity involves a kind of playfulness."

Science provides guidance on how to think and how to make decisions. The trick, obviously, is to avail yourself of the best available facts, then objectively interpret them without picking an answer that a higher authority prefers or, especially with climate change, an answer that is less painful than the likely reality. This option of scientific thinking comes at a time when, especially in the last six years, evidence (the 2016 election and its aftermath) indicates that 40 percent of US adults use various forms of magical thinking. Whether these lapses in thought have been based on lies fed us by a con man, on fundamentalist religion, on daydreams, or on conspiracy theories, they are no way to make life decisions. It is OK to be amused by Spiderman movies, to check your horoscope periodically, or to thumb through your most outrageous spot-on social media, as long as we recognize these as entertainment, not ways to make choices or decisions. As Sagan states (1996, p. 434), "[i]f we can't think for ourselves, if we're unwilling to question authority, then we're just putty in the hands of those in power." Pennock adds (p. 323) that, through skepticism and analysis of facts, science "frees the mind" by "the rejection of authority in favor of evidence," therefore it "overcomes dogma and dispels superstition." Finally, much scientific thinking is simple common sense and a rejection of nonsense ("if it sounds too good to be true, it probably is"). Everybody has the capacity to do that.

Many of the problems faced in the larger world will require collaboration with, and trust in, science and scientists. Such problems cannot be solved by the Trump practice of hiring unqualified politicians and underlings to limit or destroy scientific knowledge and then to use the ensuing chaos to enrich themselves. It will take long-term investment of our country's resources in science of all kinds, beginning in grade schools.

One of the most obvious problems facing science (because we are still experiencing it as I write) is learning all we can about COVID-19 and its related, rapidly evolving variations including long COVID, as well as about totally new viruses that appear regularly. British astronomer Martin Rees, writing in the January 2020 issue of the magazine *Nautilus*, noted some other problems that science needs to address in greater detail, including practical knowledge and answers on human diet/nutrition and on child care. He noted that drugs to counter and mitigate Alzheimer's disease have so far been ineffective, indicating that pure research on how the brain functions is a current need. There should, he said, be "no justification for snobbery of pure over applied work," as we debated during USGS work at the Nevada Test Site, and he gave the example of the invention of the zipper in 1917 by Swedish engineer Gideon Sundback as applied science of lasting benefit and "a bigger intellectual leap than most of us [scientists] ever will [do]."

The greatest scientific research is going to have to be directed to solving climate change. As Ben Ehrenreich noted in 2021 to explain the mess we are in, "*Homo sapiens* have walked the earth for at least 300,000 years, but coal-fueled industrial capitalism is less than 200 years old." Clearly a funding priority such as the mission-oriented one that John Kennedy provided to get a man to the moon will be necessary. And the excuse for such a priority will not be merely to outdo the Russians! In fact, as my work in Antarctica and with the USGS demonstrates, scientists of all countries love the exchange of ideas with researchers outside their own country, who bring different perspectives. Some people may not like to hear this, but national boundaries do not exist when doing cutting-edge science! Therefore, climate-change research will have to *include* scientists from Russia, India, China, Japan, the European Union, and all other countries, thus requiring international cooperation instead of competition, threats, and warfare. It can start with the mundane, speeding up alternative sources of energy and increasing conversations around the world on the truths and alternatives of what we are facing. Next, the carbon cycle is a massive conundrum whose fundamental answers are unknown and will require much more work and funding by all, even as some countries and industries are going all out to emit more and more carbon dioxide (CO_2) and methane (CH_4). Gathering more data and computer manipulation of those data will be necessary to understand the mind-boggling feedback loops, currently-unknown weather/climate effects, competing roles of photosynthesis, and unintended consequences of mitigation. We have hardly started on all this! In my opinion, the tipping point beyond which date nothing can be done to save civilization will happen this century, within the lifetime of people who now are relatively young. There is no technology on the near horizon that will bail us out.

In the same 2020 article in *Nautilus*, astronomer Martin Rees made the valid point that many non-scientists have the misperception that scientists are "especially clever" or "elite" in their IQ. Countering this, he noted that "to be effective in some scientific fields it's actually best not to be too bright," following the observations of ecologist E.O. Wilson, who quipped that there is a risk of boredom (in smart people) during the requisite decades of gathering information needed to make advances of significance in most scientific fields. Of course, many of the most productive scientists, especially in mathematics, chemistry, physics, and astronomy, achieved excellence and won awards early in their careers, after which they were less effective. But for many other scientists, especially those in geology, experience is paramount and may take years to attain. Hoover Mackin once told me that "a geologist is no damned good until he reaches the age of...well, how old am I, anyway?" (he was over sixty at the time!). But the real key to doing well in science is to have a love for it and a love for the truth, as well as a simple curiosity about why things happen in the natural world. In my opinion, a science profession now provides greater opportunities than any other profession, regardless of your sex, sexual orientation, race, religion, and national heritage. We need all the scientists we can get. So to all young people: give it a try!

CONCLUSIONS

This has been a story about science. Yet I have told it in the context of a life story—mine. The story addresses such things as the benefits of science, how it works, its rules, the thinking behind it, and the excitement and joy it brings the user. Among the many reasons I wrote down this story is that during the last decade, including the COVID-19 pandemic, it became clear to me that the American public and the news media do not understand how science operates.

A profession in science is unlike nearly all others, in that it makes truth and the search for facts for the benefit of society its guiding principles. All scientists have in common a fervent belief in objectivity and in seeking the truth. These pre-requirements and attitudes are fundamental, from the start to the end of any scientist's career. Scientists suggest hypotheses to explain the facts in each scientific discipline, doing so not for personal profit, ego, or fame, but instead in search of the truth. Their hypotheses may turn out to be dead wrong, but they are put forward for the right reasons. For science is never final or "correct" but instead is a path in which an individual's contribution is just one step toward what may be helpful to society. The ideas in this book have thus come from many remarkable fellow scientists and friends, whose stories I have therefore included, in tribute to and as examples of the scientific life. To them—many now deceased—I repeat the words of Bruce Springsteen in his reflective song "Ghosts," that "I make my vows to those who've come before."

The reader should contrast the values of scientists with those of folks such as used-car salesmen, TV ad men/women, and politicians. Yet scientists are under attack, and they notably rarely defend themselves, at least not against non-scientists. Nearly all scientists' communication is between themselves, notably through their publications, and they are much more comfortable with such discourse. They prefer doing and teaching science rather than communicating about it to citizens, this latter trait clearly being a major failing in retrospect. Except for renowned senior scientists who have succeeded in their careers, then stepped back and written books for the scientifically literate public, few scientists reach out to popularize what they have found. These exceptions include Loren Eiseley, Stephen Jay Gould, Carl Sagan, E.O. Wilson, and Stephen Hawking, all wonderful writers but now deceased. Making the case for science has become a lost art.

To return to our topic of the importance of scientists and how they benefit us all, I have left you, the reader, with many examples, but one more is worth emphasizing: that of increasing human life expectancy. In his 1996 book, *The Demon-Haunted World*, Carl Sagan noted that "advances in medicine and agriculture have saved vastly more lives than have been lost in all the wars in history" (p. 9), and that "diseases that once tragically carried off countless infants and children [adults too!] have been professionally mitigated and cured by science" (p. 11). Before those advances were made, living significantly beyond infancy required luck and a prayer. Vaccinations are an important part of those advances, yet now some Americans—almost exclusively in the world—have pushed senseless anti-vaccination campaigns.

It is worth returning to Noam Chomsky because of his optimism in the current time of two wars and the implications of global climate change becoming more obvious with each day. In a quote given by *Salon* and accessed on the internet on December 15, 2023 ("Noam Chomsky at 95—still speaking hard truths, still ignored by the mainstream media" by Norman Solomen), "if you assume that there is no hope, you guarantee that there will be no hope. If you assume that there is an instinct for freedom, that there are opportunities to change things, then there is a possibility that you can contribute to making a better world."

In his 1980 classic book *Cosmos*, Sagan (pp. 7–9) noted that science flourished in many early civilizations before ignorant brutal conquerors or fanatical religion stamped it out. He relates, as an example, the fascinating story of Eratosthenes in Alexandria, Egypt in the third century BC, who, among many other discoveries, recognized that the Earth's surface is curved, and he calculated to surprising accuracy its circumference using shadows from vertical sticks at noon at Alexandria and at Syene, which he measured to be 800 km to the south, along the Nile River. This discovery, of course, had major implications for discovery of new places by ship, but most of those implications were forgotten almost until Magellan circumnavigated the world (1519–1522). Sagan warned that the "present epoch is a major crossroads for our civilization and perhaps for our species" (1980, p. xxiv). More than ever, that is true now, a mere forty-five years later. My hope is that you, dear reader, will consider joining the ranks of the Sagans, Eratosthenes, my geology colleagues, and me, and make science your life or interest, or at least let science inspire your thinking and a life-long curiosity.

EPILOGUE

On November 5, 2024, Republican Donald Trump was re-elected President of the USA despite a seemingly near-flawless campaign by Democrat candidate Kamala Harris. The stories and events in this book end at that date. The vote margin was significant in the electoral count but the popular vote was close. American citizens currently are collectively responsible for the direction of the country, whereas politicians follow their lead. In the upcoming four years, however, those roles sadly may be reversed, for voters have in effect chosen to end democracy, in favor of Christian nationalistic-male authoritarian and minority-party rule. These voters have decided that four more years of deaths, corruption, recession, lies, hate, and chaos will bring on a golden age of American prosperity, a concept that surely is a repudiation of 250 years of American values. P.T. Barnum has been credited with the sayings "you can fool most of the people most of the time" and "there is a sucker born every minute," and both would seem to apply here. Yet for these voters, the wisdom of *Aesop's Fables* also pertains: "be careful of what you wish for, lest it come true." For the Republican campaign was a clinic in gaslighting. They claimed that the booming economy that Biden/Harris brought back from recession was "a disaster" and inflation brought down to a little more than 2 percent was "the worst in history." Even the fact that Democrats ended the pandemic was forgotten. This book provides a history lesson that goes back about 80 years, available to remind those who have experienced the same events that I have. Perhaps the American voter can use this book or trusted sources of their own to refresh their memory about these events, as for example during Trump's first presidency when he ignored the start of COVID-19 and deaths soon skyrocketed, the economy and stock market then cratered, and inflation hit repeated new highs.

Many terrible things have been promised by Trump himself during his 2024 campaign. These include a dictatorship instead of democracy, the end of free speech (seeking out the "enemy within" who says things Republicans disagree with), turning back the clock on women's rights, deporting millions of illegal immigrants, revenge against the FBI and military, purging civil servants, releasing his fellow criminals from prison, and items proposed in *Project 2025.* We trust that some of these are empty campaign promises. But you should remember that when Trump's unexpected 2016 victory sunk in, many of us voiced the opinion that we could survive four years of Trump unless a catastrophic world event took place, which we knew he would be unable to contain. Unfortunately, that came to pass with COVID-19, and his actions unnecessarily cost at least 200,000 Americans their lives (the actual death count was at least 1.2 million Americans). Now, as we near the start of his second administration, the next catastrophic world event has already revealed itself—climate change! Yet so far, and as with COVID-19, Republicans deny that it exists ("it's a fraud"), although as with COVID-19, deaths from it are already mounting within and especially outside our borders. A likely exponential rise in fatalities will be a gut-check for those who can recognize facts. And there will likely be many other sobering events caused by his policies for those who choose to observe things, as the economy and inflation become worse and injustices rise, or perhaps as lapses in national security allow another 911-type tragedy or more wars, or as anti-vaccination agendas lead to new pandemics, etc. These will be opportunities for Americans to make another choice, before time runs out. That is because Trump is gathering a new army of white men to his administration that are sycophants whose main motivations are self-interest and greed and whose abilities are unknown beyond servility and flattery. Concentrating all power in the hands of a single person with so many mental and character deficiencies will eventually come back to bite all Americans. The likely answers the new regime will propose for future catastrophes will be simplistic at best or ignored at worst, and they will only exacerbate each disaster. This contrasts with the start of his previous administration, when some of his employees had sense enough to provide some checks and balances. Therefore, Republican incompetence and brutality will soon become obvious, and the arrogance that results from their perceived "mandate" will result in overreach in their tactics. I am therefore hopeful that Americans will have many future chances to pick truth and science in the years before climate-change tipping points occur or a police state prevents any change back to a democracy. But reversing the current outcome will be much more difficult than the simple vote of November 5, requiring that we organize, stay engaged, and act aggressively with perseverance, energy, guts, and even optimism!

ACKNOWLEDGMENTS

My son, Scott E. Rowley, provided the most valuable support and encouragement for this book. An excellent artist, he also did the layout of the book, photos, and illustrations. Finally, he and his wife, Patty, inspired me when I thought of abandoning the project. Garrett Vice also helped with such motivation. The editor that Scott chose, Ulrike Guthrie, provided invaluable advice alongside her beautiful writing. Chet Wallace did the chapter on Saudi Arabia, including the final "note" to it, with no corrections from me. The dozens of friends and colleagues mentioned in the book all had a part in my upbringing, but they are not to be held accountable for my deficiencies. I am deeply grateful for all they did for me. Persons who took photos that I used are acknowledged in the caption of each, and here thanked. I am grateful to my wife, Dawna, for her help. I used *Wikipedia*, the online encyclopedia, to check many facts, and they are part of my family of worthy charities.

NOTES

Introduction

All comments here are based on a lifetime of memories and experiences. The book *Concrete Economics* (Cohen and DeLong, 2016) discusses the importance of federal government agencies to the economics of the U.S., including inventions by civil servants that have been important to American citizens. An example of a great civil servant is Art Allen of the U.S. Coast Guard, given in an inspiring article and book by Michael Lewis (both cited below). Michael Lewis counters the long-held Republican myth that civil servants are nonessential bureaucrats.

Chapter 1. Early awakenings

This chapter is based on memories and family interviews, notes, and recordings. My father's bachelors and masters theses are listed below.

Chapter 2. Larry Gould and undergraduate studies

This is another chapter of vivid memories, along with those of classmates. Walt Alvarez is a college classmate who became Professor of Geology at the University of California at Berkeley and achieved fame by a career of studies that included his theory of the demise of the dinosaurs by an asteroid at 66 Ma. He has many scientific publications on many topics, but I cite only two of his books for the laymen. Wally Hansen, a colleague, mentor, and close friend at the USGS, published the definitive report (Hansen et al., 1966) on the great Alaska earthquake before I met him; some of his other reports are given below (with him or me as senior author). Larry Gould, President of Carleton College when I arrived, became an inspiration to me in many ways, including his adventures in Antarctica (Byrd, 1930; Gould, 1931). The definitive biography of Larry is Hillemann (2012). Duncan Stewart, Carleton Professor of Geology when I enrolled, was a mentor, colleague, and friend; memorials to his life by Larry Gould and Eiler Henrickson, are cited. Eiler Henrickson, the other Carleton Professor of Geology when I enrolled, became another mentor, colleague, and friend. John Anderson, Professor of Geology at Kent State University, was a great mentor, friend, and colleague; many of his publications are given, including Craddock and others (1965); our memorial to John is cited as Rowley, Filkorn, and others (2018).

Chapter 3. Hoover Mackin and graduate studies

Hoover Mackin was my most important mentor, and I experienced and accumulated a lot of stories about him. Several of Hoover Mackin's most important publications are listed below under his name. A memorial to Hoover by Hal James is given as James (1974). Our dedication to Hoover, with the story of his life, is given in the 2001 Utah Geological Society guidebook (the Mackin volume), cited as Anderson, Rowley, Fisher, and others (2001). Dan Barker's report on Iron Springs is given below under his name. Earl Cook's dissertation and publications are given under his name; his 1965 report describes many of the ash-flow tuffs that he, Hoover, and Paul Williams traced and studied. Several of Paul Williams' most important publications, and his dissertation, are given under his name. Our memorial to Paul is cited as Rowley, Ford, and others (2020). Part of Robbie Gries' history came from her biography in *Wikipedia* (online). George Davis' excellent textbook (first edition) is given under his name; there have been several later editions. Several other reports that George senior authored are given also.

Chapter 4. Times are changing

This chapter is based on memories, with many historic details filled in by *Wikipedia* (online). My brother Clark contributed the best advice I ever got, and information on his life and those of other friends came from memories.

Chapter 5. The start of real life

Many strong memories are given here, along with information about long-term friends who strongly influenced my life and career. The story of Paul Williams comes from memories and from our memorial to him, cited as Rowley, Ford, and others (2020). Dick Blank's dissertation is cited under his name. Art Ford's report on the giant Dufek intrusion and two geologic maps showing it, in the Pensacola Mountains, are under his name. The geology of the Schmidt Hills and Williams Hills, also in the Pensacola Mountains, is described in Schmidt and others (1978). The evolving concepts that began as continental drift and became *plate tectonics*, covered in this chapter and several others, come from my memory, refreshed by visiting *Wikipedia* (online) and several textbooks. Details about Warren Hamilton come from memories, but his biography is from *Wikipedia* (online). Some of the books on continental drift are mentioned in the text. There are many general books on plate tectonics, some for children; the curious reader may be interested in the book *Plate Tectonics* by Oreskes and LeGrand (2003).

Chapter 6. Adventures in Antarctica and elsewhere

My Antarctica career resulted in many vivid memories, backed by careful notes in journals and field notebooks that I kept on all five field seasons, as well as many letters to family and friends, most of which I retain. We also maintained journals of most calls by short-wave radios during the five seasons. Three USGS Antarctic sketch maps (topography not controlled), at 1:500,000 scale and based on trimet aerial photos, were our only base maps used in the field. They show the topographic features mentioned here, for all five field seasons. They are cited below, giving the author as U.S. Geological Survey.

The 1970-1971 field season on The Ice

The background on Antarctica and its early explorers comes largely from books in my personal library, with some details from *Wikipedia* (online). Biographies of colleagues comes from many hours of sharing meals and storms, plus their resumes and letters. Most polar explorers wrote books to fund their expeditions, but I have not cited those books. Ernest Shackleton is Antarctica's greatest hero, so there are numerous books about him; my favorites are the adventure story by Lansing and the extensive biography by Huntford, both given. Roald Amundsen, the first to get to the South Pole, and Robert Falcon Scott, the second to get there a month later, also have many books telling their stories; another book by Huntford (in references), detailing their race to the pole, is a good introduction to this large literature. Douglas Mawson, Australia's national hero, also has many books about him; Bickel gives an account of Mawson's epic catastrophic yet inspiring journey, and Fitzsimons' biography of Mawson and the other explorers of the heroic age is noteworthy. Camping methods that we used in Antarctica, learned from previous USGS field parties, are covered in layman articles senior authored by Steve Boyer, listed under his name. Trail methods that we used are in an article for the layman by Rowley, Waitt, and others (1976). The geology of the porphyry-copper deposit found by Paul Williams, Dwight Schmidt, and others during the 1969-70 season is described by Rowley, Williams, and others (1975), Rowley, Williams, and Schmidt (1977), and Rowley, Farrar, and others (1988). The false fantasy of development of mineral resources, including oil and gas, in Antarctica has long been a topic in Antarctic Treaty negotiations, as addressed by Wright and Williams (1974), Rowley and Pride (1982), Behrendt (1983, 1990, 1991), Ford (1983), Rowley, Ford, and others (1983), and Rowley, Williams, and Pride (1983, 1991).

Between field seasons, 1971-1972

My early USGS geologic mapping project to complete the 7.5-minute quadrangles done in manuscript form but not published by Hoover Mackin (he died in 1968) resulted in three maps in which I put him as senior author (Mackin and Rowley, 1975, 1976; Mackin, Nelson, and Rowley, 1976) and three other quadrangles north of Iron Springs in the Black Mountains (Rowley, 1975, 1976) and Red Hills (Rowley and Threet, 1976), as well as a 15-minute quadrangle (Rowley, 1978) of the Black Mountains. A summary of the geology of the Iron Springs district is given by Rowley and Barker (1978) and an important outcrop that reveals Mackin's ideas is given as Rowley, Hacker, and Biek (2022). The life of Doug Morton comes largely from a memorial to him, cited as Miller, Rowley, and others (2021). Biographies of other USGS colleagues came from memories and notes.

The 1972-1973 field season on The Ice

The 1997 book about Mt. Everest by Jon Krakauer is cited. Articles in *Summit* described camping methods (Boyer and others, 1976) and trail methods (Rowley, Waitt, and others, 1976), as well as other adventures during the 1972-1973 season.

Between field seasons, 1973-1977

Reports on the 1970-1971 and 1972-1973 Antarctic field seasons include Kamenev and Orlenko (1982), Mehnert and others (1975), Rowley and Williams (1982), Farrar and others (1982), and Rowley, Kellogg, and others (1988). The affliction of groupthink, introduced to me by Ed Ruppel, was to become an important reason for incidents discussed later in my career and life, and this mental disease was pervasive during the Trump administration. The Vernal 2-degree project produced not only the 1:250,000-scale geologic map, the surficial geologic map at the same scale (Carrara, 1980), lots of 1:24,000-scale (7.5-minute) maps by three of us, and a summary geologic map (1:50,000 scale) of Dinosaur National Monument (Hansen, Rowley, and Carrara, 1983). The Vernal geologic map went through a preliminary map (Rowley, Tweto, and Hansen, 1978), a black-and-white version (Rowley, Tweto, and others, 1979) while the final version was in press, and the final color version (Rowley, Hansen, and others, 1985). I senior authored five 7.5-minute maps of Dinosaur Monument (Rowley, Dyni, and others, 1979; Rowley and Hansen, 1979a, b; Rowley, Kinney, and Hansen, 1979; and Rowley, Hansen, and Carrara, 1981). Irv Witkind and I wrote a dedication to Wally (Rowley and Witkind, 2005) in one of the annual Utah Geological Society guidebooks, giving highlights of his career and some of his publications. My son Scott and his wife Patty published a cookbook cited below.

The 1977-1978 field season on The Ice

Biographical information on Mike Thomson came from an obituary published in the Robinson College Record (2020) from Robinson College of Cambridge University and from a draft obituary (2021) by Alistair Crame and Jane Francis, sent to me by Alistair in an email in March 2021, both cited as such below. Some of the publications of Tom Laudon mentioned in this and subsequent chapters are cited under his name. Finn Ronne's last book (1979), a memoir, is listed under his name. One of the many expeditions that he led, which included wintering over at Ellsworth Station in 1957, showed him to be a troubled person, as described by John Behrendt's 1998 book; the main part of John's book describes an over-snow geophysical traverse from that base, using Sno-Cats. Several other reports by John that are mentioned in the text are cited under his name. An obituary for Tom Laudon was written by Pat Quilty for The Antarctican Society newsletter, as cited. The 1960 incident in the Crary Mountains, when their aircraft wingtip hit a mountain while in clouds, is given in John Behrendt's 2005 book and in a 2018 newsletter of The Antarctican Society, both cited under John's name.

Between field seasons, 1978-1984

Some of our reports on the geology of the Orville Coast are Laudon and others (1983), Rowley, Vennum, and others (1983), Thomson (1983), Rowley, Farrar, and others (1988), Vennum and Laudon (1988), and Kellogg and Rowley (1989). Biographies of Chet Wallace, Paul Williams, Bruce Wardlaw, and Dave Moore came from shared memories, notes, and their resumes.

Tom Steven and the Marysvale volcanic field

Two of Eugene Callaghan's reports are cited under his name. The story of pioneer Joseph A. Lott, for whom the Joe Lott Tuff Member is named, is given in the preface to a petrologic study of this tuff (Budding and others, 1987). Among several popular and scientific studies of the Kimberly (Gold Mountain) mining district are those of Herring (1981), Utley (1992), and Park and Krahulec (2009). A summary of the geology of Kimberly is given by Steven, Cunningham, and others (1979) and Steven and Morris (1987). Our work in the southern Marysvale volcanic field, largely before I joined the USGS, was published in 1975 in a book listed as Anderson and others, with chapters by Anderson and Rowley and by Fleck and others. Later, two 15-minute quadrangles of an area south of the center of the field were published as Anderson, Rowley, and others (1990a, b). Tom Steven, Skip Cunningham, and Pete Lipman had remarkable careers with the USGS, and left a long legacy of publications about the San Juan Mountains, Marysvale volcanic field, and other places, only a few of which are cited under their names in the References below (some written after they retired, thus largely free to the taxpayer), with a few given as coauthors under my name (e.g., Cunningham and others, 1994, 2004; Lipman, 2006, 2020; Lipman and others, 1970, 1978; Steven 1968; Steven and Eaton, 1975; Steven and Hail, 1989; Steven and Lipman, 1976; Steven and others, 1974; Steven and Ratte, 1960, 1965). My memorial to Tom's life is listed as Rowley, Lipman, and others (2014). One self-published book and compiled folder of Tom's poems are cited as Steven (1992) and Steven (2000). A dedication to Skip in a volume by the Utah Geological Association is listed as Rowley, Steven, and others (2009). A study of the geology of Big Rock Candy Mountain near Marysvale was done by Cunningham, Rye, and others (2005). An early study of the uranium deposits at Marysvale was done by Kerr and others (1957); both Skip and Tom senior authored more modern studies of the same uranium deposits. An early guidebook that visited mining districts in Utah (and writeups), including Marysvale and Iron Springs, is that of Shawe and Rowley (1978). An early summary of the Marysvale field is that of Rowley, Steven, and others (1979), and Skip, Tom, and I senior authored others later, some listed. The overall Richfield 2-degree sheet is by Steven and others (1990). Our general detailed summary map of the Marysvale field is that of Cunningham and others (1983), later compiled and revised by Hintze and others (2003), Rowley, Cunningham, and others (2002), and Rowley, Vice, and others (2005). Detailed precursor maps to the Cunningham and others (1983) map included Rowley (1979), Rowley, Cunningham, and others (1979), Rowley, Cunningham, and Kaplan (1981), Rowley, Steven, and Kaplan (1981), and Rowley, Williams, and others (1981). A geophysical map to accompany the Rowley, Cunningham, and others (2002) map is Campbell and others (1999). The map of the Deer Trail Mountain area is by Cunningham and Steven (1979b). The study of Kingston Canyon is that of Rowley, Steven, and Mehnert (1981) and Rowley, Biek, and Hacker (2022). A breccia dike, containing tiny blue sapphires and located in the eastern Sevier Plateau, was described by Stuart Agrell of Cambridge University and listed as Agrell and others (1999). The revised detailed quadrangle maps of the Central Uranium Area and the western part of the Monroe Peak caldera are those of Rowley, Cunningham, and others (1988a, b), whereas the entire caldera is discussed by Steven, Rowley, and Cunningham (1984) and the detailed maps of the eastern half of the caldera are those of Rowley, Williams, and Kaplan (1986a, b). Excellent reports on the Mineral Mountains were done by Sibbett and Nielson (1980) and Nielson and others (1986). The Pleistocene rhyolites of that range were described by Lipman, Rowley, and others (1978) and Mehnert and others (1978). A summary of the mineral resources of the Richfield 2-degree sheet is in Steven and Morris (1987).

Mount St. Helens

A prediction that Mount St. Helens would erupt within 100 years, if not by the end of the century, came from Crandell and others (1975) and Crandell and Mullineaux (1978), based on long-term USGS studies of the mountain. Foxworthy and Hill (1982) provided a time-line of eruptions for the first 100 days in 1980, the basis for many of the events I mention in this chapter; they also include spectacular photographs by Keith and Dorothy Stoffel, Gary Rosenquist, and others. The huge book on Mount St. Helens, edited by Lipman and Mullineaux (1981), is the best and most complete analysis of an active stratovolcano, and it contains chapters that I mention in the text. These include a timeline of the first events of the cataclysmic eruption from the incredible photos of Gary Rosenquist (Voight, 1981), eyewitness accounts by Rosenbaum and Waitt (1981), the debris-avalanche deposits (Voight and others, 1981), the directed-blast deposits (Hoblitt and others, 1981), the pyroclastic flows of 1980 by Kuntz and others (1981) and Rowley and others (1981), and temperature measurements of the pyroclastic flows and other deposits by Banks and Hoblitt (1981). Our map of the 1980 pyroclastic flows is cited as Kuntz and others (1990). Field work after 1980 at Mount St. Helens led to the recognition of high-speed ash-flow surge deposits related to the May 18 eruption, by Rowley, McLeod, and others (1985). Melanie Holmes' biography of David Johnston is as Holmes (2019). A summary of the job of USGS Information Scientist for Mount St. Helens is given as Rowley, Hait, and others (1986). A retrospective look back at the mountain after 40 years is that of Major (2020).

A trip to Saudi Arabia

This chapter is slightly abbreviated from a story published by Wallace and Rowley (2016). The final USGS geologic report of our 1984 study in Saudi Arabia is Wallace and Rowley (1985).

The 1984-1985 Field Season on The Ice

Based on memories, field notes, and letters home. Some of the information on Janet Thomson's career comes from her biography in *Wikipedia* (online).

Between field seasons, 1985-1986

Our reports on the English Coast include Laudon (1991), Kellogg and Rowley (1991), Rowley, Kellogg, and others (1991), Pankhurst and Rowley (1991), Laudon and Craddock (1992), and Laudon and Ford (1997). A large wall map at 1:500,000 scale showing the geology of the English Coast, Lassiter Coast, Orville Coast, and eastern Ellsworth Land, published by the British Antarctic Survey, is cited as Rowley, Kellogg, and others (1992). Walt Vennum summarized our studies of Cretaceous batholiths and plutons as Vennum and Rowley (1986) and Vennum and others (1995). An arm-waving paper about Antarctica is Schmidt and Rowley (1986). The Richfield 2-degree project (Marysvale volcanic field) included the map itself (Steven and others, 1990). The important background groundwater study of the Nevada Test Site is Winograd and Thordarson (1975). Our compilation of mapping needs of the U.S. is Rowley, Miller, and Miller (1986). I have pushed geologic mapping, even in letters to the editor (Rowley, Dixon, and Steven, 1997). The geologic map of Nevada is by Stewart and Carlson (1978). The geologic map of Utah is by Hintze (1980). Our map of the Newcastle quadrangle is Siders and others (1990).

The 1986-1987 field season on The Ice

The USGS sketch map of the area we visited this field season is cited as USGS (1979). Information on Charles Swithinbank comes from his biography in *Wikipedia* (online). The fathers of continental drift and plate tectonics come from my memory and class notes, refreshed by online visits to *Wikipedia*; the reader may like to read Oreskes and LeGrand (2003). Articles on the geology are cited as Rowley, Storey, and others (1987) and Storey and others (1987), and on basalts (one of many articles) in Antarctica as Rowley, Thomson, and others (1990). A quick story about a rock from Antarctica that Art Ford and I sent Guy Guthridge is cited under Guthridge (2018).

Chapter 7. Calderas in Nevada and Utah

The first map of the Caliente caldera complex is that of Ekren and others (1977). Bart's report on high-temperature ash-flow tuffs that have no caldera is Ekren and others (1984). Our map and description of the Quinn Canyon Range caldera is Ekren, Rowley, and others (2012). Our memorial to Bart Ekren is cited as Rowley, Williams, and others (2020). Study of the Caliente caldera complex, one of the largest in the U.S., began with mapping its western edge, near the town of Caliente, Nevada. Dawna Ferris, who eventually became my wife, did her Masters thesis (cited under her name) on the history and archeology of the Delamar (Ferguson) mining district, on the southwest edge of the Caliente caldera complex. I completed more than ten 7.5-minute quadrangles, although only three were published before the project was cancelled by the 1995 USGS Reduction in Force (RIF). The three quadrangles are Rowley and Shroba (1991), Rowley, Shroba, and others (1994), and Swadley and Rowley (1994). Seven others are being put together now, with two just released (Rowley, Snee, Shroba, and others, 2023; Rowley, Shroba, Swadley, and others, 2023). A study of a small gold district on the northwest edge of the caldera complex was described as Rowley, Snee, and others (1992). Field trips in which we participated included Best and others (1993), Scott and others (1996), and Rowley, Snee, and others (2001). A preliminary summary of the caldera complex, especially the geochemistry in which I relied on Dave Nealey, were published as Rowley, Nealey, and others (1995), Nealey and others (1995), and Unruh and others (1995). Snee and Rowley (2000) published a summary of some of nearly 100 Ar ages that Larry determined on the Caliente caldera complex, but this was only an abstract, and with the rapid demise of the project due to the 1995 RIF, the full data and seven maps are only in 2023 starting to be published (Nevada Bureau of Mines & Geology). Later mapping, when the Utah Geological Survey paid me as a consultant to map quadrangles, demonstrated that the caldera complex extended well into Utah (Rowley, V.S. Williams, and others, 2006; Rowley, Hacker, and others, 2008). East-west "lineaments," which I later renamed transverse zones, were first documented by Ekren and others (1976, 1977), Stewart and others (1977), and Rowley, Lipman, and others (1978). They were later analyzed, along with the geology of the Marysvale volcanic field, by Rowley, Cunningham, and others (1998); while this report languished "in press," Rowley (1998) did a more thorough analysis. The Caliente caldera complex has a pronounced east-west elongation, and my mapping showed that it was bounded by transverse zones on the northern and southern sides; the expanded texts in our two 2023 maps describe these zones. In an abstract (Rowley and Anderson, 1996), we called it a new type of caldera, named a syntectonic caldera. Hudson and others (1998) reported a paleomagnetic study of the part of the caldera in Nevada, although I do not agree with all conclusions. A later study of the paleomagnetism of the Caliente caldera complex, by Petronis and others (2014), is better. I call the part of the Colorado Plateau along Interstate-70 by the phrases "Paul Williams country" and "George Davis country" because of their excellent studies (Williams, 1964; Williams and Hackman, 1971; Davis, 1999; Davis and Reeher, 2024). Later updates on transverse zones were given by Rowley and Dixon (2001) and Rowley, Dixon, and others (2017).

Chapter 8. The rise and fall of the USGS

This chapter is largely from my memory, including notes taken at the time. Biographies of various directors, especially ones who were appointed after the RIF, came largely from *Wikipedia*. In the discussion of the eruption of Mount Pinatubo, the report by Wendel and Kumar (cited below) was helpful. One reason I survived the RIF, I assume, was because I received a national award, the Dibblee Medal, for geologic mapping. Dorothy Stout is cited for a short biographical sketch of me as well as of the festivities. The next year, the Dibblee Committee appointed me to a 3-man selection committee, chaired by Lehi Hintze, to recommend subsequent medals. Before that term expired (2 years), we selected Bart Ekren, then Tom Steven for the medal.

Chapter 9. Relevance in science: Gary Dixon and the Nevada Test Site

Gary Dixon, in collaboration with Doug Morton, Larry Snee, John Sutter, and others, was the other reason I was not RIFed—when these colleagues appointed me to join Gary as a manager of the Las Vegas project, the geology of the fastest growing city in the country at the time, yet with environmental problems that included decreasing supplies of culinary water. A year later, after a DOE manager insisted to our boss Larry Snee that delinquent reports be submitted, I was asked to run the large Nevada Test Site project by Geologic Division, working closely with a project of the same name by Water Resources Division. These two huge Test Site projects were funded by outside (DOE) money, whereas our Las Vegas project also had significant outside funding, mostly by Las Vegas Valley Water District and other water districts. Gary and I ran both projects together, the only two persons from Geologic Division in the Las Vegas office. Much of our outside funding to the Las Vegas project came from making geologic maps of 7.5-minute quadrangles that were potential sites for production water wells. Bart Ekren, Gary Dixon, and others mapped a masterful series of four 15-minute quadrangles in a caldera complex in central Nevada that the U.S. Atomic Energy Commission was considering for underground nuclear tests (one was detonated); two of the maps are cited with Bart as senior author, one map with Gary the senior, and the other with Richard Snyder the senior. We and Jeremy Workman produced two geologic and tectonic maps of the entire Death Valley regional groundwater flow system, cited under Jeremy's name. Ric Page senior-authored the Las Vegas quadrangle (Page, Lundstrom, and others, 2005). Ric also senior authored our geologic map of the southern part of the huge White River regional groundwater flow system, Nevada, which we would later continue northward as consultants; Ric's map is listed as Page, Dixon, and others (2005). The geologic map of the Nevada Test Site is cited under the name Janet Slate. Our work with hydrologists of the USGS Water Resources Division included collaboration with Randy Laczniak; one of his summary reports is listed under his name. I did mapping with Stan Hatfield of Southern Utah University, Dave Moore of the USGS, and others on two quadrangles (Navajo Lake and Brian Head, respectively, by Moore and others, 2004, and Rowley, Biek, Sable, and others, 2013) that contained Cedar Breaks National Monument. I also helped supervise the dissertation work and continued collaboration with Dave Hacker in the adjacent Pine Valley Mountains. Several publications are listed under Hacker. I retired from the USGS in August 2000, then taught for the academic year at Southern Utah University.

Chapter 10. Geologic consulting and fracture flow

I founded Geologic Mapping Inc. in January 2001. The stories here are based on memories, field notes, invoices, and publications that resulted (most publications were done on our own time, if the topics were worth publishing). Among the consulting topics that we published are ones cited as Dixon and Katzer (2002), Johnson and others

(2002), Rowley (2002), Rowley, Himmelreich, and others (2003), Morgan and others (2003), Rowley and Dixon (2004, 2010), Rowley, Dixon, and others (2004, 2009, 2016, 2017), Rowley, Dixon, and Brickey (2005), Rowley, Christensen, and others (2012), Rowley, Rutledge, and others, (2013); Rowley, Norlin, and others (2014), and Rowley, Dixon, and Layton (2018). Information on Directors of the Utah Geological Survey comes in part from Wikipedia. An obituary in *Eos* for Lee Allison is cited as Richard and Hills, the authors. Jobs trickle in as of this writing, which I accept if they sound interesting or if the request comes from former clients.

Chapter 11. Geologic mapping and the largest gravity slide on land

The importance of geologic maps has been described in many reports, one of which is cited below as Bernknopf and others (1993) and another by Rowley (2008). A letter to the editor by Rowley, Dixon, and Steven (1997) stressed the value to the taxpayer of mapping, in response to an upcoming congressional decision on whether to renew the National Cooperative Geologic Mapping Act. Reports by Earl Cook, Dick Blank, Bob Biek, Dave Hacker, and others are cited under their name. The grad student of George Davis who worked on the Miocene thrusts at the north entrance to Bryce Canyon National Park is Eric Lundin, whose report is cited under his name. Our first reports to point out the vast size of the first landslide (Markagunt gravity slide) that we described are Hacker and others (2014) and Biek and others (2014). The Panguitch 30-minute by 60-minute map (1:62,500 scale) had several preliminary versions, including Biek and others (2012); the final version was Biek and others (2015) but by then we did not report on the second great slide, the Sevier gravity slide, although we found it when this map was in press. The announcement and summary of the Thompson Field Forum are those of Biek and others (2017a, b). A large report that puts our two slides in the context of those at Iron Spings is given as Hacker and others (2017). Our big report on the two slides and the Thompson Field Forum is that of Biek and others (2019). Later work on the Sevier slide included Biek, Eaton, and others (2023) and Braunagel and others (2023). The first announcement of our recognition of the Black Mountains gravity slide, making three side-by-side gravity slides called the Marysvale gravity slide complex, was in Biek and others (2020) and in a summary of the three slides for the layman by Biek, Rowley, and Hacker (2023). The Beaver 30-minute by 60-minute map was broken into four quarters that were compiled at 1:50,000 scale, with the northwest and southwest quarters published by the Utah Geological Survey and the northeast and southeast finished and submitted, after which the overall Beaver map was submitted at 1:62,500 scale in May 2024; the five are given as Rowley, Biek, Hacker, Vice, and others in the References.

Chapter 12. Climate change

References to books by Rachel Carson, Jared Diamond, Paul Ehrlich, Donella Meadows and others, Oreskes and Conway, Elizabeth Kolbert, E.O. Wilson, Christiana Figueres and Tom Rivett-Carnac, David Lipsky, Richard Alley, and Al Gore are listed under their names. The article by Garrett Harden is under his name. Much data on climate-caused weather catastrophes came from *Wikipedia* (online). Articles by Corn (2019), Ehrenreich (2021), Mazzucato (2021), Kornei (2023), and Packer (2024) are given under their names. Noam Chomsky is quoted in several articles.

Chapter 13. Lies and insanity: Trump and the consequences

Personal memories form much of this chapter. Books by Mary Trump, Hetherington and Weiler, and John Bolton are listed under their names. A published article by Patricia Roberts-Miller is cited as Roberts-Miller (2019). An online article by Jeffrey Goldberg is cited as Goldberg (2020), and a 2016 online article by George Lakoff is also given. Other information came from online articles by Anne Applebaum (2020a, b) and Aronson and Tavris (2020), and print articles by Geoghegan (2020), Packer

(2020), Burmila (2019), Nwanevu (2020), and Shenk (2022). Noam Chomsky is quoted several times. Other articles that I collected over several years, plus verifiable facts from the internet, supply much of the other data given in the chapter.

Chapter 14. COVID-19 and Black Lives Matter

Living through these two terrible long-term events provided many horrible reminiscences as well as my written notes about them. The book *Rage* by Woodward (2020) gave the proof that Trump knew of the seriousness of the virus but downplayed it to the American public, presumably to not panic the stock market. Articles from magazines and newspapers that I cite are by Corn (2020), Dickerson (2020), Garrett (2020), Gibson and Thompson (2020), Holloway (2020), Krugman (2020), McCormack (2020), Mother Jones (2020), Yong (2020), and Fauci (2024). Several of these, including a 2021 Mother Jones Special Issue, summarized the timeline of events. Some online facts from the internet make up other data given in this chapter. A science article by Alexiane Pradelle and colleagues provides evidence that at least 17,000 persons died from mistakenly taking the drug hydroxychloroquine. Peter Hotez, in a book cited, estimated that 200,000 persons died needlessly because they refused COVID-19 vaccinations.

Chapter 15. The future of science

This provides my own predictions on where science is headed, based on information given in previous chapters. The wonderful book by Pennock (2019) is discussed, and I supply quotes from books by Orwell (1949), Sagan (1996), and Oreskes (2020), as well as data from several articles.

Conclusions

I attempt to summarize how science operates and as a method of thinking, and its benefits to society because making the case for science has become a lost art. Books by Carl Sagan (1996, 1980) are quoted, and an article by Rees (2020) supplied some advice for persons considering a science career. An article by Ehrenreich is given under his name. Noam Chomsky supplies some optimism about our future.

REFERENCES

Agrell, S.O., Chinner, G.A., and Rowley, P.D., 1999, The black skarns of Pine Canyon, Piute County, Utah: Geological Magazine [United Kingdom], v. 136, no. 4, p. 343-359.

Alley, R.B., 2014, The two-mile time machine: Princeton, New Jersey, Princeton University Press, 229 p.

Alvarez, Walter, 1997, T. Rex and the crater of doom: Princeton, New Jersey, Princeton University Press, 185 p.

Alvarez, Walter, 2009, The mountains of Saint Francis: New York, W.W. Norton & Company, 304 p.

Anderson, J.J., 1965, Bedrock geology of Antarctica—a summary of exploration, 1831-1962: American Geophysical Union Antarctic Research Series, v. 6, p. 1-70.

Anderson, J.J., 1965, Geology of northern Markaunt Plateau, Utah: Austin, University of Texas unpublished Ph.D. dissertation, 193 p.

Anderson, J.J., 1993, The Markagunt Megabreccia—large Miocene gravity slides mantling the northern Markagunt Plateau, southwestern Utah: Utah Geological Survey Miscellaneous Publication 93-2, 37 p.

Anderson, J.J., Bastien, T.W., Schmidt, P.G., Spettstoesser, J.F., and Craddock, C., 1962, Antarctica—geology of the Ellsworth Mountains: Science, v. 138, p. 824-825.

Anderson, J.J., and Rowley, P.D., 1975, Cenozoic stratigraphy of southwestern High Plateaus of Utah, *in* Anderson, J.J., Rowley, P.D., Feck, R.J., and Nairn, A.E.M., editors, Cenozoic geology of southwestern High Plateaus of Utah: Geological Society of America Special Paper 160, p. 1-51.

Anderson, J.J., Rowley, P.D., Blackman, J.T., Mehnert, H.H., and Grant, T.C., 1990a, Geologic map of the Circleville Canyon area, southern Tushar Mountains and northern Markagunt Plateau, Beaver, Garfield, Iron, and Piute Counties, Utah: U.S. Geological Survey Miscellaneous Investigations Series Map I-2000, scale 1:50,000.

Anderson, J.J., Rowley, P.D., Fisher, R.V., Dover, James, Schultz, Peter, Williams, P.L., Threet, R.L., and Osmond, John, 2001, J. Hoover Mackin—a personal tribute: *in* Erskine, M.C., Faulds, J.E., Bartley, J.M., and Rowley, P.D., editors, The geologic transition, High Plateaus to Great Basin—a symposium and field guide (The Mackin Volume): Utah Geological Association Publication no. 30 and Pacific Section of the American Association of Petroleum Geologists Guidebook GB 78, p. xii-xx.

Anderson, J.J., Rowley, P.D., Fleck, R.J., and Nairn, A.E.M., 1975, Cenozoic geology of southwestern High Plateaus of Utah: Geological Society of America Special Paper 160, 88 p.

Anderson, J.J., Rowley, P.D., Machette, M.N., Decatur, S.H., and Mehnert, H.H., 1990b, Geologic map of the Nevershine Hollow area, eastern Black Mountains, southern Tushar Mountains, and northern Markagunt Plateau, Beaver and Iron Counties, Utah: U.S. Geological Survey Miscellaneous Investigations Series Map I-1999, scale 1:50,000.

Andrews, Sarah, 2002, Fault line: New York, St. Martin's Minotaur, 307 p.

Applebaum, Anne, 2020a, Laura Ingraham's descent into despair: The Atlantic, online,

July 16, 2020.

Applebaum, Anne, 2020b, The facts just aren't getting through: The Atlantic, online, August 9, 2020.

Aronson, Elliot, and Tavris, Carol, 2020, The role of cognitive dissonance in the pandemic: The Atlantic, online, July 12, 2020.

Ball, G.W., Dixon, G.L., Johnson, Kenneth, Katzer, T.L., Rowley, P.D., Rumbaugh, James, Smith, D.L., Childress, Jack, and Squires, Robert, 2010, Hydrogeologic evaluation and ground-water modeling of the Smoke Creek Desert hydrologic basin, Washoe County, Nevada, and Lassen County, California: Unpublished report (CD) prepared for LSC Project, under direction of Watersource Consulting Engineers, Inc., Reno, Nevada; includes Plate 1, Geologic map of the Smoke Creek Desert, Nevada and California, by Dixon, G.L., Rowley, P.D., Ekren, E.B., Faulds, J.E., Ramelli, A.R., Vice, G.S., and Brickey, D.W., scale 1:100,000.

Banks, N.G., and Hoblitt, R.P., 1981, Summary of temperature studies of 1980 deposits, *in* Lipman, P.W., and Mullineaux, D.R., editors, The 1980 eruptions of Mount St. Helens, Washington: U.S. Geological Survey Professional Paper 1250, p. 295-314.

Barker, D.S., 1995, Crystallization and alteration of quartz monzonite, Iron Springs mining district, Utah—relation to associated iron deposits: Economic Geology, v. 90, p. 2197-2217.

Behrendt, J.C., 1964, Antarctic Peninsula traverse geophysical results relating to glaciological and geological studies: Research Report Series Geophysical Polar Research Center, Madison, University of Wisconsin, No. 64-1, 112 p.

Behrendt, J.C., 1983, Are there petroleum resources in Antarctica? *in* Behrendt, J.C., editor, Petroleum and mineral resources of Antarctica: U.S. Geological Survey Circular 909, p. 3-24.

Behrendt, J.C., 1990, Brief summary of geophysical and geological research in Antarctica related to the assessment of petroleum resources and potential environmental hazards to their development, *in* Splettstoesser. J.F., and Dreschhoff, G.A.M., editors, Mineral resources potential of Antarctica: Antarctic Research Series, vol. 51, Washington, D.C., American Geophysical Union, p. 163-174.

Behrendt, J.C., 1991, Scientific studies relevant to the question of Antarctica's petroleum resource potential, chap. 17, *in* Tingey, R.J., editor, The Geology of Antarctica, Oxford monographs on geology and geophysics no. 17: Oxford, Clarendon Press, p. 588-616.

Behrendt, J.C., 1998, Innocents on the ice—a memoir of Antarctic Exploration, 1957: Niwot, Colorado, University Press of Colorado, 428 p.

Behrendt, J.C., 2005, The ninth circle—a memoir of life and death in Antarctica, 1960-1962: University of New Mexico Press, 240 p.

Behrendt, J.C., 2018, Incident in the Crary Mountains, 22 November 1960: The Antarctican Society, v. 17-18 (July), no. 4, p. 4-7.

Bernknopf, R.L., Brookshire, D.S., Soller, D.R., McKee, M.J., Sutter, J.F., Matti, J.C., and Campbell, R.H., 1993, Societal value of geologic maps: U.S. Geological Survey Circular 1111, 53 p.

Best, M.G., Scott, R.B., Rowley, P.D., Swadley, WC, Anderson, R.E., Gromme, C.S., Harding, A.E., Deino, A.L., Christiansen, E.H., Tingey, D.G., and Sullivan, K.R., 1993, Oligocene-Miocene caldera complexes, ash-flow sheets, and tectonism in the central and southeastern Great Basin, *in* Lahren, M.M., Trexler, J.H., Jr., and Spinosa, Claude, editors, Crustal evolution of the Great Basin and Sierra Nevada:

Field Trip Guide, Geological Society of America, Cordilleran and Rocky Mountain Sections Meeting, p. 285-311.

Bickel, Lennard, 1977, Mawson's will: New York, Stein and Day Publishers, 237 p.

Biek, R.F., Anderson, J.J., Sable, E.G., and Rowley, P.D., 2014a, Geologic map of the Panguitch Lake quadrangle, Garfield and Iron Counties, Utah: Utah Geological Survey Map 268DM, CD, scale 1:24,000.

Biek, R.F., Anderson, J.J., Sable, E.G., and Rowley, P.D., 2014b, Geologic map of the Haycock Mountain quadrangle, Garfield County, Utah: Utah Geological Survey Map 269DM, CD, scale 1:24,000.

Biek, R.F., Eaton, J.G., Rowley, P.D., and Mattox, S.R., 2015, Interim geologic map of the western Loa 30' x 60' quadrangle, Garfield, Piute, and Wayne Counties, Utah (year 2): Utah Geological Survey Open-File Report 648, scale 1:62,500.

Biek, R.F., Eaton, J.G., Rowley, P.D., Hacker, D.B., Mattox, S.R., Bailey, Christopher, and Marchetti, D.W., 2023, Geologic map of the west half of the Loa 30' x 60' quadrangle, Garfield, Piute, and Wayne Counties, Utah: Utah Geological Survey Map 292DM, scale 1:62,500.

Biek, R.F., Hacker, D.B., and Rowley, P.D., 2014, New constraints on the extent, age, and emplacement history of the early Miocene Markagunt Megabreccia, southwest Utah—one of the World's largest subaerial gravity slides, *in* MacLean, J.S., Biek, R.F., and Huntoon, J.E., editors, Geology of Utah's far south: Utah Geological Association Publication 43, CD, p. 565-598.

Biek, R.F., Hacker, D.B., and Rowley, P.D., 2017a, Catastrophic mega-scale landslide failure of large volcanic fields (Thompson Field Forums announcement): GSA Today, v. 26, no. 11, p. 22-24.

Biek, R.F., Hacker, D.B., and Rowley, P.D., 2017b, Catastrophic mega-scale landslide failure of large volcanic fields (Thompson Field Forums report): GSA Today, v. 27, no. 12, p. 30-31.

Biek, R.F., Rowley, P.D., Anderson, J.J., Maldonado, Florian, Moore, D.W., Eaton, J.G., Hereford, Richard, and Matyjasik, Basia, 2012, Interim geologic map of the Panguitch 30' x 60' quadrangle, Garfield, Iron, and Kane Counties, Utah: Utah Geological Survey Open-File Report 599, CD, 127 p., scale 1:62,500.

Biek, R.F., Rowley, P.D., Anderson, J.J., Maldonado, Florian, Moore, D.W., Hacker, D.B., Eaton, J.G., Hereford, Richard, Sable, E.G., Filkorn, H.F., and Matyjasik, Basia, 2015, Geologic map of the Panguitch 30' x 60' quadrangle, Garfield, Iron, and Kane Counties, Utah: Utah Geological Survey Map 270DM, CD, 162 p., scale 1:62,500 (received the 2016 Charles Mankin award for year's best report, by American Association of State Geologists).

Biek, R.F., Rowley, P.D., and Hacker, D.B., 2019, The gigantic Markagunt and Sevier gravity slides resulting from mid-Cenozoic catastrophic mega-scale failure of the Marysvale volcanic field, Utah, USA: Geological Society of America Field Guide 56, 121 p.

Biek, R.F., Rowley, P.D., and Hacker, D.B., 2020, Utah's ancient mega-landslides: Utah Geological Survey, Survey Notes, v. 52, no. 2, p. 1-3.

Biek, R.F., Rowley, P.D., and Hacker, D.B., 2022, Utah's ancient mega-landslides: Utah Geological Survey Circular 132, 67 p.

Biek, R.F., Rowley, P.D., Hacker, D.B., Hayden, J.M., Willis, G.C., Hintze, L.F., Anderson, R.E., and Brown, K.D., 2009, Geologic map of the St. George and east part of the Clover Mountains 30' x 60' quadrangles, Washington and Iron Counties, Utah:

Utah Geological Survey Map 242, + 101 p. pamphlet, scale 1:100,000.

Boyer, S.J., 1975, Chemical weathering of rocks on the Lassiter Coast, Antarctic Peninsula, Antarctica: New Zealand Journal of Geology and Geophysics, v. 18, no. 4, p. 623-628.

Boyer, S.J., 1976, 75 days on "the ice:" University of Colorado Medical Center Quarterly, v. 18, no. 2, p. 1-6.

Boyer, S.J., Rowley, P.D., and Kellogg, K.S., 1976, Camping and trail methods used on the Lassiter Coast, Antarctica: Summit, v. 22, no. 4, p. 8-11, 38-39.

Boyles, J.M., Schmutzler, R.A., and Rowley, P.D., 1979, Snowmobiles in Antarctica: Arctic, v. 32. no. 3, p. 189-200.

Byrd, R.E., 1930, Little America—aerial exploration in the Antarctic, the flight to the South Pole: New York, G.P. Putnam's Sons, 422 p. (includes chapter 17, The geological sledge trip, by L.M. Gould)

Blank, H.R., Jr., 1959, Geology of the Bull Valley district, Washington County, Utah: Seattle, University of Washington unpublished Ph.D. dissertation, 177 p.

Blank, H.R., Rowley, P.D., and Hacker, D.B., 1992, Miocene monzonite intrusions and associated megabreccias of the Iron Axis region, southwestern Utah, *in* Wilson, J.R., editor, Field guide to geologic excursions in Utah and adjacent areas of Nevada, Idaho, and Wyoming; Geological Society of America, Rocky Mountain section meeting: Utah Geological Survey Miscellaneous Publication 92-3, p. 399-420.

Bolton, John, 2020, The room where it happened—a White House memoir: New York, Simon & Schuster, 588 p.

Braunagel, M.J., Griffith, W.A., Biek, R.F., Hacker, D.B., Rowley, P.D., Malone, D.H., Mayback, Danika, Rivera, T.A., Loffer, Zachary, and Smith, Z.D., 2023, Structural relationships across the Sevier gravity slide of southwest Utah and implications for catastrophic translation and emplacement processes of long runout landslides: Geochemistry, geophysics, geosystems, v. 24, e2022GCo10783, https://doi.org/10.1029/2022GCo10783.

Budding, K.E., Cunningham, C.G., Zielinski, R.A., Steven, T.A., and Stern, C.R., 1987, Petrology and chemistry of the Joe Lott Tuff Member of the Mount Belknap Volcanics, Marysvale volcanic field, west-central Utah: U.S. Geological Survey Professional Paper 1354, 47 p.

Burmila, Ed, 2019, State under siege: The New Republic, November 2019 issue, p. 30-35.

Callaghan, Eugene, 1939, Volcanic sequence in the Marysvale region in southwest-central Utah: EOS (American Geophysical Union Transactions), 20th Annual Meeting, Washington, D.C., 1939, pt. 3, p. 438-452.

Callaghan, Eugene, 1973, Mineral resource potential of Piute County, Utah, and adjoining area: Utah Geological and Mineralogical Survey Bulletin 102, 135 p.

Campbell, D.L., Steven, T.A., Cunningham, C.G., and Rowley, P.D., 1999, Aeromagnetic and gravity maps of the central Marysvale volcanic field, southwestern Utah: U.S. Geological Survey Geologic Investigation Series Map I-2645-B, 50 p., scale 1:100,000.

Carrara, P.E., 1980, Surficial geologic map of the Vernal 1° x 2° quadrangle, Colorado and Utah: U.S. Geological Survey Miscellaneous Investigations Series Map I-1204, scale 1:250,000.

Carrara, P.E., 1981, Evidence for a former large ice sheet in the Orville Coast-Ronne Ice Shelf area, Antarctica: Journal of Glaciology, v. 27, no. 97, p. 487-491.

Carson, Rachel, 1962, Silent spring: Boston, Houghton Mifflin Company, 368 p.

Cohen, S.S., and DeLong, J.B., 2016, Concrete economics—the Hamilton approach to economic growth and policy: Harvard Business Review Press, 223 p.

Cook, Earl, 1976, Man, energy, society: San Francisco, W.H. Freeman and Company, 478 p.

Cook, E.F., 1954, Geology of the Pine Valley Mountains, Utah: Seattle, University of Washington unpublished Ph.D. dissertation, 236 p.

Cook, E.F., 1957, Geology of the Pine Valley Mountains, Utah: Utah Geological and Mineralogical Survey Bulletin 58, 111 p.

Cook, E.F., 1960, Geologic atlas of Utah, Washington County: Utah Geological and Mineralogical Survey Bulletin 70, 119 p.

Cook, E.F., 1965, Stratigraphy of Tertiary volcanic rocks in eastern Nevada: Nevada Bureau of Mines Report 11, 61 p.

Corn, David, 2019, Weight of the world—when you can see disaster unfolding and nobody listens: Mother Jones, July/August issue, p. 18-25.

Corn, David, 2020, Covid pro quo—Donald Trump's killer corruption: Mother Jones, July/August issue, p. 6-10.

Craddock, C., Bastien, T.W., Rutford, R.H., and Anderson, J.J., 1965, *Glossopteris* discovered in West Antarctica: Science, v. 148, p. 634-637.

Crame, Alistair, and Francis, Jane, 2021, Draft obituary, M.R.A. Thomson, 1942-2020: 314/2021.

Crandell, D.R., and Mullineaux, D.R., 1978, Potential hazards from future eruptions of Mount St. Helens volcano, Washington: U.S. Geological Survey Bulletin 1383-C, 26 p.

Crandell, D.R., Mullineaux, D.R., and Rubin, Meyer, 1975, Mount St. Helens volcano—recent and future behavior: Science, v. 187, no. 4175, p. 438-441.

Cunningham, C.G., Austin, G.W., Naeser, C.W., Rye, R.O., Ballantyne, G.H., Stamm, R.G., and Barker, C.E., 2004, Formation of a paleothermal anomaly and disseminated gold deposits associated with the Bingham Canyon porphyry Cu-Au-Mo system, Utah: Economic Geology, v. 99, p. 789-806.

Cunningham, C.G., Naeser, C.W., Marvin, R.F., Luedke, R.G., and Wallace, A.R., 1994, Ages of selected intrusive rocks and associated ore deposits in the Colorado Mineral Belt: U.S. Geological Survey Bulletin 2109, 31 p.

Cunningham, C.G., Rasmussen, J.D., Steven, T.A., Rye, R.O., Rowley, P.D., Romberger, S.B., and Selverstone, Jane, 1998, Hydrothermal uranium deposits containing molybdenum and fluorite in the Marysvale volcanic field, west-central Utah: Mineralium Deposita, v. 33, p. 477-494.

Cunningham, C.G., Rowley, P.D., Steven, T.A., and Rye, R.O., 2007, Geologic evolution and mineral resources of the Marysvale volcanic field, west-central Utah, in Willis, G.C., Hylland, M.D., Clark, D.L., and Chidsey, T.C., Jr., editors, Central Utah—diverse geology of a dynamic landscape: Utah Geological Association Publication 36, p. 144-161.

Cunningham, C.G., Rye, R.O., Rockwell, B.W., Kunk, M.J., and Councell, T.B., 2005, Origin and natural destruction of a hydrothermal replacement alunite deposit at Big Rock Candy Mountain, Utah—mineralogy, spectroscopic remote sensing, stable isotope, and argon age: Chemical Geology, v. 215, p. 317-337.

Cunningham, C.G., and Steven, T.A., 1979a, Mount Belknap and Red Hills calderas and associated rocks, Marysvale volcanic field, west-central Utah: U.S. Geological Survey Bulletin 1468, 34 p.

Cunningham, C.G., and Steven, T.A., 1979b, Geologic map of the Deer Trail Mountain – Alunite Ridge mining area, west-central Utah: U.S. Geological Survey Miscellaneous Investigations Series Map I-1177, scale 1:24,000.

Cunningham, C.G., Steven, T.A., Rowley, P.D., Glassgold, L.B., and Anderson, J.J., 1983, Geologic map of the Tushar Mountains and adjoining areas, Marysvale volcanic field, Utah: U.S. Geological Survey Miscellaneous Investigations Series Map I 1430-A, scale 1:50,000.

Cunningham, C.G., Unruh, D.M., Steven, T.A., Rowley, P.D., Naeser, C.W., Mehnert, H.H., Hedge, C.E., and Ludwig, K.R., 1998, Geochemistry of volcanic rocks in the Marysvale volcanic field, west-central Utah, *in* Friedman, J.D., and Huffman, A.C., Jr., editors, Laccolith complexes of southeastern Utah—time of emplacement and tectonic setting—workshop proceedings: U.S. Geological Survey Bulletin 2158, p. 223-232.

Davis, G.H., 1984, Structural geology of rocks and regions: New York, John Wiley & Sons, 492 p.

Davis, G.H., 1999, Structural geology of the Colorado Plateau region of southern Utah, with special emphasis on deformation bands: Geological Society of America Special Paper 342, 157 p.

Davis, G.H. and 19 coauthors, 2023, Structural analysis and chronological constraints on progressive deformation within the Rincon Mountains, Arizona—implications for development of metamorphic core complexes: Geological Society of America Memoir 222, 125 p.

Davis, G.H., and Reeher, L.J., 2025, Transverse faulting in the Paradox Basin, Colorado Plateau, U.S.A.—active intermittent faulting over a 300 million year history of strain accommodation and fluid flow: Geological Society of America Bulletin v.137, p. 1069-1064, https://doi.org/10.1130/B37485.1.

Diamond, Jared, 2005, Collapse—how societies choose to fail or succeed: New York, Viking, 576 p.

Dickerson, John, 2020, A presidential guide to crisis management—what Trump should have learned from his predecessors: The Atlantic, July/August issue, p. 9-12.

Dixon, G.L., Hedlund, D.C., and Ekren, E.B., 1972, Geologic map of the Pritchards Station quadrangle, Nye County, Nevada: U.S. Geological Survey Miscellaneous Geologic Investigations Map I-728, scale 1:48,000.

Dixon, G.L., and Katzer, Terry, 2002, Geology and hydrology of the lower Virgin River Valley in Nevada, Arizona, and Utah: Mesquite, Nevada, Virgin Valley Water District Report VVWD-01, 125 p., scale 1:100,000, with CD.

Dixon, Gary, and Rowley, Peter, 2007, Hydrogeology of Spring Valley, *in* Coache, Robert, editor, Regional tour of the carbonate system guidebook: Reno, Nevada Water Resources Association, p. 35-36. (This is one of many of ours in this guidebook volume.)

Dixon, G.L., Rowley, P.D., Burns, A.G., Watrus, J.M., and Ekren, E.B., 2007, Geology of

White Pine and Lincoln Counties and adjacent areas, Nevada and Utah—the geologic framework of regional groundwater flow systems: Las Vegas, Southern Nevada Water Authority, Doc. No. HAM-ED-0001, variously paginated + CD-ROM, plus 9 plates of 1:250,000-scale digital geologic maps, hydrogeologic maps, geologic cross sections, hydrogeologic cross sections, and explanation of map units; 25,000 mi^2.

Ehrenreich, Ben, 2021, Suicide prevention—why we must do everything differently to ensure the planet's survival: The New Republic, April 2021, p. 14-21.

Ehrlich, P.R., 1968, The population bomb: New York, Ballantine Books Inc., 223 p.

Ekren, E.B., Bucknam, R.C., Carr, W.J., Dixon, G.L., and Quinlivan, W.D., 1976, East-trending structural lineaments in central Nevada: U.S. Geological Survey Professional Paper 986, 16 p.

Ekren, E.B., Hinrichs, E.N., and Dixon, G.L., 1972, Geologic map of The Wall quadrangle, Nye County, Nevada: U.S. Geological Survey Miscellaneous Geologic Investigations Map I-719, scale 1:48,000.

Ekren, E.B., McIntyre, D.H., and Bennett, E.H., 1984, High-temperature, large-volume lava-like ash-flow tuffs without calderas in southwest Idaho: U.S. Geological Survey Professional Paper 1272, 76 p.

Ekren, E.B., Orkild, P.P., Sargent, K.A., and Dixon, G.L., 1977, Geologic map of Tertiary rocks, Lincoln County, Nevada: U.S. Geological Survey Miscellaneous Investigations Series Map I-1041, scale 1:250,000.

Ekren, E.B., Rogers, C.L., and Dixon, G.L., 1973, Geologic map of the Reveille Range quadrangle, Nye County, Nevada: U.S. Geological Survey Miscellaneous Geologic Investigations Map I-806, scale 1:48,000.

Ekren, E.B., Rowley, P.D., Dixon, G.L., Page, W.R., Kleinhampl, F.J., Ziony, J.I., Brandt, J.M., and Patrick, B.G., 2012, Geology of the Quinn Canyon Range and vicinity, Nye and Lincoln Counties, Nevada: Las Vegas, Nevada, Southern Nevada Water Authority Doc. No. HAM-ED-0004, 61 p., scale 1:100,000.

Farrar, Edward, McBride, S.L., and Rowley, P.D., 1982, Ages and tectonic implications of Andean plutonism in the southern Antarctic Peninsula, *in* Craddock, Campbell, editor, Antarctic geoscience: Madison, University of Wisconsin Press, p. 349-356.

Fauci, Anthony, 2024, The first three months—what I saw inside the government's response to COVID-19: The Atlantic, v. 334, no. 1, p. 22-28.

Ferris, D.E., 1991, The best of times, the worst of times—a biocultural analysis of the Ferguson district, 1892-1909: Las Vegas, University of Nevada, Las Vegas Master's thesis, 296 p.

Figueres, Christiana, and Rivett-Carnac, Tom, 2020, The future we choose—surviving the climate crisis: New York, Alfred A. Knopf, 210 p.

Fitzsimons, Peter, 2011, Mawson and the ice men of the heroic age—Scott, Shackleton and Amundsen: Australia, William Heinemann, 737 p.

Fleck, R.J., Anderson, J.J., and Rowley, P.D., 1975, Chronology of mid-Tertiary volcanism in High Plateaus region of Utah, *in* Anderson, J.J., and others, Cenozoic geology of southwestern High Plateaus of Utah: Geological Society of America Special Paper 160, p. 53-62.

Ford, A.B., 1976, Stratigraphy of the layered gabbroic Dufek intrusion, Antarctica: U.S. Geological Survey Bulletin 1405-D, 36 p.

Ford, A.B., 1983, The Dufek intrusion of Antarctica and a survey of its minor metals and possible resources, *in* Behrendt, J.C., ed., Petroleum and mineral resources of Antarctica: U.S. Geological Survey Circular 909, p. 51-75.

Ford, A.B., Schmidt, D.L., and Boyd, W.W., Jr., 1978, Geologic map of the Davis Valley quadrangle and part of the Cordiner Peaks quadrangle, Pensacola Mountains, Antarctica: U.S. Geological Survey U.S. Antarctic Research Program Map A-10, scale 1:250,000.

Ford, A.B., Schmidt, D.L., Boyd, W.W., Jr., and Nelson, W.H., 1978, Geologic map of the Saratoga Table quadrangle, Pensacola Mountains, Antarctica: U.S. Geological Survey U.S. Antarctic Research Program Map A-9, scale 1:250,000.

Foxworthy, B.L., and Hill, Mary, 1982, Volcanic eruptions of 1980 at Mount St. Helens—the first 100 days: U.S. Geological Survey Professional Paper 1249, 124 p.

Garrett, Laurie, 2020, Grim reapers—how Trump and Xi set the stage for the coronavirus pandemic: The New Republic, May 2020 issue, p. 12-23.

Geoghegan, Thomas, 2020, Educated fools—why democratic leaders still misunderstand the politics of social class: The New Republic, January-February 2000 issue.

Gibson, Dave, and Thompson, Laura, 2020, Failed state—how the president's denial of reality made a dangerous virus even more deadly: Mother Jones, July/August issue, p. 12-25.

Goldberg, Jeffrey, 2020, Trump—Americans who died in war are "losers" and "suckers": The Atlantic, online, September 3, 2020.

Gore, Al, 2009, Our choice—a plan to solve the climate crisis: New York, Melcher Media, 416 p.

Gould, L.M., 1931 (republished in 1984 by Carleton), Cold—the record of an Antarctic sledge journey: New York, Brewer Warren & Putnam (Northfield, Minnesota, Carleton College), 213 p.

Gould, L.M., 1971, Memorial to Duncan Stewart (1905-1969): Geological Society of America, 5 p.

Guthridge, Guy, 2018, A rock from Antarctica: The Antarctican Society, v. 17-18 (July), no. 4, p. 4.

Hacker, D.B., 1998, Catastrophic gravity sliding and volcanism associated with the growth of laccoliths—examples from early Miocene hypabyssal intrusions of the Iron Axis magmatic province, Pine Valley Mountains, southwest Utah: Kent, Ohio, Kent State University unpublished Ph.D. dissertation, 258 p.

Hacker, D.B., Biek, R.F., and Rowley, P.D., 2014, Catastrophic emplacement of the gigantic Markagunt gravity slide, southwest Utah (USA)—implications for hazards associated with sector collapse of volcanic fields: Geology, v. 42, no. 11, p. 943-946.

Hacker, D.B., Holm, D.K., Rowley, P.D., and Blank, H.R., 2002, Associated Miocene laccoliths, gravity slides, and volcanic rocks, Pine Valley Mountains and Iron Axis region, southwestern Utah, *in* Lund, W.R., editor, Field guide to geologic excursions in southwestern Utah and adjacent areas of Arizona and Nevada, prepared for the Geological Society of America, Rocky Mountain Section Meeting in Cedar City, Utah, May 7-9, 2002: U.S. Geological Survey Open-File Report 02-172, p. 236-283.

Hacker, D.B., Rowley, P.D., and Biek, R.F., 2017, Catastrophic collapse features in volcanic terrains—styles and links to subvolcanic magma systems, *in* Breitkreuz, C., and Rocchi, S., editors, Physical geology of shallow magmatic systems: Advances in Volcanology Series, Springer International Publishing, p. 215-248.

Hansen, W.R., and others, 1966, The Alaska earthquake, March 27, 1964—field investigations and reconstruction effort: U.S. Geological Survey Professional Paper 541, 111 p.

Hansen, W.R., Carrara, P.E., and Rowley, P.D., 1980, Geologic map of the Haystack Rock quadrangle, Moffat County, Colorado: U.S. Geological Survey Geologic Quadrangle Map GQ-1535, scale 1:24,000.

Hansen, W.R., Carrara, P.E., and Rowley, P.D., 1981, Geologic map of the Crouse Reservoir quadrangle, Uintah and Daggett Counties, Utah: U.S. Geological Survey Geologic Quadrangle Map GQ-1554, scale 1:24,000.

Hansen, W.R., Carrara, P.E., and Rowley, P.D., 1982, Geologic map of the Canyon of Lodore North quadrangle, Moffat County, Colorado: U.S. Geological Survey Geologic Quadrangle Map GQ-1568, scale 1:24,000.

Hansen, W.R., and Rowley, P.D., 1980, Geologic map of the Stuntz Reservoir quadrangle, Utah-Colorado: U.S. Geological Survey Geologic Quadrangle Map GQ-1530, scale 1:24,000.

Hansen, W.R., and Rowley, P.D., 1980, Geologic map of the Hells Canyon quadrangle, Moffat County Colorado: U.S. Geological Survey Geologic Quadrangle Map GQ-1536, scale 1:24,000.

Hansen, W.R., Rowley, P.D., and Carrara, P.E., 1983, Geologic map of Dinosaur National Monument and vicinity, Utah and Colorado: U.S. Geological Survey Miscellaneous Investigations Series Map I-1407, scale 1:50,000.

Hardin, Garrett, 1968, The tragedy of the commons: Science, v. 162, p. 1243-1248.

Hatfield, S.C., Rowley, P.D., Sable, E.G., Maxwell, D.J., Cox, B.V., McKell, M.D., and Kiel, D.E., 2000, Geology of Cedar Breaks National Monument, Utah, *in* Sprinkel, D.A., Chidsey, T.C., Jr., and Anderson, P.B., editors, Geology of Utah's parks and monuments: Utah Geological Association Publication 28, p. 139-154.

Hatfield, S.C., Rowley, P.D., Sable, E.G., Maxwell, D.J., Cox, B.V., McKell, M.D., and Kiel, D.E., 2024, Geology of Cedar Breaks National Monument, Utah, *in* Sprinkel, D.A., Chidsey, T.C., Jr., Anderson, P.B., and Willis, G.C., editors, Geology of Utah's parks and monuments, 4th edition: Utah Geological Association Publication 28, p. 201-216.

Hatfield, S.C., Rowley, P.D., Sable, E.G., Maxwell, D.J., Cox, B.V., McKell, M.D., and Kiel, D.E., 2000, Geologic road log of Cedar Breaks National Monument, Utah, *in* Anderson, P.B., and Sprinkel, D.A., editors, Geologic road, trail, and lake guides to Utah's parks and monuments: Utah Geological Association Publication 29, CD-ROM.

Henrickson, E.L., 1971, Memorial of Duncan Stewart VII (1905-1969): American Mineralogist, v. 56, p. 720-728.

Herring, D.F., 1981, Kimberly, Utah—from lode to dust: Eugene, Oregon, Herring Publishers, 106 p.

Hetherington, Marc, and Weiler, Jonathan, 2008, Prius or pickup?—how the answers to four simple questions explain America's great divide: Boston, Houghton Mifflin Harcourt, 259 p.

Hillemann, Eric, 2012, A beacon so bright—the life of Laurence McKinley Gould: Northfield, Minnesota, Carleton College, 658 p.

Hintze, L.F., 1980, Geologic map of Utah: Utah Geological and Mineralogical Survey, scale 1:500,000.

Hintze, L.F., Davis, F.D., Rowley, P.D., Cunningham, C.G., Steven, T.A., and Willis, G.C., 2003, Geologic map of the Richfield 30' x 60' quadrangle, southeast Millard County and parts of Beaver, Piute, and Sevier Counties, Utah: Utah Geological Survey Map 195, scale 1:100,000.

Hoblitt, R.P., Miller, C.D., and Vallance, J.W., 1981, Origin and stratigraphy of the deposit produced by the May 18 directed blast, *in* Lipman, P.W., and Mullineaux, D.R., editors, The 1980 eruptions of Mount St. Helens, Washington: U.S. Geological Survey Professional Paper 1250, p. 401-420.

Holloway, Kali, 2020, The Whitelash next time: The Nation, September 21/28, p. 10-11.

Holmes, Arthur, 1965, Principles of physical geology, 2nd edition: New York, The Ronald Press Company.

Holmes, Melanie, 2019, A hero on Mount St. Helens—the life and legacy of David A. Johnston: Urbana, Illinois, University of Illinois Press, 222 p.

Hotez, P.J., 2023, The deadly rise of anti-science—a scientist's warning: Baltimore, Johns Hopkins University Press, 218 p.

Hudson, M.R., Rosenbaum, J.G., Gromme, C.S., Scott, R.B., and Rowley, P.D., 1998, Paleomagnetic evidence for counterclockwise rotation in a broad sinistral shear zone, Basin and Range province, southeastern Nevada and southwestern Utah, *in* Faulds, J.E., and Stewart, J.H., eds., Accommodation zones and transfer zones—the regional segmentation of the Basin and Range province: Geological Society of America Special Paper 323, p. 149-180.

Huntford, Roland, 1980, Scott and Amundsen: New York, G.P. Putnam's Sons, 665 p.

Huntford, Roland, 1986, Shackleton: New York, Atheneum, 774 p.

James, H.L., 1974, Joseph Hoover Mackin, 1905-1968: Biographical Memoir, v. XLV, National Academy of Sciences of the United States of America, p. 249-262.

Johnson, Michael, Dixon, G.L., Rowley, P.D., Katzer, T.C., and Winters, Michael, 2002, Hydrology and ground-water conditions of the Tertiary Muddy Creek Formation in the lower Virgin River basin of southeastern Nevada and adjacent Arizona and Utah, *in* Lund, W.R., editor, Field guide to geologic excursions in southwestern Utah and adjacent areas of Arizona and Nevada, prepared for the Geological Society of America, Rocky Mountain Section Meeting in Cedar City, Utah, May 7-9, 2002: U.S. Geological Survey Open-File Report 02-172, p. 284-315.

Kamenev, E.N., and Orlenko, E.M., 1982, Metamorphism of sedimentary formations on the Lassiter Coast, *in* Craddock, Campbell, editor, Antarctic Geoscience: Madison, University of Wisconsin Press, p. 357-361.

Keizer, Garret, 2018, Nihilist nation, the empty core of the Trump mystique: The New Republic, November 2018, p. 26-33.

Kellogg, K.S., 1980, Paleomagnetic evidence for oroclinal bending of the southern Antarctic Peninsula: Geological Society of America Bulletin, pt. I, v. 91, p. 414-420.

Kellogg, K.S., and Reynolds, R.L., 1978, Paleomagnetic results from the Lassiter Coast, Antarctica, and a test for oroclinal bending of the Antarctic Peninsula: Journal of Geophysical Research, v. 83, no. B5, p. 2293-2299.

Kellogg, K.S., and Rowley, P.D., 1989, Structural geology and tectonics of the Orville Coast region, southern Antarctic Peninsula, Antarctica: U.S. Geological Survey Professional Paper 1498, 25 p.

Kellogg, K.S., and Rowley, P.D., 1991, Tectonic evolution of the south-eastern Antarctic Peninsula, *in* Thomson, M.R.A., Crame, J.A., and Thomson, J.W., eds., Geological evolution of Antarctica: Cambridge, England, Cambridge University Press, p. 461-466.

Kerr, P.F., Brophy, G.P., Dahl, H.M., Green, Jack, and Woolard, L.E.,1957, Marysvale, Utah, uranium area—geology, volcanic relations, and hydrothermal alteration: Geological Society of America Special Paper 64, 212 p.

Kolbert, Elizabeth, 2006, Field notes from a catastrophe—man, nature, and climate change: New York, Bloomsbury, 225 p.

Kornei, Katherine, 2023, The mental toll of climate change: Eos, v. 104, no. 5, p. 28-33.

Krakauer, Jon, 1997, Into thin air—a personal account of the Mt. Everest disaster: New York, Villard Books, 416 p.

Krugman, Paul, 2020, The cult of selfishness is killing America: The New York Times, July 27, 2020.

Kuntz, M.A., Rowley, P.D., and MacLeod, N.S., 1990, Geologic maps of pyroclastic-flow and related deposits of the 1980 eruptions of Mount St. Helens, Washington: U.S. Geological Survey Miscellaneous Investigations Series Map I-1950, scale 1:12,000.

Kuntz, M.A., Rowley, P.D., MacLeod, N.S., Reynolds, R.L., McBroome, L.A., Kaplan, A.M., and Lidke, D.J., 1981, Petrography and particle-size distribution of pyroclastic-flow, ash-cloud, and surge deposits, *in* Lipman, P.W., and Mullineaux, D.R., editors, The 1980 eruptions of Mount St. Helens, Washington: U.S. Geological Survey Professional Paper 1250, p. 525-539.

Lansing, Alfred, 1959, Endurance—Shackleton's incredible journey: London, Hodder and Stoughton, 287 p.

Laudon, T.S., 1972, Stratigraphy of eastern Ellsworth Land, *in* Adie, R.J., editor, Antarctic geology and geophysics: Oslo, Universitetsforlaget, p. 215-223.

Laudon, T.S., 1982, Geochemistry of Mesozoic and Cenozoic igneous rocks, eastern Ellsworth Land, *in* Craddock, Campbell, editor, Antarctic Geoscience: Madison, University of Wisconsin Press, p. 775-785.

Laudon, T.S., 1991, Petrology of sedimentary rocks from the English Coast, eastern Ellsworth Land, *in* Thomson, M.R.A., Crame, J.A., and Thomson, J.W., editors, Geological evolution of Antarctica: Cambridge, England, Cambridge University Press, p. 455-460.

Laudon, T.S., and Craddock, C., 1992, Petrologic comparison of Paleozoic rocks from the English Coast, eastern Ellsworth Land, and the Ellsworth Mountains, *in* Yoshida, Y., Kaminuma, K., and Shiraishi, K., editors, Recent progress in Antarctic earth science: Tokyo, Terra Scientific Publishing Company, p. 341-345.

Laudon, T.S., and Ford, A.B., 1997, Provenance and tectonic setting of sedimentary rocks from eastern Ellsworth Land based on geochemical indicators and sandstone modes, *in* Ricci, C.A., editor, The Antarctic region—geological evolution and processes: Siena, Terra Antarctica Publication, p. 417-427.

Laudon, T.S., Lackey, L.L., Quilty, P.D., and Otway, P.M., 1969, Geology of eastern Ellsworth Land, *in* Bushnell, V.C., and Craddock, Campbell, editors, Geologic maps of Antarctica: New York, American Geographical Society, Antarctic Map Folio Series 12, plate 3.

Laudon, T.S., Thomson, M.R.A., Williams, P.L., Milliken, K.L., Rowley, P.D., and Boyles,

J.M., 1983, The Jurassic Latady Formation, southern Antarctic Peninsula, *in* Oliver, R.L., James, P.R., and Jago, J.B., editors, Antarctic earth science: Canberra, Australian Academy of Science, p. 308-314.

Laczniak, R.J., Cole, J.C., Sawyer, D.A., and Trudeau, D.A., 1996, Summary of hydrogeologic controls on ground-water flow at the Nevada Test Site, Nye County, Nevada: U.S. Geological Survey Water-Resources Investigations Report 96-4109, 59 p.

Lewis, Michael, 2018, The fifth risk: New York, W.W. Norton & Company, 255 p.

Lewis, Michael, 2019, Portrait of an inessential worker: Bloomberg Opinion Today, October 2.

Lipman, P.W., 2000, The central San Juan caldera cluster—regional geologic framework, *in* Bethke, P.M., and Hay, R.L., editors, Ancient Lake Creede—its volcano-tectonic setting, history of sedimentation, and relation to mineralization in the Creede mining district: Geological Society of America Special Paper 346, p. 9-70.

Lipman, P.W., 2006, Geologic map of the central San Juan caldera cluster, southwestern Colorado: U.S. Geological Survey Geologic Investigation Series I-2799, 4 sheets, 34 p., scale 1:50,000.

Lipman, P.W., 2020, Geologic map of the Bonanza caldera area, northeastern San Juan Mountains, Colorado: U.S. Geological Survey Scientific Investigations Map 3394, 2 sheets, 73 p., scale 1:50,000.

Lipman, P.W., Doe, B.R., Hedge, C.E., and Steven, T.A., 1978, Petrologic evolution of the San Juan volcanic field, southwestern Colorado—Pb and Sr isotope evidence: Geological Society of America Bulletin, v. 89, p. 59-82.

Lipman, P.W., and Mullineaux, D.R., editors, 1981, The 1980 eruptions of Mount St. Helens, Washington: U.S. Geological Survey Professional Paper 1250, 844 p.

Lipman, P.W., Rowley, P.D., Mehnert, H.H., Evans, S.H., Jr., Nash, W.P., and Brown, F.H., 1978, Pleistocene rhyolite of the Mineral Mountains, Utah—geothermal and archeological significance, *with sections on* Fission-track dating, by Izett, G.A., and Naeser, C.W., *and on* Obsidian-hydration dating, by Irving Friedman: U.S. Geological Survey Journal of Research, v. 6, no. 1, p. 133-147.

Lipman, P.W., Steven, T.A., and Mehnert, H.H., 1970, Volcanic history of the San Juan Mountains, Colorado, as indicated by potassium-argon dating: Geological Society of America Bulletin, v. 81, p. 2327-2352.

Lipsky, David, 2023, The parrot and the igloo—climate and the science of denial: New York, W.W. Norton & Company, Inc., 480 p.

Lund, W.R., Knudsen, T.R., Vice, G.S., and Shaw, L.M., 2008, Geologic hazards and adverse construction conditions, St. George—Hurricane metropolitan area, Washington County, Utah: Utah Geological Survey Special Study 127, 105 p., DVD.

Lundin, E.R., 1989, Thrusting of the Claron Formation, the Bryce Canyon region, Utah: Geological Society of America Bulletin, v. 101, p. 1038-1050.

Mackin, J.H., 1947, Some structural features of the intrusions in the Iron Springs district: Utah Geological Society Guidebook 2, 62 p.

Mackin, J.H., 1948, Concept of the graded river: Geological Society of America Bulletin, v. 59, p. 463-512.

Mackin, J.H., 1960, Structural significance of Tertiary volcanic rocks in southwestern Utah: American Journal of Sciences, v. 58, p. 81-131.

Mackin, J.H., 1954, Geology and iron deposits of the Granite Mountain area, Iron County, Utah: U.S. Geological Survey Mineral Investigations Field Studies Map MF-14, scale 1:12,000.

Mackin, J.H., 1968, Iron ore deposits of the Iron Springs district, southwestern Utah, *in* Ridge, J.D., editor, Ore deposits of the United States, 1933-1967 (Graton-Sales volume): New York, American Institute of Mining and Metallurgical Petroleum Engineers, v. 2, p. 992-1019.

Mackin, J.H., and Ingerson, F.E., 1960, An hypothesis for the origin of ore-forming fluid: U.S. Geological Survey Professional Paper 400, p. B1-B2.

Mackin, J.H., Nelson, W.H., and Rowley, P.D., 1976, Geologic map of the Cedar City NW quadrangle, Iron County, Utah: U.S. Geological Survey Geologic Quadrangle Map GQ-1295, scale 1:24,000.

Mackin, J.H., and Rowley, P.D., 1975, Geologic map of the Avon SE quadrangle, Iron County, Utah: U.S. Geological Survey Geologic Quadrangle Map GQ 1294, scale 1:24,000.

Mackin, J.H., and Rowley, P.D., 1976, Geologic map of The Three Peaks quadrangle, Iron County, Utah: U.S. Geological Survey Geologic Quadrangle Map GQ-1297, scale 1:24,000.

Major, J.J., 2020, Mount St. Helens at 40: Science, v. 368, no. 6492, p. 704-705.

Mazzucato, Mariana, 2021, Saving the climate in a triple crisis—a moon shot model for the transformation of capitalism: The New Republic, April 2021, p. 22-29.

McCormack, Win, 2020, Self-interest, wrongly understood—Covid-19 and the limits of individual liberty: The New Republic, September 2020 issue, p. 68.

Meadows, D.H., Meadows, D.L., and Randers, Jorgen, 1992, Beyond the limits: Post Mills, Vermont, Chelsea Green Publishing Company, 300 p.

Meadows, D.H., Meadows, D.L., Randers, Jorgen, and Behrens, W.W., III, 1972, The limits to growth: New York, Universe Books, 205 p.

Mehnert, H.H., Rowley, P.D., and Lipman, P.W., 1978, K-Ar ages and geothermal implications of young rhyolites in west-central Utah: Isochron/West, no. 21, p. 3-7.

Mehnert, H.H., Rowley, P.D., and Schmidt, D.L., 1975, K-Ar ages of plutonic rocks in the Lassiter Coast area, Antarctic Peninsula: U.S. Geological Survey Journal of Research, v. 3, no. 2, p. 233-236.

Miller, F.K., Rowley, P.D., Lee, Cin-Ty, Snee, L.W., Sutter, J.F., Nichols, Lindsay, and Tilley, S.E., 2021, Memorial to Douglas M. Morton (1935-2020): Geological Society of America Memorials, v. 50, p. 5-11 (online).

Moore, D.W., Nealey, L.D., Rowley, P.D., Hatfield, S.C., Maxwell, D.J., and Mitchell, Eric, 2004, Geologic map of the Navajo Lake quadrangle, Kane and Iron Counties, Utah: Utah Geological Survey Map 199, scale 1:24,000.

Morgan, M.L., Siddoway, C.S., Rowley, P.D., Temple, Jay, Keller, J.W., Archuleta, B.H., and Himmelreich, J.W., Jr., 2003, Geologic map of the Cascade quadrangle, El Paso County, Colorado: Colorado Geological Survey Open-File Report 03-18, 45 p., scale 1:24,000.

Mother Jones, 2020, The plague—how Trump bungled America's coronavirus response: Mother Jones special issue, 36 p.

Nealey, L.D., Rowley, P.D., Unruh, D.M., Budahn, J.R., Snee, L.W., Mehnert, H.H., and Anderson, R.E., 1995, Preliminary geochemistry of Miocene ash-flow tuffs in and near the Caliente caldera complex, southeastern Nevada and southwestern Utah, *in* Scott, R.B., and Swadley, WC, editors, Geologic studies in the Basin and Range-Colorado Plateau transition, southeastern Nevada, southwestern Utah, and northwestern Arizona: U.S. Geological Survey Bulletin 2056-C, p. 89-110.

Nielson, D.L., Evans, S.H., Jr., and Sibbett, B.S., 1986, Magmatic, structural, and hydrothermal evolution of the Mineral Mountains intrusive complex, Utah: Geological Society of America Bulletin, v. 97, p. 765-777.

Nwanevu, Osita, 2020, Make it stop: The New Republic, March 2020 issue, p. 12-19.

Oreskes, Naomi, and Conway, E.M., 2010, Merchants of Doubt: New York, Bloomsbury Press, 355 p.

Oreskes, Naomi, and Conway, E.M., 2023, The big myth: New York, Bloomsbury Publishing, 564 p.

Oreskes, Naomi, and LeGrand, Homer, 2003, Plate tectonics—an insider's history of the modern theory of the earth: CRC Press, 452 p.

Packer, George, 2020, Underlying conditions—in America, the coronoviris has revealed a sick and unequal society incapable of self-government: The Atlantic, June 2020 issue, p. 9-12 (online version in June 2020 was titled "We are living in a failed state").

Packer, George, 2024, The valley—searching for the future in the most American city: The Atlantic, v. 334, no. 1, p. 52-93.

Page, W.R., Dixon, G.L., Rowley, P.D., and Brickey, D.W., 2005, Geologic map of parts of the Colorado, White River, and Death Valley groundwater flow systems, Nevada, Utah, and Arizona: Nevada Bureau of Mines & Geology Map 150, 21 p., scale 1:250,000.

Page, W.R., Lundstrom, S.C., Harris, A.G., Langenheim, V.E., Workman, J.B., Mahan, S.A., Paces, J.B., Dixon, G.L., Rowley, P.D., Burchfiel, B.C., Bell, J.W., and Smith, E.I., 2005, Geologic and geophysical maps of the Las Vegas 30' x 60' quadrangle, Clark and Nye Counties, Nevada, and Inyo County, California: U.S. Geological Survey Scientific Investigation Map SIM-2814, scale 1:100,000.

Pankhurst, R.J., and Rowley, P.D., 1991, Rb-Sr study of Cretaceous plutons from southern Antarctic Peninsula and eastern Ellsworth Land, Antarctica, *in* Thomson, M.R.A., Crame, J.A., and Thomson, J.W., eds., Geological evolution of Antarctica: Cambridge, England, Cambridge University Press, p. 387-394.

Park, Gerald, and Krahulec, Ken, 2009, Gold Mountain mining district, Piute County, Utah, *in* Tripp, B.T., Krahulec, Ken, and Jordan, J.L., editors, Geology and geologic resources and issues of western Utah: Utah Geological Association Publication 38, CD, p. 177-186.

Pennock, R.T., 2019, An instinct for truth—curiosity and the moral character of science: Cambridge, Massachusetts, The MIT Press, 415 p.

Petronis, M.S., Holm, D.K., Geissman, J.W., Hacker, D.B., and Arnold, B.J., 2014, Paleomagnetic results from the eastern Caliente-Enterprise zone, southwestern Utah—implications for initiation of a major Miocene transfer zone: Geosphere, v. 10, p. 534-563.

Quilty, Pat, 2018, Thomas S. Laudon, 1932-2017, The Antarctican Society, v. 17-18 (January), no. 2, p. 9-10.

Rees, Martin, 2020, Lessons for a young scientist: Nautilus, January 2020 issue, p. 38-43.

Richard, S.M., and Hills, D.J., 2017, Merle Lee Allison (1948-2016): Eos, v. 98, no. 8, p. 9.

Roberts-Miller, Patricia, 2019, Deconstructing Trump: The Washington Spectator, March 1, 2019.

Robinson College, 2020, Dr. Michael R.A. Thomson (1942-2020): Robinson College Record, 2020, Robinson College, Cambridge University [England], p. 100-104.

Ronne, Finn, 1979, Antarctica, my destiny: New York, Hastings House Publishers, 278 p.

Rosenbaum, J.G., and Waitt, R.B., Jr., 1981, Summary of eyewitness accounts of the May 18 eruption, *in* Lipman, P.W., and Mullineaux, D.R., editors, The 1980 eruptions of Mount St. Helens, Washington: U.S. Geological Survey Professional Paper 1250, p. 53-68.

Rowley, A.C., 1938, Flow of liquid through sealed circular weirs: Cambridge, Massachusetts Institute of Technology unpublished B.S. thesis, 52 p.

Rowley, A.C., 1939, Diffusion of water vapor into various gases in a wetted-wall column: Cambridge, Massachusetts Institute of Technology unpublished M.S. thesis, 44 p. + figures.

Rowley, P.D., 1968, Geology of the southern Sevier Plateau, Utah: Austin, University of Texas at Austin unpublished Ph.D. dissertation, 340 p.

Rowley, P.D., 1975, Geologic map of the Enoch NE quadrangle, Iron County, Utah: U.S. Geological Survey Geologic Quadrangle Map GQ 1301, scale 1:24,000.

Rowley, P.D., 1976, Geologic map of the Enoch NW quadrangle, Iron County, Utah: U.S. Geological Survey Geologic Quadrangle Map GQ 1302, scale 1:24,000.

Rowley, P.D., 1978, Geologic map of the Thermo 15-minute quadrangle, Beaver and Iron Counties, Utah: U.S. Geological Survey Geologic Quadrangle Map GQ-1493, scale 1:62,500.

Rowley, P.D., 1979, Geologic map of the Marysvale SE quadrangle, Piute County, Utah: U.S. Geological Survey Miscellaneous Field Studies Map MF-1115, scale 1:24,000.

Rowley, P.D., 1998, Cenozoic transverse zones and igneous belts in the Great Basin, western United States—their tectonic and economic implications, *in* Faulds, J.E., and Stewart, J.H., editors, Accommodation zones and transfer zones—the regional segmentation of the Basin and Range province: Geological Society of America Special Paper 323, p. 195-228.

Rowley, P.D., 2002, Water resources technical report—hydrogeological assessment of the Gunlock part of the Navajo/Kayenta aquifer and its bearing on the Gunlock well field, Washington County, Utah, Appendix D *in* Bureau of Land Management and Bureau of Indian Affairs, Environmental assessment for the Santa Clara pipeline, Utah, UT 045-98-02, p. D1-D32.

Rowley, P.D., 2008, The value of geologic mapping—a perspective from a consulting geologist (abs.): Newsletter of the Utah Geological Association, v. 40, no. 11, p. 2, http://utahgeology.org (newsletter also includes "Presentation of the 2008 Lehi Hintze Award for Outstanding Contributions to the Geology of Utah to Peter D. Rowley" on p. 1).

Rowley, P.D., and Anderson, R.E., 1996, The syntectonic caldera—a new caldera type bounded by synchronous linear faults: Geological Society of America Abstracts with Programs, v. 28, no. 7, p. A449.

Rowley, P.D., Anderson, R.E., Hacker, D.B., Boswell, J.T., Maxwell, D.J., Cox, D.P., Willden, Ronald, and Adair, D.H., 2007, Interim geologic map of the Goldstrike quadrangle and east part of the Docs Pass quadrangle, Washington County, Utah: Utah Geological Survey Open-File Report 510, 30 p., scale 1:24,000.

Rowley, P.D., and Barker, D.S., 1978, Geology of the Iron Springs mining district, Utah, *in* Shawe, D.R., and Rowley, P.D., editors, Utah Geological Association Publication 7, p. 49-58.

Rowley, P.D., Biek, R.F., and Hacker, D.B., 2022, Basin-range uplift and canyon cutting in 3 million years, Kingston Canyon, Piute County, southwestern Utah: Geology of the Intermountain West, v. 9, p. 1-11, https://doi:10.31711/giw.v9.pp1-11.

Rowley, P.D., Biek, R.F., and Hacker, D.B., 2022, An unconformity in the Pole Creek area (Sevier Plateau) west of Antimony, western Garfield County, Utah, and its bearing on the Sevier gravity slide: Geology of the Intermountain West, v. 9, p. 13-24, https://doi:10.31711/giw.v9.pp13-24.

Rowley, P.D., Biek, R.F., Hacker, D.B., Vice, G.S., McDonald, R.E., Maxwell, D.J., Cunningham, C.G., Steven, T.A., Anderson, J.J., Ekren, E.B., Machette, M.N., and Wardlaw, B.R., in press, Interim geologic map of the northeast quarter of the Beaver 30' x 60' quadrangle, Beaver and Piute Counties, Utah: Utah Geological Survey Contract Deliverable, scale 1:50,000.

Rowley, P.D., Biek, R.F., Hacker, D.B., Vice, G.S., McDonald, R.E., Maxwell, D.J., Fasselin, R., Dustin, J.A., Cunningham, C.G., Steven, T.A., Anderson, J.J., Ekren, E.B., Machette, M.N., and Wardlaw, B.R., Smith, Z.D., Kirby, S.M., Knudsen, T.R., Kleber, E.J., Hiscock, A.I., Malone, D.H., Rivera, T.A., and Jicha, B.R., in review, Interim geologic map of the Beaver 30' x 60' quadrangle, Beaver, Piute, Iron, and Garfield Counties, Utah; Utah Geological Survey Contract Deliverable, xx p., 3 plates, scale 1:62,500.

Rowley, P.D., Biek, R.F., Hacker, D.B., Vice, G.S., McDonald, R.E., Maxwell, D.J., Smith, Z.D., Cunningham, C.G., Steven, T.A., Anderson, J.J., Ekren, E.B., Machette, M.N, Wardlaw, B.R., and Loffer, Z.J., in press, Interim geologic map of the southeastern quarter of the Beaver 30' x 60' quadrangle, Beaver, Garfield, Iron, and Piute Counties, Utah: Utah Geological Survey Contract Deliverable, scale 1:50,000.

Rowley, P.D., Biek, R.F., Hacker, D.B., Vice, G.S., McDonald, R.E., Maxwell, D.J., Smith, Z.D., Cunningham, C.G., Steven, T.A., Anderson, J.J., Ekren, E.B., Machette, M.N., and Wardlaw, B.R., 2019, Interim geologic map of the southwestern quarter of the Beaver 30' x 60' quadrangle, Beaver, Iron, and Garfield Counties, Utah: Utah Geological Survey Open-File Report 686DM, CD, 18 p., scale 1:100,000.

Rowley, P.D., Biek, R.F., Hacker, D.B., Vice, G.S., McDonald, R.E., Maxwell, D.J., Smith, Z.D., Cunningham, C.G., Steven, T.A., Anderson, J.J., Ekren, E.B., Machette, M.N., Wardlaw, B.R., Kirby, S.M., Knudsen, T.R., Kleber, E.J., and Hiscock, A.I., 2020, Interim geologic map of the northwestern quarter of the Beaver 30' x 60' quadrangle, Beaver and Piute Counties, Utah: Utah Geological Survey Open-File Report 729DM, CD, scale 1:62,500, https://doi.org/10.34191/OFR-729DM.

Rowley, P.D., Biek, R.F., Sable, E.G., Boswell, J.T., Vice, G.S., Hatfield, S.C., Maxwell, D.J., and Anderson, J.J., 2013, Geologic map of the Brian Head quadrangle, Iron County, Utah: Utah Geological Survey Map 263DM, CD, 38 p., scale 1:24,000.

Rowley, P.D., Christensen, E.F., Dixon, G.L., Burns, A.G., Watrus, J.M., and Collins, C.A., 2012, Fault derived conduit and barrier groundwater flow at Sawyer Spring, eastern margin of the Pine Valley Mountains, Washington County, Utah, *in* Hylland, M.D., and Harty, K.M., editors, Selected topics in engineering and environmental geology *in* Utah: Utah Geological Association Publication 41, CD, p. 99-114.

Rowley, P.D., Cunningham, C.G., Anderson, J.J., and Steven, T.A., 1979, Geologic map of the Marysvale SW quadrangle, Piute County, Utah: U.S. Geological Survey Miscellaneous Field Studies Map MF-1116, scale 1:24,000

Rowley, P.D., Cunningham, C.G., Anderson, J.J., Steven, T.A., Workman, J.B., and Snee, L.W., 2002, Geology and mineral resources of the Marysvale volcanic field, southwestern Utah, *in* Lund, W.R., editor, Field guide to geologic excursions in southwestern Utah and adjacent areas of Arizona and Nevada, prepared for the Geological Society of America, Rocky Mountain Section Meeting in Cedar City, Utah, May 7-9, 2002: U.S. Geological Survey Open-File Report 02-172, p. 131-170.

Rowley, P.D., Cunningham, C.G., and Kaplan, A.M., 1981, Geologic map of the Monroe SE quadrangle, Sevier and Piute Counties, Utah: U.S. Geological Survey Miscellaneous Field Studies Map MF-1331, scale 1:24,000.

Rowley, P.D., Cunningham, C.G., Steven, T.A., Mehnert, H.H., and Naeser, C.W., 1988a, Geologic map of the Antelope Range quadrangle, Sevier and Piute Counties, Utah: Utah Geological and Mineral Survey Map 106, 2 sheets, scale 1:24,000.

Rowley, P.D., Cunningham, C.G., Steven, T.A., Mehnert, H.H., and Naeser, C.W., 1988b, Geologic map of the Marysvale quadrangle, Piute County, Utah: Utah Geological and Mineral Survey Map 105, 2 sheets, scale 1:24,000.

Rowley, P.D., Cunningham, C.G., Steven, T.A., Mehnert, H.H., and Naeser, C.W., 1998, Cenozoic igneous and tectonic setting of the Marysvale volcanic field and its relation to other igneous centers in Utah and Nevada, *in* Friedman, J.D., and Huffman, A.C., Jr., editors, Laccolith complexes of southeastern Utah—time of emplacement and tectonic setting—workshop proceedings: U.S. Geological Survey Bulletin 2158, p. 167-202.

Rowley, P.D., Cunningham, C.G., Steven, T.A., Workman, J.B., Anderson, J.J., and Theissen, K.M., 2002, Geologic map of the central Marysvale volcanic field, southwestern Utah: U.S. Geological Survey Investigation Series Map I-2645-A, scale 1:100,000.

Rowley, P.D., and Dixon, G.L., 2001, The Cenozoic evolution of the Great Basin area, U.S.A.—New interpretations based on regional geologic mapping, *in* Erskine, M.C., Faulds, J.E., Bartley, J.M., and Rowley, P.D., editors, The geologic transition, High Plateaus to Great Basin—A symposium and field guide (The Mackin Volume): Utah Geological Association Publication no. 30 and Pacific Section of the American Association of Petroleum Geologists Guidebook GB 78, p. 169-188.

Rowley, P.D., and Dixon, G.L., 2004, The role of geology in increasing Utah's ground-water resources from faulted terranes—lessons from the Navajo Sandstone, Utah, and the Death Valley flow system, Nevada-California, *in* Spangler, Larry, editor, Ground Water in Utah—resource, protection, and remediation: Utah Geological Association Publication 31, p. 27-41.

Rowley, P.D., and Dixon, G.L., 2010, Geology and hydrogeology of Anderson Junction Reservoir, Anderson Junction well field, and Cottam well field, Washington County, Utah: St. George, Utah, Washington County Water Conservancy District WCWCD-03, 22 p., scale 1 inch = 500 feet.

Rowley, P.D., Dixon, G.L., D'Agnese, F.A., O'Brien, G.M., and Brickey, D.W., 2004, Geology and hydrology of the Sand Hollow Reservoir and well field area, Washington County, Utah: St. George, Utah, Washington County Water Conservancy District WCWCD-01, 14 p. + CD-ROM, scale 1:12,000.

Rowley, P.D., Dixon, G.L., and Brickey, D.W., 2005, Geology along the route of the Lake Powell water pipeline, Utah and Arizona: St. George, Utah, Washington County Water Conservancy District WCWCD-02, 17 p. + CD-ROM, scale 1:75,000.

Rowley, P.D., Dixon, G.L., Burns, A.G., and Collins, C.A., 2009, Geology and hydrogeology of the Snake Valley area, western Utah and eastern Nevada, *in* Tripp, B.T., Krahulec, Ken, and Jordan, J.L., editors, Geology and geologic resources and issues of western Utah: Utah Geological Association Publication 38, CD, p. 251-269 + Plate 1, Geologic map, cross sections, and explanation of geologic units of the Snake Valley area, scale 1:250,000.

Rowley, P.D., Dixon, G.L., and Layton, M.E., 2018, Hydrogeology of the Ash Creek drainage basin, Washington and Iron Counties, Utah, with implications from fracture-flow analysis on amounts, sources, and destinations of groundwater, *in* Emerman, S.H., Bowen, B., Schamel, S., and Simmons, S., editors, Geofluids of Utah: Utah Geological Association Publication 47, CD, p. 23-53.

Rowley, P.D., Dixon, G.L., Mankinen, E.A., Pari, K.T., McPhee, D.K., McKee, E.H., Burns, A.G., Watrus, J.M., Ekren, E.B., Patrick, W.G., and Brandt, J.M., 2017, Geology and geophysics of White Pine and Lincoln Counties, Nevada, and adjacent parts of Nevada and Utah—The geologic framework of regional groundwater flow systems: Nevada Bureau of Mines & Geology Report 56, 146 p., scale 1:250,000.

Rowley, P.D., Dixon, G.L., and Steven, T.A., 1997, The value to the country of geologic mapping, *in* Letters: GSA Today, v. 7, no. 10, p. 17-19.

Rowley, P.D., Dixon, G.L., Watrus, J.M., Burns, A.G., Mankinen, E.A., McKee, E.H., Pari, K.T., Ekren, E.B., and Patrick, W.G., 2016, Geology, selected geophysics, and hydrogeology of the White River and parts of the Great Salt Lake Desert regional groundwater flow systems, Utah and Nevada, *in* Comer, J.B., Inkenbrandt, P.C., Krahulec, K.A., and Pinnell, M.L., editors, Resources and geology of Utah's West Desert: Utah Geological Association Publication 45, CD, p. 167-200.

Rowley, P.D., Dyni, J.R., Hansen, W.R., and Pipiringos, G.N., 1979, Geologic map of the Indian Water Canyon quadrangle, Moffat County Colorado: U.S. Geological Survey Geologic Quadrangle Map GQ-1516, scale 1:24,000.

Rowley, P.D., Farrar, E., Carrara, P. E., Vennum, W.R., and Kellogg, K.S., 1988, Porphyry-type copper deposits and potassium-argon ages of plutonic rocks of the Orville Coast and eastern Ellsworth Land, Antarctica, *in* Rowley, P.D., and Vennum, W.R., editors, Studies of the geology and mineral resources of the southern Antarctic Peninsula and eastern Ellsworth Land, Antarctica: U.S. Geological Survey Professional Paper 1351, p. 35-49.

Rowley, P.D., Filkorn, H.F., Lassen, P.L., and Spurney, J.C., 2018, Memorial to John J. Anderson (1930-2017): Geological Society of America Memorials, v. 47, June 2018, p. 15-20 (online).

Rowley, P.D., Ford, A.B., Thorman, C.H., Wallace, C.A., and K.L. and L.W. Pierce, 2020, Memorial to Paul L. Williams (1932-2020): Geological Society of America Memorials, v. 49, p. 45-50 (online).

Rowley, P.D., Ford, A.B., Williams, P.L., and Pride, D.E., 1983, Metallogenic provinces of Antarctica, *in* Oliver, R.L., James, P.R., and Jago, J.B., editors, Antarctic earth science: Canberra, Australian Academy of Science, p. 414-419.

Rowley, P.D., Hait, M.H., Jr., Russell-Robinson, S.L., Buchanan-Banks, J.M., and Cashman, K.V., 1986, The role of the U.S. Geological Survey in providing information to news media about the 1980 eruptions of Mount St. Helens: U.S. Geological Survey Open-File Report 86-509, 15 p.

Rowley, P.D., Hacker, D.B., and Biek, R.F., 2022, A site bearing on the origin of iron deposits in the Iron Springs mining district, Iron County, Utah Geology of the Intermountain West, v. 9, p. 25-37, https://doi:10.31711/giw.v9.pp25-37.

Rowley, P.D., Hacker, D.B., Maxwell, D.J., Maxwell, J.D., and Boswell, J.T., 2008, Interim geologic map of the Utah part of the Deer Lodge Canyon, Prohibition Flat, Uvada, and Pine Park quadrangles (east part of the Caliente 30' x 60' quadrangle), Iron and Washington Counties, Utah: Utah Geological Survey Open-File Report 530, 20 p., scale 1:24,000.

Rowley, P.D., and Hansen, W.R., 1979a, Geologic map of the Plug Hat Rock quadrangle, Moffat County, Colorado: U.S. Geological Survey Geologic Quadrangle Map GQ-1514, scale 1:24,000.

Rowley, P.D., and Hansen, W.R., 1979b, Geologic map of the Split Mountain quadrangle, Uintah County, Utah: U.S. Geological Survey Geologic Quadrangle Map GQ-1515, scale 1:24,000.

Rowley, P.D., Hansen, W.R., and Carrara, P.E., 1981, Geologic map of the Island Park quadrangle, Uintah County, Utah: U.S. Geological Survey Geologic Quadrangle Map GQ-1560, scale 1:24,000.

Rowley, P.D., Hansen, W.R., Tweto, Ogden, and Carrara, P.E., 1985, Geologic map of the Vernal 1° x 2° quadrangle, Colorado, Utah, and Wyoming: U.S. Geological Survey Miscellaneous Investigations Series Map I-1526, scale 1:250,000.

Rowley, P.D., Hereford, Richard, and Williams, V.S., 1987, Geologic map of the Adams Head-Johns Valley area, southern Sevier Plateau, Garfield County, Utah: U.S. Geological Survey Miscellaneous Investigations Series Map I-1798, scale 1:50,000.

Rowley, P.D., Himmelreich, J.W., Jr., Kupfer, D.H., and Siddoway, C.S., 2003, Geologic map of the Cheyenne Mountain quadrangle, El Paso County, Colorado: Colorado Geological Survey Open-File Report 02-5, 30 p., scale 1:24,000.

Rowley, P.D., Katzer, T.C., Kuntz, M.A., Vice, G.S., Morton, D.M., and Faulds, J.E., 2017, Memorial to Gary L. Dixon (1943-2017): Geological Society of America Memorials, v. 46, p. 17-21 (online).

Rowley, P.D., Kellogg, K.S., Vennum, W.R., Laudon, T.S., Thomson, J.W., O'Neill, J.M., and Lidke, D.J., 1991, Tectonic setting of the English Coast, eastern Ellsworth Land, Antarctica, *in* Thomson, M.R.A., Crame, J.A., and Thomson, J.W., editors, Geological evolution of Antarctica: Cambridge, England, Cambridge University Press, p. 467-474.

Rowley, P.D., Kellogg, K.S., Vennum, W.R., Waitt, R.B., and Boyer, S.J., 1988, Geology of the southern Black Coast, Antarctic Peninsula, *in* Rowley, P.D., and Vennum, W.R., editors, Studies of the geology and mineral resources of the southern Antarctic Peninsula and eastern Ellsworth Land, Antarctica: U.S. Geological Survey Professional Paper 1351, p. 1-20.

Rowley, P.D., Kellogg, K.S., Williams, P.L., Willan, C.F.H., and Thomson, J.W., 1992, Geological map, southern Palmer Land and eastern Ellsworth Land: British Antarctic Survey Series BAS 500G, sheet 6, scale 1:500,000.

Rowley, P.D., Kinney, D.M., and Hansen, W.R., 1979, Geologic map of the Dinosaur Quarry quadrangle, Uintah County, Utah: U.S. Geological Survey Geologic Quadrangle Map GQ-1513, scale 1:24,000.

Rowley, P.D., Kuntz, M.A., and MacLeod, N.S., 1981, Pyroclastic-flow deposits, *in* Lipman, P.W., and Mullineaux, D.R., editors, The 1980 eruptions of Mount St. Helens, Washington: U.S. Geological Survey Professional Paper 1250, p. 489-512.

Rowley, P.D., Lipman, P.W., Cunningham, C.G., Casadevall, T.J., Shawe, D.R., and Rye, R.O., 2014, Memorial to Thomas A. Steven (1917-2013): Geological Society of America Memorials, v. 43, p. 13-18 (online).

Rowley, P.D., Lipman, P.W., Mehnert, H.H., Lindsey, D.A., and Anderson, J.J., 1978, Blue Ribbon lineament, an east trending structural zone within the Pioche mineral belt of southwestern Utah and eastern Nevada: U.S. Geological Survey Journal of Research, v. 6, no. 2, p. 175-192.

Rowley, P.D., MacLeod, N.S., Kuntz, M.A., and Kaplan, A.M., 1985, Proximal bedded deposits related to pyroclastic flows of May 18, 1980, Mount St. Helens, Washington: Geological Society of America Bulletin, v. 96, p. 1373-1383.

Rowley, P.D., Mehnert, H.H., Naeser, C.W., Snee, L.W., Cunningham, C.G., Steven, T.A., Anderson, J.J., Sable, E.G., and Anderson, R.E., 1994, Isotopic ages and stratigraphy of Cenozoic rocks of the Marysvale volcanic field and adjacent areas, west-central Utah: U.S. Geological Survey Bulletin 2071, 35 p.

Rowley, P.D., Miller, D.M., and Miller, F.K., compilers, 1986, An overview of geologic mapping needs in the United States: U.S. Geological Survey Open-File Report 86-572, 82 p.

Rowley, P.D., Nealey, L.D., Unruh, D.M., Snee, L.W., Mehnert, H.H., Anderson, R.E., and Gromme, C.S., 1995, Stratigraphy of Miocene ash-flow tuffs in and near the Caliente caldera complex, southeastern Nevada and southwestern Utah, *in* Scott, R.B., and Swadley, WC, editors, Geologic studies in the Basin and Range-Colorado Plateau transition, southeastern Nevada, southwestern Utah, and northwestern Arizona: U.S. Geological Survey Bulletin 2056-B, p. 43-88.

Rowley, P.D., Norlin, Keith, Maxwell, J.A., and Maxwell, D.J., 2014, The Minersville gravity slide and metallic mineral potential at Black Mountain, northern Black Mountains, Beaver County, Utah, *in* MacLean, J.S., Biek, R.F., and Huntoon, J.E., editors, Geology of Utah's far south: Utah Geological Association Publication 43, CD, p. 599-616; includes 1:6,000-scale geologic map as Plate 1.

Rowley, P.D., and Pride, D.E., 1982, Metallic mineral resources of the Antarctic Peninsula (Review paper), *in* Craddock, Campbell, editor, Antarctic geoscience: Madison, University of Wisconsin Press, p. 859-870.

Rowley, P.D., Rutledge, E.F., Maxwell, D.J., Dixon, G.L., and Wallace, C.A., 2013, Geology of the Sulphurdale geothermal-resource area, Beaver and Millard Counties, Utah: Utah Geological Survey Open-File Report 609, CD, 27 p., scale 1:12,000.

Rowley, P.D., and Shroba, R.R., 1991, Geologic map of the Indian Cove quadrangle, Lincoln County, Nevada: U.S. Geological Survey Geologic Quadrangle Map GQ-1701, scale 1:24,000.

Rowley, P.D., Shroba, R.R., Simonds, F.W., Burke, K.J., Axen, G.J., and Olmore, S.D., 1994, Geologic map of the Chief Mountain quadrangle, Lincoln County, Nevada: U.S. Geological Survey Geologic Quadrangle Map GQ-1731, scale 1:24,000.

Rowley, P.D., Shroba, R.R., Swadley, W C, Simonds, F.W., Snee, L.W., and Axen, G.J., 2023, Geologic map of the Caliente NW quadrangle, Lincoln County, Nevada: Nevada Bureau of Mines & Geology Open-File Report 2023-10, 25 p., scale 1:24,000.

Rowley, P.D., Snee, L.W., Anderson, R.E., Nealey, L.D., Unruh, D.M., and Ferris, D.E., 2001, Field trip to the Caliente caldera complex, east-striking transverse zones, and nearby mining districts in Nevada-Utah—implications for petroleum, ground-water, and mineral resources, *in* Erskine, M.C., Faulds, J.E., Bartley, J.M., and Rowley, P.D., editors, The geologic transition, High Plateaus to Great Basin—a symposium and field guide (The Mackin Volume): Utah Geological Association Publication no. 30 and Pacific Section of the American Association of Petroleum Geologists Guidebook GB 78, p. 401-418.

Rowley, P.D., Snee, L.W., Mehnert, H.H., Anderson, R.E., Axen, G.J., Burke, K.J., Simonds, F.W., Shroba, R.R., and Olmore, S.D., 1992, Structural setting of the Chief mining district, eastern Chief Range, Lincoln County, Nevada, chapter H, *in* Thorman, C.H., editor, Application of structural geology to mineral and energy resources of the central and western United States: U.S. Geological Survey Bulletin 2012, p. H1-H17.

Rowley, P.D., Snee, L.W., Swadley, W C, Shroba, R.R., and Simonds, F.W., 2023, Geologic map of the Chokecherry Mountain quadrangle, Lincoln County, Nevada: Nevada Bureau of Mines & Geology Open-File Report 2023-09, 31 p., scale 1:24,000.

Rowley, P.D., Storey, B.C., Ford, A.B., and Wever, H.E., 1987, Geologic studies of the Black Coast, Antarctic Peninsula: Antarctic Journal of the U.S., v. 22, no. 5, p. 1-3.

Rowley, P.D., Steven, T.A., Anderson, J.J., and Cunningham, C.G., 1979, Cenozoic stratigraphic and structural framework of southwestern Utah: U.S. Geological Survey Professional Paper 1149, 22 p.

Rowley, P.D., Steven, T.A., Briskey, J.A., Bethke, P.M., and Rye, R.O., 2009, Dedication to Charles G. ("Skip") Cunningham: *in* Tripp, B.T., Krahulec, Ken, and Jordan, J.L., editors, Geology and geologic resources and issues of western Utah: Utah Geological Association Publication 38, CD, p. vii-xii.

Rowley, P.D., Steven, T.A., and Mehnert, H.H., 1981, Origin and structural implications of upper Miocene rhyolites in Kingston Canyon, Piute County, Utah: Geological Society of America Bulletin, pt. 1, v. 92, p. 590-602.

Rowley, P.D., Steven, T.A., and Kaplan, A.M., 1981, Geologic map of the Monroe NE quadrangle, Sevier County, Utah: U.S. Geological Survey Miscellaneous Field Studies Map MF-1330, scale 1:24,000.

Rowley, P.D., Thomson, J.W., Smellie, J.L., Laudon, T.S., La Prade, K.E., and LeMasurier, W.E., comps., 1990, Alexander Island, Palmer Land, and Ellsworth Land, chapter C *of* LeMasurier, W.E., and Thomson, J.W., editors, Volcanoes of the Antarctic plate and southern oceans: Washington, D.C., American Geophysical Union, Antarctic Research Series, v. 48, p. 257-301.

Rowley, P.D., and Threet, R.L., 1976, Geologic map of the Enoch quadrangle, Iron County, Utah: U.S. Geological Survey Geologic Quadrangle Map GQ 1296, scale 1:24,000.

Rowley, P.D., Tweto, Ogden, and Hansen, W.R., 1978, Preliminary geologic map of the Vernal 1° x 2° quadrangle, Colorado, Utah, and Wyoming: U.S. Geological Survey Open-File Report 78 573, scale 1:250,000.

Rowley, P.D., Tweto, Ogden, Hansen, W.R., and Carrara, P.E., 1979, Geologic map of the Vernal 1° x 2° quadrangle, Colorado, Utah, and Wyoming: U.S. Geological Survey Miscellaneous Field Studies Map MF 1163, scale 1:250,000.

Rowley, P.D., and Vennum, W.R., editors, 1988, Studies of the geology and mineral resources of the southern Antarctic Peninsula and eastern Ellsworth Land, Antarctica: U.S. Geological Survey Professional Paper 1351, 49 p.

Rowley, P.D., Vennum, W.R., Kellogg, K.S., Laudon, T.S., Carrara, P.E., Boyles, J.M., and Thomson, M.R.A., 1983, Geology and plate tectonic setting of the Orville Coast and eastern Ellsworth Land, Antarctica, *in* Oliver, R.L., James, P.R., and Jago, J.B., editors, Antarctic earth science: Canberra, Australian Academy of Science, p. 245-250.

Rowley, P.D., Vice, G.S, McDonald, R.E., Anderson, J.J., Machette, M.N., Maxwell, D.J., Ekren, E.B., Cunningham, C.G., Steven, T.A., and Wardlaw, B.R., 2005, Interim geologic map of the Beaver 30' x 60' quadrangle, Beaver, Piute, Iron, and Garfield Counties, Utah: Utah Geological Survey Open-File Report 454, 32 p., scale 1:100,000.

Rowley, P.D., Waitt, R.B., Jr., Boyer, S.J., Kellogg, K.S., and Vennum, W.R., 1976, Expedition to the Lassiter Coast, Antarctica: Summit, v. 22, no. 1, p. 18-27.

Rowley, P.D., and Williams, P.L., 1982, Geology of the northern Lassiter Coast and southern Black Coast, Antarctic Peninsula, *in* Craddock, Campbell, editor, Antarctic geoscience: Madison, University of Wisconsin Press, p. 339-348.

Rowley, P.D., Williams, P.L., Anderson, J.J., and Kaplan, A.M., 1981, Geologic map of the Marysvale NE quadrangle, Piute County, Utah: U.S. Geological Survey Miscellaneous Field Studies Map MF-1329, scale 1:24,000.

Rowley, P.D., Williams, P.L., Anderson, R.E., Thorman, C.H., and Vice, G.S., 2020, Memorial to E. Bartlett Ekren (1923-2019): Geological Society of America Memorials, v. 49, p. 23-27 (online).

Rowley, P.D., Williams, P.L., and Kaplan, A.M., 1986a, Geologic map of the Koosharem quadrangle, Sevier and Piute Counties, Utah: U.S. Geological Survey Geologic Quadrangle Map GQ 1590, scale 1:24,000.

Rowley, P.D., Williams, P.L., and Kaplan, A.M., 1986b, Geologic map of the Greenwich quadrangle, Piute County, Utah: U.S. Geological Survey Geologic Quadrangle Map GQ-1589, scale 1:24,000.

Rowley, P.D., Williams, P.L., and Pride, D.E., 1983, Mineral occurrences of Antarctica, *in* Behrendt, J.C., editor, Petroleum and mineral resources of Antarctica: U.S. Geological Survey Circular 909, p. 25-49.

Rowley, P.D., Williams, P.L., and Pride, D.E., 1991, Metallic and non-metallic mineral resources of Antarctica, chapter 18, *in* Tingey, R.J., editor, The Geology of Antarctica, Oxford monographs on geology and geophysics no. 17: Oxford, Clarendon Press, p. 617-651.

Rowley, P.D., Williams, P.L., and Schmidt, D.L., 1977, Geology of an Upper Cretaceous copper deposit in the Andean Province, Lassiter Coast, Antarctic Peninsula: U.S. Geological Survey Professional Paper 984, 36 p.

Rowley, P.D., Williams, P.L., Schmidt, D.L., Reynolds, R.L., Ford, A.B., Clark, A.H., Farrar, E., and McBride, S.L., 1975, Copper mineralization along the Lassiter coast of the Antarctic Peninsula: Economic Geology, v. 70, no. 5, p. 982-987.

Rowley, P.D., Williams, V.S., Vice, G.S., Maxwell, D.J., Hacker, D.B., Snee, L.W., and Mackin, J.H., 2006, Interim geologic map of the Cedar City 30' x 60' quadrangle, Iron and Washington Counties, Utah: Utah Geological Survey Open-File Report 476DM, CD-ROM, scale 1:100,000.

Rowley, P.D., and Witkind, I.J., 2005, Dedication to Wallace R. Hansen, *in* Dehler, C.M., Pederson, J.L., Sprinkel, D.A., and Kowallis, B.J., editors, Uinta Mountain geology: Utah Geological Association Publication 33, p. 1-4

Rowley, Scott, and Rowley, Patty, 2005, Rowley's favorite recipes: self-published, 94 p.

Sagan, Carl, 1996, The demon-haunted world—science as a candle in the dark: New York, Ballantine Books, 457 p.

Sagan, Carl, 1980, Cosmos: New York, Ballantine Books Trade Paperbacks edition (2013), 396 p.

Schmidt, D.L., and Rowley, P.D., 1986, Continental rifting and transform faulting along the Jurassic Transantarctic rift, Antarctica: Tectonics, v. 5, p. 279-291.

Schmidt, D.L., Williams, P.L., and Nelson, W.H., 1978, Geologic map of the Schmidt Hills quadrangle and part of the Gambacorta Peak quadrangle, Pensacola Mountains, Antarctica: U.S. Geological Survey U.S. Antarctic Research Program Map A-8, scale 1:250,000.

Scott, R.B., Rowley, P.D., Snee, L.W., Anderson, R.E., Harding, A.E., Unruh, D.M., Nealey, L.D., Hudson, M.R., Swadley, W C, and Ferris, D.E., 1996, Synchronous Oligocene and Miocene extension and magmatism in the vicinity of caldera complexes in southeastern Nevada, in Thompson, R.A., Hudson, M.R., and Pillmore, C.L., editors, Geologic excursions to the Rocky Mountains and beyond, Field trip guidebook for the 1996 annual meeting, Geological Society of America, Denver, Colorado, October 28-31: Colorado Geological Survey Spec. Pub. 44, 36 p.

Scott, W.E., and Shroba, R.R., 1985, Surficial geologic map of an area along the Wasatch fault zone in the Salt Lake Valley, Utah: U.S. Geological Survey Open-File Report 85-448, scale 1:24,000.

Shawe, D.R., and Rowley, P.D., editors, 1978, Guidebook to mineral deposits of southwestern Utah, field excursion C 2, International Association on the Genesis of Ore Deposits: Utah Geological Association Publication 7, 75 p.

Shenk, Timothy, 2022, Struggle for the soul of the GOP—is the Republican party compatible with democracy? The New Republic, May 2022, p. 51-55.

Sibbett, B.S., and Nielson, D.L., 1980, Geology of the central Mineral Mountains, Beaver County, Utah: Salt Lake City, Earth Science Laboratory, University of Utah Research Institute DOE/ET/28392-40, 42 p., scale 1:24,000.

Siders, M.A., Rowley, P.D., Shubat, M.A., Christenson, G.E., and Galyardt, G.L., 1990, Geologic map of the Newcastle quadrangle, Iron County, Utah: U.S. Geological Survey Geologic Quadrangle Map GQ-1690, scale 1:24,000.

Simon, D.B., 2003-2004, A case study—the Holocene "Downtown fault" in Salt Lake City, Utah—technical controversies and geopolitics: unpublished abstract for a national presidential address given at 16 sections, coast-to-coast, of the Association of Engineering Geologists.

Simon, D.B., and Shlemon, R.J., 1999, The Holocene "Downtown fault" in Salt Lake City, Utah (abs.): Association of Engineering Geologists, 42nd Annual Meeting, Program with Abstracts Volume, 1999, p. 85.

Simon, D.B., Bartlett, S.F., and Shlemon, R.J., 1999, Holocene ground failure in downtown Salt Lake City, Utah (abs.): Geological Society of America, Cordilleran Section, Abstracts with Program, v. 31, no. 6, p. A-95.

Simon, D.B., Black, D.K., Hanson, J.R., and Rowley, P.D., 2015, Surface-fault-rupture-hazard investigation for a portion of the Southern Parkway Northern Extension (State Route 7), Fort Pearce section, Washington fault zone, Washington County, Utah, *in* Lund, W.R., editor, Geologic mapping and paleoseismic investigations of the Washington fault zone, Washington County, Utah, and Mohave County, Arizona, Paleoseismicity of Utah, Volume 27: Utah Geological Survey Miscellaneous Publication 27, p. 107-160.

Slate, J.L., Berry, M.E., Rowley, P.D., Fridrich, C.J., Morgan, K.S., Workman, J.B., Young, O.D., Dixon, G.L., Williams, V.S., McKee, E.H., Ponce, D.A., Hildenbrand, T.G., Swadley, W C, Lundstrom, S.C., Ekren, E.B., Warren, R.G., Cole, J.C., Fleck, R.J., Lanphere, M.A., Grunwald, D.J., Laczniak, R.J., Menges, C.M., Yount, J.C., and Jayko, A.S., 1999, Digital geologic map of the Nevada Test Site and vicinity, Nye, Lincoln, and Clark Counties, Nevada, and Inyo County, California: U.S. Geological Survey Open-File Report 99-554-A, CD-ROM, scale 1:100,000.

Snee, L.W., and Rowley, P.D., 2000, New $^{40}Ar/^{39}Ar$ dates from the Caliente caldera complex, Nevada-Utah—at least 10 million years of Tertiary volcanism in one of the World's largest caldera complexes: Geological Society of America Abstracts with Programs, v. 32, no. 7, p. A-461.

Snyder, R.P., Ekren, E.B., and Dixon, G.L., 1972, Geologic map of the Lunar Crater quadrangle, Nye County, Nevada: U.S. Geological Survey Miscellaneous Geologic Investigations Map I-700, scale 1:48,000.

Steven, T.A., 1968, Ore deposits in the central San Juan Mountains, Colorado, *in* Ridge, J.D., editor, Ore deposits of the United States, 1933-1967 (Graton-Sales volume): American Institute of Mining and Metallurgical Petroleum Engineers, v. 1, p. 706-713.

Steven, Tom, 1992, Echoes from my mind: Lakewood, Colorado, TAS Enterprises, 116 p.

Steven, Thomas, 2000, Bigfoot, the poetry of Thomas Steven, 1994-2000: a binder perhaps 100 pages long.

Steven, T.A., Cunningham, C.G., and Machette, M.N., 1981, Integrated uranium systems in the Marysvale volcanic field, west-central Utah, *in* Goodell, P.C., and Waters, A.C., editors: American Association of Petroleum Geologists, Studies in Geology no. 13, p. 111-122.

Steven, T.A., Cunningham, C.G., Naeser, C.W., and Mehnert, H.H., 1979, Revised stratigraphy and radiometric ages of volcanic rocks and mineral deposits in the Marysvale area, west-central Utah: U.S. Geological Survey Bulletin 1469, 40 p.

Steven, T.A., and Eaton, G.P., 1975, Environment of ore deposition in the Creede mining district, San Juan Mountains, Colorado—the geologic, hydrologic, and geophysical setting: Economic Geology, v. 70, no. 6, p. 1023-1037.

Steven, T.A., and Hail, W.J., Jr., 1989, Geologic map of the Montrose 30' x 60' quadrangle, southwestern Colorado: U.S. Geological Survey Miscellaneous Investigations Series Map I-1939, scale 1:100,000.

Steven, T.A., and Lipman, P.W., 1976, Calderas of the San Juan volcanic field, southwestern Colorado: U.S. Geological Survey Professional Paper 958, 35 p.

Steven, T.A., Lipman, P.W., Hail, W.J., Jr., Barker, Fred, and Luedke, R.G., 1974, Geologic map of the Durango quadrangle, Colorado: U.S. Geological Survey Miscellaneous Geologic Investigations Map I-764, scale 1:250,000.

Steven, T.A., and Morris, H.T., 1987, Summary mineral resource appraisal of the Richfield 1° x 2° quadrangle, west-central Utah: U.S. Geological Survey Circular 916, 24 p.

Steven, T.A., Morris, H.T., and Rowley, P.D., 1990, Geologic map of the Richfield 1° x 2° quadrangle, west-central Utah: U.S. Geological Survey Miscellaneous Investigations Series Map I-1901, scale 1:250,000.

Steven, T.A., and Ratte, J.C., 1960, Geology and ore deposits of the Summitville district, Colorado: U.S. Geological Survey Professional Paper 343, 70 p.

Steven, T.A., and Ratte, J.C., 1965, Geology and structural control of ore deposition in the Creede district, San Juan Mountains, Colorado: U.S. Geological Survey Professional Paper 487, 87 p.

Steven, T.A., Rowley, P.D., and Cunningham, C.G., 1984, Calderas of the Marysvale volcanic field, west-central Utah: Journal of Geophysical Research, v. 89, no. B10,

p. 8751-8764.

Stewart, J.H., and Carlson, J.E., 1978, Geologic map of Nevada: U.S. Geological Survey, scale 1:500,000.

Stewart, J.H., Moore, W.J., and Zietz, I., 1977, East-west patterns of Cenozoic igneous rocks, aeromagnetic anomalies, and mineral deposits, Nevada and Utah: Geological Society of America Bulletin, v. 88, p. 67-77.

Storey, B.C., Wever, H.E., Rowley, P.D., and Ford, A.B., 1987, The geology of the central Black Coast, eastern Palmer Land: British Antarctic Survey Bulletin, no. 77, p. 145-155.

Stout, D.L., 1995, Peter D. Rowley receives 1995 Dibblee Medal: GSA Today, v. 5, no. 8, p. 157-159.

Swadley, W.C., and Rowley, P.D., 1994, Geologic map of the Pahroc Spring SE quadrangle, Lincoln County, Nevada: U.S. Geological Survey Geologic Quadrangle Map GQ-1752, scale 1:24,000.

Thomson, Janet, 1986, A summer on the English Coast: Home and Country, August 1986.

Thomson, M.R.A., 1983, Late Jurassic ammonites from the Orville Coast, Antarctica, *in* Oliver, R.L., James, P.R., and Jago, J.B., editors, Antarctic earth science: Canberra, Australian Academy of Science, p. 315-319.

Trump, M.L., 2020, Too much and never enough—how my family created the world's most dangerous man: New York, Simon & Schuster, 240 p.

Unruh, D.M., Nealey, L.D., Rowley, P.D., Snee, L.W., Mehnert, H.H., and Anderson, R.E., 1995, Strontium and neodymium isotopic survey of ash-flow tuffs and related rocks from the Caliente caldera complex, southeastern Nevada and southwestern Utah, in Scott, R.B., and Swadley, WC, editors, Geologic studies in the Basin and Range-Colorado Plateau transition, southeastern Nevada, southwestern Utah, and northwestern Arizona: U.S. Geological Survey Bulletin 2056-D, p. 111-129.

U.S. Geological Survey, 1967, Antarctica sketch map, Ellsworth Land (East part) – Palmer Land (South Part), showing the area between 73° to 76°30' South, 59° to 75° W., scale 1:500,000.

U.S. Geological Survey, 1967, Antarctica sketch map, Bryan Coast – Ellsworth Land, showing the area between 72° to 75°30' South, 75° to 90°W., scale 1:500,000.

U.S. Geological Survey, 1979, Antarctica sketch map, Palmer Land (North part), showing the area between 68° to 73° South, 60° to 70° W., scale 1:500,000.

Utley, M.G., 1992, The ghosts of Gold Mountain—a history of the Gold Mountain mining district, Piute County, Utah: self-published, 272 p.

Vennum, W.R., 1978, Igneous and metamorphic petrology of the southwestern Dana Mountains, Lassiter Coast, Antarctic Peninsula: U.S. Geological Survey Journal of Research, v. 6, no. 1, p. 95-106.

Vennum, W.R., 1980, Evaporite encrustations and sulphide oxidation products from the southern Antarctic Peninsula: New Zealand Journal of Geology and Geophysics, v. 23, p. 499-505.

Vennum, W.R., and Laudon, T.S., 1988, Igneous petrology of the Merrick Mountains, eastern Ellsworth Land, Antarctica, *in* Rowley, P.D., and Vennum, W.R., editors, Studies of the geology and mineral resources of the southern Antarctic Peninsula and eastern Ellsworth Land, Antarctica: U.S. Geological Survey Professional

Paper 1351, p. 21-34.

Vennum, W.R., and Meyer, C.E., 1979, Plutonic garnets from the Werner batholith, Lassiter Coast, Antarctic Peninsula: American Mineralogist, v. 64, p. 268-273.

Vennum, W.R., and Rowley, P.D., 1986, Reconnaissance geochemistry of the Lassiter Coast intrusive suite, southern Antarctic Peninsula: Geological Society of America Bulletin, v. 97, no. 12, p. 1521-1533.

Vennum, W.R., Rowley, P.D., and Laudon, T.S., 1995, Intrusive igneous rocks of eastern Ellsworth Land, West Antarctica—the southwestward extension of the Lassiter Coast intrusive suite: American Geophysical Union, Contributions to Antarctic Research IV, Antarctic Research Series, v. 67, p. 75-101.

Voight, Barry, 1981, Time scale for the first moments of the May 18 eruption, *in* Lipman, P.W., and Mullineaux, D.R., editors, The 1980 eruptions of Mount St. Helens, Washington: U.S. Geological Survey Professional Paper 1250, p. 69-86.

Voight, Barry, Glicken, Harry, Janda, R.J., and Douglass, P.M., 1981, Catastrophic rockslide avalanche of May 18, *in* Lipman, P.W., and Mullineaux, D.R., editors, The 1980 eruptions of Mount St. Helens, Washington: U.S. Geological Survey Professional Paper 1250, p. 347-378.

Wallace, C.A., and Rowley, P.D., 2016, A visit to Saudi Arabia in 1984: The Geologic Division Retirees Newsletter, spring 2016, no. 72, p. 11-17.

Wallace, C.A., and Rowley, P.D., 1985, Reconnaissance study of the Murdama group and related formations, main Murdama basin, Kingdom of Saudi Arabia: Deputy Ministry for Mineral Resources, U.S. Geological Survey Administrative Report, 72 p.

Wegener, A.L., 1924, The origin of continents and oceans, 3rd edition, translated by J.F.A. Skerl: London, Methuen.

Wendel, JoAnna, and Kumar, Mohi, 2016, Pinatubo, 25 years later—eight ways the eruption broke ground: Eos, v. 97, no. 15, p. 16-22.

Williams, P.L., 1960, A stained slice method for rapid determination of the phenocryst composition of volcanic rocks: American Journal of Science, v. 258, p. 148-152.

Williams, P.L., 1964, Geology, structure, and uranium deposits of the Moab quadrangle, Colorado and Utah: U.S. Geological Survey Miscellaneous Investigations Series I-360, scale 1:250,000.

Williams, P.L., 1967, Stratigraphy and petrography of the Quichapa Group, southwestern Utah and southeastern Nevada: Seattle, University of Washington unpublished Ph.D. dissertation, 139 p.

Williams, P.L., and Cole, J.C., 2007, Geologic map of the Albuquerque 30' x 60' quadrangle, north-central New Mexico: U.S. Geological Survey Scientific Investigations Map 2946, 31 p., scale 1:100,000.

Williams, P.L., and Hackman, R.J., 1971, Geology, structure, and uranium deposits of the Salina quadrangle, Utah: U.S. Geological Survey Miscellaneous Geologic Investigations Map I-591, scale 1:250,000.

Williams, P.L., Schmidt, D.L., Plummer, C.C., and Brown, L.E., 1972, Geology of the Lassiter Coast area, Antarctic Peninsula—preliminary report, *in* Adie, R.J., editor, Antarctic Geology and Geophysics: Oslo, Universitetsforlaget, p. 143-148.

Wilson, E.O., 2016, Half-earth—our planet's fight for life: New York, Liveright Publishing Corporation, 272 p.

Woodward, Bob, 2020, Rage: New York, Simon & Schuster, 480 p.

Winograd, I.J., and Thordarson, William, 1975, Hydrogeologic and hydrochemical framework, south-central Great Basin, with special reference to the Nevada Test Site: U.S. Geological Survey Professional Paper 712-C, 126 p.

Workman, J.B., Menges, C.M., Page, W.R., Ekren, E.B., Rowley, P.D., and Dixon, G.L., 2002, Tectonic map of the Death Valley ground-water model area, Nevada and California: U.S. Geological Survey Miscellaneous Field Studies Map MF-2381-B, scale 1:250,000.

Workman, J.B., Menges, C.M., Page, W.R., Taylor, E.M., Ekren, E.B., Rowley, P.D., Dixon, G.L., Thompson, R.A., and Wright, L.A., 2003, Geologic map of the Death Valley ground-water model area, Nevada and California: U.S. Geological Survey Miscellaneous Field Studies Map MF-2381-A, scale 1:250,000.

Wright, N.A., and Williams, P.L., 1974, Mineral resources of Antarctica: U.S. Geological Survey Circular 705, 29 p.

Yong, Ed, 2020, Anatomy of an American failure: The Atlantic, September, p. 32-47.

INDEX

(Includes pages containing definitions of technical terms)

D

E

f

G

H

I

N

O

P

Q

R

S

T

U

V

W

Y

Z

www.ingramcontent.com/pod-product-compliance
Lightning Source LLC
LaVergne TN
LVHW051058210926
848486LV00024B/420/J
9798996046706